Angiotensin Receptors

Angiotensin Receptors

Edited by

Juan M. Saavedra

National Institute of Mental Health
National Institutes of Health
Bethesda, Maryland

and

Pieter B. M. W. M. Timmermans

The DuPont Merck Pharmaceutical Company
Wilmington, Delaware

Springer Science+Business Media, LLC

Library of Congress Cataloging-in-Publication Data

Angiotensin receptors / edited by Juan M. Saavedra and Pieter B.M.W.M.
  Timmermans.
      p.   cm.
  Includes bibliographical references and index.
  ISBN 978-1-4613-6049-0   ISBN 978-1-4615-2464-9 (eBook)
  DOI 10.1007/978-1-4615-2464-9
  1. Angiotensin--Receptors.   I. Saavedra, Juan M.   II. Timmermans,
P. B. M. W. M.
  [DNLM: 1. Receptors, Angiotensin.   QU 55 A588 1994]
  QP572.A54A55   1994
  615'.71--dc20
  DNLM/DLC
  for Library of Congress   94-5810
                                                                CIP

ISBN 978-1-4613-6049-0

© 1994 Springer Science+Business Media New York
Originally published by Plenum Press, New York in 1994
Softcover reprint of the hardcover 1st edition 1994

Contributors

R. Wayne Alexander • Cardiology Division, Emory University School of Medicine, Atlanta, Georgia 30322

Keshwar Baboolal • Department of Medicine, Stanford University, Stanford, California 94305

Serge P. Bottari • Cardiovascular Research Laboratories, CIBA-GEIGY Limited, CH-4002 Basel, Switzerland. Present address: Centre d'Etudes Nucléaires de Grenoble DBMS/BRCE, INSERM U 244, 38041 Grenoble Cedex, France

Véronique Brechler • Cardiovascular Research Laboratories, CIBA-GEIGY Limited, CH-4002 Basel, Switzerland. Present address: Laboratoire de Biochimie Moléculaire de l'Hypertension, Institut de Recherche Clinique de Montréal, H3W 1R7 Montréal, Quebec, Canada

Hans R. Brunner • Division of Hypertension and Cardiovascular Research Group, University Hospital, 1010 Lausanne, Switzerland

John C. Burnett, Jr. • Department of Cardiovascular Diseases, Mayo Clinic, Rochester, Minnesota 55905

Michel Burnier • Division of Hypertension and Cardiovascular Research Group, University Hospital, 1010 Lausanne, Switzerland

Joseph M. Capasso • Department of Anatomy, University of South Dakota School of Medicine, Vermillion, South Dakota 57069

David P. Chan • Division of Cardiology, Children's Hospital, Ohio State University, Columbus, Ohio 43205

Andrew T. Chiu • The Du Pont Merck Pharmaceutical Company, Wilmington, Delaware 19880-0400

Marc de Gasparo • Cardiovascular Research Laboratories, CIBA-GEIGY Limited, CH-4002 Basel, Switzerland

Jeremy J. Edmunds • Department of Medicinal Chemistry, Parke-Davis Pharmaceutical Research, Division of Warner-Lambert Company, Ann Arbor, Michigan 48105

Eric A. Espiner • Departments of Cardiology and Endocrinology, The Princess Margaret Hospital, Christchurch, New Zealand

Mechthild Falkenhahn • German Institute for High Blood Pressure Research and Pharmacological Institute, University of Heidelberg, D-6900 Heidelberg, Germany

Michael Andrew Fitzpatrick • Departments of Cardiology and Endocrinology, The Princess Margaret Hospital, Christchurch, New Zealand

Pascal Furet • Cardiovascular Research Laboratories, CIBA-GEIGY Limited, CH-4002 Basel, Switzerland

Peter Gohlke • German Institute for High Blood Pressure Research and Pharmacological Institute, University of Heidelberg, D-6900 Heidelberg, Germany

Steven Goldman • University of Arizona Heart Center and Tucson VA Medical Center, Tucson, Arizona 85723

William J. Greenlee • Merck Research Laboratories, Rahway, New Jersey 07065

Kathy K. Griendling • Cardiology Division, Emory University School of Medicine, Atlanta, Georgia 30322

Deng-Fu Guo • Department of Biochemistry, Vanderbilt University School of Medicine, Nashville, Tennessee 37232-0146

Robert Gyurko • Department of Physiology, College of Medicine, University of Florida, Gainesville, Florida 32610

Frank M. J. Heemskerk • Section on Pharmacology, Laboratory of Clinical Science, National Institute of Mental Health, National Institutes of Health, Bethesda, Maryland 20892

John C. Hodges • Department of Medicinal Chemistry, Parke-Davis Pharmaceutical Research, Division of Warner-Lambert Company, Ann Arbor, Michigan 48105

Tadashi Inagami • Department of Biochemistry, Vanderbilt University School of Medicine, Nashville, Tennessee 37232-0146

Bruno Kamber • Cardiovascular Research Laboratories, CIBA-GEIGY Limited, CH-4002 Basel, Switzerland

Joan A. Keiser • Cardiovascular Pharmacology, Parke-Davis Pharmaceutical Research, Division of Warner-Lambert Company, Ann Arbor, Michigan 48105

Birgitta Kimura • Department of Physiology, College of Medicine, University of Florida, Gainesville, Florida 32610

Bernard Lassègue • Cardiology Division, Emory University School of Medicine, Atlanta, Georgia 30322

Nigel R. Levens • Cardiovascular Research Laboratories, CIBA-GEIGY Limited, CH-4002 Basel, Switzerland

Kaj P. Metsärinne • Minerva Foundation Institute for Medical Research, and IVth Department of Medicine, Helsinki University Hospital, SF-00250 Helsinki, Finland

Timothy W. Meyer • Department of Medicine, Stanford University, Stanford, California 94305

Masato Mizukoshi • Department of Biochemistry, Vanderbilt University School of Medicine, Nashville, Tennessee 37232-0146

Eugene Morkin • University of Arizona Heart Center and Tucson VA Medical Center, Tucson, Arizona 85723

James J. Morton • Medical Research Council Blood Pressure Unit, Western Infirmary, Glasgow G11 6NT, Scotland

Liisa Näveri • Department of Pharmacology and Toxicology, University of Helsinki, 00170 Helsinki, Finland

Robert L. Panek • Cardiovascular Pharmacology, Parke-Davis Pharmaceutical Research, Division of Warner-Lambert Company, Ann Arbor, Michigan 48105

M. Ian Phillips • Department of Physiology, College of Medicine, University of Florida, Gainesville, Florida 32610

Miriam Rademaker • Departments of Cardiology and Endocrinology, The Princess Margaret Hospital, Christchurch, New Zealand

Thomas E. Raya • University of Arizona Heart Center and Tucson VA Medical Center, Tucson, Arizona 85723

Juan M. Saavedra • Section on Pharmacology, Laboratory of Clinical Sciences, National Institute of Mental Health, National Institutes of Health, Bethesda, Maryland 20892

Ronald D. Smith • The Du Pont Merck Pharmaceutical Company, Wilmington, Delaware 19880-0400

Monika Stoll • German Institute for High Blood Pressure Research and Pharmacological Institute, University of Heidelberg, D-6900 Heidelberg, Germany

Christer Strömberg • National Agency for Medicines, Pharmacological Department, 00301 Helsinki, Finland

Pieter B.M.W.M. Timmermans • The Du Pont Merck Pharmaceutical Company, Wilmington, Delaware 19880-0400

Thomas Unger • German Institute for High Blood Pressure Research and Pharmacological Institute, University of Heidelberg, D-6900 Heidelberg, Germany

Mohan Viswanathan • Section of Pharmacology, Laboratory of Clinical Science, National Institute of Mental Health, National Institutes of Health, Bethesda, Maryland 20892

Bernard Waeber • Division of Hypertension and Cardiovascular Research Group, University Hospital, 1010 Lausanne, Switzerland

Ruth R. Wexler • The Du Pont Merck Pharmaceutical Company, Wilmington, Delaware 19880-0400

Steven Whitebread • Cardiovascular Research Laboratories, CIBA-GEIGY Limited, CH-4002 Basel, Switzerland

Pancras C. Wong • The Du Pont Merck Pharmaceutical Company, Wilmington, Delaware 19880-0400

Preface

Angiotensin II is considered to be one of the most important hormones in the control of blood pressure. Inhibition of angiotensin II synthesis with angiotensin-converting enzyme inhibitors is one of the preferred treatments for human hypertension and heart failure. The converting enzyme inhibitors, however, are not devoid of side effects, and their mechanism of action is not selective for the angiotensin system, since they affect bradykinin metabolism and prostaglandin actions. Recently, it was discovered that angiotensin II receptors were not homogeneous, and could be classified in at least two subtypes: AT_1 and AT_2. Nonpeptidic ligands, selective for each receptor subtype, were quickly developed. The discovery of selective, nonpeptidic, orally effective blockers of angiotensin receptors offered in theory a more selective approach to the therapy of hypertension and heart failure. This in turn stimulated research on the molecular biology, physiology, and pharmacology of angiotensin receptors and in the medicinal chemistry of angiotensin receptor blockers. Using the selective AT_1 blocker losartan, it was found that stimulation of the AT_1 subtype mimicked the vasoconstrictive and water retention actions of angiotensin II. It was quickly found that the effects of AT_1 blockade were not substantially different, physiologically and pathophysiologically, from those of converting enzyme inhibitors. In experimental models, and in preliminary clinical investigations, administration of losartan resulted in clinical benefits similar to those of the converting enzyme inhibitors captopril and enalapril. This evidence substantiates the possible role of AT_1 blockers such as losartan as a new approach to the therapy of cardiovascular disease.

At the same time, the renewed interest in the field of angiotensin receptors generated new hypotheses and raised new, challenging questions. It is now clear that the angiotensin receptors are more heterogeneous than was considered only a few years ago. The function of the newly discovered AT_2 receptor subtype is unknown. However, AT_2 receptors are abundant in developing tissues and in the immature brain, and many groups are devoted to the attempt to clarify the proposed roles of angiotensin as a growth factor and as a central neurotransmitter.

In this volume, we have tried to compile some of the most recent discoveries in the emerging field of angiotensin receptors, from their molecular biology to the clinical application of selective receptor blockade. We include chapters on the molecular biology of the recently cloned AT_1 receptors; the pharmacology and medicinal chemistry of the two main angiotensin receptor subtypes currently known, the AT_1 and the AT_2 subtypes; and a classification of receptor subtypes defined by ligand selectivity, biochemical receptor characteristics, and genetic analysis.

Two new areas of interest receive particular attention in this volume: the newly discovered AT_2 receptors and the brain angiotensin receptor subtypes. New aspects of the physiology and pathophysiology of angiotensin receptor subtypes are also included, and establish without doubt the beneficial effects of AT_1 blockade in models of hypertension, heart failure, and renal disease. Finally, clinical studies demonstrate that in humans the blockade of AT_1 receptors with losartan is significantly devoid of side effects and appears to be equally active as converting enzyme inhibitors in hypertensive patients with chronic nephropathy and in patients with congestive heart failure.

Advances in the molecular biology, pharmacology, and medicinal chemistry of angiotensin receptor subtypes over the last few years offer the clinician a new direction in the treatment of cardiovascular disease. The new developments, and those to come, promise safer, more selective treatments for cardiovascular disease and a better understanding of a system with a multiplicity of effects and great therapeutic importance.

Contents

Chapter 3

Defining Angiotensin Receptor Subtypes **49**

Andrew T. Chiu, Ronald D. Smith, and Pieter B.M.W.M. Timmermans

Chapter 4

Medicinal Chemistry of Angiotensin II Antagonists **67**

William J. Greenlee and Ruth R. Wexler

Chapter 5

The Angiotensin II AT_2 Receptor Subtype **95**

*Marc de Gasparo, Nigel R. Levens, Bruno Kamber, Pascal Furet,
Steven Whitebread, Véronique Brechler, and Serge P. Bottari*

Chapter 9

Chapter 10

Chapter 11

Angiotensin II Receptor Subtypes and Growth **205**

Mohan Viswanathan and Juan M. Saavedra

Chapter 12

**Inhibiting the Effects of Angiotensin II on Cardiovascular
Hypertrophy in Experimental Hypertension** **221**

James J. Morton

Chapter 15

Angiotensin II Receptor Antagonism in an Ovine Model of Heart Failure: Comparison with ACE and Renin Inhibition 287

Michael Andrew Fitzpatrick, Miriam Rademaker, and Eric A. Espiner

Chapter 21

Angiotensin Receptors

1

Angiotensin II Receptor
Molecular Cloning, Functions, and Regulation

Tadashi Inagami, Masato Mizukoshi, and Deng-Fu Guo

1. INTRODUCTION

The diversity of specific physiological effects elicited by angiotensin II[1,2] (ANG II) poses a simple but important question: Are they mediated by a single or multiple type of receptors? While workers on ANG II action expected more than one subtype as suggested by the presence of the dithiothreitol-sensitive and -insensitive receptors, exceeding instability of the receptor protein thwarted numerous attempts at obtaining direct evidence for its multiplicity. Thus, the development of the new types of ANG II receptor antagonists,[3–5] which turned out to be isoform specific, was the important turning point that provided indirect but important tools to reveal the problem of ANG II receptor properties.

On the other hand, alternatives to the orthodox approaches to receptor protein purification were sought for cloning the receptor cDNA. One such recourse produced a report that the *mas* oncogene product expressed in frog oocytes responds to ANG II by opening a chloride channel.[6] The contention that this oncogen product is an ANG II receptor has not been validated. An important methodological breakthrough was made in molecular biology[7] for an expression cloning that did not require the purification of a receptor protein. A review on the early phase of ANG II receptor has been published.[8–11]

Tadashi Inagami, Masato Mizukoshi, and Deng-Fu Guo • Department of Biochemistry, Vanderbilt University School of Medicine, Nashville, Tennessee 37232-0146.

Angiotensin Receptors, edited by Juan M. Saavedra and Pieter B.M.W.M. Timmermans. Plenum Press, New York, 1994.

2. CLONING OF ANGIOTENSIN II RECEPTOR cDNA

Realizing that cloning of individual isoforms of ANG II receptors was required to clarify the molecular basis for diverse responses of various cells and tissues to ANG II, Sasaki et al.[12] and Murphy et al.[13] adapted the expression cloning method to the ANG II receptor. mRNA from cultured bovine adrenal zona glomerulosa cells[12] or rat aortic vascular smooth muscle cells[13] with a high abundance of ANG II receptor expression was used to prepare cDNA, which was inserted into the expression vector pCDM8 and then expressed and selected in mammalian cells (COS-7 cells). Cells transfected with plasmid-containing cDNA for the receptor were identified by their binding to the [^{125}I]angiotensin II analogue, Sar[1], Ile[8]-ANG II (Sarile). Sixty-seven thousand clones were used to transfect 1 million COS-7 cells, which were autoradiographed in 12 wells. More than 10 positive clones were isolated.

Iwai et al.[14] obtained rat renal AT_1 cDNA from spontaneously hypertensive rats (SHR) by plaque hybridization using bovine adrenal AT_1 cDNA as a probe. The SHR renal receptor and vascular smooth muscle cell receptor[13] were found to have an identical amino acid sequence. This amino acid sequence differed from the bovine adrenal AT_1 amino acid sequence by 8% (Fig. 1). The difference could have been due to either different species or organ-dependent AT_1 subtypes. To resolve this question, AT_1 receptor from rat adrenal (AT_{1B}) was cloned[15,16] and was found to be different from the receptor found in vascular and renal tissue (AT_{1A}). The cDNA for AT_1 cloned from rat pituitary was identical to that expressed dominantly in the rat adrenal (AT_{1B}).[17] It was also possible to demonstrate the presence of the two similar but distinct genes AT_{1A} and AT_{1B} in the rat genomic DNA library[18] Since the entire coding regions of AT_{1A} and AT_{1B} were contained in single exons, their structures were readily determined. (Note that AT_{1B} defined here for the rat adrenal/pituitary AT_1 receptor seems to be different from AT_{1B} of rat kidney mesangial cells by Ernsberger et al.[19]) Two AT_1 subtypes were also found in mouse genomic DNA.[20,21]

To determine the structure of human AT_1 receptor, we used the property that the entire coding region is contained in a single exon. The coding region for AT_1 was cloned from a human genomic DNA library and its amino acid sequence was deduced.[22] In addition, we and others also cloned human AT_1 cDNA from a human liver cDNA library.[23–25] An AT_1 receptor has also been cloned from a cDNA library of rabbit kidney cortex,[26] but no evidence for divergence to AT_{1A} and AT_{1B} is available in this species.

Southern blot analysis of human genomic DNA consistently produced evidence for the presence of single AT_1 gene. However, similar analysis of rodent's gene strongly indicated the presence of two closely related AT_1 genes, both in rat and mouse.[18,20,24,27,28] These observations were further supported by experiments using somatic cell hybrids.[27] Rat AT_{1A} was localized on chromosome 17 and AT_{1B} on chromosome 2. Human chromosome 3 bears the single AT_1 gene in the human genome. Homology (synteny conservation) between human chromosome 3 and rodent chromosome 2 indicates that human AT_1 may be homologous to rat AT_{1B}.

Human AT_1 gene was further localized to chromosome 3q21–3q25 by Curnow *et al.*[28] Analysis and comparison of the frequency of interspecies and intersubtype difference in the amino acid sequences indicated that AT_{1A}/AT_{1B} differentiation in rodents occurred much earlier than divergence of rat and mouse.[20]

3. STRUCTURE OF AT₁ RECEPTOR

The amino acid sequences encoded by the coding regions of AT_{1A} and AT_{1B} show 96% identity with each other, but marked differences exist in the noncoding regions, suggesting a possible difference in the mechanism of receptor regulation. Indeed, estradiol[17] was found to regulate AT_{1B} but not AT_{1A} mRNA. Expressions of AT_{1A} and AT_{1B} were different from tissue to tissue.[15–17,29]

The amino acid sequences encoded by the open reading frame of AT_1 cDNA or genomic DNA of all species consist of 359 amino acid residues with relative molecular weight of approximately 41,000, which is in reasonable agreement with that of the deglycosylated form of the ANG II receptor isolated from the adrenal gland (mol. wt. 35,000).[30,31]

Hydropathy analysis of the amino acid sequences indicates that AT_1 has seven transmembrane domains characteristic to the superfamily of the G-protein–coupled receptors (Fig. 2). The amino acid sequence showed 20–30% sequence identity with other G-protein–coupled receptors. Although previous studies had suggested that the product of the *mas* oncogene might encode a G-protein–coupled receptor for ANG II,[6] the amino acid sequence of AT_1 showed only 9% sequence identity.

In all species determined to date, the AT_1 receptors possess four cysteine residues, each distributed in each of four extracellular domains, suggesting the presence of two disulfide bridges. One of these, spanning the second and third extracellular domains, is highly conserved in the superfamily of the seven transmembrane domain receptors. It is likely then that a second disulfide bridge connects the first and last extracellular domain resulting in a well-like structure (Fig. 3), and disruption of this structure by interruption of the disulfide bridges is the likely explanation for the sensitivity of the AT_1 receptor subtype to sulfhydryl reagents such as dithiothreitol.

```
                    ▼                                      I
Rat AT₁ₐ     MALNSSAEDG  IKRIQDDCPK  AGRHSYIFVM  IPTLYSIIFV  VGIFGNSLVV   50
Rat AT₁ᵦ     -T----T---  ----------  ----------  ----------  ----------
Mouse AT₁ₐ   ------T---  ---------R  ----------  ----------  ----------
Mouse AT₁ᵦ   -I----I---  ----------  ----N-----  ----------  ----------
Rabbit AT₁   -M----T---  ----------  ----N-----  ----------  --------A-
Bovine AT₁   -I----T---  ----------  ----N---I-  ----------  ----------
Human AT₁    -I----T---  ----------  ----------  ----------  ----------

                                    II
Rat AT₁ₐ     IVIYFYMKLK  TVASVFLLNL  ALADLCFLLT  LPLWAVYTAM  EYRWPFGNHL  100
Rat AT₁ᵦ     ----------  ----------  ----------  ----------  ----------
Mouse AT₁ₐ   ----------  ----------  ----------  ----------  ----------
Mouse AT₁ᵦ   ----------  ----------  ----------  ----------  --Q-------
Rabbit AT₁   ----------  ----------  ----------  ----------  --------Y-
Bovine AT₁   ----------  ----------  ----------  ----------  --------Y-
Human AT₁    ----------  ----------  ----------  ----------  --------Y-

                        III
Rat AT₁ₐ     CKIASASVTF  NLYASVFLLT  CLSIDRYLAI  VHPMKSRLRR  TMLVAKVTCI  150
Rat AT₁ᵦ     --------S-  ----------  ----------  ----------  ----------
Mouse AT₁ₐ   --------S-  ----------  ----------  ----------  ----------
Mouse AT₁ᵦ   --------S-  ----------  ----------  ----------  ----------
Rabbit AT₁   --------S-  ----------  ----------  ----------  ----------
Bovine AT₁   --------S-  ----------  ----------  ----------  ----------
Human AT₁    --------S-  ----------  ----------  ----------  ----------

                    IV
Rat AT₁ₐ     IIWLMAGLAS  LPAVIHRNVY  FIENTNITVC  AFHYESRNST  LPIGLGLTKN  200
Rat AT₁ᵦ     ----------  -----Y----  ----------  ------Q---  ----------
Mouse AT₁ₐ   ----------  ----------  ----------  ------Q---  ----------
Mouse AT₁ᵦ   ----------  ----------  ----------  ------Q---  ----------
Rabbit AT₁   ----L-----  ---I-----F  ----------  ------Q---  ----------
Bovine AT₁   ----L-----  --TI-----F  ----------  ------Q---  --V-------
Human AT₁    ----L-----  ---I-----F  ----------  ------Q---  ----------

                    V
Rat AT₁ₐ     ILGFLFPFLI  ILTSYTLIWK  ALKKAYEIQK  NKPRNDDIFR  IIMAIVLFFF  250
Rat AT₁ᵦ     ----V-----  ----------  ------K---  -T--------  ----------
Mouse AT₁ₐ   ----------  ----------  ------K---  ----------  ----------
Mouse AT1ᵦ   ----V---V-  ----------  ------K---  -T--------  ----------
Rabbit AT₁   ----------  ----------  ----------  --------K   ----------
Bovine AT₁   ----------  ----------  T---------  ----K----K  --L-------
Human AT₁    ----------  ----------  ----------  --------K   ----------

                    VI                                      VII
Rat AT₁ₐ     FSWVPHQIFT  FLDVLIQLGV  IHDCKISDIV  DTAMPITICI  AYFNNCLNPL  300
Rat AT₁ᵦ     ----------  ---------I  -R--E-A---  ----------  ----------
Mouse AT₁ₐ   ----------  ----------  ------A---  ----------  ----------
Mouse AT₁ᵦ   ---------S  ----------  ----E-A-V-  ----------  ----------
Rabbit AT₁   ----------  ----------  ----R-A---  ----------  ----------
Bovine AT₁   ----------  -M-------L  -R----E---  ---------L  ----------
Human AT₁    ---I------  ---------I  -R--R-A---  ----------  ----------

Rat AT₁ₐ     FYGFLGKKFK  KYFLQLLKYI  PPKAKSHSSL  STKMSTLSYR  PSDNMSSSAK  350
Rat AT₁ᵦ     ----------  ----------  --T----AG-  ----------  ----------
Mouse AT₁ₐ   ----------  ----------  ----------  ----------  -------A--
Mouse AT₁ᵦ   ----------  R---------  ----------  ----------  ---------R
Rabbit AT₁   F---------  ----------  -------N-   ----------  ----V---S-
Bovine AT₁   ----------  ----------  -------N-   ----------  --E-GN--T-
Human AT₁    ----------  R---------  -------N-   ----------  ----V---T-

Rat AT₁ₐ     KPASCFEVE*                                                 359
Rat AT₁ᵦ     -S--F-----
Mouse AT₁ₐ   -----S----
Mouse AT₁ᵦ   -S-Y------
Rabbit AT₁   --VP------
Bovine AT₁   ---P-I----
Human AT₁    ---P------
```

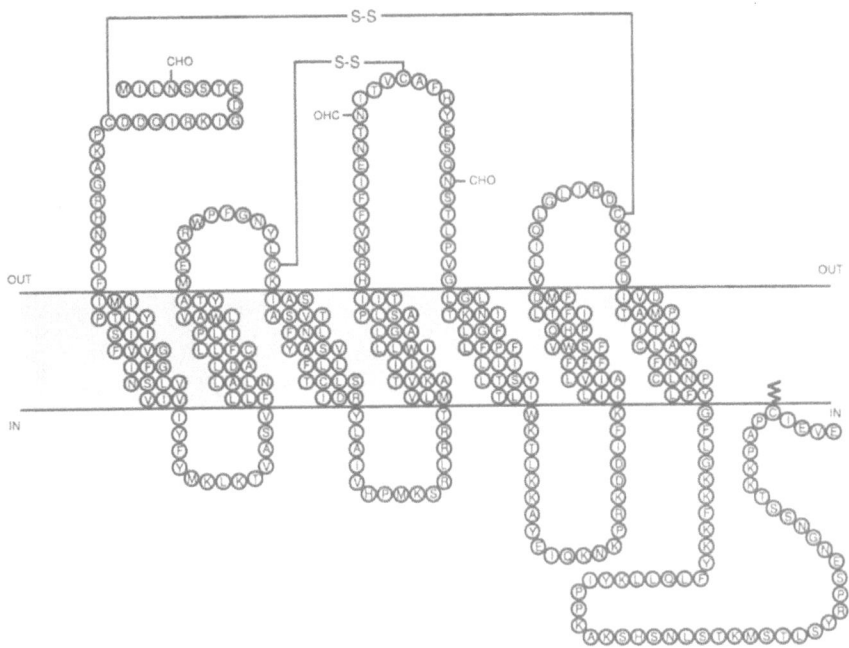

Figure 2. The 359 amino acid residues in bovine AT_1 receptor are arranged in the configuration with seven transmembrane domains. Note a characteristic short third cytosolic loop and two putative disulfide bridges in the extracellular domains which may confer sensitivity to reducing agents such as dithiothreitol.

4. FUNCTIONAL PROPERTIES OF CLONED AND EXPRESSED AT_1 RECEPTORS

The receptor proteins have been expressed either transiently or permanently by transfecting mammalian cells with expression plasmids containing cDNA inserts for the receptor. In all cases, these cells have shown saturable binding with $[^{125}I]$-ANG II with a single high-affinity binding site that is inhibited by losartan (an AT_1-specific antagonist) but not AT_2-specific antagonists[12–17,21,23–26] (Fig. 4). This is in contrast to cells transfected with *mas*

Figure 1. Amino acid sequences of AT_1 receptors deduced from the base sequences of cloned cDNA. Rat AT_{1A} sequence is from refs. 13,14; rat AT_{1B} from refs. 15–18; mouse AT_{1A} and AT_{1B} from refs. 20,21; rabbit AT_1 from ref. 26; bovine AT_1 from ref. 12; and human AT_1 from refs. 22–25. Bars with Roman numerals indicate putative transmembrane regions; ▼ indicates potential glycosylation sites; hyphens in the amino acid sequence indicate identical sequence. All sequences agreed with each other by more than 90%.

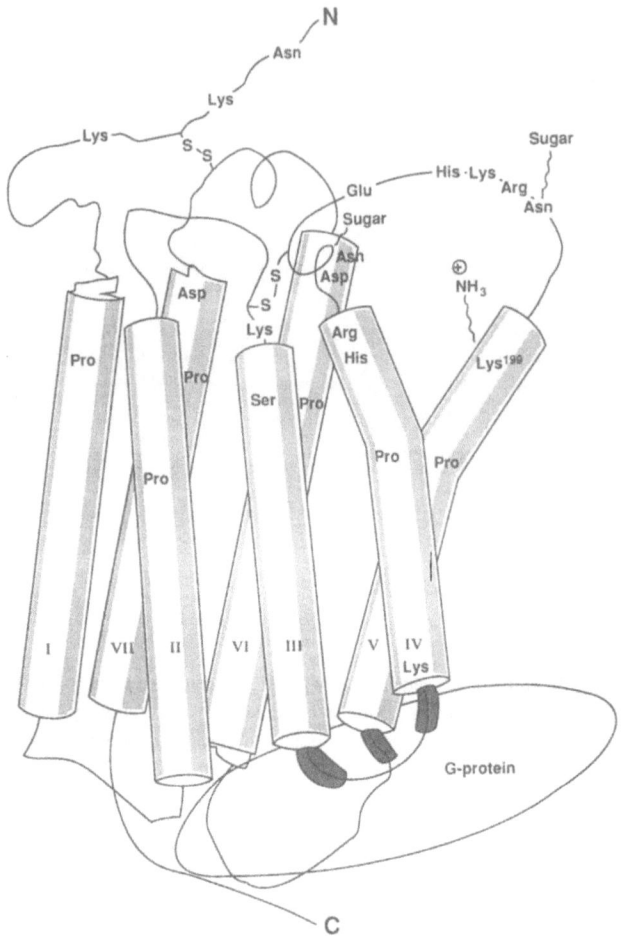

Figure 3. Putative three-dimensional structure by computer graphics of rat AT$_{1A}$ receptor. Lys[199] was found to be a key anchoring site for a negatively charged group in ligands. Courtesy of Dr. Inouye of Green Cross Corp.

oncogene DNA in which the ligand binding was not observed.[6] GTP analogues shifted the receptor to a low-affinity form. In COS-7 cells transiently expressing AT$_{1A}$, ANG II elicited generation of inositol-1,4,5-*tris*-phosphate (IP$_3$) and a transient increase in intracellular calcium, which is consistent with activation of phospholipase C.[12–14,16,17] In CHO cells permanently expressing rat AT$_1$, ANG II has also been shown to couple to a G$_i$ and inhibit adenylyl cyclase activity as well as to activate the IP$_3$ mediated Ca^{2+} release presumably via the Gq-mediated phospholipase C activation.[32] In these cells, ANG II also stimu-

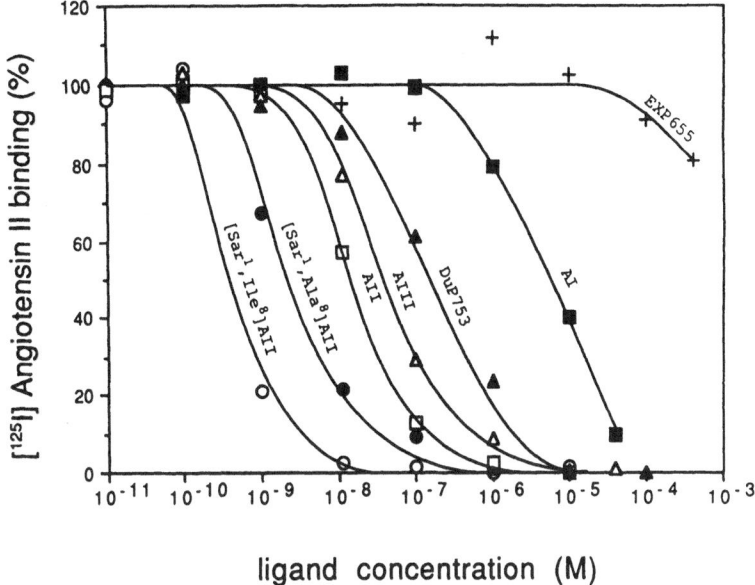

Figure 4. Binding of [^{125}I]-ANG II to the cloned bovine adrenal AT_1 receptor expressed transiently in COS-7 cells. Losarton competed with [^{125}I]-ANG II, whereas AT_2-specific EXP 655 did not compete. (From Sasaki *et al.*[12] Reprinted with permission.)

lated a nifedipine-sensitive influx of calcium.[33] Since untransfected COS-7 or CHO cells do not express detectable amounts of endogenous ANG II receptor, these observations indicate that the AT_1 receptor has the potential for coupling with multiple G proteins, which may explain at least part of the diversity of ANG II actions in different tissues and cells.

5. FUNCTIONAL DOMAINS OF AT₁ₐ RECEPTOR

The elucidation of the mechanism of action of AT_1 receptor requires identification of the site of ANG II binding, structures involved in the ensuing changes in the conformation of the receptor, and the sites involved in the interaction with G-proteins. Further, structural features required for the regulation by internalization or possible uncoupling of G-protein interaction must be identified. Clarifications of the regulation of gene expression requires information on various regulatory elements in the promotor regions.

Approaches toward these objectives have begun using site-directed mutagenesis. The octapeptide ANG II should bind over a sizable binding domain. However, the importance of the anionic group of the carboxy-terminal phenylalanine has been noted. Mutagenesis of Lys[199] (shown in Fig. 3) to Gln resulted in a profound (eightfold) reduction in the affinity of [125I]-ANG II binding to the transiently expressed receptor. The binding of [125I]-Sar[1], Ile[8]–ANG II was even more drastically reduced (20-fold), indicating that the positively charged ε-amino group of Lys[199] in the fifth transmembrane domain provides an anchoring site for the negatively charged carboxylate anion in the carboxy-terminal of ANG II.[34] Two putative disulfide bridges between the second and third extracellular loop and the amino-terminal and the third extracellular domain also seem to play important roles. Both appear to be essential to receptor function since mutation of each of the four Cys residues to Gly drastically reduce the binding affinity of [125]-ANG II. On the other hand elimination of the three Asn residues in the extracellular domains, potential *N*-glycosylation sites, by mutating it to Ala had no effect for receptor binding functions or its expression.[34]

Cytosolic domains needed for binding of G-proteins were localized in the second cytosolic loop and carboxy-terminal tail from studies with multiple mutations of polar groups in these domains as shown in Fig. 5. In Mut-1, the highly conserved Asp[125]-Arg[126]-Tyr[127] sequence was mutated to Gly-Gly-Ala. In Mut-2, Lys[135]-Ser[136]-Arg[137]—Arg[139]-Arg[140] were mutated to Gln-Gly-Gln—Gln-Gln. Either of these mutations abolished interaction with G-protein as mediated by the lack of shift to a low-affinity (to ANG II) state upon exposure to GTP-γ-S (Fig. 6) and also by the absence of stimulation of IP₃ formation on exposure to ANG II. Another important site for the G-protein coupling is the carboxy-terminal portion. Mutant M-7, in which 51 residues in the carboxy-terminal were deleted by inserting a termination codon, also lost the ability to shift to the low-affinity state by GTP-γ-S and response to ANG II in increasing IP₃ formation.[35] Interestingly, this is in contrast to α or β-adrenergic receptors, which contain a very large third cytosolic loop that seems to play the key role in their interaction with various G-proteins.[36]

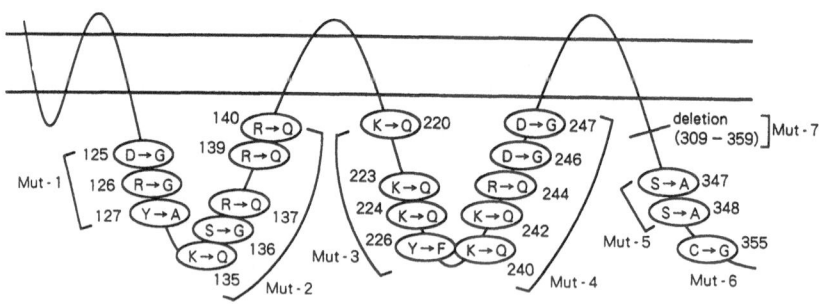

Figure 5. Illustration of multiple-site-directed mutagenesis of rat AT₁A receptor. Mut-1—Mut-7 illustrate mutated sites in each of the mutants. Mut-1, Mut-2, and Mut-7 have lost the ability to interact with G-protein almost completely. (From Ohyama *et al.*[35] Reprinted with permission.)

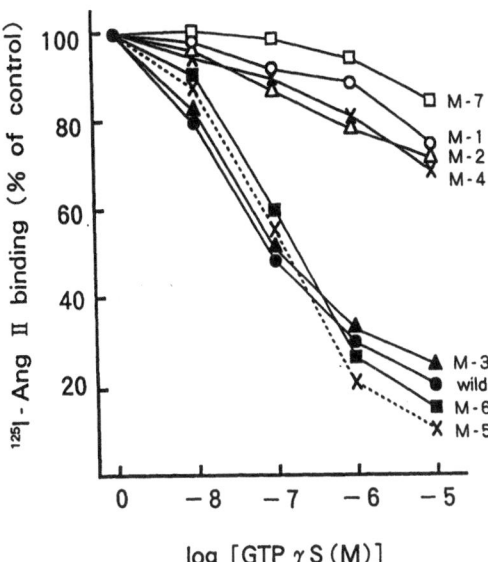

Figure 6. Results of elimination of putative G-protein coupling domains in Mut-1, Mut-2, and Mut-7 leading to the loss of the ability to shift to low-affinity state. ● = Unmutated rat receptor; ○ = Mut-1; –△– = Mut-2; –▲– = Mut-3; –X– = Mut-4; – –X– – = Mut-5; –■– = Mut-6; –□– = Mut-7. Sites of mutation are given in Fig. 5. Taken from Ref. 35.

Some areas in the transmembrane domains are more highly conserved than those in the cytosolic or extracellular domains. Among them are several prolyl residues, Asp^{74} and Cys^{76}. The conservation suggests that these residues may play universally important roles in signal transmission from the ligand to the interior of the receptor D-F Guo *et al.* (unpublished observation). Indeed mutation of Cys^{76} to Ala completely abolished the Gq activation.

6. REGULATION OF AT_1 RECEPTOR AND ITS GENE EXPRESSION

The regulation of the AT_1 receptor occurs at several levels: (1) the well-known downregulation of receptor proteins by ANG II (homologous down-regulation), (2) regulation of gene expression, and (3) regulation of the signal transduction mechanisms involving the G-protein coupling.

6.1. Downregulation of Receptor Molecule

The receptor internalization by ANG II takes place rapidly, which results in a rapid-onset desensitization (tachyphylaxis). Cloned AT_1 receptor transiently expressed in COS-7 cells undergoes the internalization. This process is stimulated only by ANG II but is not activated by phorbol esters or other agents that interact or mimic the process of ANG II signaling[37] (also S. Chaki *et al.*, unpublished results). In hepatocytes it has been shown to be affected specifically by ANG II but not mediated by a phorbol ester.[36] Interestingly, mutagenesis studies have shown that a carboxyl tail region of rat AT_{1A} receptor is the sole determinant of the internalization because its deletion completely abolished the internalization of bound radioiodinated ANG II (S. Chaki, unpublished observations). The identification of the specific residues involved in this region will clarify the mechanism involved in the process of the internalization.

6.2. Regulation of Gene Expression

In regulating the expression of rat AT_{1A} gene, the 5′-noncoding region is expected to play decisive roles. It contains a TATA box,[37,38] three glucocorticoid-responsive elements,[37] one of which is overlapped with an AP-1 recognition element.[37] In addition, a cyclic AMP (cAMP) responsive element is present (Fig. 7).

Regulation of gene expression was studied *in vitro* with rat mesangial and aortic smooth muscle cells in culture. Makita *et al.*[39] found that treatment of the mesangial cells with ANG II or dibutylyl cAMP (as well as forskolin) markedly suppressed AT_{1A} mRNA. These effects appear to be mediated by neither protein kinase C nor a calmodulin-mediated process since staurosporin, depletion of protein kinase C by a prolonged treatment with a phorbol ester TPA, and treatment with the calmodulin inhibitor W-7 did not interfere with the extensive downregulation of AT_{1A} mRNA by ANG II. Interestingly, elevation of intracellular Ca^{2+} by the calcium ionophore A23187 failed to mimic the downregulating effect of ANG II. These observations indicate that the homologous downregulation by ANG II of its AT_1 receptor gene expression is not mediated by the elevated cytosolic Ca^{2+} or protein kinase C activation, as expected from the well-known downstream effects of ANG II binding. However, the phorbol ester TPA downregulates the AT_1 mRNA independent of ANG II as shown by Makita *et al.*[40] and Uno *et al.*[37] This latter mechanism may be explained by the activation of a phorbol ester-sensitive protein kinase C, whereas the homologous downregulation of AT_1 mRNA cannot be explained without invoking another mechanism, which may exert a negative modulatory effect in the 5′ noncoding region of the AT_{1A} gene.[37] Several glucocorticoid-responsive elements (GRE) also exist in the 5′-regulatory region and account for stimulation by aldosterone or dexamethasone. Interestingly, one of the GRE

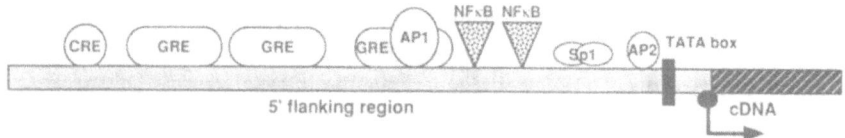

Figure 7. Illustration of the promotor regions structure in the 5′ flanking sequence of rat AT_{1A} gene. Note the presence of AMP-responsive elements (CRE), glucocorticoid-responsive elements (GRE), and domains (AP-1) which binds immediate early gene products. Initiation of mRNA synthesis occurs downstream of a TATA box

sites and an AP-1 site overlap each other. This results in a synergistic effect of the phorbol ester TPA and dexamethasone, producing a marked increase in the AT_1 mRNA level. However, such a synergism does not occur between ANG II and dexamethasone, again indicating that the effect of ANG II is not exerted through protein kinase C. The positive effect of aldosterone on the vascular smooth muscle is interesting since this effect is compounded by the stimulation of aldosterone production by ANG II in the adrenal. *In vitro* studies also showed that cAMP or forskolin inhibits the AT_1 receptor gene expression,[39] which can be a negative result of cAMP-responsive element in the upstream regions of the 5'-noncoding region of the gene.[37] A preliminary report on the genomic structure and the promotor region sequence with a TATA box had been reported.[40]

The cloning of AT_{1A} and AT_{1B} cDNA permitted the determination of tissue distribution of AT_1 mRNA by Northern blot analysis.[12–14,16] For a quantitative estimation of AT_1 expressed at low levels, various polymerase chain reaction (PCR) methods[40] and competitive PCR methods[41,42] have been developed. The presence of the two closely related subtypes, AT_{1A} and AT_{1B}, made their differential determinations necessary. The Northern blot analysis can be used for this purpose, using a probe for noncoding regions where base sequences are sufficiently different between AT_{1A} and AT_{1B}.[16,37] Since the coding sequences are similar to each other and the entire coding sequence is located in a single exon, specific competitive PCR methods were developed using primer sequences from noncoding sequence differences (specific) for the two subtypes.[17,29,41,42] Studies with these methods revealed that the liver, kidney, and adrenal are the major sites of AT_1 gene expression, whereas the heart (both ventricles and atria) and vasculature express it at lower levels.[15,40,41] AT_{1A} is the dominant form expressed in the liver, kidney, vasculature, lung, ovary, and heart, whereas AT_{1B} is expressed in greater quantities in the adrenal, anterior pituitary, and uterus.[16,17,28,41] The expression of these receptors in the brain varies from region to region. The cerebellum expresses more AT_{1B} than AT_{1A}, whereas the hypothalamus has more AT_{1A} than AT_{1B}.[29] Among the circumventricular organs with high concentrations of AT_1 receptors, the subfornical organ and organum vasculosum of lamina terminalis has more AT_{1B} than AT_{1A}, the area postrema expresses AT_{1A} and AT_{1B} in comparable quantities, and the median eminence expresses more AT_{1B} than AT_{1A}.

In view of the alleged sensitivity of spontaneously hypertensive rats (SHR) to ANG II, we examined the possibility of an elevated expression level in SHR tissues. However, no difference was seen in the expression levels of AT_{1A} in the kidney, liver, adrenal, and whole brain between SHR and the normotensive control [Wistar-Kyoto (WKY)] rats.[14] In the heart, a significant increase in AT_{1B} was observed by Iwai *et al.*[42] A positive response (upregulation) was seen in the adrenal expression to low salt feeding, whereas it caused no change in the kidney, liver, or brain.[14]

In vivo adrenal AT_{1A} response to manipulations that are expected to change ANG II is interesting. Bilateral nephrectomy or infusion of captopril or losartan for 4 days reduced adrenal AT_{1A} as well as that in the brainstem without affecting those in the kidney or aorta.[41] Another AT_1 antagonist, TCV 116, was found to downregulate adrenal AT_{1B} but not AT_{1A}. In the heart and aorta, both AT_{1A} and AT_{1B} were downregulated by TCV 116.[43] Prolonged infusion of ANG II (2 weeks) to a level to elicit mild hypertension (elevation of systolic blood pressure by 20 mmHg) significantly upregulated adrenal AT_{1A} but did not affect those in the kidney, aorta, or brainstem. Later studies by Iwai *et al.*[42] confirmed the downregulation of AT_{1A} in the adrenal in nephrectomized rats but AT_{1B} was found upregulated. AT_{1A} in the liver was also downregulated.

In rats, an increase in the cardiac afterload by aortic coarctation just above the right renal artery also caused a marked increase in the left ventricular AT_{1A} expression level in 5 days (M. Mizukoshi *et al.*, unpublished observation). In these animals, neither the plasma renin nor the ventricular tissue ANG II levels were elevated. Thus, parallel rises in the ventricular AT_1 expression and left ventricular hypertrophy appear to be linked by a causal relationship.

Although the AT_1 receptor expression levels are relatively stable during these *in vitro* manipulations, more detailed studies are needed to resolve some of the disagreements in results from different laboratories.

7. FUTURE DIRECTIONS

The cloning of AT_1 receptor has already resolved many important questions and confirmed the hypotheses postulated without using purified receptors. The use of cDNAs has made it possible to determine the level of AT_1 gene expression and also to transiently or permanently express native or mutated AT_1 receptors. These developments promise further clarification of the mechanism of the binding of ANG II or its antagonist, mechanisms of generation of signals in the receptor molecule and its transmission to G-proteins, and mechanisms of regulation of the receptor protein and mRNA levels. While AT_1 belongs to the superfamily of receptors with seven transmembrane domains, it has its own uniqueness, particularly as a receptor for a small peptide ligand. Rapid and interesting developments are expected.

ACKNOWLEDGMENT. I wish to express my sincere appreciation of my talented and diligent colleagues and collaborators. The editorial assistance of Mrs. Tina Stack is gratefully acknowledged. This work was supported by research grants HL35323 and HL14192, and training grant HL07323 from the National Institutes of Health.

REFERENCES

1. Peach MJ: Molecular actions of angiotensin. *Biochem Pharmacol* 30:2745–2751, 1981.
2. Douglas JC: Angiotensin receptor subtypes of the kidney cortex. *Am J Physiol* 253 (Renal Fluid Electrolyte Physiol 22): F1–F7, 1987.
3. Whitebread S, Mele M, Kamber B, et al: Preliminary biochemical characterization of two angiotensin II receptor subtypes. *Biochem Biophys Res Commun* 163:284–291, 1989.
4. Chiu AT, Herblin WF, McCall DE, et al: Identification of angiotensin II receptor subtypes. *Biochem Biophys Res Commun* 165:196–203, 1989.
5. Chang RSL, Lotti VJ: Two distinct angiotensin II receptor binding sites in rat adrenal revealed by new selective nonpeptide ligands. *Mol Pharmocol* 29:347–351, 1990.
6. Jackson TR, Blair LAC, Marshal J, et al: The *mas* oncogene encodes an angiotensin receptor. *Nature* 335:437–440, 1988.
7. Seed B: An LFA-3 cDNA encodes a phospholipid-linked membrane protein homologous to its receptor CD2. *Nature* 329:840–842, 1987.
8. Inagami T, Sasaki K, Iwai N. et al: Cloning and characterization of angiotensin II receptor and its regulation in bovine adrenocortical cells and rat kidney. *J Vascular Med Biol* 3:192–196, 1991.
9. Inagami T, Iwai N. Sasaki K, et al: Cloning, expression and regulation of angiotensin II receptors. *J Hypertension* 10:713–716, 1992.
10. Inagami T, Harris RC: Molecular insights into angiotensin II receptor subtypes. *News Physiol Sci* 8:215–218, 1993.
11. Bernstein KE, Alexander WR: Counterpoint: Molecular analysis of the angiotensin II receptor. *Endocrine Rev* 13:381–386, 1992.
12. Sasaki K, Yamano Y, Bardhan S, et al: Cloning and expression of a complementary DNA encoding a bovine adrenal angiotensin II type 1 receptor. *Nature* 351:230–233, 1991.
13. Murphy TJ, Alexander RW, Griendling KK, et al: Isolation of a cDNA encoding the vascular type-1 angiotensin II receptor. *Nature* 351:233–236, 1991.
14. Iwai N, Yamano Y, Chaki S, et al: Rat angiotensin II receptor: cDNA sequence and regulation of the gene expression. *Biophys Res Commun* 177:299–304, 1991.
15. Iwai N, Inagami T: Identification of two subtypes in the rat type 1 angiotensin II receptor. *FEBS Lett* 298:257–260, 1992.
16. Sandberg K, Ji H, Clanr JJI, et al: Cloning and expression of a novel angiotensin II receptor subtype. *J Biol Chem* 267:9544–9458, 1992.
17. Kakar SS, Selter SJC, Devor DC, et al: Angiotensin II type-a receptor subtype cDNAs: Differential tissue expression and hormonal regulation. *Biochem Biophys Res Commun* 183:1090–1096, 1992.
18. Elton TS, Stephan CC, Taylor GR, et al: Isolation of two distinct type I angiotensin II receptor genes. *Biochem Biophys Res Commun* 184:1067–1073, 1992.
19. Ernsberger P, Zhou J, Damon TH, et al: Angiotensin II receptor subtypes in cultured rat renal mesangial cells. *Am J Physiol* 263 (Renal Fluid Electrolyte Physiol 32):F411–F416, 1992.
20. Yoshida H, Kakiuchi J, Guo D-F, et al: Analysis of the evolution of angiotensin II type II

receptor gene in mammals (mouse, rat, bovine and human). *Biochem Biophys Res Commun* 186:1042–1049, 1992.

21. Sasamura H, Hein L, Krieger JE, et al: Cloning, characterization and expression of two angiotensin receptor (AT-1) isoforms from the mouse genome. *Biochem Biophys Res Commun* 185:253–259, 1992.

22. Furuta H, Guo D-F, Inagami T: Molecular cloning and sequencing of the gene encoding human angiotensin II type 1 receptor. *Biochem Biophys Res Commun* 183:8–13, 1992.

23. Takayanagi R, Ohnaka K, Sakai Y, et al: Molecular cloning, sequencing analysis and expression of a cDNA encoding human type-1 angiotensin II receptor. *Biochem Biophys Res Commun* 183:910–916, 1992.

24. Bergsma D-Z, Ellis C, Kumar C, et al: Cloning and characterization of a human angiotensin II type 1 receptor. *Biochem Biophys Res Commun* 183:989–995, 1992.

25. Mauzy CA, Hwang O, Egloff AM, et al: Cloning, expression and characterization of a gene encoding the human angiotensin II type 1A receptor. *Biochem Biophys Res Commun* 186:277–284, 1992.

26. Burns KD, Inagami T, Harris RC: Cloning, sequencing and expression of a rabbit kidney cortex angiotensin II receptor. *Am J Physiol* 246:F645–F654, 1993.

27. Szpirer C, Riviere M, Szpirer J, et al: The human and rat angiotensin II receptor genes: Chromosomal localization of the type 1 genes. *J Hypertens* 11:919–925, 1993.

28. Curnow KM, Pascoe L, White PC: Genetic analysis of the human type-1 angiotensin II receptor. *Mol Endocrinol* 6:1113–1118, 1992.

29. Kakar SS, Kristen KR, Neill JD: Differential expression of angiotensin II receptor subtype mRNAs (AT-1A and AT1-B) in the brain. *Biochem Biophys Res Commun* 185:688–692, 1992.

30. Carson MD, Leach-Harper CM, Baukal AJ, et al: Physiochemical characterization of photo affinity-labeled angiotensin II receptors. *Mol Endocrinol* 1:147–153, 1987.

31. Rondeau JJ, McNecoll N, Meloche S, et al: Hydrodynamic properties of the angiotensin II receptor from bovine adrenal zona glomerulosa. *Biochem J* 268:443–448, 1980.

32. Chang RSL, Lotti V, Keegan ME: Inactivation of angiotensin II receptors in bovine adrenal cortex by dithiothreitol. *Biochem Pharmacol* 31:1903–1906, 1982.

33. Ohnishi J, Ishido M, Shibata T, et al: The rat angiotensin II AT_{1A} receptor couples with three different signal transduction pathways. *Biochem Biophys Res Commun* 186:1094–1011, 1992.

34. Yamano Y, Ohyama K, Chaki S, et al: Identification of amino acid residues of rat angiotensin II receptor for ligand binding by site directed mutagenesis. *Biochem Biophys Res Commun* 187:1426–1431, 1992.

35. Ohyama K, Yamano Y, Chaki S, et al: Domains for G-protein coupling in angiotensin II receptor type 1: Studies by site-directed mutagenesis. *Biochem Biophys Res Commun* 189:677–683, 1992.

36. Bouscarel B, Wilson PB, Blackmore PF, et al: Agonist-induced down-regulation of the angiotensin II receptor in primary culture of rat hepatocytes. *J Biol Chem* 263:14920–14929, 1988.

37. Uno S, Guo DF, Ohi H, et al: Sequence analysis of the rat angiotensin II type 1A receptor promoter. *J Clin Invest* (submitted).

38. Langford K, Frenzel K, Martin BM, et al: The genomic organization of the rat AT_1 angiotensin receptor. *Biochem Biophys Res Commun* 183:1025–1032, 1992.

39. Makita N, Iwai N, Inagami T: Two distinct pathways in the down-regulation of type-1 angiotensin II receptor gene in rat glomerular mesangial cells. *Biochem Biophys Res Commun* 185:142–146, 1992.

40. Makita N, Fukunaga M, Iwai N, et al: Two distinct pathways in the down-regulation of

type-1 angiotensin II receptor gene in rat aortic smooth muscle cells (abstract). *Circulation* 86 (Suppl I):90, 1992.

41. Iwai N, Inagami T: Regulation of the expression of the rat angiotensin II receptor mRNA. *Biochem Biophys Res Commun* 182:1094–1099, 1992.

42. Iwai N, Inagami T, Ohnishi N, et al: Differential regulation of rat AT_{1A} and AT_{1B} receptor mRNA. *Biochem Biophys Res Commun* 188:298–303, 1992.

43. Kitami Y, Okura T, Marumoto K, et al: Differential gene expression and regulation of type-1 angiotensin II receptor subtypes in the rat. *Biochem Biophys Res Commun* 188:446–452, 1992.

2

Molecular Biology of Angiotensin II Receptors

Bernard Lassègue, Kathy K. Griendling, and R. Wayne Alexander

1. INTRODUCTION

Angiotensin II (ANG II), a central component of the renin–angiotensin–aldosterone system, helps to sustain blood pressure by affecting multiple targets.[1,2] ANG II stimulates glomerulosa cells of the adrenal cortex to release aldosterone, which is responsible for reabsorption of sodium and water in the kidney, and increases heart rate and contractile force and constricts blood vessel smooth muscle cells, thus increasing blood pressure. In the kidney, ANG II contracts glomerular mesangial cells and smooth muscle in efferent arterioles, resulting in a decrease in filtration. It also stimulates epithelial cells of the proximal tubules and promotes reabsorption of sodium and water. In addition, ANG II increases pituitary secretion of vasopressin, a vasoconstrictor and antidiuretic hormone. In the brain, ANG II induces thirst and salt appetite. Finally, ANG II increases catecholamine release by stimulating central sympathetic activity and secretion by adrenal chromaffin cells.

Much evidence indicates that ANG II may also participate in the development of pathological states. It can promote muscle cell growth and cardiovascular hypertrophy and can contribute to neointimal proliferation of smooth muscle after vessel injury.[3,4] The potential role of ANG II as a growth factor led to its implication in the pathogenesis of hypertension and atherosclerosis.[5]

Bernard Lassègue, Kathy K. Griendling, and R. Wayne Alexander • Cardiology Division, Emory University School of Medicine, Atlanta, Georgia 30322.

Angiotensin Receptors, edited by Juan M. Saavedra and Pieter B.M.W.M. Timmermans. Plenum Press, New York, 1994.

In order to elucidate the pathways underlying the striking multiplicity of effects of this single octapeptide hormone, many biochemical studies endeavored to isolate ANG II receptors, the cellular effectors that bind ANG II and transduce its physiological effects, and to dissect their signaling mechanisms. Considerable progress has resulted from the synthesis of nonpeptidic antagonists of ANG II and the subsequent characterization of receptor diversity. This was rapidly followed by the long-awaited successful cloning of one type of ANG II receptor. In this chapter, we will summarize these recent advances and attempt to clarify a rapidly expanding literature.

2. EARLY EVIDENCE OF ANG II RECEPTOR DIVERSITY

Experiments performed with diverse analogues of the angiotensin peptides for more than 15 years have suggested the existence of variations in the ANG II receptor molecule. For example, in the unanesthetized rat, ANG II and des-Asp[1]-ANG II (common peptidic analogues of ANG II are listed in Table I) displayed similar potencies in increasing serum aldosterone. Saralasin, a peptidic ANG II antagonist, blocked the effect of the former, but inhibited only

Table I. Common Peptidic Analogues of Human ANG II and ANG III and Their Usual Agonist Activity

Peptide	Amino acid no.										Agonist +/−
	1	2	3	4	5	6	7	8	9	10	
ANG I	Asp	Arg	Val	Tyr	Ile	His	Pro	Phe	His	Leu	+
ANG II	Asp	Arg	Val	Tyr	Ile	His	Pro	Phe			+
ANG III		Arg	Val	Tyr	Ile	His	Pro	Phe			+
ANG IV			Val	Tyr	Ile	His	Pro	Phe			+
Sar[1]–ANG II	Sar	Arg	Val	Tyr	Ile	His	Pro	Phe			+
Saralasin	Sar	Arg	Val	Tyr	Ile	His	Pro	Ala			−
Sarile	Sar	Arg	Val	Tyr	Ile	His	Pro	Ile			−
Sar[1]Gly[8]–ANG II	Sar	Arg	Val	Tyr	Ile	His	Pro	Gly			−
Sar[1]Leu[8]-ANG II	Sar	Arg	Val	Tyr	Ile	His	Pro	Leu			−
Sar[1]Thr[8]-ANG II	Sar	Arg	Val	Tyr	Ile	His	Pro	Thr			−
Sar[1]Val[5]Ala[8]ANG II	Sar	Arg	Val	Tyr	Val	His	Pro	Ala			−
Des-Asp[1]–ANG II		Arg	Val	Tyr	Ile	His	Pro	Phe			+
Des-Phe[8]–ANG II	Asp	Arg	Val	Tyr	Ile	His	Pro				+
Des-Asp[1]Ile[8]–ANG II		Arg	Val	Tyr	Ile	His	Pro	Ile			−

	1	2	3	4	5	6	7				
ANG III		Arg	Val	Tyr	Ile	His	Pro	Phe			+
Ile[7]-ANG III		Arg	Val	Tyr	Ile	His	Pro	Ile			−
Val[4]Ile[7]-ANG III		Arg	Val	Tyr	Val	His	Pro	Ile			−

80% of the effect of the latter. Furthermore, des-Asp[1]-ANG II was about 10 times less potent than ANG II in increasing arterial pressure.[6] Similarly, while both saralasin and des-Asp[1]-Ile[8]-ANG II were able to increase renal blood flow in anesthetized dogs with partial occlusion of the vena cava, only saralasin inhibited ANG II-induced contraction of isolated rabbit glomeruli.[7] In isolated rat hypothalamoneurohypophysial preparations, des-Phe[8]-ANG II, which has no vasopressor activity, and ANG II were equally potent in releasing vasopressin[8]

Interpretation of the above experiments is complicated by possible metabolism of peptides injected into whole animals. However, *in vitro* experiments also suggested the existence of different classes of binding sites. In rat hepatic membranes, while [125I]-ANG II bound to two populations of sites with different affinities, [125I]-saralasin only bound to the low-affinity site.[9] Similarly, saralasin preferentially bound to one of two sites present in a rabbit ventricle particulate fraction, and the GTP analogue Gpp(NH)p selectively reduced the affinity of the high-affinity site for its ligand.[10] Even more striking was the presence of two sites in rat liver membranes apparently coupled to different effector systems.[11] ANG II binding to the high-affinity site and stimulation of glycogen phosphorylase with corresponding potency were both inhibited by dithiothreitol. In contrast, binding to the low-affinity site and inhibition of glucagon-stimulated adenylate cyclase with corresponding potency were not affected by dithiothreitol.[11] Paradoxically, the presence of two sites and their differential sensitivity to dithiothreitol were not confirmed in rat cultured hepatocyte membranes,[12] perhaps due to differences in the source and preparation of the membranes. Two different ANG II receptors also seem to be expressed in the kidney cortex.[13] The receptor of glomerular mesangial cells has high affinity for both ANG II and ANG III, is downregulated by ANG II, and appears to be primarily coupled to phospholipase C activation. In contrast, the receptor of tubular epithelial cells has lower affinity for ANG II and ANG III, is upregulated by ANG II, and appears primarily coupled to inhibition of adenylate cyclase.[13] Such experiments were the basis for the widespread consensus among investigators in the field that there was heterogeneity in the ANG II receptor population.

3. DISCRIMINATION OF AT₁ AND AT₂ SITES WITH SPECIFIC LIGANDS

Diversity in the ANG II receptor population was finally proven unequivocally when new nonpeptidic ANG II antagonists were developed. Chang and Lotti[14] showed that displacement of [125I]-ANG II binding in rat adrenals by ANG II or ANG III only revealed one binding site, whereas competition with the nonpeptidic antagonists DuP 89 (losartan, the prototype AT₁ antagonist, is the potassium salt of DuP 89; see below) or WL 19 was clearly biphasic, suggesting the existence of two sites, each responsible for about 50% of total binding capacity. The dose–response relationship of each inhibitor in the presence of a high concentration of the other showed that one site preferentially bound DuP 89 and the other WL 19. This was the first direct evidence that two sites existed.

This discovery was quickly confirmed in similar studies. Two binding sites were also found in rat adrenal cortex and medulla, rat and human adrenal glomerulosa, and rat uterus (Table II), each characterized by affinity for a different antagonist. Other tissues, such as rabbit adrenal, liver, or vascular smooth muscle only had losartan-sensitive sites, whereas human uterus and PC12W cells exclusively contained WL 19-sensitive sites (Table II).

Table II. Relative Distribution of AT_1 and AT_2 Receptors
Determined by Binding Studies[a]

Species	Tissue	Percent AT_1	Percent AT_2	References
Rat	Whole adrenal	50	50	17
	Adrenal cortex	80	20	14, 177, 178
	Adrenal medulla	20	80	14, 17
	Adrenal glomerulosa	60	40	15, 20
	PC12W cells (adrenal chromaffin tumor)	0	100	16
	Kidney cortex	>90	<10	178
	Uterus	60	40	15
	Ovarian follicular granulosa cells	0	100	166
	Heart	>90	<10	178
	Aorta	60–70	30–36	19, 178
	Vascular smooth muscle	100	0	15, 23
	Liver	100	0	16, 18
	Brain	40	56	178
Mouse	R3T3 cells	0	100	21
Rabbit	Adrenal	100	0	18
	Adrenal cortex	>90	<10	178
	Kidney cortex	>90	<10	178
	Uterus	40	60	18
	Heart	64	32	178
	Aorta	>90	<10	178
	Brain	>90	<10	178
Monkey	Adrenal cortex	>90	<10	178
	Kidney cortex	45	58	178
	Heart	71	28	178
	Aorta	61	37	178
	Brain	85	12	178
Human	Adrenal glomerulosa	80	20	15
	Adrenal fasciculata-reticularis	>95	<5	79
	Uterus	0	100	15, 20, 23
	Vascular smooth muscle	100	0	15, 20, 23
Bovine	Cerebellar cortex	0	100	20

[a]Detailed distribution of receptors subtypes in brain nuclei can be found elsewhere.[31–33]

An interesting difference between the two binding sites, confirming the early binding observations, was their sensitivity to sulfhydryl-reducing reagents such as dithiothreitol and mercaptoethanol. These reagents usually greatly reduced the affinity and binding capacity of the losartan-sensitive site and increased the affinity of the other site.[14–17]

Another interesting divergent feature of the binding to the two sites was rapidly discovered: their sensitivity to GTP and nonhydrolyzable analogues. In tissues that contain only losartan-sensitive sites, such as rat liver, the affinity for ANG II was decreased in the presence of Gpp(NH)p or GTPγS.[17–20] In contrast, in tissues that have only WL 19-sensitive sites, such as bovine cerebellar cortex or PC12W cells, the affinity for ANG II was not affected by guanine nucleotide analogues.[17,19–21] In tissues that contain both types of binding sites, the effect of GTP analogues was seen when WL 19- but not losartan-sensitive sites were blocked with a high concentration of antagonist.[18,20] These results indicate that only one of the ANG II binding sites, the losartan-sensitive one, appears to interact with guanine nucleotide binding proteins (G-proteins). Furthermore, these binding sites have specific physiological and biochemical properties (see Sections 4, 17).

Together, these characteristics serve as a basis for classification. The accepted current nomenclature[22] recommends the name AT_1 (previously AII-1, AII-B, or AII_α) for the receptor blocked by losartan and AT_2 (previously AII-2, AII-A, or AII_β) for the PD 123319-sensitive site. A summary of the most common site-specific antagonists is presented in Table III.

4. PHYSIOLOGICAL EFFECTS MEDIATED BY AT_1 AND AT_2 RECEPTORS

In order to assess the importance of these newly characterized sites, a correlation with separate physiological functions was actively pursued. It became rapidly evident that all of the functions of ANG II known at the time were exclusively mediated by AT_1 receptors. Thus, ANG II-stimulated inositol phosphate formation in rat hepatocytes and vascular smooth muscle cells, calcium mobilization in rat cardiomyocytes and vascular smooth muscle cells, and contraction of rat mesangial cells and rabbit aorta were totally blocked (via an apparent competitive mechanism) by AT_1 and unaffected by AT_2 antagonists.[18,23–26] ANG II-induced inhibition of adenylate cyclase in rat liver membranes was also sensitive to AT_1 but not AT_2 antagonists.[27] In addition, ANG II-induced protein synthesis and [3H]thymidine incorporation in rat vascular smooth muscle cells were abolished by AT_1 antagonists.[26] Even in AT_2-rich tissues such as human uterus, ANG II-induced contraction was reduced by losartan and unaffected by the AT_2 antagonist PD 123319.[18]

Table III. Specific Ligands Used to Distinguish
AT_1 and AT_2 Receptors in Binding Studies[a]

AT₁			AT₂		
Compound	K_D	References	Compound	K_D	References
Losartan = Ex 89	10–100 nM	14–19, 23,	PD 123319	30 nM	18
= DuP 89		27, 31, 33,			
= DuP 753		60, 178			
= MK 954					
= PD 134756					
EXP 3174	1 nM	26, 172	CGP 42112A	0.24–1 nM	15, 23, 31
(= metabolite					
of losartan)					
EXP 9270	90 nM	64	EXP 326	100 μM	177
EXP 6803	100–160 nM	18, 64	EXP 655	30–300 nM	14, 19, 33
			= PD 123177		
SKF 108566	3 nM	49	WL 19 = PD 121981	18–78 nM	17, 178
TCV 116	ND	65	Ex 169	310 nM	23
GR 117289C	28 nM	64	p-aminoPhe⁶-ANG II	10 nM	16, 173

[a]All are nonpeptidic, except CGP 42112A, which is a hexapeptide. All are antagonists, except p-aminoPhe^6-ANG II.

Similarly, in animal studies, only AT₁ antagonists reduced blood pressure in renal hypertensive rats and inhibited the following ANG II-induced effects: pressor and tachycardiac response in control and pithed rats; aldosterone release in rat adrenal cortex and conscious rats; epinephrine secretion from rat adrenal gland; and water drinking and regulation of proximal tubule transport.[14,18,24,28] Losartan also inhibited the renal actions and drinking behavior induced in rats by intracerebroventricular injection of renin[29] and the pressor response to electrical stimulation of the subfornical organ.[30] In accord with these observations, histological and binding studies indicate that AT₁ receptors predominate in the forebrain, circum- and paraventricular nuclei, the hypothalamus and autonomic control centers of the medulla oblongata, and areas involved in the regulation of blood pressure, fluid and electrolyte balance, and pituitary hormone secretions.[31–35] Furthermore, AT₁ but not AT₂ antagonists reversed the firing pattern of individual neurons excited by ANG II.[36] AT₂ receptors were also present in different brain structures, but there is a marked species variation and their function is as yet unknown.[32–34]

Clearly, the discovery of AT₂ binding sites provided little explanation for the diversity of ANG II effects. The potential function of this site is under current investigation, and only recently has information emerged concerning its possible function and signaling mechanism (see Section 17). Recent work has centered on understanding the molecular basis for AT₁ receptor expression, coupling, and function.

5. CLONING OF THE AT₁ RECEPTOR

A more detailed understanding of the function of this receptor requires characterization of its signaling mechanisms at the molecular level. For over 10 years, different groups attempted to isolate the ANG II receptor using protein purification strategies. Numerous techniques were used, such as photoaffinity labeling and affinity chromatography,[37–44] which permitted characterization of two membrane-bound receptors and one cytoplasmic protein, but complete purification of the particulate proteins remained elusive, perhaps because of the instability of the receptor and its peptide ligands in the presence of detergents.[37,45]

The method of expression cloning in the mammalian COS-7 cell proved more fruitful, and a cDNA encoding an ANG II-binding protein was cloned simultaneously by two groups from rat vascular smooth muscle[46,47] and bovine adrenal gland.[48] These very similar clones had an open reading frame of 1077 bp (base pairs) corresponding to 359-amino acid, 41-kDa proteins (Fig. 1). Northern hybridization of mRNA from different tissues revealed a distribution consistent with that obtained in pharmacological studies of AT₁ receptors. Furthermore, ANG II induced inositol phosphate accumulation and calcium mobilization in the transfected COS-7 cells, indicating that the cDNAs coded functional receptors coupled to phospholipase C activation.[46,48] A cDNA for the human AT₁ receptor, encoding a protein 95% similar to its rat and bovine homologues, has also been cloned.[49–51] The cDNA for a turkey AT receptor has been cloned by reverse transcription of adrenal cortex mRNA and amplification (RT-PCR), using primers based on the smooth muscle sequence.[52] The protein coded by this clone is 78% similar to the rat AT₁ receptor, but has different affinities for losartan and PD 123177. At this time the function of this protein has not been confirmed by physiological data.

Several interesting features are suggested by the primary sequence of the AT₁ receptor.[46,48] A hydropathy plot of the amino acid sequence revealed seven hydrophobic domains thought to confer to the protein the seven transmembrane segment geometry common to many adenylate cyclase-stimulating and calcium-mobilizing, G-protein–coupled receptors, such as the muscarinic and β-adrenergic receptors.[53] Each of the four putative extracellular domains contains a cysteine residue. These are hypothesized to form two disulfide bonds and stabilize the tertiary structure of the receptor. This would explain the sensitivity of the receptor to sulfhydryl-reducing reagents noted above.[11] The putative extracellular sequences also feature three consensus *N*-glycosylation sites, consistent with reports from purification experiments showing that chemical or enzymatic deglycosylation of the receptor shifted its molecular mass from 65 to 40 kDa,[44] a figure in close agreement with the 41 kDa predicted molecular mass. Variations in the glycosylation of the receptor between tissues and species may explain the heterogeneity in molecular mass (up to 79 kDa) observed in purification studies.[54]

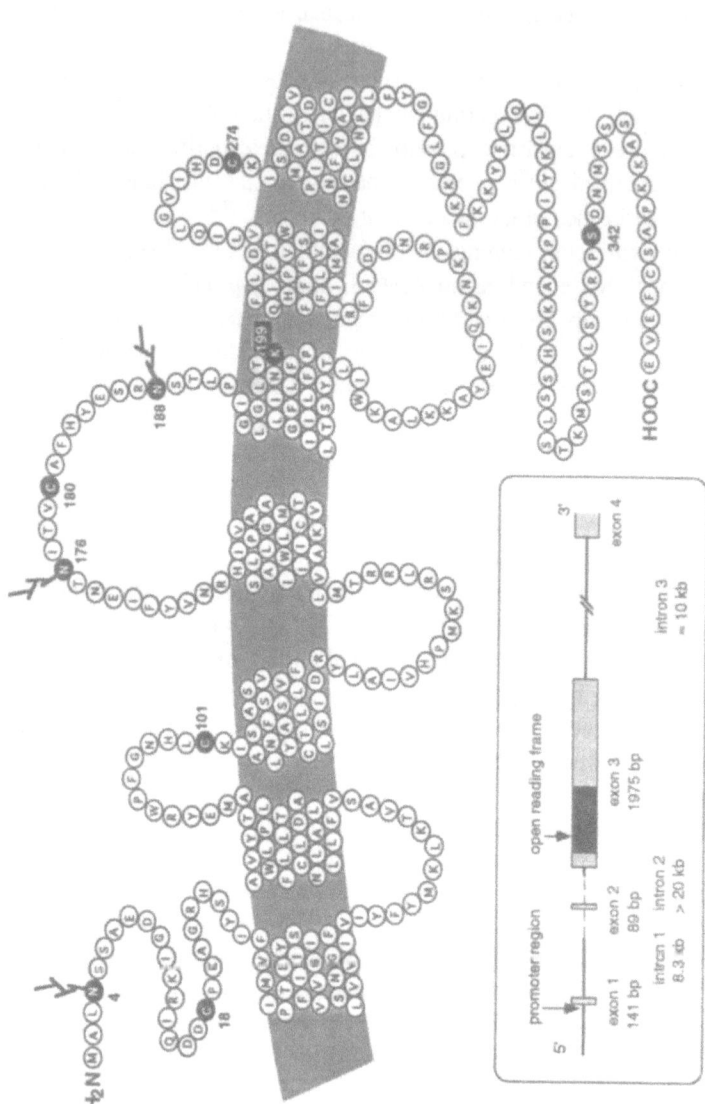

Figure 1. The rat AT$_{1A}$ receptor: gene organization and inferred protein conformation. The cDNA-deduced sequence is composed of 359 amino acids, corresponding to a 40,889 Da protein. Seven hydrophobic domains constitute putative transmembrane domains (gray layer). Three potential glycosylation sites are indicated: Asn 4, 176, and 188. Each extracellular domain contains one Cys residue (18, 101, 180, and 274) which may participate in ANG II binding, along with Lys 199. Ser 342 in the carboxy-terminal region is a potential site of phosphorylation by protein kinase A. Inset: gene organization. The gene includes at least three introns and four exons (boxes). The sequence around exon 2 is not entirely known (dotted line). The open reading frame (black box) is included in exon 3.

6. STRUCTURE–ACTIVITY RELATIONSHIP OF THE AT₁ RECEPTOR

To date, two studies designed to determine which amino acids are critical to the normal functioning of the AT_1 protein describe site-directed mutagenesis of the cDNA expressed transiently in COS-7 cells. The first report concentrated on mutations that may affect ANG II binding.[55] The replacement of any of the four cysteine residues from the extracellular domains by glycine markedly decreased affinity for [^{125}I]-ANG II. This confirms the results mentioned above demonstrating inhibition of radioligand binding by sulfhydryl-reducing reagents.[11] Mutation of the putative glycosylation sites and deletion of the intracellular carboxy-terminal tail did not affect binding. Mutations of positively charged lysine residues presumed to electrostatically bind the negatively charged ANG II were also performed. Replacement of lysine at position 199 in the fifth intramembrane domain (Fig. 1) by uncharged glutamine induced a marked inhibition of [^{125}I]-ANG II binding, suggesting that the ligand might be tightly associated with the receptor inside the plasma membrane.

The second study focused on the domains likely to interact with cytoplasmic G-proteins.[56] In each mutant a group of polar amino acids was replaced by neutral homologues. Mutations affecting either end of the second cytoplasmic loop, deletion of the cytoplasmic tail, and to a lesser degree mutation of the carboxy side of the third cytoplasmic loop, all inhibited the GTPγS-induced shift in agonist binding and inositol phosphate production. These results indicate which areas of the receptor may be important to signal transduction. Further studies will undoubtedly scrutinize the molecular basis of receptor coupling to the many cytoplasmic effectors activated by ANG II (see Section 13).

7. ORGANIZATION OF THE AT₁ RECEPTOR GENE

Another area of intense current interest is the organization of the AT_1 receptor gene and identification of the factors controlling its transcription. Fragments of the very long AT_1 gene were isolated from rat genomic libraries probed with segments of the previously cloned cDNA.[57,58] The gene encoding the AT_1 receptor is comprised of at least four exons and three introns (Fig. 1, inset) and spans over 80 kb.[57,59] The third exon includes the entire coding region and segments of the 5′ and 3′ untranslated flanking regions.[57,58] The absence of introns in the coding region was also later found in the mouse and human genes.[50,60] The region 5′ of the transcription initiation site includes the following consensus sequences: a putative TATA box; a cap site to allow binding of the mRNA to the small ribosomal subunit; putative transcription factor binding sites including several AP-1 and SP-1 binding sites; a cAMP-responsive element; and an HNF-1 binding element.[57,59] The presence of two bands on Northern blots of mRNA from different rat tissues suggested the existence of two transcripts (2.3 and 3.5 kb), possibly resulting from alternative splicing of the hnRNA.

To examine the possible existence of AT_1 subtypes, rat and mouse genomes were digested with several restriction endonucleases. Southern blots of the digests were hybridized with probes annealing to segments resistant to the digestion. Two bands were present in each sample, suggesting the existence of at least two rodent genes.[58,60] In contrast, a single copy of the AT_1 gene seemed to be present in the human genome.[49,50] Specific features of the human gene included the presence of an intron–exon junction just upstream of the translation initiation site, two putative polyadenylation sites, and six AUUUA motifs common to labile mRNA, consistent with a possible posttranscriptional regulation.[50]

8. CLONING OF AT_1 SUBTYPES: AT_{1A} AND AT_{1B}

Stimulated by the evidence for more than one AT_1 gene in the mouse and rat genomes, several groups simultaneously isolated variants of the rat and mouse AT_1 receptors.[58,60–63] In these reports, rat and mouse cDNA or genomic libraries were screened with fragments of the AT_1 cDNA or amplified using the polymerase chain reaction with primers based on the AT_1 sequence. In every case two clones were isolated. The first clone encoded a protein virtually identical to the rat vascular and very close to the bovine sequence. This clone was called AT_{1A}. The second clone that encoded a protein 90–94% similar to previous rat and bovine sequences was called AT_{1B}. The main features of the new clone are:

1. The protein is a 41-kDa molecule that is encoded by a single exon.[58,60]
2. The major differences in protein sequence between subtypes AT_{1A} and AT_{1B} are located in the carboxy terminal of the molecule,[60] the portion most likely to be modified by intracellular signals and thus important in regulation of function.
3. Potentially important differences in the primary protein sequence include two additional putative protein kinase C phosphorylation sites,[63] the absence of two other possible protein kinase C phosphorylation sites,[58,60] and the absence of a possible palmitoylation site.[58]

Significantly, there was only a 35% homology in the 5′ and 3′ untranslated regions between AT_{1A} and AT_{1B}, and the restriction maps of the genes were notably different.[58,61–63]

Pharmacological studies of the expressed cDNAs indicated that AT_{1A} and AT_{1B} coded for proteins with virtually identical properties[60–63]; ANG II had high affinity for both subtypes and was displaced exclusively by AT_1 antagonists. Binding to both subtypes was sensitive to GTPγS.[64] The only reported differences were a lower affinity of AT_{1B} for ANG I[62] (tenfold, not seen elsewhere[64]), and a decreased response to ANG II at concentrations above 1 μM,[63] similar to the observed decrease in ANG II-induced aldosterone secretion in adrenal glomerulosa cells. Functionally, like AT_{1A} receptors, AT_{1B} receptors expressed in COS-7 cells are coupled to phospholipase C, as evidenced by their ability to mobilize calcium.[62,63]

The majority of published reports used the terms AT_{1A} for the previously isolated rat vascular smooth muscle and bovine adrenal gland receptors and AT_{1B} for the new clones. In one case, the authors used the terminology AT_3 for a newly cloned rat adrenal cortex receptor.[63] However, this receptor is very similar to AT_{1B} by sequence, pharmacology, and tissue distribution (see Section 9); the term AT_{1B} seems, therefore, more appropriate.

9. TISSUE DISTRIBUTION OF THE AT_{1A} AND AT_{1B} SUBTYPES IN THE RAT

To determine possible functional differences between these receptors, relative tissue distributions were examined. Since there is no notable pharmacological difference between the subtypes, mRNA expression has been measured. However, since AT_1 probes used in previous studies hybridized to the nearly identical coding regions of AT_{1A} and AT_{1B}, the strategy was adopted of using PCR amplification of the untranslated divergent portions of the two subtypes. Both subtypes were expressed more or less equally in spleen, liver, and kidney.[62,63] However, AT_{1A} seemed to be predominant in vascular smooth muscle, heart, lung, ovary, and hypothalamus.[62,63,65,66] This preponderance in muscle suggests that AT_{1A} may be responsible for contraction.[66] The AT_{1B} subtype seemed to be predominant in anterior pituitary, adrenal gland, and uterus.[61–63,65,66] These data imply that this subtype might mediate aldosterone secretion from the adrenal and ACTH and prolactin secretion from the anterior pituitary.[66] In addition, the preferential expression of the AT_{1B} subtype in the subfornical organ and organum vasculosum of the lamina terminalis[66] raises the possibility that this subtype may be responsible for the induction of water drinking.

10. OTHER AT_1 RECEPTOR SUBTYPES

Long before the introduction of specific AT_1 and AT_2 antagonists, the research group of Douglas and co-workers extensively characterized the binding and function of two ANG II receptors in the kidney.[13,67–69] These investigators originally proposed the names type A and type B for these receptors (this classification has no connection with receptors distinguished by nonpeptidic antagonists also called A and B by some authors before the introduction of the currently accepted nomenclature corresponding to AT_2 and AT_1, respectively). Douglas's type A receptors predominate in the glomerular mesangium and are characterized by high affinity for both ANG II and ANG III, coupling to phospholipase C and calcium mobilization, and downregulation by ANG II. In contrast, Douglas's type B receptors are present in the tubular epithelium, have a low affinity for ANG III, are coupled to adenylate cyclase and phospholipase A_2, and are upregulated by prolonged exposure to ANG II. Since both of Douglas's receptors are coupled to G-proteins, they probably both belong to the AT_1 type. However, the terms AT_{1A} and AT_{1B}, used in recent papers from this group,[70,71] should probably be avoided since they have been used by most other laboratories to refer to the cloned receptors described in the previous section.

Recent experiments using nonpeptidic antagonists have been performed by Douglas's group in rat mesangial cell membranes.[70] The results indicate that Douglas's type A receptor is similar but not identical to the cloned AT_1 receptor. Both AT_1 and AT_2 receptors are known to have high affinity for ANG II (1–10 nM[46,48,49,51] and 0.6–3.5 nM,[21,72] respectively) and ANG III (5–30 nM[46,48,49,51] and 1.1–1.4 nM,[21,72] respectively). The AT_{1A} and AT_{1B} subtypes have similar relative affinities for ANG II and ANG III.[60–62] In experiments from Douglas's laboratory, different ligands displayed the following order of potency (K_d) for type A receptor: ANG II (1 nM) ≥ losartan (8 nM) > ANG III (129 nM) > > PD 123319 (24 μM) > CGP 42112A (167 μM). Binding to type B receptors was similar to AT_2 (high affinity for the PD compound, low affinity for losartan), except that they had low affinity for the CGP compound [1 μM vs. 1 nM for AT_2 (Table III)]. Binding to both types was sensitive to GTPγS and pertussis toxin. Another recent report by the same group,[71] indicated that both types of receptors coupled to activation of adenylate cyclase and phospholipase C. However, only type A coupled to protein synthesis. In view of the results described in this and previous sections, it appears that ANG II receptors of mesangial cells may be unique. Definitive classification will require cloning of these receptor subtypes.

11. NON-AT₁ AND NON-AT₂ ANG II BINDING SITES

Different authors have described ANG II binding sites that are not blocked by AT_1 or AT_2 antagonists. The data are still too recent to determine whether these sites correspond to a single new receptor. In three successive studies,[73–75]

Chaki and Inagami found that mouse neuroblastoma Neuro-2A cells differentiated with PGE_1 express a high-affinity (12 nM) ANG II binding site. [^{125}I]-ANG II binding was inhibited by low concentrations of ANG II ($IC_{50} = 7$ nM) and high concentrations of ANG III ($IC_{50} \approx 30$ μM), and was insensitive to micromolar concentrations of AT_1 and AT_2 antagonists. Binding was also inhibited by millimolar concentrations of dithiothreitol. The affinity for ANG II was not affected by stable analogues of GTP.[73] Stimulation of this putative receptor did not increase inositol phosphates or cAMP[75] but transiently increased cGMP (during the first minute of stimulation).[74,75] This transient fivefold increase in cGMP seemed to result from activation of a soluble guanylate cyclase, since it was inhibited by methylene blue. Coupling of the receptor to the cyclase was not direct, however, since it was dependent on calcium entry through a channel different from the L and N types.[75]

Le Noble *et al.*[76] studied ANG II-induced angiogenesis in chick embryo chorioallantoic membrane. In this model, the effect of ANG II could not be inhibited by micromolar concentrations of the AT_1 antagonist losartan or the AT_2 antagonist PD 123319. The response to ANG II was blocked by micromolar concentrations of the peptidic AT_2 antagonist CGP 42112A, which usually has nanomolar affinity for AT_2 sites (Table III). Binding assays revealed a single site having high affinity for ANG II (3 nM). ANG II binding was inhibited by CGP 42112A ($IC_{50} = 724$ nM) and losartan ($IC_{50} = 59$ μM) and not by PD 123319 ($IC_{50} > 100$ μM). Therefore, this receptor does not seem to belong to either AT_1 or AT_2 groups. Recent evidence suggests that avian ANG II receptors may differ pharmacologically from mammalian ones; the cloning of an avian ANG II receptor[52] may help to resolve this issue.

12. REGULATION OF EXPRESSION OF THE AT$_1$ mRNA

It has long been observed that the response to ANG II is altered in certain pathophysiological conditions, particularly those in which circulating levels of ANG II are affected. With the immunologic and molecular biologic tools now available, it has become possible to test the likelihood that the molecular basis of these differences resides in differential expression of AT_1 mRNA. Several groups have assessed this possibility by examining the factors that regulate AT_1 mRNA expression. Some studies have focused directly on specific regulation of the ANG II receptor by ANG II itself and other related biochemical stimuli, while others have examined receptor expression in animal models of vascular and renal disease.

Of the factors shown to be most important in the regulation of AT_1 mRNA expression, ANG II itself is among the most potent. ANG II decreased AT_1 mRNA by 50% in 6 hr in vascular smooth muscle cells,[77] by 70% in 6 hr in mesangial cells,[78] and by 20–50% in 24–96 hr in adrenal fasciculata reticularis.[79] Administration of an AT_1 antagonist for 1 week in the rat reduced AT_{1A} and AT_{1B} mRNAs by 66% in the heart and 52–33% in the aorta.[65] Dibutyryl cAMP and forskolin also decreased AT_1 mRNA in vascular smooth muscle[77] and mesangial cells.[78] In the latter model, two possible mechanisms contributed to downregulation of AT_1 receptors. The first pathway was dependent on ANG II stimulation; independent of calcium, protein kinase C, and activation of an inhibitory guanine nucleotide binding protein (G_i); and required protein synthesis. The second pathway was dependent on cAMP, independent of ANG II, and did not require protein synthesis.[78] These results suggest possible mechanisms by which ANG II receptor activation might become desensitized and raise the possibility of cross talk between signaling pathways.

In animal models in which the renin–angiotensin system (RAS) was compromised, AT_1 mRNA expression was also altered, although not necessarily in the direction predicted from the studies on cultured cells or isolated tissues described above. When the RAS was stimulated by a 4-week low-sodium diet (0.03% sodium vs. 3.15% sodium)[47] or a 2-week infusion of ANG II,[80] AT_1 mRNA expression was increased 2.3-fold in the adrenal gland.[47] Conversely, suppressing the RAS by a 24-hr bilateral nephrectomy[80] or administration of an AT_1 antagonist for several days[65,80] decreased AT_1 mRNA levels by 85%[80] and AT_{1B} mRNA levels by 65%[65] in the adrenal glands.[80] This is consistent with an upregulation by ANG II of its receptor in the adrenal gland, in contrast with the smooth muscle mentioned above. These results suggest that expression of the AT_1 receptor in the adrenal is upregulated by RAS activity.

AT_1 receptor expression has also been compared in different organs of Wistar-Kyoto (WKY) and spontaneously hypertensive rats (SHR). In these rats, hypertension does not appear to be primarily caused by changes in RAS activity. There was no apparent difference in mRNA levels between the two strains in kidney, liver, adrenal glands, or whole brain of 6- or 16-week-old animals.[47] However, AT_{1A} and AT_{1B} mRNAs were higher in specific areas of SHR brains compared to WKY. The increases were 3- and 20-fold in the hypothalamus and brainstem, respectively.[81] This difference was maintained when neuronal cells were cultured from the hypothalamus and brainstem of 1-day-old SHR and WKY. The expression of AT_{1A} and AT_{1B} mRNAs were fourfold higher in SHR than in WKY cultures. Furthermore, the AT_1 protein was also increased in cultures from SHR.[82] These results raise the possibility that hypertension in the SHR may result in part from an increased expression of AT_1 receptors in the specific brain nuclei that control water and electrolyte balance.

Since estrogens have been shown to regulate ANG II receptors in the anterior pituitary, the effect of estrogens on AT_1 receptor expression was examined in ovariectomized rats.[62] Four weeks after the operation, a large increase in AT_{1B} but not AT_{1A} was observed. This effect was reversed by administration of estrogen but not by progesterone,[62] confirming the regulatory function of estrogens on AT receptors in the pituitary.

Regulation of AT_1 expression during ontogeny of the rat has also been observed. In the liver, AT_1 mRNA was increased fivefold between day 15 of gestation and 5 days after birth, when adult levels were attained.[47,83] In the kidney, AT_1 mRNA levels are also higher in the newborn than in the adult.[47,83] Distribution within the kidney also changes with development. *In situ* hybridization studies indicated that AT_1 mRNA-expressing cells are widely distributed in the renal cortex of newborn rats and limited in the adult to glomeruli, arteries, and vasa recta.[83]

13. SIGNALING THROUGH THE AT_1 RECEPTOR

The binding of ANG II to its receptors at the surface of the plasma membrane activates specific signaling systems. These mechanisms had been studied extensively long before receptor classes were recognized. In retrospect, it is apparent that most, if not all, studies investigated ANG II signaling through AT_1 receptors. This receptor has been shown to be coupled to at least five effector systems that will be briefly reviewed in this section: phospholipases C, D, and A_2; adenylate cyclase; and calcium channels (Fig. 2). Additional evidence exists for coupling to other ion channels such as potassium and sodium.

Hormone coupling to different effectors is usually mediated by different receptor subtypes, each being linked to a specific effector. This is the case for α-adrenergic receptors. The α_1 subtype is coupled to phospholipase C activation, whereas the α_2 subtype mediates adenylate cyclase inhibition. To determine whether this is also true of ANG II receptors, the cloned rat AT_{1A} subtype was stably expressed in CHO cells.[84] In that system, the ANG II receptor appeared to be coupled to three effectors. First, ANG II induced a rapid and transient intracellular calcium increase that was reduced by the phospholipase C inhibitor neomycin. Second, the agonist induced a small and sustained elevation in intracellular calcium that was dependent on the presence of extracellular calcium and was inhibited by nicardipine, a calcium entry blocker. Finally, high ANG II concentrations were able to reduce forskolin-induced cAMP elevation.[84] These results indicate that the AT_{1A} receptor, at least, can couple to three different effectors in CHO cells.

13.1. Coupling to Phospholipase C

Coupling of a variety of receptors to phospholipase C has been extensively studied in a number of systems and reviewed elsewhere.[85] ANG II is coupled

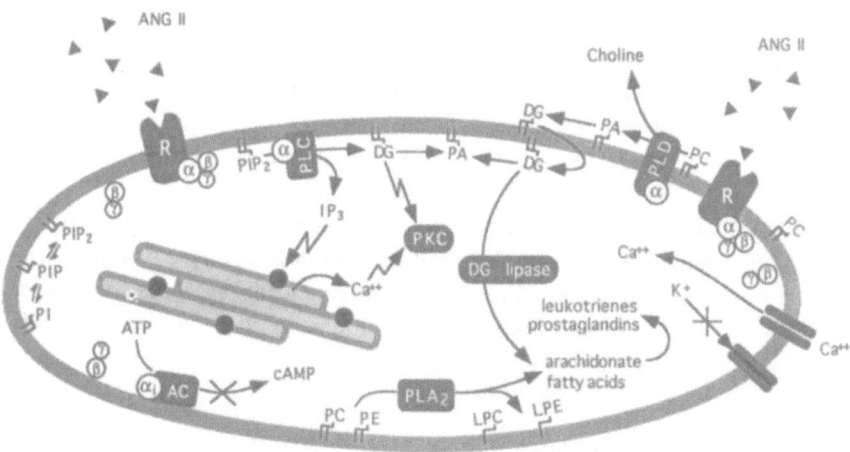

Figure 2. Signaling pathways coupled to the AT$_1$ receptor. ANG II binds to the AT$_1$ receptor (R), which activates phospholipases C (PLC), D (PLD), and A$_2$ (PLA$_2$) and inhibits adenylate cyclase (AC). The receptor is coupled to the effector enzymes by G-proteins (the trimers represented by α, β, and γ). α$_i$ inhibits adenylate cyclase, reducing the conversion of ATP to cAMP. Phospholipase C hydrolyzes phosphatidylinositol bisphosphate (PIP$_2$), a metabolite of phosphatidylinositol (PI) to produce inositol trisphosphate (IP$_3$) and diacylglycerol (DG). IP$_3$ binds to a specific receptor (black disk) that induces the release of calcium from an intracellular compartment. Calcium and DG synergistically activate protein kinase C (PKC). Phospholipase D hydrolyzes phosphatidylcholine (PC) into phosphatidic acid (PA) and choline. PA and DG can be interconverted; the latter can be degraded to fatty acids and glycerol by DG lipase. Phospholipase A$_2$ hydrolyzes PC and phosphatidylethanolamine (PE) into the corresponding lysophospholipids (LPC and LPE) and fatty acids. One fatty acid frequently released by PLA$_2$ and DG lipase is arachidonic acid, a precursor of leukotrienes and prostaglandins. ANG II also triggers the opening of calcium channels through which extracellular calcium can enter the cytoplasm and closes potassium channels, inhibiting diffusion of cytoplasmic potassium into the extracellular fluid.

to a rapid and transient activation of phospholipase C through a G-protein. In most systems this protein is insensitive to pertussis toxin[86–89] and, as has been shown for the muscarinic receptor, may belong to the G$_q$ family of G-proteins.[90] Although several phospholipase Cs have been isolated and cloned, it is still not entirely clear which subtype is involved in coupling to ANG II receptors. Activation of phospholipase C hydrolyzes phosphatidylinositol 4,5-bisphosphate, a minor plasma membrane phospholipid to generate the two second messengers: inositol 1,4,5-trisphosphate and diacylglycerol.[91,92] Inositol trisphosphate diffuses in the cell cytoplasm, binds to a specific intracellular receptor, and mobilizes calcium from internal stores.[85,91] Calcium itself acts as an intracellular messenger and activates calmodulin-dependent protein kinases. Together with calcium, diacylglycerol (the other product of phospholipase C stimulation) activates the serine/threonine protein kinase, protein kinase C.[93,94]

Both types of kinases then mediate specific cellular functions such as contraction or growth.

13.2. Coupling to Phospholipase D

ANG II can also activate phospholipase D.[95,96] As with phospholipase C, phospholipase D coupling to a number of receptors is also frequently mediated by G-proteins,[97,98] which in some cases is pertussis toxin- or cholera toxin-sensitive.[99–101] Phospholipase D may also be activated by protein kinase C or tyrosine phosphorylation, but these are not necessarily the primary activation pathways,[95,102] and their involvement in receptor-mediated phospholipase D activation remains to be demonstrated. Phospholipase D hydrolyzes phosphatidylcholine, a major phospholipid of the plasma membrane, and releases phosphatidic acid and choline. The preferential location of phosphatidylcholine in the external leaflet of the plasma membrane may explain the extracellular release of choline.[95] Although the possible direct function of phosphatidic acid as a second messenger is still controversial,[103] phosphatidic acid is converted to diacylglycerol by phosphatidate phosphohydrolase and may thus prolong diacylglycerol accumulation.[96] In vascular smooth muscle, phospholipase D activation appears to be more intense in growing cells or proliferating diseased vessels; it may, therefore, be involved in cell growth.[104,105]

13.3. Coupling to Phospholipase A$_2$

Phospholipase A$_2$ coupling to the ANG II receptor has been demonstrated in cells from proximal tubules,[67] vascular smooth muscle,[106] and mesangium.[107] Phospholipase A$_2$ hydrolyzes the bond in position 2, attaching fatty acids to the glycerol backbone of phospholipids such as phosphatidylcholine and phosphatidylethanolamine, thus releasing free fatty acids and lysophospholipids. The fatty acid from position 2 in phospholipids is frequently arachidonic acid. This fatty acid is the rate-limiting precursor for the generation of prostaglandins and leukotrienes. In proximal tubular cells, ANG II-stimulated phospholipase A$_2$ releases arachidonic acid[108] that can be further metabolized by cytochrome P$_{450}$ epoxygenase.[109,110] ANG II induces prostaglandin synthesis in vascular smooth muscle[106] and mesangial cells.[107] In the latter cell type, phospholipase A$_2$ activation is inhibited by pertussis toxin[111] and may be mediated by protein kinase C.[112]

13.4. Coupling to Adenylate Cyclase

The negative coupling of ANG II receptors to adenylate cyclase has been extensively studied. ANG II inhibits adenylate cyclase in several cell types, therefore attenuating the generation of the second messenger cAMP. In rat liver membranes, ANG II inhibits basal and GTP- or glucagon-stimulated adenylate cyclase activity.[113-115] Similar effects were observed in Leydig cells,[116] cultured bovine adrenal cell membranes,[117] and rat and rabbit heart myocytes or membranes.[118,119] Furthermore, the effect of ANG II was abolished by ADP-ribosylation of the G_i G-protein by pertussis toxin.[88,89,115-119] A major form of G_i that can couple ANG II receptors to adenylate cyclase in hepatocytes is G_{i3}.[120]

cAMP activates protein kinase A, which mediates such functions as decreasing proximal tubule epithelium sodium transport.[67] By inhibiting adenylate cyclase, ANG II opposes the effects of protein kinase A.

ANG II has also been reported to increase cAMP in bovine adrenocortical cells,[121-123] cultured fetal fibroblasts,[124,125] and vascular smooth muscle cells.[126] However, the effect of ANG II alone was small or was limited to potentiation of the accumulation induced by other agonists, such as forskolin or cholera toxin. Furthermore, this effect of ANG II seemed to be indirect, possibly mediated by protein kinase C activation and a calcium–calmodulin pathway.[123,125-127] Direct, positive coupling of adenylate cyclase to AT receptors thus remains to be conclusively demonstrated.

13.5. Coupling to Calcium Channels

ANG II is known to induce a sustained calcium influx from the extracellular medium.[128] Both voltage-dependent and receptor-operated channels may be involved.[129] Coupling to voltage-dependent calcium entry channels was shown to be mediated by a pertussis toxin-sensitive G-protein in adrenocortical cells.[130] In vascular smooth muscle, coupling of ANG II to calcium channels was mediated by a pertussis and cholera toxin-insensitive G-protein.[131] In the heart, ANG II may stimulate both T- and L-type voltage dependent calcium channels,[25,132,133] the activity of some of which is increased by protein kinase C.[134,135] Opening of calcium channels induces a small sustained elevation of cytoplasmic calcium concentration that may prolong the effect of the initial spike in calcium produced by mobilization of internal stores.

13.6. Coupling to Other Ion Channels

ANG II can inhibit potassium channels, possibly through a G-protein.[136-138] The resulting depolarization may be involved in determining the level of cellular activation. ANG II may also depolarize cells through the opening of sodium channels, secondary to protein kinase C activation.[139]

13.7. Temporal Sequence of Signaling Events: Example of Smooth Muscle

The signaling sequence in smooth muscle is similar to that of adrenal granulosa or glomerular mesangial cells. These biochemical events were initially studied in cultured vascular smooth muscle cells labeled with radioactive phospholipid precursors.[91,92] Within seconds of exposure to ANG II, phosphatidylinositol bisphosphate was hydrolyzed and diacylglycerol and inositol trisphosphate were increased. These observations indicate that a phosphoinositide-specific phospholipase C is rapidly activated by ANG II.[91,92] This activation was apparently not dependent on calcium, since it was not affected by an increase of cytoplasmic calcium concentration induced by an ionophore.[91] During the same time frame, a large and transient increase in cytoplasmic calcium concentration was observed, thought to result from inositol trisphosphate-induced mobilization of calcium from intracellular stores.[91] A cytoplasmic acidification, also taking place during that initial period of activation, was thought to be a consequence of the expulsion of calcium from the cytoplasm by the plasma membrane calcium-ATPase.[140]

Formation of inositol trisphosphate by activation of phospholipase C appeared to be the predominant intracellular signal only during the first minute of exposure to the agonist. At later times, diacylglycerol accumulation became the major signal.[92] After two minutes of exposure to ANG II, a sustained accumulation of phosphatidic acid was observed, soon followed by an increase of diacylglycerol. During this period, choline was also released from the cells. These events appeared to be a consequence of phospholipase D-mediated phosphatidylcholine hydrolysis, an observation that was confirmed by following the accumulation of phosphatidylethanol, a specific marker for phospholipase D.[95] The observed sustained diacylglycerol accumulation thus most likely derives from phosphatidic acid hydrolysis by phosphatidate phosphohydrolase.[96] As expected from diacylglycerol formation, protein kinase C activation was also sustained.[94] Diacylglycerol may then be metabolized by diacylglycerol lipase.[141,142] The late phase of signaling was also characterized by cytoplasmic alkalinization and an increase in intracellular sodium, due at least in part to activation of the sodium–proton exchanger.[143,144] An activation of phospholipase $A_2$112 and a prolonged calcium influx,[91,128] accompanied in some instances by oscillations in cytoplasmic calcium concentration,[144–146] were also observed.

The immediate consequence of these intracellular signals is the activation of protein kinases, including protein kinase C, tyrosine kinases,[147] and calcium–calmodulin-dependent protein kinase. These enzymes, in turn, phosphorylate a number of specific proteins including vimentin,[148] the myosin light chain,[149] nuclear lamins,[150] and MAP kinase.[147] It is the phosphorylation of these and other cellular proteins that ultimately mediates cellular functions.

The two phases of activation in vascular smooth muscle were also differentially regulated. The first phase was independent of cytoplasmic calcium and was attenuated by protein kinase C.[91,92,94,151] The second phase was enhanced by alkalinization,[94,143] dependent on calcium influx,[95] and independent of protein kinase C.[92,95] However, the development of the second phase of signaling, characterized by calcium influx and a sustained generation of diacylglycerol by activation of phospholipase D, required internalization of the ligand–receptor complex.[95,152,153] These results indicate that the two phases of smooth muscle contraction observed *in vitro* correspond to different biochemical signaling mechanisms.

13.8. Desensitization of the AT₁ Receptor

ANG II has long been shown to downregulate its own receptor. In rats in which the plasma concentration of ANG II was elevated there was a decreased number of ANG II receptors in vascular smooth muscle, glomeruli, or platelets.[154–157] This was confirmed in cultured vascular smooth muscle cells and hepatocytes exposed to ANG II.[154,158]

An important factor in the regulation of ANG II signaling is the availability of receptors at the cell surface. In rat vascular smooth muscle cells in culture, radioactive ANG II was internalized with a half-life of 1.5 min at 37°C; about 90% of the receptor was internalized within 10 min.[152,159,160] Similar results were found in cultured proximal tubule cells[161] and neuroblastoma cells N1E-115.[162] The internalization of the receptor–agonist complex was time-dependent. It was inhibited by decreasing the temperature (almost absent at 4°C) and incubation with agents that inhibit endocytosis such as phenylarsine oxide, colchicine, or cytochalasin D.[152,160,161] Experiments using ANG II–gold complexes and electron microscopy showed that the agonist was originally bound to the whole cell surface. Migration in the plane of the membrane then occurred, concentrating the receptor–agonist complex in coated pits. The complexes were then internalized in small vesicles. After 1 hr, the gold label was associated with lysosomes.[159] Recovery of the surface receptor after removal of the agonist occurred with a half-life of 15 min. About 25% of the internalized receptors were recycled to the plasma membrane, the rest were degraded, and complete recovery of surface receptors required protein synthesis.[154,160,162,163] Internalization of the agonist–receptor complex thus rapidly desensitized the cells to further agonist exposure. However, the receptor–ligand complex remained functional after internalization.[95,152]

14. SUBTYPES OF AT₂ BINDING SITES

The selective sensitivity of AT₁ receptors to guanine nucleotides has been used as a characteristic feature to distinguish them from AT₂ binding sites. However, recent data indicate that binding sites from specific rat brain nuclei that have typical AT₂ affinity for selective antagonists may also be sensitive to GTP analogues. [^{125}I-Sar¹]-ANG II binding to locus coeruleus, ventral thalamic, and medial geniculate nuclei showed the typical high affinity for CGP 42112A and PD 123177 and low affinity for losartan, characteristic of AT_2 binding sites.[164] Nonetheless, affinity for the radioligand was decreased by preincubation with GTPγS or pertussis toxin. Furthermore, the locus coeruleus was more sensitive than the other nuclei to the GTP analogue.[164] These results indicate the existence of a new subtype of AT_2 binding site that may be coupled to a G-protein.

Another report supporting the existence of subtypes of the AT_2 binding site in the rat was published by Speth.[165] Binding of [^{125}I]-CGP 42112 or [^{125}I]-sarile to fractions of the brain was enhanced by β-mercaptoethanol, a typical feature of AT_2 sites as described above. However, binding of the same ligands to adrenal gland homogenates was not sensitive to β-mercaptoethanol. This suggests the existence of two different AT_2 binding sites in rat brain and adrenal glands. Definitive classification of these binding sites will require development of specific antagonists and cloning of the AT_2 subtype(s).

As a first step toward characterizing the molecular features of this binding site, rat ovarian follicular granulosa cells that express exclusively AT_2 binding sites were incubated with [^{125}I]-ANG II. The radioligand was cross-linked to the receptor with disuccinimidyl suberate. After solubilization, proteins were separated by electrophoresis. The labeled receptor appeared as an M_r 79-kDa band on the autoradiogram. Binding of the radioligand was specific since it was inhibited by unlabeled agonist and did not occur after downregulation of the receptor with 8-bromo-cAMP.[166] Further characterization awaits purification or cloning.

15. TISSUE DISTRIBUTION AND ONTOGENY OF AT_2 BINDING SITES

Distribution of AT_2 binding sites was studied using autohistoradiography of nonspecific radioactive AT ligands followed by displacement with specific antagonists. Binding was detected in the rat fetus by day 11 of gestation and was maximal by days 19–21[167] but decreased markedly within 12 hr of birth. These highly abundant binding sites were almost exclusively of the AT_2 type and were most abundant in the buccal and gastrointestinal submucosal undifferentiated mesenchyme, the connective tissue and the choroid around the retina, the subdermal mesenchyme surrounding developing cartilage, and the diaphragm.[167] In addition, AT_2 sites were present in skeletal muscle and adrenal medulla,[168] the fetal aorta,[19] and in brain areas of newborn rats involved in the control and learning of motor activity, sensory areas, and some limbic system structures.[31,34] AT_2 sites in the anterior cerebral artery were also present in higher number in 2-week- compared with 8-week-old rats.[169] Together, these results suggest that AT_2 binding sites may have some function in development.

16. REGULATION OF AT_2 BINDING SITES

In cultured rat ovarian granulosa cells that only express AT_2[166] sites, ANG II binding was downregulated by serum or by incubation with 8-bromo-cAMP.[170] ANG II also downregulated its own binding site, an effect that could be mimicked by a calcium ionophore but not by phorbol esters.[170] To investigate the factors controlling the high expression of AT_2 binding sites in fetal tissue, fibroblasts were isolated from rat fetuses.[124] In primary culture, the proportion of binding sites classified AT_2 rapidly shifted from 90 to ≈30% after six days. Actinomycin D prevented this transition by increasing the number of AT_2 binding sites, suggesting that the expression of this subtype is controlled posttranscriptionally.[124] Further investigation of the effect of growth on AT_2 expression was studied by radioligand binding in the R3T3 cell line that selectively expresses this subtype.[21] Growing R3T3 cells express very few AT_2 sites. Both fetal bovine serum and bFGF markedly downregulated the binding site in quiescent cells: binding was almost abolished after a 24 hr incubation with 100 ng/ml bFGF. Incubation with AT_2 agonists or antagonists increased receptor numbers almost threefold after 24 hr, an effect that required protein synthesis.[21] Thus, consistent with its potential role in development, AT_2 binding site expression appears to be regulated by growth-related factors.

17. FUNCTION OF AT_2 BINDING SITES: SIGNALING

Recent observations indicate that central AT_2 binding sites may, along with AT_1 receptors, participate in the control of cardiovascular function. The rapid inhibition of the effects of intracerebroventricular ANG II by an AT_1 antagonist was more complete in the presence of an AT_2 antagonist.[171] The AT_2 antagonist alone caused a slow and sustained inhibition of the effect of ANG II.[172] The molecular basis of this effect has not been determined.

At this time, the only second messenger that has been functionally associated with the AT_2 binding site is cGMP. Neuron cultures from 1-day-old rats express AT receptors, about 80% belonging to the AT_2 subtype.[173] In these cells, 100 nM ANG II induced a 62% decrease in cGMP in 10 min.[173,174] The effect of ANG II was abolished by AT_2 antagonists and unaffected by losartan.[173] ANG II seemed to have no effect on soluble or particulate guanylate cyclase,[174] but the alteration in cGMP was abolished by calcium channel blockers and a nonselective phosphodiesterase inhibitor.[174] In PC12W cells and rat adrenal glomerulosa cells, AT_2 stimulation by ANG II also inhibited basal-

and atrial natriuretic peptide-stimulated particulate guanylate cyclase activity.[175] This inhibition was blocked by an inhibitor of phosphotyrosine phosphatase but not by an inhibitor of serine–threonine phosphatases.[175] Finally, in murine neuroblastoma N1E-115 cells, ANG II induced a dose-dependent transient three-fold increase in cGMP in 15 sec.[176] This effect was inhibited by losartan and also partially by blockade of the AT_2 binding sites.[176] Whether or not AT_2 sites utilize cGMP as a true intracellular second messenger remains controversial.

ANG II may induce internalization of the AT_2 receptor as described above in the case of the AT_1 subtype. However, this is still controversial since it was observed in murine neuroblastoma N1E-115 cells[162] with a slower time course than for the AT_1 subtype, but not in R3T3 cells.[21]

18. CONCLUSION

The past few years have witnessed a tremendous expansion of our knowledge of angiotensin II receptors. The new nonpeptidic antagonists, along with the cloning of the AT_1 receptors, have made it clear that although multiple subtypes of AT receptors exist, this heterogeneity does not explain the diversity of signal generation and tissue specificity. The AT receptors cloned to date couple to multiple intracellular signaling pathways, but the detailed mechanisms of coupling to different species of G-proteins is unknown. Knowledge of the factors controlling receptor–effector coupling could provide insight into control of diverse physiological functions. In addition, more data are needed to determine whether AT_{1A} and AT_{1B} receptors have specific functions. Pharmacological studies suggest a greater heterogeneity of receptor population than has so far been evident from cloning studies. Isolation or cloning of these molecules and/or the development of new selective pharmacological antagonists will provide definitive evidence concerning their structure and function. An additional area of major uncertainty surrounds the AT_2 binding site. Isolation of this protein will determine not only its status as a receptor but also its signaling pathways and its potential function in growth and development. Resolving these issues will provide a much-needed insight into the role of AT receptors in the development and maintenance of pathophysiological diseases such as hypertension and atherosclerosis. Understanding these proteins at the molecular level will direct the evolution of new therapeutic strategies for the treatment of vascular and renal disease.

NOTE ADDED IN PROOF. Since this chapter was written, two independent research groups simultaneously cloned an AT_2 receptor from different rat cDNA libraries.[179,180] Both clones encode indentical 363 amino acid proteins with seven putative transmembrane regions. This new protein is 33% identical to the AT_1 subtype and has features typical of an AT_2 receptor including binding specificity, effect of dithiothreitol on binding, and absence of coupling to phospholipase C. Tissue distribution and developmental expression of the mRNA is also in good agreement with AT_2 binding data. This receptor appears capable of mediating ANG II-induced inhibition of phosphotyrosine phosphatase.

REFERENCES

1. Bottari SP, de Gasparo M, Steckelings UM, et al: Angiotensin II receptor subtypes: Characterization, signaling mechanisms, and possible physiological implications. *Frontiers Neuroendocrinol* 2:123–171, 1993.

2. Peach MJ: Renin–angiotensin system: Biochemistry and mechanisms of action. *Physiol Rev* 57:313–370, 1977.

3. Dzau VJ, Gibbons GH, Pratt RE: Molecular mechanisms of vascular renin–angiotensin system in myointimal hyperplasia. *Hypertension* 18:II-100–II-105, 1991.

4. Schelling P, Fisher H, Ganten D: Angiotensin and cell growth: A link to cardiovascular hypertrophy? *J Hypertens* 9:3–15, 1991.

5. Krieger JE, Dzau VJ: Molecular biology of hypertension. *Hypertension* 18:I-3–I-17, 1991.

6. Campbell WB, Pettinger WA: Organ specificity of angiotensin II and Des-aspartyl angiotensin II in the conscious rat. *J Pharmacol Exp Ther* 198:450–456, 1976.

7. Caldicott WJH, Taub KJ, Margulies SS, et al: Angiotensin receptors in glomeruli differ from those in renal arterioles. *Kidney Int* 19:687–693, 1981.

8. Schiavone MT, Santos RAS, Brosnihan KB, et al: Release of vasopressin from the rat hypothalamoneurohypophysial system by angiotensin-(1-7) heptapeptide. *Proc Natl Acad Sci USA* 85:4095–4098, 1988.

9. Campanile CP, Crane JK, Peach MJ, et al: The hepatic angiotensin II receptor. I. Characterization of the membrane-binding site and correlation with physiological response in hepatocytes. *J Biol Chem* 257:4951–4958, 1982.

10. Wright GB, Alexander RW, Ekstein LS, et al: Characterization of the rabbit ventricular myocardial receptor for angiotensin II. Evidence for two sites of different affinities and specificities. *Mol Pharmacol* 24:213–221, 1983.

11. Gunther S: Characterization of angiotensin II receptor subtypes in rat liver. *J Biol Chem* 259:7622–7629, 1984.

12. Bouscarel B, Blackmore PF, Exton JH: Characterization of the angiotensin II receptor in primary cultures of rat hepatocytes. Evidence that a single population is coupled to two different responses. *J Biol Chem* 263:14913–14919, 1988.

13. Douglas JG: Angiotensin II receptor subtypes of the kidney cortex. *Am J Physiol* 253:F1–F7, 1987.

14. Chang RS, Lotti VJ: Two distinct angiotensin II receptor binding sites in rat adrenal revealed by new selective nonpeptide ligands. *Mol Pharmacol* 37:347–351, 1990.

15. Chiu AT, Herblin WF, McCall DE, et al: Identification of angiotensin II receptor subtypes. *Biochem Biophys Res Commun* 165:196–203, 1989.

16. Whitebread S, Mele M, Kamber B, et al: Preliminary biochemical characterization of two angiotensin II receptor subtypes. *Biochem Biophys Res Commun* 163:284–291, 1989.

17. Speth RC, Kim KH: Discrimination of two angiotensin II receptor subtypes with a selective agonist analogue of angiotensin II, *p*-aminophenylalanine[6] angiotensin II. *Biochem Biophys Res Commun* 169:997–1006, 1990.

18. Dudley DT, Panek RL, Major TC, et al: Subclasses of angiotensin II binding sites and their functional significance. *Mol Pharmacol* 38:370–377, 1990.

19. Viswanathan M, Tsutsumi K, Correa FM, et al: Changes in expression of angiotensin receptor subtypes in the rat aorta during development. *Biochem Biophys Res Commun* 179:1361–1367, 1991.

20. Bottari SP, Taylor V, King IN, et al: Angiotensin II AT$_2$ receptors do not interact with guanine nucleotide binding proteins. *Eur J Pharmacol* 207:157–163, 1991.

21. Dudley DT, Summerfelt RM: Regulated expression of angiotensin II (AT$_2$) binding sites in R3T3 cells. *Regul Pept* 44:199–206, 1993.

22. Bumpus FM, Catt KJ, Chiu AT, et al: Nomenclature for angiotensin receptors. A report of the Nomenclature Committee of the Council for High Blood Pressure Research. *Hypertension* 17:720–721, 1991.

23. Criscione L, Thomann H, Whitebread S, et al: Binding characteristics and vascular effects of various angiotensin II antagonists. *J Cardiovasc Pharmacol* 4:s56–s59, 1990.

24. Wong PC, Hart SD, Zaspel AM, et al: Functional studies of nonpeptide angiotensin II receptor subtype-specific ligands: DuP 753 (AII-1) and PD 123177 (AII-2). *J Pharmacol Exp Ther* 255:584–592, 1990.

25. Kem DC, Johnson EI, Capponi AM, et al: Effect of angiotensin II on cytosolic free calcium in neonatal rat cardiomyocytes. *Am J Physiol* 261:C77–C85, 1991.

26. Sachinidis A, Ko Y, Weisser P, et al: EXP3174, a metabolite of losartan (MK 954, DuP 753) is more potent than losartan in blocking the angiotensin II-induced responses in vascular smooth muscle cells. *J Hypertens* 11:155–162, 1993.

27. Bauer PH, Chiu AT, Garrison JC: DuP 753 can antagonize the effects of angiotensin II in rat liver. *Mol Pharmacol* 39:579–585, 1991.

28. Bernstein KE, Alexander RW: Counterpoint: Molecular analysis of the angiotensin II receptor. *Endocrinol Rev* 13:381–386, 1992.

29. Barbella Y, Cierco M, Israel A: Effect of losartan, a nonpeptide angiotensin II receptor antagonist, on drinking behavior and renal actions of centrally administered renin. *Proc Soc Exp Biol Med* 202:401–406, 1993.

30. Li Z, Bains JS, Ferguson AV: Functional evidence that the angiotensin antagonist losartan crosses the blood–brain barrier in the rat. *Brain Res Bull* 30:33–39, 1993.

31. Tsutsumi K, Saavedra JM: Characterization and development of angiotensin II receptor subtypes (AT$_1$ and AT$_2$) in rat brain. *Am J Physiol* 261:R209–R216, 1991.

32. Aldred GP, Chai SY, Song K, et al: Distribution of angiotensin II receptor subtypes in the rabbit brain. *Regul Pept* 44:119–130, 1993.

33. Barnes JM, Steward LJ, Barber PC, et al: Identification and characterization of angiotensin II receptor subtypes in human brain. *Eur J Pharmacol* 230:251–258, 1993.

34. Millan MA, Jacobowitz DM, Aguilera G, et al: Differential distribution of AT$_1$ and AT$_2$ angiotensin II receptor subtypes in the rat brain during development. *Proc Natl Acad Sci USA* 88:11440–11444, 1991.

35. Bunnemann B, Iwai N, Metzger R, et al: The distribution of angiotensin II AT$_1$ receptor subtype mRNA in the rat brain. *Neurosci Lett* 142:155–158, 1992.

36. Barnes KL, McQueeney AJ, Ferrario CM: Receptor subtype that mediates the neuronal effects of angiotensin II in the rat dorsal medulla. *Brain Res Bull* 31:195–200, 1993.

37. Weber V, Monnot C, Bihoreau C, et al: The difficult challenge of cloning the angiotensin II receptor. *Horm Res* 34:101–104, 1990.

39. Elton TS, Dion LD, Bost KL, et al: Purification of an angiotensin II binding protein by using antibodies to a peptide encoded by angiotensin II complementary RNA. *Proc Natl Acad Sci USA* 85:2518–2522, 1988.

39. Guillemette G, Escher E: Analysis of the adrenal angiotensin II receptor with the photoaffinity labeling method. *Biochemistry* 22:5591–5596, 1983.

40. Kwok YG, Moore GJ: Comparison of angiotensin receptors in isolated smooth muscle tissues by photoaffinity labeling. *Eur J Pharmacol* 115:53–58, 1985.

41. Guillemette G, Guillon G, Marie J, et al: High yield photoaffinity labeling of angiotensin II receptors. *Mol Pharmacol* 30:544–551, 1986.

42. Carson MC, Harper CM, Baukal AJ, et al: Physicochemical characterization of photoaffinity-labeled angiotensin II receptors. *Mol Endocrinol* 1:147–153, 1987.

43. Rondeau JJ, McNicoll N, Escher E, et al: Hydrodynamic properties of the angiotensin II receptor from bovine adrenal zona glomerulosa. *Biochem J* 268:443–448, 1990.

44. Desarnaud F, Marie J, Lombard C, et al: Deglycosylation and fragmentation of purified rat liver angiotensin II receptor: Application to the mapping of hormone-binding domains. *Biochem J* 289:289–297, 1993.

45. Murphy TJ, Takeuchi K, Alexander RW: Molecular cloning of AT_1 angiotensin receptors. *Am J Hypertens* 5:236s–242s, 1992.

46. Murphy TJ, Alexander RW, Griendling KK, et al: Isolation of a cDNA encoding the vascular type-1 angiotensin II receptor. *Nature* 351:233–236, 1991.

47. Iwai N, Yamano Y, Chaki S, et al: Rat angiotensin II receptor: cDNA sequence and regulation of the gene expression. *Biochem Biophys Res Commun* 177:299–304, 1991.

48. Sasaki K, Yamano Y, Bardhan S, et al: Cloning and expression of a complementary DNA encoding a bovine adrenal angiotensin II type-1 receptor. *Nature* 351:230–233, 1991.

49. Bergsma DJ, Ellis C, Kumar C, et al: Cloning and characterization of a human angiotensin II type 1 receptor. *Biochem Biophys Res Commun* 183:989–995, 1992.

50. Furuta H, Guo DF, Inagami T: Molecular cloning and sequencing of the gene encoding human angiotensin II type 1 receptor. *Biochem Biophys Res Commun* 183:8–13, 1992.

51. Takayanagi R, Ohnaka K, Sakai Y, et al: Molecular cloning, sequence analysis and expression of a cDNA encoding human type-1 angiotensin II receptor. *Biochem Biophys Res Commun* 183:910–916, 1992.

52. Carsia RV, McIlroy PJ, Kowalski KI, et al: Isolation of turkey adrenocortical cell angiotensin II (AII) receptor partial cDNA: Evidence for a single-copy gene expressed predominantly in the adrenal gland. *Biochem Biophys Res Commun* 191:1073–1080, 1993.

53. Dohlman HG, Caron MG, Lefkowitz RJ: A family of receptors coupled to guanine nucleotide regulatory proteins. *Biochemistry* 26:2657–2664, 1987.

54. Catt KJ, Carson MC, Hausdorff WP, et al: Angiotensin II receptors and mechanisms of action in adrenal glomerulosa cells. *J Steroid Biochem* 27:915–927, 1987.

55. Yamano Y, Ohyama K, Chaki S, et al: Identification of amino acid residues of rat angiotensin II receptor for ligand binding by site-directed mutagenesis. *Biochem Biophys Res Commun* 187:1426–1431, 1992.

56. Ohyama K, Yamano Y, Chaki S, et al: Domains for G-protein coupling in angiotensin II receptor type I: Studies by site-directed mutagenesis. *Biochem Biophys Res Commun* 189:677–683, 1992.

57. Langford K, Frenzel K, Martin BM, et al: The genomic organization of the rat AT_1 angiotensin receptor. *Biochem Biophys Res Commun* 183:1025–1032, 1992.

58. Elton TS, Stephan CC, Taylor GR, et al: Isolation of two distinct type I angiotensin II receptor genes. *Biochem Biophys Res Commun* 184:1067–1073, 1992.

59. Takeuchi K, Murphy TJ, Nakamura Y, et al: Molecular cloning of the rat vascular AT_1 angiotensin II receptor gene. *Hypertension* 20:410, 1992.

60. Sasamura H, Hein L, Krieger JE, et al: Cloning, characterization, and expression of two angiotensin receptor (AT-1) isoforms from the mouse genome. *Biochem Biophys Res Commun* 185:253–259, 1992.

61. Iwai N, Inagami T: Identification of two subtypes in the rat type I angiotensin II receptor. *Febs Lett* 298:257–260, 1992.

62. Kakar SS, Sellers JC, Devor DC, et al: Angiotensin II type-1 receptor subtype cDNAs: Differential tissue expression and hormonal regulation. *Biochem Biophys Res Commun* 183:1090–1096, 1992.

63. Sandberg K, Ji H, Clark AJL, et al: Cloning and expression of a novel angiotensin II receptor subtype. *J Biol Chem* 267:9455–9458, 1992.

64. Chiu AT, Dunscomb JH, McCall DE, et al: Characterization of angiotensin AT_{1A} receptor isoform by its ligand binding signature. *Regul Pept* 44:141–147, 1993.

65. Kitami Y, Okura T, Marumoto K, et al: Differential gene expression and regulation of type-1 angiotensin II receptor subtypes in the rat. *Biochem Biophys Res Commun* 188:446–452, 1992.

66. Kakar SS, Riel KK, Neill JD: Differential expression of angiotensin II receptor subtype mRNAs (AT-1A and AT-1B) in the brain. *Biochem Biophys Res Commun* 185:688–692, 1992.

67. Douglas JG, Romero M, Hopfer U: Signaling mechanisms coupled to the angiotensin receptor of proximal tubular epithelium. *Kidney Int Suppl* 30:s43–s47, 1990.

68. Brown GP, Douglas JG, Krontiris-Litowitz J: Properties of angiotensin II receptors of isolated rat glomeruli: Factors influencing binding affinity and comparative binding of angiotensin analogues. *Endocrinology* 106:1923–1929, 1980.

69. Brown GP, Douglas JG: Angiotensin II binding sites in rat and primate isolated renal tubular basolateral membranes. *Endocrinology* 112:2007–2014, 1983.

70. Ernsberger P, Zhou J, Damon TH, et al: Angiotensin II receptor subtypes in cultured rat mesangial cells. *Am J Physiol* 263:F411–F416, 1992.

71. Madhun ZT, Ernsberger P, Ke FC, et al: Signal transduction mediated by angiotensin II receptor subtypes expressed in rat renal mesangial cells. *Regul Pept* 44:149–157, 1993.

72. Heemskerk FM, Zorad S, Seltzer A, et al: Characterization of brain angiotensin II AT_2 receptor subtype using [^{125}I]-CGP 42112A. *Neuroreport* 4:103–105, 1993.

73. Chaki S, Inagami T: Identification and characterization of a new binding site for angiotensin II in mouse neuroblastoma neuro-2A cells. *Biochem Biophys Res Commun* 182:388–394, 1992.

74. Chaki S, Inagami T: A newly found angiotensin II receptor subtype mediates cyclic GMP formation in differentiated neuro-2A cells. *Eur J Pharmacol* 225:355–356, 1992.

75. Chaki S, Inagami T: New signaling mechanism of angiotensin II in neuroblastoma neuro-2A cells: Activation of soluble guanylyl cyclase via nitric oxide synthesis. *Mol Pharmacol* 43:603–608, 1993.

76. Le Noble FA, Schreurs NH, van Straaten HW, et al: Evidence for a novel angiotensin II receptor involved in angiogenesis in chick embryo chorioallantoic membrane. *Am J Physiol* 264:R460–R465, 1993.

77. Lassègue B, Griendling KK, Murphy TJ, et al: Regulation of angiotensin II receptor expression in vascular smooth muscle cells. *FASEB J* 6:A1859, 1992.

78. Makita N, Iwai N, Inagami T, et al: Two distinct pathways in the downregulation of type-1 angiotensin II receptor gene in rat glomerular mesangial cells. *Biochem Biophys Res Commun* 185:142–146, 1992.

79. Naville D, Lebrethon MC, Kermabon AY, et al: Characterization and regulation of the angiotensin II type-1 receptor (binding and mRNA) in human adrenal fasciculata-reticularis cells. *FEBS Lett* 321:184–188, 1993.

80. Iwai N, Inagami T: Regulation of the expression of the rat angiotensin II receptor mRNA. *Biochem Biophys Res Commun* 182:1094–1099, 1992.

81. Raizada MK, Sumners C, Lu D: Angiotensin II type 1 receptor mRNA levels in the brains of normotensive and spontaneously hypertensive rats. *J Neurochem* 60:1949–1952, 1993.

82. Raizada MK, Lu D, Tang W, et al: Increased angiotensin II type-1 receptor gene expression in neuronal cultures from spontaneously hypertensive rats. *Endocrinology* 132:1715–1722, 1993.

83. Tufro-McReddie A, Harrison JK, Everett AD, et al: Ontogeny of type 1 angiotensin II receptor gene expression in the rat. *J Clin Invest* 91:530–537, 1993.

84. Ohnishi J, Ishido M, Shibata T, et al: The rat angiotensin II AT_{1A} receptor couples with three different signal transduction pathways. *Biochem Biophys Res Commun* 186:1094–1101, 1992.

85. Berridge MJ, Irvine RF: Inositol phosphates and cell signaling. *Nature* 341:197–205, 1989.

86. Socorro L, Alexander RW, Griendling KK: Cholera toxin modulation of angiotensin II-stimulated inositol phosphate production in cultured vascular smooth muscle cells. *Biochem J* 265:799–807, 1990.

87. Enyedi P, Mucsi I, Hunyady L, et al: The role of guanyl nucleotide binding proteins in the formation of inositol phosphates in adrenal glomerulosa cells. *Biochem Biophys Res Commun* 140:941–947, 1986.

88. Lynch CJ, Prpic V, Blackmore PF, et al: Effect of islet-activating pertussis toxin on the binding characteristics of Ca^{2+}-mobilizing hormones and on agonist activation of phosphorylase in hepatocytes. *Mol Pharmacol* 29:196–203, 1986.

89. Hausdorff WP, Sekura RD, Aguilera G, et al: Control of aldosterone production by angiotensin II is mediated by two guanine nucleotide regulatory proteins. *Endocrinology* 120:1668–1678, 1987.

90. Berstein G, Blank JL, Smrcka AV, et al: Reconstitution of agonist-stimulated phosphatidylinositol 4,5-bisphosphate hydrolysis using purified m_1 muscarinic receptor, Gq/11, and phospholipase C-β_1. *J Biol Chem* 267:8081–8088, 1992.

91. Brock TA, Alexander RW, Ekstein LS, et al: Angiotensin increases cytosolic free calcium in cultured vascular smooth muscle cells. *Hypertension* 7:I-105–109, 1985.

92. Griendling KK, Rittenhouse SE, Brock TA, et al: Sustained diacylglycerol formation from inositol phospholipids in angiotensin II-stimulated vascular smooth muscle cells. *J Biol Chem* 261:5901–5906, 1986.

93. Nishizuka Y: The role of protein kinase C in cell surface signal transduction and tumour promotion. *Nature* 308:693–698, 1984.

94. Griendling KK, Tsuda T, Berk BC, et al: Angiotensin II stimulation of vascular smooth muscle. *J Cardiovasc Pharmacol* 14:S27–39, 1989.

95. Lassègue B, Alexander RW, Clark M, et al: Angiotensin II-induced phosphatidylcholine hydrolysis in cultured vascular smooth-muscle cells. Regulation and localization. *Biochem J* 276:19–25, 1991.

96. Lassègue B, Alexander RW, Clark M, et al: Phosphatidylcholine is a major source of phosphatidic acid and diacylglycerol in angiotensin II-stimulated vascular smooth muscle cells. *Biochem J* 292:509–317, 1993.

97. Martin TW, Michaelis K: P2-purinergic agonists stimulate phosphodiesteratic cleavage of phosphatidylcholine in endothelial cells. Evidence for activation of phospholipase D. *J Biol Chem* 264:8847–8856, 1989.

98. Anthes JC, Eckel S, Siegel MI, et al: Phospholipase D in homogenates from HL-60 granulocytes: Implications of calcium and G protein control. *Biochem Biophys Res Commun* 163:657–664, 1989.

99. Agwu DE, McPhail LC, Chabot MC, et al: Choline-linked phosphoglycerides. *J Biol Chem* 264:1405–1413, 1989.

100. Kanaho Y, Kanoh H, Nozawa Y: Activation of phospholipase D in rabbit neutrophils by fMet-Leu-Phe is mediated by a pertussis toxin-sensitive GTP-binding protein that may be distinct from a phospholipase C-regulating protein. *Febs Lett* 279:249–252, 1991.

101. Qian Z, Drewes LR: Muscarinic acetylcholine receptor regulates phosphatidylcholine phospholipase D in canine brain. *J Biol Chem* 264:21720–21724, 1989.

102. Garland LG: New pathways of phagocyte activation: The coupling of receptor-linked phospholipase D and the role of tyrosine kinase in primed neutrophils. *FEMS Microbiol Immunol* 105:229–238, 1992.

103. Billah MM, Anthes JC: The regulation and cellular functions of phosphatidylcholine hydrolysis. *Biochem J* 269:281–291, 1990.

104. Ollerenshaw JD, Lassègue B, Alexander RW, et al: Intracellular signaling in arteries and vascular smooth muscle cells in culture. In Mulvany MJ, Aalkjaer C, Heagerty AM, et al (eds): *Resistance Arteries, Structure and Function.* Amsterdam, Elsevier, 1991, pp 73–76.

105. Ohanian J, Ollerenshaw J, Collins P, et al: Agonist-induced production of 1,2-diacylglycerol and phosphatidic acid in intact resistance arteries. Evidence that accumulation of diacylglycerol is not a prerequisite for contraction. *J Biol Chem* 265:8921–8928, 1990.

106. Alexander RW, Gimbrone MAJ: Stimulation of prostaglandin E synthesis in cultured human umbilical vein smooth muscle cells. *Proc Natl Acad Sci USA* 73:1617–1620, 1976.

107. Schlondorff D, DeCandido S, Satriano JA: Angiotensin II stimulates phospholipases C and A_2 in cultured rat mesangial cells. *Am J Physiol* 253:C113–C120, 1987.

108. Morduchowicz GA, Sheikh-Hamad D, Dwyer BE, et al: Angiotensin II directly increases rabbit renal brush-border membrane sodium transport: Presence of local signal transduction system. *J Membr Biol* 122:43–53, 1991.

109. Madhun ZT, Goldthwait DA, McKay D, et al: An epoxygenase metabolite of arachidonic acid mediates angiotensin II-induced rises in cytosolic calcium in rabbit proximal tubule epithelial cells. *J Clin Invest* 88:456–461, 1991.

110. Burns KD, Homma T, Harris RC: The intrarenal renin–angiotensin system. *Semin Nephro* 13:13–30, 1993.

111. Pfeilschifter J, Bauer C: Pertussis toxin abolishes angiotensin II-induced phosphoinositide hydrolysis and prostaglandin synthesis in rat renal mesangial cells. *Biochem J* 236:289–294, 1986.

112. Lang U, Vallotton MB: Effects of angiotensin II and of phorbol ester on protein kinase C activity and on prostacyclin production in cultured rat aortic smooth-muscle cells. *Biochem J* 259:477–483, 1989.

113. Jard S, Cantau B, Jakobs KH: Angiotensin II and α-adrenergic agonists inhibit rat liver adenylate cyclase. *J Biol Chem* 256:2603–2606, 1981.

114. Crane JK, Campanile CP, Garrison JC: The hepatic angiotensin II receptor. II. Effect of guanine nucleotides and interaction with cyclic AMP production. *J Biol Chem* 257:4959–4965, 1982.

115. Pobiner BF, Hewlett EL, Garrison JC: Role of N_i in coupling angiotensin receptors to inhibition of adenylate cyclase in hepatocytes. *J Biol Chem* 260:16200–16209, 1985.

116. Khanum A, Dufau ML: Angiotensin II receptors and inhibitory actions in Leydig cells. *J Biol Chem* 263:5070–5074, 1988.

117. Bégeot M, Langlois D, Penhoat A, et al: Variations in guanine-binding proteins (Gs, Gi) in cultured bovine adrenal cells. Consequences on the effects of phorbol ester and angiotensin II on adrenocorticotropin-induced and cholera toxin-induced cAMP production. *Eur J Biochem* 174:317–321, 1988.

118. Allen IS, Gaa ST, Rogers TB: Changes in expression of a functional G_i protein in cultured rat heart cells. *Am J Physiol* 255:C51–C59, 1988.

119. Anand-Srivastava MB: Angiotensin II receptors negatively coupled to adenylate cyclase in rat myocardial sarcolemma. Involvement of inhibitory guanine nucleotide regulatory protein. *Biochem Pharmacol* 38:489–496, 1989.

120. Pobiner BF, Northup JK, Bauer PH, et al: Inhibitory GTP-binding regulatory protein G_{i3} can

couple angiotensin II receptors to inhibition of adenylyl cyclase in hepatocytes. *Mol Pharmacol* 40:156–167, 1991.

121. Brami B, Vilgrain I, Chambaz EM: Sensitization of adrenocortical cell adenylate cyclase activity to ACTH by angiotensin II and activators of protein kinase C. *Mol Cell Endocrinol* 50:131–137, 1987.

122. Rainey WE, Byrd EW, Sinnokrot RA, et al: Angiotensin II activation of cAMP and corticosterone production in bovine adrenocortical cells: Effects of nonpeptide angiotensin II antagonists. *Mol Cell Endocrinol* 81:33–41, 1991.

123. Langlois D, Bégeot M, Berthelon M-C, et al: Angiotensin II potentiates agonist-induced 3′,5′-cyclic adenosine monophosphate production by cultured bovine adrenal cells through protein kinase C and calmodulin pathways. *Endocrinology* 131:2189–2195, 1992.

124. Johnson MC, Aguilera G: Angiotensin-II receptor subtypes and coupling to signaling systems in cultured fetal fibroblasts. *Endocrinology* 129:1266–1274, 1991.

125. Johnson MC, Aguilera G: Studies on the mechanism of the novel stimulatory effect of angiotensin II on adenylate cyclase in rat fetal skin fibroblasts. *Endocrinology* 131:2404–2412, 1992.

126. Kubalak SW, Webb JG: Angiotensin II enhancement of hormone-stimulated cAMP formation in cultured vascular smooth muscle cells. *Am J Physiol* 264:H86–H96, 1993.

127. Bird IM, Mason JI, Oka K, et al: Angiotensin-II stimulates an increase in cAMP and expression of 17α-hydroxylase cytochrome P_{450} in fetal bovine adrenocortical cells. *Endocrinology* 132:932–934, 1993.

128. Smith JB: Angiotensin-receptor signaling in cultured vascular smooth muscle cells. *Am J Physiol* 250:F759–769, 1986.

129. Blayney LM, Gapper PW, Newby AC: Vasoconstrictor agonists activate G-protein-dependent receptor-operated calcium channels in pig aortic microsomes. *Biochem J* 282:81–84, 1992.

130. Hescheler J, Rosenthal W, Hinsch KD, et al: Angiotensin II-induced stimulation of voltage-dependent Ca^{2+} currents in an adrenal cortical slice. *EMBO J* 7:619–624, 1988.

131. Ohya Y, Sperelakis N: Involvement of a GTP-binding protein in stimulating action of angiotensin II on calcium channels in vascular smooth muscle cells. *Circ Res* 68:763–771, 1991.

132. Baker KM, Singer HA, Aceto JF: Angiotensin II receptor-mediated stimulation of cytosolic-free calcium and inositol phosphates in chick myocytes. *J Pharmacol Exp Ther* 251:578–585, 1989.

133. Baker KM, Booz GW, Dostal DE: Cardiac actions of angiotensin II: Role of an intracardiac renin–angiotensin system. *Annu Rev Physiol* 54:227–241, 1992.

134. Dosemeci A, Dhallan RS, Cohen NM, et al: Phorbol ester increases calcium current and simulates the effects of angiotensin II on cultured neonatal rat heart myocytes. *Circ Res* 62:347–357, 1988.

135. Cohen CJ, McCarthy RT, Barret PQ, et al: Ca^{2+} channels in adrenal glomerulosa cells: K^+ and angiotensin II increase T-type Ca^{2+} channel current. *Proc Natl Acad Sci USA* 85:2412–2416, 1988.

136. Quinn SJ, Cornwall MC, Williams GH: Electrophysiological responses to angiotensin II of isolated rat adrenal glomerulosa cells. *Endocrinology* 120:1581–1589, 1987.

137. Brauneis U, Vassilev PM, Quinn SJ, et al: ANG II blocks potassium currents in zona glomerulosa cells from rat, bovine, and human adrenals. *Am J Physiol* 260:E772–E779, 1991.

138. Hoyer J, Popp R, Meyer J, et al: Angiotensin II, vasopressin and GTP[γ-S] inhibit inward-rectifying K^+ channels in porcine cerebral capillary endothelial cells. *J Membr Biol* 123:55–62, 1991.

139. Moorman JR, Kirsch GE, Lacerda AE, et al: Angiotensin II modulates Na^+ channels in neonatal rat. *Cir Res* 65:1804–1809, 1989.

140. Berk BC, Brock TA, Gimbrone MAJ, et al: Early agonist-mediated ionic events in cultured vascular smooth muscle cells. Calcium mobilization is associated with intracellular acidification. *J Biol Chem* 262.5065–5072, 1987.

141. Chuang M, Dell KR, Severson DL: Protein kinase C does not regulate diacylglycerol metabolism in aortic smooth muscle cells. *Mol Cell Biochem* 96:69–77, 1990.

142. Severson DL, Hee-Cheong M: Diacylglycerol metabolism in isolated aortic smooth muscle cells. *Am J Physiol* 256:C11–C17, 1989.

143. Griendling KK, Berk BC, Alexander RW: Evidence that Na^+/H^+ exchange regulates angiotensin II-stimulated diacylglycerol accumulation in vascular smooth muscle cells. *J Biol Chem* 263:10620–10624, 1988.

144. Vallotton MB, Capponi AM, Johnson EI, et al: Mode of action of angiotensin II and vasopressin on their target cells. *Horm Res* 34:105–110, 1990.

145. Quinn SJ, Williams GH, Tillotson DL: Calcium oscillations in single adrenal glomerulosa cells stimulated by angiotensin II. *Proc Natl Acad Sci USA* 85:5754–5758, 1988.

146. Johnson EM, Theler JM, Capponi AM, et al: Characterization of oscillations in cytosolic free Ca^{2+} concentration and measurement of cytosolic Na^+ concentration changes evoked by angiotensin II and vasopressin in individual rat aortic smooth muscle cells. Use of microfluorometry and digital imaging. *J Biol Chem* 266:12618–12626, 1991.

147. Tsuda T, Kawahara Y, Shii K, et al: Vasoconstrictor-induced protein–tyrosine phosphorylation in cultured vascular smooth muscle cells. *FEBS Lett* 285:44–48, 1991.

148. Tsuda T, Griendling KK, Alexander RW: Angiotensin II stimulates vimentin phosphorylation via a Ca^{2+}-dependent, protein kinase C-independent mechanism in cultured vascular smooth muscle cells. *J Biol Chem* 263:19758–197563, 1988.

149. Tsuda T, Griendling KK, Ollerenshaw JD, et al: Angiotensin II- and endothelin-induced protein phosphorylation in cultured vascular smooth muscle cells. *J Vascular Res* 30:241–249, 1993.

150. Tsuda T, Alexander RW: Angiotensin II stimulates phosphorylation of nuclear lamins via a protein kinase C-dependent mechanism in cultured vascular smooth muscle cells. *J Biol Chem* 265:1165–1170, 1990.

151. Pfeilschifter J, Ochsner M, Whitebread S, et al: Down-regulation of protein kinase C potentiates angiotensin II-stimulated polyphosphoinositide hydrolysis in vascular smooth muscle cells. *Biochem J* 262:285–291, 1989.

152. Griendling KK, Delafontaine P, Rittenhouse SE, et al: Correlation of receptor sequestration with sustained diacylglycerol accumulation in angiotensin II-stimulated cultured vascular smooth muscle cells. *J Biol Chem* 262:14555–14562, 1987.

153. Linas SL, Marzec-Calvert R, Ullian ME: K depletion alters angiotensin II receptor expression in vascular smooth muscle cells. *Am J Physiol* 258:C849–C854, 1990.

154. Gunther S, Gimbrone MA Jr, Alexander RW: Regulation by angiotensin II of its receptors in resistance blood vessels. *Nature* 287:230–232, 1980.

155. Schiffrin EL, Gutkowska J, Genest J: Effect of angiotensin II and deoxycorticosterone infusion on vascular angiotensin II receptors in rats. *Am J Physiol* 246:H608–H614, 1984.

156. Kitamura E, Kikkawa R, Fujiwara Y, et al: Effect of angiotensin II infusion on glomerular angiotensin II receptor in rats. *Biochim Biophys Acta* 885:309–316, 1986.

157. Mann JFE, Leidig M, Ritz E: Human angiotensin II receptors are regulated by angiotensin II. *Clin Exp Hypertens [a]* 10:151–168, 1988.

158. Bouscarel B, Wilson PB, Blackmore PF, et al: Agonist-induced downregulation of the angiotensin II receptor in primary cultures of rat hepatocytes. *J Biol Chem* 263:14920–14924, 1988.

159. Anderson KM, Murahashi T, Dostal DE, et al: Morphological and biochemical analysis of angiotensin II internalization in cultured rat aortic smooth muscle cells. *Am J Physiol* 264:C179–C188, 1993.

160. Ullian ME, Linas SL: Role of receptor cycling in the regulation of angiotensin II surface

receptor number and angiotensin II uptake in rat vascular smooth muscle cells. *J Clin Invest* 84:840–846, 1989.

161. Schelling JR, Hanson AS, Marzec R, et al: Cytoskeleton-dependent endocytosis is required for apical type 1 angiotensin II receptor-mediated phospholipase C activation in cultured rat proximal tubule cells. *J Clin Invest* 90:2472–2480, 1992.

162. Reagan LP, Ye X, Maretzski CH, et al: Down-regulation of angiotensin II receptor subtypes and desensitization of cyclic GMP production in neuroblastoma N1E-115 cells. *J Neurochem* 60:24–31, 1993.

163. Ullian ME, Linas SL: Angiotensin II surface receptor coupling to inositol trisphosphate formation in vascular smooth muscle cells. *J Biol Chem* 265:195–200, 1990.

164. Tsutsumi K, Saavedra JM: Heterogeneity of angiotensin II AT_2 receptors in the rat brain. *Mol Pharmacol* 41:290–297, 1992.

165. Speth RC: [^{125}I]CGP 42112 binding reveals differences between rat brain and adrenal AT_2 receptor binding sites. *Regul Pept* 44:189–197, 1993.

166. Pucell AG, Hodges JC, Sen I, et al: Biochemical properties of the ovarian granulosa cell type 2-angiotensin II receptor. *Endocrinology* 128:1947–1959, 1991.

167. Grady EF, Sechi LA, Griffin CA, et al: Expression of AT_2 receptors in the developing rat fetus. *J Clin Invest* 88:921–933, 1991.

168. Feuillan PP, Millan MA, Aguilera G: Angiotensin II binding sites in the rat fetus: Characterization of receptor subtypes and interaction with guanyl nucleotides. *Regul Pept* 44:159–169, 1993.

169. Tsutsumi K, Saavedra JM: Characterization of AT_2 angiotensin II receptors in rat anterior cerebral arteries. *Am J Physiol* 261:H667–H670, 1991.

170. Pucell AG, Bumpus FM, Husain A: Regulation of angiotensin II receptors in cultured rat ovarian granulosa cells by follicle-stimulating hormone and angiotensin II. *J Biol Chem* 263:11954–11961, 1988.

171. Toney GM, Porter JP: Functional role of brain AT_1 and AT_2 receptors in the central angiotensin II pressor response. *Brain Res* 603:57–63, 1993.

172. Widdop RE, Gardiner SM, Kemp PA, et al: Central administration of PD 123319 or EXP-3174 inhibits effects of angiotensin II. *Am J Physiol* 264:H117–H125, 1993.

173. Sumners C, Tang W, Zelezna B, et al: Angiotensin II receptor subtypes are coupled with distinct signal-transduction mechanisms in neurons and astrocytes from rat brain. *Proc Natl Acad Sci USA* 88:7567–7571, 1991.

174. Sumners C, Myers LM: Angiotensin II decreases cGMP levels in neuronal cultures from rat brain. *Am J Physiol* 260:C79–C87, 1991.

175. Bottari SP, King IN, Reichlin S, et al: The angiotensin AT_2 receptor stimulates protein tyrosine phosphatase activity and mediates inhibition of particulate guanylate cyclase. *Biochem Biophys Res Commun* 183:206–211, 1992.

176. Zarahn ED, Ye X, Ades AM, et al: Angiotensin-induced cyclic GMP production is mediated by multiple receptor subtypes and nitric oxide in N1E-115 neuroblastoma cells. *J Neurochem* 58:1960–1963, 1992.

177. Chiu AT, McCall DE, Nguyen TT, et al: Discrimination of angiotensin II receptor subtypes by dithiothreitol. *Eur J Pharmacol* 170:117–118, 1989.

178. Chang RS, Lotti VJ: Angiotensin receptor subtypes in rat, rabbit and monkey tissues: Relative distribution and species dependency. *Life Sci* 49:1485–1490, 1991.

179. Kambayashi Y, Bardhan S, Takahashi K, et al: Molecular cloning of a novel angiotensin II receptor isoform involved in phosphotyrosine phosphatase inhibition. *J Biol Chem* 268:24543–24546, 1993.

180. Mukoyama M, Nakajima M, Horiuchi M, et al: Expression cloning of type 2 angiotensin II receptor reveals a unique class of seven-transmembrane receptors. *J Biol Chem* 268:24539–24542, 1993.

3

Defining Angiotensin Receptor Subtypes

Andrew T. Chiu, Ronald D. Smith,
and Pieter B.M.W.M. Timmermans

1. HISTORICAL PERSPECTIVE

Historically, suggestions of multiple receptor subtypes for angiotensin II (ANG II) were based on a number of experimental observations involving the differences in the concentration–response relationships in isolated smooth muscle preparations,[1] differences in membrane binding properties in various tissues such as in rat liver[2] and kidney,[3] and differences in the coupling to various second messenger systems.[4] Early in our evaluation of a series of nonpeptide ANG II receptor antagonists, which led to the development of DuP 753 (losartan), it was noted that 20–30% of the labeled ANG II-specific binding could not be displaced by these compounds, in contrast to saralasin and unlabeled ANG II, which completely inhibited this binding.[5] With the patent disclosure of two novel series of nonpeptide antagonists exemplified by losartan[6] and by PD 123177,[7] it was found that losartan selectively inhibited one population of ANG II binding sites, whereas PD 123177 selectively inhibited the other losartan-resistant site.[8] A similar observation was independently established by using losartan and a novel peptide ligand, CGP 42112A, which also showed selectivity for the losartan-resistant site.[9] Subsequently, a unified nomenclature for these two binding sites or receptors (AT_1 and AT_2) was recommended by a committee established by the American Heart Association Council for High

Andrew T. Chiu, Ronald D. Smith, and Pieter B.M.W.M. Timmermans • The Du Pont Merck Pharmaceutical Company, Wilmington, Delaware 19880-0400.

Angiotensin Receptors, edited by Juan M. Saavedra and Pieter B.M.W.M. Timmermans. Plenum Press, New York, 1994.

Blood Pressure Research in 1990[10] and has now gained widespread acceptance. Accordingly, the ANG II binding sites or receptors that are specifically
blocked by losartan and dithiothreitol (DTT) are designated the AT_1 receptor
subtype and those blocked by PD 123177 or CGP 42112A and enhanced by
DTT are designated the AT_2 receptor subtype. The committee also made pro-
vision for the subdivision of the recognized classes into AT_{1A}, AT_{1B}, AT_{2A},
AT_{2B}, and so forth if evidence clearly justifies it. Additional subtypes such as
AT_3, $AT_4 \ldots AT_n$, may also be necessary if the established criteria cannot be
applied (see Table I).

2. DEFINING AT RECEPTOR SUBTYPES BY SELECTIVE ANTAGONISTS OR LIGANDS

From a practical standpoint, receptor subtypes are best defined by the use
of highly discriminative antagonists or agonists that can be applied as pharmacological markers for most experimental conditions. ANG II and its homologues and analogues have to be used first to define a binding site or response
that is associated with an angiotensin receptor. However, these peptides do not
have the specificity and stability that permit them to be the discriminatory
markers for the various receptor subtypes.

Currently, a number of highly selective AT_1 or AT_2 nonpeptide receptor antagonists/ligands have been developed.[11-15] Losartan is the prototypic
marker for the AT_1 receptor subtype. Other AT_1-selective nonpeptide receptor
antagonists include EXP3174, DuP 532, L-158,809, SK&F 108566, TCV-116,

Table I. Biochemical Definition of AT Receptor Subtypes

	AT_1	AT_2
Previous names	AII-1; AII-B, AIIα	AII-2; AII-A; AIIβ
Receptor isoforms	rAT_{1A}, rAT_{1B}, hAT ...	AT_{2A}, AT_{2B}
Affinity order	Saralasin > ANG II > ANG III >>> ANG I	ANG III ≥ ANG II ≥ saralasin >> ANG I
Selective antagonists	Losartan; EXP3174; DuP 532; L-158,809; GR117289; SK&F 108566; TCV-166	PD123177; PD123319; PD121981; PD124125; CGP42112A
Selective radioligands	[^{3}H]losartan, [^{125}I]-EXP985	(3-[^{125}I]-Tyr)CGP42112A
Sensitivity to –SH reagents	Inactivation	Enhancement
Coupling to G-protein(s)	Yes/No	Yes/No
Signal transduction	↑ Ca^{2+}/↑ IP_3 ↓ Adenyl cyclase ↑ Prostaglandins ↓ K+ current	↓ cGMP/↑ cGMP ↓ Adenyl cyclase ↑ Prostaglandins ↑ K+ current
Cloned receptor	359 Amino acids 7 Trans. domains	Unknown
Molecular size	~65–80 kDa	~120 kDa?

and GR117289 (Fig. 1). As a class, these losartanlike compounds show a high selectivity for the AT_1 receptor, showing about a 10,000-fold greater affinity than for the AT_2 receptor. Losartan is the least potent among this group, having

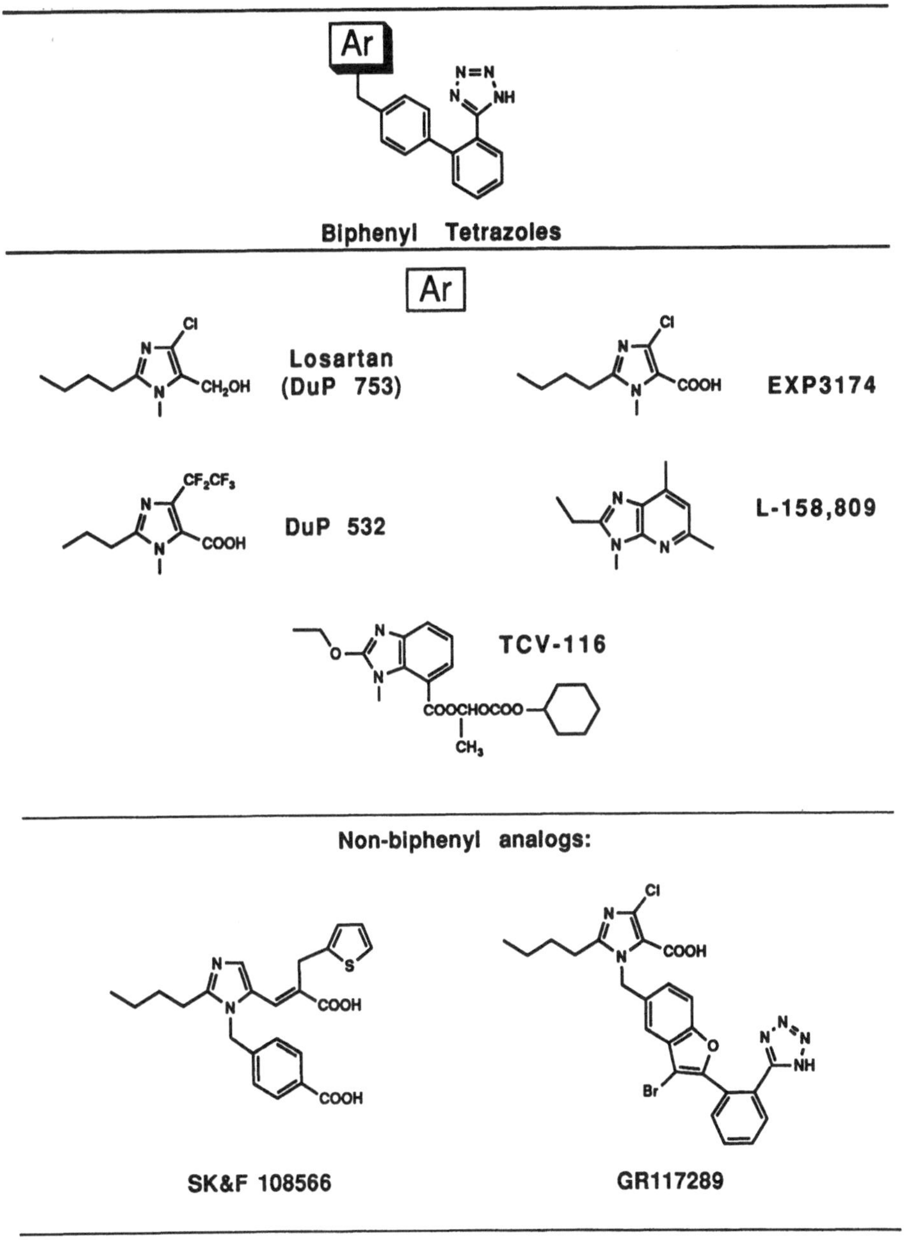

Biphenyl   Tetrazoles

Non-biphenyl   analogs:

Figure 1. AT_1-selective receptor antagonists.

an affinity constant (K_B) of 3×10^{-9} M, whereas EXP3174, DuP 532, L-158,809, SK&F 108566, TCV-116, and GR117289 are more potent nonpeptide ANG II receptor antagonists possessing K_B values of about 10^{-10} M. Virtually all of the known physiological effects of ANG II are blocked by losartan and are therefore mediated via the AT_1 receptor subtype (Table II).

The AT_2 receptor has been defined by those angiotensin binding sites or responses that are not blocked by losartan or analogues but are sensitive to inhibition by either PD 123177 (EXP655) or CGP 42112A. Two classes of AT_2-selective ligands are now known: (1) the peptidic agents, such as CGP 42112A and pNH$_2$Phe^6-ANG II, and (2) the nonpeptidic agents, such as PD 123177, PD 123319, and other congeners (Table I). Judging from their binding affinities, these peptidic and nonpeptidic agents are about 1000- to 3500-fold more selective for the AT_2 than for AT_1 receptors. It is unclear whether these AT_2 selective agents are in fact antagonists or agonists for the AT_2 receptors. Part of the complexity is due to a lack of understanding of the function(s) of this newly discovered AT_2 receptor. Their use as pharmacological markers is even more complicated especially in *in vivo* studies since their pharmacokinetic data are currently unknown. In view of the above limitations, the use of these agents must be carefully scrutinized without preconceived assumptions.

There is a third group of heterogeneous angiotensin "receptors" (Table III) that share one common characteristic; that is, they are not blocked by either the AT_1 or AT_2 receptor blockers. These receptors/binding sites can be activated by ANG II or bind ANG II with high or reasonable affinity and some selectivity. For example, binding sites for ANG II with distinct pharmacological profiles have been reported in the aorta of the domestic fowl,[16] amphibian cardiac

Table II. AT$_1$ Receptor-Mediated Responses

AT$_1$-mediated response	Species	Reference
(+)^a Vascular contraction	Rat, rabbit	74–76
(+) Nonvascular contraction	Rat, guinea pig, human	74,77
(+) Vasoconstriction	Rat	78,79
	Monkey, dog, human	80,81
(+) Aldosterone secretion	Rat, dog	80,82,83
	Ovine, human	81,84
(+) Catecholamine release	Rat, dog	85,86
(−) Renin release	Rat, monkey	87,88
	Dog, human	80,81
(+) Vasopressin release	Rat	89
(+) Cell growth	Rat, human	38,90
Protein synthesis		
Thymidine incorporation		
(+) Drinking	Rat	91

^{a}(+), Stimulation; (−), inhibition.

Table III. Novel Angiotensin AT_n "Receptors" or Binding Sites

Occurrence	Response[a]	Inhibited by	Resistant to[b]	Reference
Mas oncogene	(+) Ca^{2+} transients	[D-Arg, D-Pro², D-Trp⁷,⁹, Leu¹¹] substance P	LO, PD	23
Xenopus laevis oocytes	(+) $[Ca^{2+}]_i$	Saralasin CGP (400 nM)[c]	LO, PD	18
Turkey adrenal cortical cells	(+) Aldosterone secretion	Sarile	LO, PD	92
Chicken aorta	Relaxation, ANG II binding	Unknown?	LO, PD	93
Amphibian cardiac	Sarile binding	Sarile, saralasin CGP (1.2 µM)[c]	LO, PD	17
Mycoplasma hyorrhinis	ANG II binding $K_d = 5$ nM	ANG I, aprotinin bactracin	LO, CGP, ANG III, saralasin	22
Mouse neuro-blastoma	ANG II binding $K_d = 12$ nM	Unknown	LO, PD, ANG III	20
Neuro-2A	(+) cGMP	Sarile	LO, PD	21
Rabbit/rat liver cytosolic protein	ANG II binding $K_d = 6.7$ nM	Saralasin	LO, PD	94,95

[a](+), Stimulation.
[b]LO, losartan; PD, PD123177 or PD123319; CGP, CGP42112A.
[c]IC_{50}.

membranes[17] and amphibian follicular cells,[18,19] differentiated mouse neuroblastoma cells,[20,21] mycoplasma hyorrhinis,[22] and *mas* oncogene product,[23] all of which are insensitive to inhibition by losartan, PD 123177, or CGP 42112A. The latter agent may give variable and inconsistent results, possibly because of its partial agonistic effect. Since there are no specific ligands or blockers to distinguish this group of "receptor subtype(s)," their individual biological relevance cannot be critically addressed. For the sake of reference, these angiotensin receptors may be tentatively referred to as AT_n "receptor" subtype.

3. DEFINING AT RECEPTOR SUBTYPES BY BIOCHEMICAL CHARACTERISTICS

The presence or absence of a coupling to a GTP-binding protein has been used to distinguish the subtypes of ANG II receptor.[24,25] The AT_1 receptors are generally thought to involve a GTP-binding protein for the receptor–second messenger coupling, whereas AT_2 may not. The evidence supporting this contention is based primarily on the ability of GTP or its nonhydrolyzable analogues [GTPγS or Gpp(NH)p] to exert an inhibitory effect on the binding of ANG II to its binding sites. The marked GTP effect on the binding of ANG II was most carefully characterized in rat liver membranes, where the ANG II receptor was subsequently identified as AT_1.[2,26,27] In contrast, the binding of ANG II to tissues expressing exclusively AT_2 receptors (human myometrium, bovine cerebral cortex, or PC12W cells) was not affected by these GTP analogues,[24,28] suggesting an absence of a G-protein in AT_2 receptor coupling. Based on these results, it was concluded that AT_2 receptors are not G-protein linked and do not belong to the G-protein-linked superfamily of receptors.

The foregoing evidence would be strongly suggestive if the GTPγS effects were unequivocally predictive. However, it remains prudent at this time not to assume the lack of a GTP effect as a conclusive marker for receptor differentiation. It has now become known that the α subunit of Gq-like proteins have much lower affinity for GTP than most other G-proteins.[29] Recently, Fluharty and co-workers[30,31] reported that two biochemically distinct populations of AT_2 receptors were isolated from N1E-115 cells. One of these AT_2 receptor subtypes was apparently coupled to PLC-α through an association with a Gq-like protein. For this preparation, GTPγS was ineffective at reducing [^{125}I]-ANG II binding until the concentration exceeded 200 μM. On the other hand, not all AT_1 receptor preparations are sensitive to GTPγS[32] and not all AT_2 receptor preparations are insensitive to it either. Tsutsumi and Saavedra[33] reported that the [^{125}I]-Sar^1-ANG II binding to AT_2 receptors in the ventral thalamic and medial geniculate nuclei and in the locus coeruleus was sensitive to GTPγS and to pertussis toxin pretreatment, whereas in the inferior olive the binding was insensitive to both treatments. They suggested the nomenclature of AT_{2A} for the former and AT_{2B} for the latter receptor subtype (Table IV). Ernsberger and co-workers[25] also reported that two ANG II receptor subtypes were found in rat renal mesangial cell membranes, one being sensitive to losartan and the other to PD 123319. Both classes of binding sites appeared to be inhibited by GTPγS and pertussis toxin treatment. These authors suggested the use of G-protein coupling as the criterion for receptor classification. On this basis, the losartan-sensitive site was referred to as AT_{1A} and the PD 123319-sensitive, G-protein-coupled site was referred to as AT_{1B}, whereas the previously pub-

Table IV. Suggested Subclassification of AT_2 "Receptor" Based on Differences in Certain Biochemical Properties

Property	AT_2	
	AT_{2A}	AT_{2B}
PD-sensitive	+	+
GTPγS-sensitive	+	−
Pertussis toxin-sensitive	+	−
(−)Adenylcyclase	+	?

(−), Inhibition; +, positive effect; −, no effect.

lished PD 123319-sensitive, non–G-protein-coupled site was referred to as AT_2 (Table IV). These findings are quite similar to those reported by Tsutsumi and Saavedra[33] in which the AT_{2A} designation appears to be equivalent to AT_{1B}. As will be discussed in Section 4, the nomenclature of AT_{1B} has been used in the molecular cloning of ANG II receptor isoforms designated for a second losartan-sensitive and PD 123319-insensitive type of receptor that shows a distinct difference from AT_{1A} on a genetic level. Whether the heterogeneity of AT_2 receptors, one coupled to a G-protein and the other not, represents two distinguishable AT_2 receptor subtypes or isoforms in different conformational states remains to be further delineated. It is anticipated that the current speculations will be resolved either by successful cloning of these receptors or by elucidation of their biochemical or pharmacological functions with the help of specific and discriminatory antagonists when they become available.

In most cases, ANG II receptors were shown to couple to many of the traditional signal–transduction pathways in various sensitive target tissues.[34] For example, in adrenal cortex, liver, and vascular smooth muscle, angiotensin receptors are coupled to phospholipase C or D, activating the conversion of phosphatidylinositol to 1,4,5-inositol trisphosphate and diacylglycerol.[26,35,36] In some of these cells, ANG II may directly couple to the Ca^{2+} channel, thus transducing an increased influx of extracellular Ca^{2+},[37,38] or to phospholipase AT_2, leading to the release of arachidonate and its metabolic products.[39–41] Furthermore, ANG II receptors found in liver, adrenal, and kidney are shown also to inhibit adenylate cyclase.[2,3,42] In spite of the multiplicity of biochemical actions, all of these ANG II receptors are blocked by losartan. Therefore, they all belong to the AT_1 receptor subtype. Some of the specific examples are listed in Table V. The fact that AT_1 receptors are capable of interacting with multiple biochemical pathways suggests that a possible heterogeneity may exist among the AT_1 receptors. Losartan may well be a nonselective antagonist for the different AT_1 receptor isoforms. On the other hand, one could invoke the possibility that all of these AT_1 receptors may share one common (or similar) extracellular recognition site but differ only in the subtle changes in the intracellular domains where coupling to various second messenger systems may be dependent on the intrinsic properties of each particular tissue.

In contrast to the above findings, none of these traditional transducing pathways (PLC, PLA_2, Ca^{2+}, adenyl cyclase) was found to be coupled to the AT_2 receptor subtype, which was exclusively expressed in Swiss R3T3 cells, rat ovarian granulosa cells, PC12W cells, and some other cell types.[43–46] In these cultured cell lines, ANG II did not affect the basal or stimulated inositol phosphate production, intracellular Ca^{2+} mobilization, adenyl cyclase or guanyl cyclase activity, or prostaglandin production. On the other hand, there are a number of studies suggesting a functional linkage of AT_2 receptors (Table VI). Sumners and co-workers[47,48] reported the coupling of AT_2 receptors to a

Table V. AT$_1$ Receptor-Mediated Signal Transduction
Pathways Sensitive to Blockade by Losartan

Signal[a]	Species	Tissue	Losartan (M)	Reference
(+) [Ca^{2+}]$_i$	Rat	Aortic smooth muscle cells	2×10^{-8b}	76
	Rat	Hepatocytes	10^{-5}	26
	Mouse/rat	NG108-15 cells	10^{-6}	96
	Human	Omental endothelial cells	1.9×10^{-8b}	77
(−) K$^+$ current	Rat	Fetal neuronal cells	10^{-7}	48
(−) K$^+$ efflux	Rat	Adrenal glomerulosa cells	1×10^{-7b}	83
(+) IP/IP$_3$	Rat	Mesangial cells	3.8×10^{-9b}	36
	Rat	Liver clone 9 cells	1.9×10^{-8b}	75
	Rat	Neonatal astrocytic glia	10^{-6}	47
	Mouse	N1E-115 cells	10^{-9}	49
	Rat	Adrenal glomerulosa cell	10^{-5}	83
(+) cAMP	Rat	(−) Hepatocytes	10^{-5}	26
	Rat	(+) Fetal fibroblasts	10^{-6}	97
	Rat	(−) Adrenal glomerulosa cells	10^{-5}	42
	Mouse	(−) Anterior pituitary 7315c cells	2.8×10^{-8b}	98
(+) cGMP	Mouse	(+) N1E-115 cells	10^{-9}	49
(+) PGE$_2$/PGI$_2$	Rat	C$_6$ glioma cells	10^{-7}	41
	Pig	Aortic smooth muscle cell	10^{-6}	39
(+) PGE$_2$	Human	Astrocytes	10^{-6}	51
	Human/rat	Glomerular mesangial	10^{-7}	99

[a](+), Stimulation; (−), inhibition.
[b]IC$_{50}$ values.

reduction in basal cGMP levels and also to an increase in the net outward current (I$_{no}$) in different neuronal cultures from neonatal rat brain. In N1E-115 neuroblastoma cells, Zarahn and co-workers[49] found that ANG II stimulated an increase in cGMP production, which was mediated predominantly through AT$_1$ and partly through AT$_2$ receptor subtype. Interestingly, this effect was attenuated by the nitric oxide synthetase inhibitor, N-monomethyl-L-arginine, suggesting an involvement of NO in ANG II-induced cGMP production. Chen and Re[50] also observed that, in insulin-treated human SHSY5Y neuroblastoma cells, the ANG II-induced increase in [^{3}H]thymidine incorporation can be partially blocked by either losartan or PD 123177, suggesting that both AT receptor subtypes may have a functional role in cell growth. In human astrocytes, Jaiswal and co-workers[51] reported that AT$_1$ receptors are involved in the release of prostaglandins (PGs) and in mobilization of calcium, whereas the AT$_2$ receptors are coupled to the release of PGs only through a calcium-independent mechanism. More recently, Bottari et al.[52] suggested that the AT$_2$ receptors found in membrane preparations of rat adrenal glomerulosa or PC12W cells may signal through activation of a phosphotyrosine phosphatase, resulting in an inhibition of an elevated particulate guanylate cyclase activity that was

Table VI. Reported AT_2 Receptor-Mediated Functions

Function[a]	Species	Tissue	Blocked by[b]	Reference
(+) K^+ current	Rat, neonate	Neuronal cells	PD	48
(−) cGMP	Rat, neonate	Neuronal cells	PD	47
(+) cGMP	Mouse	N1E-115 cells	PD/LO	49
(+) P-tyrosine phosphatase	Rat	PC12W cells, adrenal	???	52
(−) Collagenase	Rat	Cardiac fibroblast	PD	100
(+) Thymidine incorporation		SHSY5Y neuroblastoma cells	PD/LO	50
(+) PGs	Human	Cultured cells	PD/LO	51
(+) Dilation	Rat	Brain arterioles	PD/LO	101
(+) Angiogenesis	Chick	Chorioallantoic membrane	PD/LO	102
(+) LH & PL	Rat	*In vivo*	PD/LO	103
Drinking	Rat	*In vivo*	PD/LO	89

[a](+), Stimulation; (−), inhibition.
[b]LO, losartan; PD, PD123177 or PD123319.

effected through ANP receptors. However, the significance of such modulation is rather unclear especially in view of the lack of overall effects on the intact cell system.[43–46] In spite of the increasing number of reports describing the functional activity mediated through AT_2 receptors (Table VI), there is still no definitive demonstration of a physiological or pathological role for these sites. Nevertheless, the predominance of AT_2-binding sites found in the rat fetus[53–55] or very young rats[56,57] and its disappearance after birth further fuels the speculation of its involvement in growth and development processes.

4. DEFINING RECEPTOR SUBTYPES BY GENETIC CODINGS

A detailed knowledge of the angiotensin receptor gene family coupled with the availability of specific agonists or antagonists may in the future allow us to define the various receptor subtypes by their distinctive genetic codings. At present, our knowledge of the angiotensin receptor gene family is in its infancy. Just recently, cDNA clones that encode the rat vascular[58] and SHR kidney[59] and the bovine adrenal capsular ANG II receptor[60] have been successfully isolated and sequenced. Each of these ANG II receptors was deduced to contain 359 amino acids and displayed very high sequence homology. Hydropathic modeling of the deduced protein suggested that they belong to the class of G-protein-coupled seven transmembrane receptors. The biochemical and pharmacological properties of these expressed proteins were found to portray the characteristics of the AT_1 ANG II receptor subtype. Subsequently, two structurally related but separate AT_1 receptor genes were detected from rat genomic DNA libraries[61–63]; the one that was identical to those previously reported was designated AT_{1A} and the other AT_{1B}. The amino acid sequence of these two receptor types has 96% identity. The differences between the two cloned receptors involved 17–18 amino acids primarily in the intracellular domains.[61–63] The adrenal and pituitary gland express primarily AT_{1B} mRNA, whereas vascular smooth muscle and lung express primarily AT_{1A} mRNA.[62] By means of restriction analysis, AT_{1A} gene was found to localize on rat chromosome 17 and AT_{1B} on chromosome 2.[64] Two separate AT_1 genes analogous to the rat were also found in mouse, whereas a single class of AT_1 gene was found in bovine, human, Rhesus monkey, dog, rabbit, and chicken.[65]

The human ANG II-AT_1 receptor has been cloned from lymphocyte and hepatic cDNA libraries.[66,67] Within the coding region the human receptor nucleotide sequence was 86% and 91% homologous to the rat and bovine sequences, respectively. Again the differences are concentrated in the C-terminal portion of the molecule and some in the extracellular and intracellular domains. Therefore it is inappropriate to classify the human receptor as either the AT_{1A} or AT_{1B} subtype.

All four receptor types (rat AT_{1A}, AT_{1B}, human AT_1, and bovine AT_1) have been introduced via expression vector constructs into ANG II-receptor-negative cell lines, offering the opportunity for examination of their pharmacological properties. All expressed receptors, with the exception of bovine AT_1, display ANG II binding affinity in the nanomolar range. The relative potency or affinity of the peptide homologues and analogues is essentially the same: sarile > saralasin > ANG II > ANG III >> ANG I, and all are blocked by AT_1-receptor-specific antagonists (e.g., losartan), but not by AT_2-receptor-blocking agents (e.g., PD 123177). The expressed bovine AT_1 receptor has essentially the same ligand-binding specificity, except the potency of all ligands is 3- to 30-fold lower. Since this observation has not been confirmed by other investigators, it is not known if the bovine AT_1 receptor has altered biochemical characteristics. If ligand specificity is the criterion for defining receptor subtype, then the rat AT_{1A}, AT_{1B}, and human AT_1 receptors are essentially indistinguishable from each other and should only be considered as multi-isoforms from a protein structural vantage point.

The functional significance of the AT_1 receptor isoforms is beginning to unfold. These receptor isoforms are potentially capable of coupling to G-protein-mediated signaling pathways (Table VII). When the AT_{1A} receptor gene was transfected to Chinese Hamster ovary cells, the expressed receptor was found to couple with all three signal transduction mechanisms known to be induced by ANG II, namely (1) activation of phospholipase C resulting in IP_3 generation with a subsequent release of intracellular Ca^{2+}, (2) activation of dihydropyridine-sensitive voltage-dependent Ca^{2+} channels, and (3) inhibition of adenylate cyclase activity.[68] All AT_1 receptor isoforms have been shown to

Table VII. Signal Transduction Pathways Utilized by Various Cloned AT_1 Receptors

	AT_1 receptor isoforms[a]			
	rAT_{1A}	rAT_{1B}	hAT_1	bAT_1
PLC	+	+	+	+
Ca^{2+} channels	+	ND	ND	+
$\downarrow$ Adenylcyclase	+	ND	ND	ND
Receptor internalization	+	+	ND	ND

[a]rAT_{1A}, rAT_{1B} = rat AT_1; hAT_1 = human AT_1; bAT_1 = bovine AT_1; + = positive effect; ND = not determined.

couple with the IP_3/Ca^{2+} pathways.[62,66,69–71] Although there are structural differences in the intracellular domains that may be involved in receptor desensitization, both AT_{1A} and AT_{1B} appeared to be equally sensitive to ANG II-induced receptor internalization.[72] Since there is only one AT_1 gene in human and other species, this AT_1 receptor must be capable of interacting with all known signal transduction pathways. Any variation in its mechanism of action is more likely due to the intrinsic nature or the environment of individual cell or tissue.

At the level of the receptor, these different isoforms are not distinguishable among themselves. However, on the genetic level, major differences are observed in the 5′- and 3′-untranslated regions.[58,60] The presence of two AT_1 receptor gene loci thus allows for differential regulation of the receptor protein. It has been reported that estrogen treatment suppressed the rat AT_{1B} but not AT_{1A} mRNA levels in the anterior pituitary gland,[62] whereas sodium deprivation increased the expression level of AT_1 mRNA in rat adrenal (where the AT_{1B} isoform predominates) but not in kidney, liver, and brain.[71] Because of this differential genetic regulation of isoreceptors, the nomenclature of rAT_{1A} and rAT_{1B} (the prefix "r" is used to designate the species from which the AT receptor is found) should be used for the purpose of differentiating the two isoforms in rat. Similarly, human AT_1 could be referred to as hAT_1, bovine AT_1 as bAT_1, and so forth.

Attempts to clone the AT_2 receptor have thus far proved unsuccessful. Northern blot analysis of total RNA using AT_1 receptor cDNA under low stringency failed to produce a hybrid in cells or tissues where the AT_2 receptor is exclusively expressed.[43,73] These findings further support that the AT_2 receptor is structurally distinct from the AT_1 receptor at the level of the gene.

Based on the previous observations, distinct structural diversity among the angiotensin receptor gene family should be expected. Possible candidate members of this gene family may include the mammalian proto-oncogene *mas*, the intracellular ANG II-binding protein, and other atypical "ANG II receptors" found in animal and microbial kingdoms. In each case, the biological relevance cannot be fully understood unless specific antagonists or agonists are available for their functional determinations.

5. CONCLUDING REMARKS

The discovery and availability of specific nonpeptide ANG II receptor antagonists has intensified our search for new functional roles of ANG II through previously unrecognized receptor pathways. Numerous prohypertensive actions of ANG II on the blood vessels, adrenal, brain, and kidney are clearly mediated through the AT_1 receptor subtype. The cloning of the mammalian AT_1 receptors has shown the existence of various isoforms of AT_1 receptor that are indistinguishable by the currently available ligands and by their signal transduction mechanisms. Potential functional roles of the AT_2 receptor subtype have been suggested by numerous investigators, but the physiological significance of these remains to be delineated. Novel AT receptor subtypes have also been detected and their biological relevance has to be determined. The future definition of ANG II receptor subtypes will undoubtedly be refined and broadened as our understanding, derived from composite studies based on ligand-binding profiles, tissue distribution, structural stability, second messengers, functional responses, molecular cloning, and sequencing of the receptor proteins, gains further depth and new dimensions.

ACKNOWLEDGMENTS. The author wishes to thank Dr. Jo Anne Saye for her critical review and Ms. Anne Robichaud for the preparation of this manuscript.

REFERENCES

1. Papdimitriou A, Worcel M: Dose–response curves for angiotensin II and synthetic analogues in three types of smooth muscle: Existence of different forms of receptor sites for angiotensin II. *Br J Pharmacol* 50:291–297, 1974
2. Gunther S: Characterization of angiotensin receptor subtypes in rat liver. *J Biol Chem* 259:7622–7629, 1984.
3. Douglas JG: Angiotensin receptor subtypes of the kidney cortex. *Am J Physiol* 253:F1–F7, 1987.
4. Garcia-Sainz JA: Angiotensin II receptors: One type coupled to two signals of receptor subtypes? *Trends Pharmacol Sci* 8:48–49, 1987.
5. Chiu AT, Duncia JV, McCall DE, et al: Nonpeptide angiotensin II receptor antagonists. III. Structure–function studies. *J Pharmacol Exp Ther* 250:867–874, 1989.
6. Duncia JV, Chiu AT, Carini DJ, et al: The discovery of potent nonpeptide angiotensin II receptor antagonists: A new class of potent antihypertensives. *J Med Chem* 33:1312–1329, 1990.
7. Blankley CJ, Hodges JC, Klutchko SR, et al: Synthesis and structure–activity relationships

of a novel series of non-peptide angiotensin II receptor binding inhibitors specific for the AT_2 subtype. *J Med Chem* 34:3248–3260, 1991.

8. Chiu AT, Herblin WF, Ardecky RJ, et al: Identification of angiotensin II receptor subtypes. *Biochem Biophys Res Commun* 165(1):196–203, 1989.

9. Whitebread S, Mele M, Kamber B, et al: Preliminary biochemical characterization of two angiotensin II receptor subtypes. *Biochem Biophys Res Commun* 163:284–291, 1989.

10. Bumpus FM, Catt KJ, Chiu AT, et al: Nomenclature for angiotensin receptors, *Hypertension* 17(5):720–723, 1991.

11. Smith RD, Chiu AT, Wong PC, et al: Nonpeptide angiotensin II receptor antagonists: A new class of antihypertensive agents. In Hansson L (ed): *1992 Hypertension Annual.* Current Science Ltd., London, 1992, p. 35–50.

12. Siegl PKS, Chang RSL, Mantlo NB, et al: *In vivo* pharmacology of L-158,809, a new highly potent and selective nonpeptide angiotensin II receptor antagonist. *J Pharmacol Exp Ther* 262(1):139–144, 1992.

13. Middlemiss D, Drew GM, Ross BC, et al: Bromobenzofurans: A new class of potent, non-peptide antagonists of angiotensin II. *Biorg Med Chem Lett* 1(12):711–716, 1991.

14. Weinstock J, Keenan RM, Samanen J, et al: 1-(Carboxybenzyl) imidazole-5-acrylic acids: Potent and selective angiotensin II receptor antagonists. *J Med Chem* 34:1514–1517, 1991.

15. Shibouta Y, Inada Y, Ojima M, et al: Pharmacological profiles of TCV-116, a highly potent and long acting angiotensin II (AII) receptor antagonist. *J Hypertens* 10(Suppl 4):S143, 1992.

16. Stallone JN, Nishimura H, Khosla MC: Angiotensin II vascular receptors in fowl aorta: Binding specificity and modulation by divalent cations and guanine nucleotides. *J Pharmacol Exp Ther* 252:1076–1082, 1989.

17. Sandberg K, Hong J, Millan MA, et al: Amphibian myocardial angiotensin II receptors are distinct from mammalian AT_1 and AT_2 receptor subtypes. *FEBS Lett* 284(2):281–284, 1991.

18. Hong J, Sandberg K, Catt KJ: Novel angiotensin II antagonists distinguish amphibian from mammalian angiotensin II receptors expressed in xenopus laevis oocytes. *Mol Pharmacol* 39:120–123, 1990.

19. Fluharty SJ, Reagan LP, White MM: Endogenous and expressed angiotensin II receptors on Xenopus oocytes. *J Neurochem* 56:1307–1311, 1991.

20. Chaki S, Inagami T: Identification and characterization of a new binding site for angiotensin II in mouse neuroblastoma neuro-2A cells. *Biochem Biophys Res Commun* 182(1):388–394, 1992.

21. Chaki S, Inagami T: A newly found angiotensin II receptor subtype mediates cyclic GMP formation in differentiated neuro-2A cells. *Eur J Pharmacol* 225(4):355–356, 1992.

22. Bergwitz C, Madoff S, Abou-Samra AB, et al: Specific, high-affinity binding sites for angiotensin II on mycoplasma hyorrhinis. *Biochem Biophys Res Commun* 179(3):1391–1399, 1991.

23. Hanley MR: Molecular and cell biology of angiotensin receptors. *J Cardiovasc Pharmacol* 18(Suppl 2):S7–S13, 1991.

24. Botarri SP, Taylor V, King IN, et al: Angiotensin II AT_2 receptors do not interact with guanine nucleotide binding proteins. *Eur J Pharmacol* 207:157–163, 1991.

25. Ernsberger P, Zhou J, Damon T, et al: Angiotensin II receptor subtypes in cultured rat renal mesangial cells. *Am J Physiol* 263:F411–F416, 1992.

26. Bauer PH, Chiu AT, Garrison JC: DuP 753 can antagonize the effects of angiotensin II in rat liver. *Mol Pharmacol* 39:579–585, 1991.

27. Crane JK, Campanile CP, Garrison JC: The hepatic angiotensin II receptor. II. Effect of guanine nucleotides and interaction with cyclic AMP production. *J Biol Chem* 257:4959–4965, 1982.

28. Speth RC, Kim KH: Discrimination of two angiotensin II receptor subtypes with a selective

agonist analogue of angiotensin II, *p*-aminophenylalanine angiotensin II. *Biochem Biophys Res Commun* 169(3):997–1006, 1990.

29. Smrcka AV, Hepler JR, Brown KO, et al: Regulation of polyphosphoinositide-specific phospholipase C activity by purified Gq. *Science* 251:804–807, 1991.

30. Mah SJ, Ades AM, Mir R, et al: Association of solubilized angiotensin II receptors with phospholipase C-α in murine neuroblastoma NIE-115 cells. *Mol Pharmacol* 42:217–226, 1992.

31. Siemens IR, He PF, Fluharty SJ: Biochemical characterization of two distinct angiotensin AT₂ receptor populations in murine neuroblastoma cells. *J Neurochem* 62, 1994 (in press).

32. Chiu AT, Leung KH, Smith RD, et al: Defining angiotensin receptor subtypes. In Sumners C, Razaida M, Phillips I (eds): *Cellular and Molecular Biology of Angiotensin Receptors.* CRC Press, Orlando, FL, 1993, pp. 245–271.

33. Tsutsumi K, Saavedra JM: Heterogeneity of angiotensin II AT₂ receptors in the rat brain. *Mol Pharmacol* 41(2):290–297, 1992.

34. Peach MJ: Molecular actions of angiotensin. *Biochem Pharmacol* 30:2745–2751, 1981.

35. Peach MJ, Dostal DE: The angiotensin II receptor and the actions of angiotensin II. *J Cardiovasc Pharmacol* 16(Suppl 4):S25–S30, 1990.

36. Pfeilschifter J: Angiotensin II B-type receptor mediates phosphoinositide hydrolysis in mesangial cells. *Eur J Pharmacol* 184:201–202, 1990.

37. Kojima I, Shibata H, Ogata E: Pertussis toxin blocks angiotensin II-induced calcium influx but not inositol triphosphate production in adrenal glomerulosa cells. *FEBS Lett* 204:347–351, 1986.

38. Chiu AT, Roscoe WA, McCall DE, et al: Angiotensin II-1 receptors mediate both vasoconstrictor and hypertrophic responses in rat aortic smooth muscle cells. *Receptor* 1(3):133–140, 1991.

39. Leung KH, Chang RSL, Lotti VJ, et al: AT₁ receptors mediate the release of prostaglandins in porcine smooth muscle cells and rat astrocytes. *Am J Hypertens* 5:648–656, 1992.

40. Leung KH, Roscoe WA, Smith RD, et al: DuP 753, a nonpeptide angiotensin II receptor antagonist, does not have a direct stimulatory effect on prostacyclin and thromboxane synthesis. *FASEB J* 5(6):A1767, 1991.

41. Jaiswal N, Diz DI, Tallant EA, et al: Characterization of angiotensin receptors mediating prostaglandin synthesis in C6 glioma cells. *Am J Physiol* 260(5, Part 2):1000R–1006R, 1991.

42. Balla T, Baukal AJ, Eng S, et al: Angiotensin II receptor subtypes and biological responses in the adrenal cortex and medulla. *Mol Pharmacol* 40:401–406, 1991.

43. Webb ML, Liu ECK, Cohen RB, et al: Molecular characterization of angiotensin II type II receptors in rat pheochromocytoma cells. *Peptides* 13:499–508, 1992.

44. Pucell AG, Hodges JC, Sen I, et al: Biochemical properties of the ovarian granulosa cell type 2-angiotensin II receptor. *Endocrinology* 128(4):1947–1959, 1991.

45. Dudley DT, Hubbell SE, Summerfelt RM: Characterization of angiotensin II binding sites (AT₂) in R3T3 cells. *Mol Pharmacol* 40:360–367, 1991.

46. Leung KH, Roscoe WA, Smith RD, et al: Characterization of biochemical responses of angiotensin II (AT₂) binding sites in the rat pheochromocytoma PC12W cells. *Eur J Pharmacol Mol Pharmacol Section* 227:63–70, 1992.

47. Sumners C, Tang W, Zelezna B, et al: Angiotensin II receptor subtypes are coupled with distinct signal transduction mechanisms in cultured neurons and astrocyte glia from rat brain. *Proc Natl Acad Sci USA* 88:7567–7571, 1991.

48. Kang J, Sumners C, Posner P: Modulation of net outward current in cultured neurons by angiotensin II: Involvement on AT₁ and AT₂ receptors. *Brain Res* 580(1-2):317–324, 1992.

49. Zarahn ED, Ye X, Ades AM, et al: Angiotensin induced cyclic cGMP production is mediated by multiple receptor subtypes and nitric oxide in N1E-115 neuroblastoma cells. *J Neurochem* 58:1960–1963, 1992.

50. Chen L, Re RN: Angiotensin and the regulation of neuroblastoma cell growth. *Am J Hypertens* 4(5), Part 2):82A, 1991.

51. Jaiswal N, Tallant EA, Diz DI, et al: Subtype 2 angiotensin receptors mediate prostaglandin synthesis in human astrocytes. *Hypertension* 17:1115–1120, 1991.

52. Bottari SP, King IN, Reichlin S, et al: The angiotensin AT2 receptor stimulates protein tyrosine phosphatase activity and mediates inhibition of particulate guanylate cyclase. *Biochem Biophy Res Commun* 183(1):206–211, 1992.

53. Grady EF, Sechi LA, Griffin CA, et al: Expression of AT2 receptors in the developing rat fetus. *J Clin Invest* 88:921–933, 1991.

54. Viswanathan M, Tsutsumi K, Correa FMA, et al: Changes in expression of angiotensin receptor subtypes in the rat aorta during development. *Biochem Biophys Res Commun* 179(3):1361–1367, 1991.

55. Tsutsumi K, Stromberg C, Viswanathan M, et al: Angiotensin II receptor subtypes in fetal tissues of the rat: Autoradiography, guanine nucleotide sensitivity, and association with phosphoninositide hydrolysis. *Endocrinology* 129(2):1075–1082, 1991.

56. Tsutsumi K, Saavedra JM: Characterization and development of type-1 and type-2 angiotensin II receptors in rat brain. *Am J Physiol* 261(1, 30–1):209R–216R, 1991.

57. Millan MA, Jacobowitz DM, Aguilera G, et al: Differential distribution of AT1 and AT2 angiotensin II receptor subtypes in the rat brain during development. *Proc Natl Acad Sci USA* 88:11440–11444, 1991.

58. Murphy TJ, Alexander RW, Griendling KK, et al: Isolation of a cDNA encoding the vascular type-1 angiotensin II receptor. *Nature* 351:233–236, 1991.

59. Iwai N, Yamano Y, Chaki S, et al: Rat angiotensin II receptor: cDNA sequence and regulation of the gene expression. *Biochem Biophys Res Commun* 177(1):299–304, 1991.

60. Sasaki K, Yamano Y, Bardhan S, et al: Cloning and expression of a complementary DNA encoding a bovine adrenal angiotensin II type-I receptor. *Nature* 351:230–232, 1991.

61. Iwai N, Inagami T: Identification of two subtypes in the rat type I angiotensin II receptor. *FEBS Lett* 298(2,3):257–260, 1992.

62. Kakar SS, Sellers JC, Devor DC, et al: Angiotensin II type 1 receptor subtype cDNAs: Differential tissue expression and hormonal regulation. *Biochem Biophys Res Commun* 183:1090–1096, 1992.

63. Mauzy CA, Egloff AM, Wu LH, et al: Cloning of a new subtype of rat angiotensin II type 1 receptor gene. *FASEB J* 6(4):A1577, 1992.

64. Lewis JL, Serikawa T, Warnock DG: Chromosomal localization of type 1A and 1B angiotensin II receptor in the rat. *Hypertension* 20:411, 1992.

65. Elton TS, Stephan CC, Taylor GR, et al: Isolation of two distinct type 1 angiotensin II receptor genes. *Hypertension* 20:411, 1992.

66. Bergsma DJ, Ellis C, Kumar C, et al: Cloning and characterization of a human angiotensin II type 1 receptor. *Biochem Biophys Res Commun* 183(3):989–995, 1992.

67. Furuta H, Guo DF, Inagami T: Molecular cloning and sequencing of the gene encoding human angiotensin II type 1 receptor. *Biochem Biophys Res Commun* 183(1):8–13, 1992.

68. Ohnishi J, Ishido M, Shibata T, et al: The rat angiotensin II AT1A receptor couples with three different signal transduction pathways. *Biochem Biophys Res Commun* 186:1094–1101, 1992.

69. Sandberg K, Ji H, Clark AJL, et al: Cloning and expression of a novel angiotensin II receptor subtype. *J Biol Chem* 267(14):9455–9458, 1992.

70. Ishido M, Kondoh M, Ohnishi J, et al: Establishment of chinese hamster ovary cell lines stably expressing the cloned human type 1 angiotensin II receptor and characterization of the expressed receptor. *Biomed Res* 13:349–356, 1992.

71. Inagami T, Sasaki K, Iwai N, et al: Cloning and characterization of angiotensin II receptor

and its regulation in bovine adrenocortical cell and rat kidney. *J Vasc Med Biol* 3:192–196, 1991.

72. Sasamura H, Hein L, Krieger JE, et al: Molecular evidence for two angiotensin (AT-1) receptor isoforms: Tissue distribution and functional implications. *Hypertension* 20(3):416, 1992.

73. Takayanagi R, Ohnaka K, Sakai Y, et al: Molecular cloning, sequence analysis and expression of a cDNA encoding human type-1 angiotensin II receptor. *Biochem Biophys Res Commun* 183(2):910–916, 1992.

74. Rhaleb NE, Rouissi N, Nantel F, et al: DuP 753 is a specific antagonist for the angiotensin receptor. *Hypertension* 17:480–484, 1991.

75. Dudley DT, Panek RL, Major TC, et al: Subclasses of angiotensin II binding sites and their functional significance. *Mol Pharmacol* 38:370–377, 1990.

76. Chiu AT, McCall DE, Price WA, et al: Nonpeptide angiotensin II receptor antagonists. VII. Cellular and biochemical pharmacology of DuP 753, an orally active antihypertensive agent. *J Pharmacol Exp Ther* 252:711–718, 1990.

77. Chiu AT, McCall DE, Price WA, et al: In vitro pharmacology of DuP 753, a nonpeptide AII receptor antagonist. *Am J Hypertens* 4(4, Pt 2):282S–287S, 1991.

78. Wong PC, Price WA, Chiu AT, et al: Nonpeptide angiotensin II receptor antagonists. VIII. Characterization of functional antagonism displayed by DuP 753, an orally active antihypertensive agent. *J Pharmacol Exp Ther* 252:719–725, 1990.

79. Wong PC, Hart SD, Duncia JV, et al: Nonpeptide angiotensin II receptor antagonists. XIII. Studies with DuP 753 and EXP3174 in dogs. *Eur J Pharmacol* 202:323–330, 1991.

80. Clark KL, Robertson MJ, Drew GM: Effects of the non-peptide angiotensin receptor antagonist, DuP 753, on basal renal function and on the renal affects of angiotensin II in the anaesthetised dog. *Br J Pharmacol* 104(Oct Suppl):78, 1991.

81. Christen Y, Waeber B, Nussberger J, et al: Oral administration of DuP 753, a specific angiotensin II antagonist, to normal male volunteers: Inhibition of pressor response to exogenous angiotensin I and II. *Circulation* 83(4):1333–1342, 1991.

82. Chang RSL, Lotti VJ: Two distinct angiotensin II receptor binding sites in rat adrenal revealed by new selective nonpeptide ligands. *Mol Pharmacol* 29:347–351, 1990.

83. Hajnoczky G, Csordas G, Bago A, et al: Angiotensin II exerts its effect on aldosterone production and potassium permeability through receptor subtype AT_1 in rat adrenal glomerulosa cells. *Biochem Pharmacol* 43(5):1009–1012, 1992.

84. Fitzpatrick MA, Rademaker MT, Espiner EA: Acute effects of the angiotensin II antagonist, DuP 753, in hear failure. *J Am Coll Cardiol* 19(3):146A, 1992.

85. Wong PC, Hart SD, Zaspel A, et al: Functional studies of nonpeptide angiotensin II receptor subtype-specific ligands: DuP 753 (AII-1) and PD 123177 (AII-2). *J Pharmacol Exp Ther* 255(2):584–592, 1990.

86. Wong PC, Hart SD, Timmermans PBMWM: Effect of angiotensin II antagonism on canine renal sympathetic nerve function. *Hypertension* 17(6, Pt 2):1127–1134, 1991.

87. Koepke JP, Bovy PR, McMahon EG, et al: Central and peripheral actions of a nonpeptidic angiotensin II receptor antagonist. *Hypertension* 15(6, Pt 2):841–847, 1991.

88. Gibson RE, Thorpe HH, Cartwright ME, et al: Angiotensin II receptor subtypes in the renal cortex of rat and rhesus monkey. *Am J Physiol* 261(3):F512–F518, 1991.

89. Hogarty DC, Phillips MI: Vasopressin release by central angiotensin II is mediated through an angiotensin type-1 receptor and the drinking response is mediated by both AT-1 and AT-2 receptors. *Soc Neuro Sci* 17:1188, 1991.

90. Bakris GL, Akerstrom V, Re RN: Insulin, angiotensin II antagonism and converting enzyme inhibition: Effect on human mesangial cell mitogenicity and endothelin. *Hypertension* 3:326, 1991.

91. Wong PC, Price WA, Chiu AT, et al: Nonpeptide angiotensin II receptor antagonists: Studies with EXP9270 and DuP 753. *Hypertension* 15:823–834, 1990.
92. Kocsis JF, Carsia RV, Chiu AT, et al: Properties of angiotensin II receptors of domestic turkey adrenocortical cells. *Gen Comp Endocrinol* (submitted).
93. Nishimura H, Walker OE, Patton CM: Novel vascular angiotensin receptor subtypes and signal pathway in fowl. *Hypertension* 20(3):435, 1992.
94. Bandyopadhyay SK, Rosenberg E, Kiron RMA, et al: Purification and properties of an angiotensin-binding protein from rabbit liver particles. *Arch Biochem Biophys* 263:272, 1988.
95. Schelhorn TM, Burkard MR, Rauch AL, et al: Differentiation of the rat liver cytoplasmic AII binding protein from the membrane angiotensin II receptor. *Hypertension* 16:36, 1990.
96. Tallant EA, Diz DI, Khosla MC, et al: Identification and regulation of angiotensin II receptor subtypes on NG108-15 cells. *Hypertension* 17(6, Pt 2):1135–1143, 1991.
97. Johnson C, Aguilera G: Angiotensin-II receptor subtypes and coupling to signaling systems in cultured fetal fibroblasts. *Endocrinology* 129(3):1266–1274, 1991.
98. Crawford KW, Frey EA, Cote TE: Angiotensin II receptor recognized by DuP 753 regulates two distinct guanine nucleotide-binding protein signaling pathways. *Mol Pharmacol* 41(1):154–162, 1992.
99. Chansel D, Czekalski S, Pham P, et al: Characterization of angiotensin II receptor subtypes in human glomeruli and mesangial cells. *Am J Physiol* 262(3):F432–F441, 1992.
100. Matsubara L, Brilla CG, Weber KT: Angiotensin II-mediated inhibition of collagenase activity in cultured cardiac fibroblasts. *FASEB J* 6(4):A941, 1992.
101. Brix J, Haberl RL: The AT_2-receptor mediates endothelium-dependent dilation of rat brain arterioles. *FASEB J* 6(4):A1264, 1992.
102. LeNoble FAC, Schreurs, N, VanStraaten HWM, et al: Angiotensin-II induced angiogenesis is not mediated through the AT_1 receptor. *FASEB J* 6(4):A937, 1992.
103. Stephenson KN, Steele MK: Brain angiotensin II receptor subtypes and the control of luteinizing hormone and prolactin secretion in female rats. *J Neuroendocrinol* 4(4):441–447, 1992.

4

Medicinal Chemistry of Angiotensin II Antagonists

William J. Greenlee and Ruth R. Wexler

1. INTRODUCTION

The renin–angiotensin system (RAS) has been demonstrated to be a key element in blood pressure regulation and sodium balance.[1] The therapeutic and commercial success of angiotensin-converting enzyme (ACE) inhibitors, such as captopril and enalapril for the treatment of hypertension, has stimulated the search for additional modes of intervention in the RAS.[2–4] Since the active hormone of the RAS is angiotensin II (ANG II), antagonism of ANG II at the level of its receptor represents the most direct way of selectively inhibiting the RAS independently of the source of ANG II. Although potent peptide ANG II receptor antagonists such as saralasin[5] have been used as pharmacological tools for the past two decades, they have not proved suitable for development as therapeutic agents because of their poor oral bioavailability, short duration, and partial agonist activity.[6,7] The concept that nonpeptide ANG II receptor antagonists would lack the disadvantages associated with peptide ANG II receptor antagonists and would also be potentially more selective than ACE inhibitors for blocking the RAS was an attractive but elusive strategy until 1982 when Furukawa and co-workers at Takeda Chemical Industries disclosed the first nonpeptidic ANG II antagonists.[8,9] Several compounds from the Takeda series were reported to inhibit the ANG II-induced contractile response in rabbit aorta and the pressor response in the ANG II-infused rat, but no information on

William J. Greenlee • Merck Research Laboratories, Rahway, New Jersey 07065.   *Ruth R. Wexler* • The Du Pont Merck Pharmaceutical Company, Wilmington, Delaware 19880-0402.

Angiotensin Receptors, edited by Juan M. Saavedra and Pieter B.M.W.M. Timmermans. Plenum Press, New York, 1994.

selectivity was provided.[8,9] Two compounds from the Takeda patent, S-8307 and S-8308 (CV-2961), were studied thoroughly by the Du Pont group and confirmed to have extremely weak antihypertensive effects, but were specific and competitive for the ANG II receptor without agonist activity.[10,11] These early leads demonstrated the feasibility of nonpeptide ANG II receptor antagonists, and hence served as the foundation for the recent discovery of potent, selective, long-acting, and orally active nonpeptidic ANG II receptor antagonists, of which losartan (DuP 753, MK-954) is the prototype.

2. ANGIOTENSIN II RECEPTOR SUBTYPES

The availability of peptide and nonpeptide antagonists has permitted the discovery of ANG II receptor heterogeneity, and it is now widely recognized that there are at least two distinct ANG II receptor subtypes.[12–17] The initial reports of ANG II receptor heterogeneity introduced a variety of designations for the observed subtypes.[12,13,18–21] The recent convention for the nomenclature of ANG II receptor subtypes is based primarily on the selective displacement of ANG II with nonpeptidic receptor antagonists.[22] AT_1 receptors are selective in their recognition of agents such as losartan,[23] DuP 532,[24] L-158,809,[25–27] and SK&F 108566.[28,29] AT_1 receptors are inactivated by dithiothreitol (DTT) and other sulfhydryl-reducing agents[12,30] and are G-protein coupled.[31] To date, all actions of ANG II that affect blood pressure appear to be blocked by AT_1-selective antagonists such as losartan, and hence most of the current pharmaceutical effort toward developing nonpeptide ANG II antagonists is focused on AT_1-selective agents. Thus, AT_1-selective receptor antagonists are the focus of this chapter.

The other ANG II receptor subtype, now known as AT_2, is inhibited by PD 123177, structural analogues such as PD 123319[31] and PD 121981,[13,31–33] and also by peptide analogues such as CGP 42112A.[18] Although there have been reports of functional activity mediated through AT_2 sites, the pharmacological role and a coupling mechanism for the AT_2 receptor have not yet been elucidated. Recently, reports have appeared that describe receptors or binding sites that do not appear to be either AT_1 or AT_2 sites.[34,35]

3. PEPTIDIC ANTAGONISTS OF ANG II

Peptidic receptor antagonists of ANG II, first reported over 20 years ago, were the first effective blockers of the RAS. Early work in this area provided valuable information about structure–activity relationships and yielded potent and specific antagonists of ANG II.[36–38] Although one of these antagonists, saralasin (Sarenin) has been studied extensively in animal models and in man.[5,39] lack of oral bioavailability and partial agonist activity have blocked saralasin, or other peptidic antagonists, from successful development as therapeutic agents. Recent reviews of ANG II antagonists provide excellent summaries of structure–activity relationships of ANG II peptide analogues[40,41] and highlight the development of other potent antagonists such as sarmesin and [Sar[1], Ile[8]]-ANG II (Fig. 1).

Peptide analogues of ANG II have continued to provide important insights into how ANG II interacts with its two binding sites, the AT_1 and AT_2 receptors. Moreover, hypothetical overlays of early nonpeptidic antagonist lead structures with the carboxy-terminus of ANG II have been useful in the development of more potent nonpeptidic antagonists (Section 3.3). Peptidic antagonists that define structure–activity relationships at the carboxy-terminus of ANG II have provided information useful in nonpeptidic antagonist design as have both amino-terminal-truncated analogues of ANG II and conformationally restricted analogues. While peptide analogues of ANG II have not yet been discovered which are orally active drug candidates, recent progress in the discovery of low-molecular-weight peptide antagonists of CCK[42] and endothelin[43,44] and recent success in obtaining orally active peptide-based renin inhibitors[45] leave open the possibility that optimization of binding affinity of shortened ANG II antagonist peptides could provide orally active agents.

3.1. Carboxy-Terminal Modifications

It is generally accepted that a free carboxy-terminus is required for full interaction of ANG II with its receptor. Not only does ANG II bind 100-fold less tightly, but carboxy-terminal esters and alcohols also bind poorly.[46] Dimerization of [Sar[1],Ile[8]]-ANG II through the carboxy-terminus as exemplified by CGP 37534-decreased AT_1 binding affinity by 100-fold (AT_1 $IC_{50} = 80$ nM; rat vascular smooth muscle cells), presumably due to the lack of a free carboxy-terminus.[47] Although the carboxy-terminal-truncated analogues 1 ($pA_2 = 7.6$) and 2 ($pA_2 = 7.0$) are modestly potent, competitive, reversible receptor antagonists in the rabbit aorta, they are antagonists of ANG II-induced pressor response in rats at doses only six- to tenfold higher than inhibitory doses of

Angiotensin II	H-Asp-Arg-Val-Tyr-Ile-His-Pro-Phe-OH
Saralasin	Sar-Arg-Val-Tyr-Val-His-Pro-Ala-OH
Sarmesin	Sar-Arg-Val-Tyr(OMe)-Ile-His-Pro-Phe-OH
[Sar[1],Ile[8]]-AII	Sar-Arg-Val-Tyr-Ile-His-Pro-Ile-OH

Figure 1. Angiotensin peptide analogues [Sar = CH_3NHCH_2CO-].

saralasin.[48,49] ANG II analogues with more extensive carboxy-terminal truncations (hexapeptides, pentapeptides) show extremely low binding affinity.[49]

Analogues of [Sar[1]]-ANG II have been reported in which Phe[8] has been replaced by aromatic amino acids with increased bulk or conformational restraint. For example, a highly hydrophobic analogue of ANG II, [Sar[1], (2′,3′,4′,5′,6′-Br$_5$)Phe[8]]-ANG II, is a potent antagonist of ANG II both *in vitro* and *in vivo*. Its prolonged duration of action may be due to slow dissociation from the receptor.[50] The (pentafluoro)Phe[8] analogue of ANG II retains agonist activity, but is also a slowly reversible antagonist.[51] Analogues in which Phe[8] is replaced with the sterically demanding amino acids diphenylalanine (Dip), or (biphenyl-4-yl)alanine (Bip), or with 2-aminoindane-2-carboxylic acid (2-Ind) have also been reported. In a rat uterus contraction assay, both [Sar[1], Dip[8]]-ANG II and [Sar[1], D-Dip[8]]-ANG II are potent agonists, while [Sar[1], Bip[8]]-ANG II, [Sar[1], D-Bip[8]]-ANG II, and [Sar[1], 2-Ind[8]]-ANG II are potent antagonists.[52] Both [Dmp[8]]-ANG II and [Sar[1], Dmp[8]]-ANG II (DMP = 2,6-dimethylphenylalanine) are potent ANG II antagonists in rabbit aorta.[53]

That ANG II analogues in which Phe[8] is replaced by amino acids with exceedingly large side chains bind well to ANG II receptors suggests that if nonpeptidic antagonists (in fact) derive their high binding potency by mimicking the carboxy-terminus of ANG II, they may do so (in part) by utilizing auxiliary binding sites.

3.2. Amino-Terminal Modifications

Recent work on peptide antagonists which are truncated at the amino-terminus has helped define the minimum peptide sequence required for binding and agonist activity (Fig. 2). The success of this work also supports the hypothesis that nonpeptidic antagonists are mainly mimics of the carboxy-

CGP-37534	Sar-Arg-Val-Tyr-Val-His-Pro-Ile-NH-CH$_2$ ⌐ Sar-Arg-Val-Tyr-Val-His-Pro-Ile-NH-CH$_2$ ⌐
1	Sar-Arg-Val-Tyr-Ile-His-Pro-NH$_2$
2	Sar-Arg-Val-Tyr-Ile-His-Pro-N(CH$_3$)$_2$
3	Ape-Tyr-Ile-His-Pro-Ile-OH
4	Ac-Tyr-Val-His-Pro-Ile-OH
CGP-37346	⌐CH$_2$NHCH$_2$CO-Asn-Arg-Val-Ile-His-Pro-Phe-OH ⌐CH$_2$NHCH$_2$CO-Asn-Arg-Val-Ile-His-Pro-Phe-OH

Figure 2. Angiotensin peptide analogues: carboxy-terminal and amino-terminal modifications [Sar = CH$_3$NHCH$_2$CO–; Ape = 5-aminopentanoyl].

terminal portion of ANG II. For example, a series of amino-terminally modified hexa- and pentapeptides, including the pentapeptide **3** (pA$_2$ = 6.6), are antagonists of ANG II in rabbit aorta and lack agonist activity.[54] The pentapeptide **4** has only modest potency as an AT$_1$ antagonist (AT$_1$ IC$_{50}$ = 13.5 µM; rat vascular smooth muscle cells), but has nanomolar potency for the AT$_2$ receptor (AT$_2$ IC$_{50}$ = 30 nM; human uterus).[47]

In general, extension of the amino-terminus of ANG II does not affect binding potency, suggesting that the carboxy-terminus may bind more deeply inside the receptor. Thus, dimerization at the amino-terminus[47] yielded the potent antagonist CGP-37346 (AT$_1$ IC$_{50}$ = 3.4 nM; rat vascular smooth muscle cells).[47]

3.3. Conformationally Restricted Analogues of ANG II

Conformations of ANG II have been proposed on the basis of analogue studies or spectroscopic evidence (see reference 55 for an extensive bibliography). Models of proposed low-energy conformations of ANG II include "folded" conformations in which the amino-terminal Asp-Arg sequence lies near the carboxy-terminus in space. Spectroscopic evidence has been offered also to support a clustering of the aromatic side chains of Tyr^4, His^6, and Phe^8 in solution, a situation that might also be reflected also in the bound conformation of ANG II.[56]

Recently attempts have been made to gain information about the bound conformation of ANG II by preparing conformationally restricted analogues of ANG II. Introduction of a conformationally constrained amino acid in place of Ile^5, yields **5**, a potent ANG II agonist (Fig. 3).[57] Cyclic ANG II analogues have been prepared by connecting the side chains of homocysteine (Hcy) residues at positions 3 and 5 via a disulfide bridge. The analogue [Hcy^{3,5}]-ANG II (**6**) has excellent binding affinity (IC$_{50}$ = 0.9 nM; rat uterus) and high

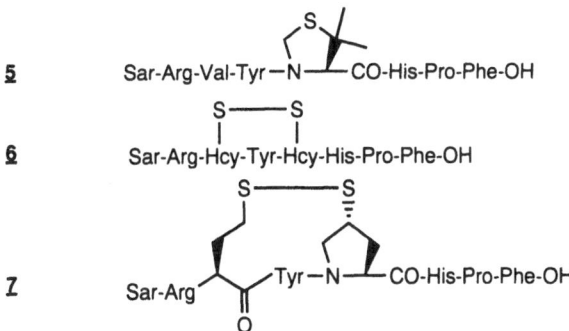

Figure 3. Conformationally restricted ANG II analogues [Sar = CH$_3$NHCH$_2$CO–].

contractile activity on rabbit aortic rings ($pD_2 = 8.48$).[58] A cyclization of this type was used to prepare a highly potent antagonist [Sar[1], Hcy[3,5], Ile[8]]-ANG II ($IC_{50} = 2.1$ nM; rat uterus), which model building suggests contains a three-residue turn in its bioactive conformation.[58,59] Further conformational restraint has been introduced by incorporation of 4-mercapto-proline in place of one Hcy residue, yielding a potent cyclized ligand **7** that binds to both AT_1 (AT_1 IC_{50} = 1.3 nM; rat liver) and AT_2 receptors (AT_2 $IC_{50} = 0.65$ nM; rabbit uterus).[60]

Putative conformations of ANG II have been of utility in the development of nonpeptidic antagonists. Hypothetical overlays of nonpeptidic antagonists with the carboxy-terminus of ANG II have been used to propose modifications of nonpeptidic antagonist lead structures that in at least two cases (Section 4.1) have resulted in antagonists with improved binding affinity.[55,61–63]

4. DISCOVERY AND STRUCTURE–ACTIVITY RELATIONSHIPS OF AT₁-SELECTIVE NONPEPTIDIC ANTAGONISTS

Due to the clear link between the AT_1 receptor and the control of blood pressure (Section 2),[16] most current effort in the development of nonpeptidic ANG II antagonists is focused on AT_1-selective agents. However, the widespread occurrence of the AT_2 receptor and its near-exclusive presence in specific tissues[16] suggest that important roles may be discovered that could make AT_2-selective agents interesting drug candidates. "Balanced" (nonselective AT_1/AT_2) antagonists (Section 4.9) may also prove to be of interest.

AT_1-selective antagonists (both peptidic and nonpeptidic) have been evaluated in radioligand binding assays using a number of protocols. Because of the variation in tissue, competing radioligand, buffer, presence (or absence) of plasma protein and proteinase inhibitors, ionic strength, and other differences in assays reported for the determination of binding potencies, direct comparison of antagonists reviewed in this chapter is difficult. Reference should be made to the original manuscripts for details of the individual assays. Recent publications have made comparisons of several leading antagonists candidates in a single assay.[41,64,65]

4.1. Development of Losartan and SK&F 108,566

The potent, nonpeptide ANG II receptor antagonists with high AT_1 selectivity, of which losartan is the prototype, can all be traced back to the Takeda series of 1-benzyl-imidazole-5-acetic acid derivatives (Fig. 4).[8,9] Both losartan and SK&F 108566 were derived from the benzylimidazole by using two different molecular models of putative active conformations of ANG II to align the Takeda derivatives, with the C-terminal region of ANG II.[23,28,55] The Du Pont group overlapped the Takeda lead with a solution conformer of

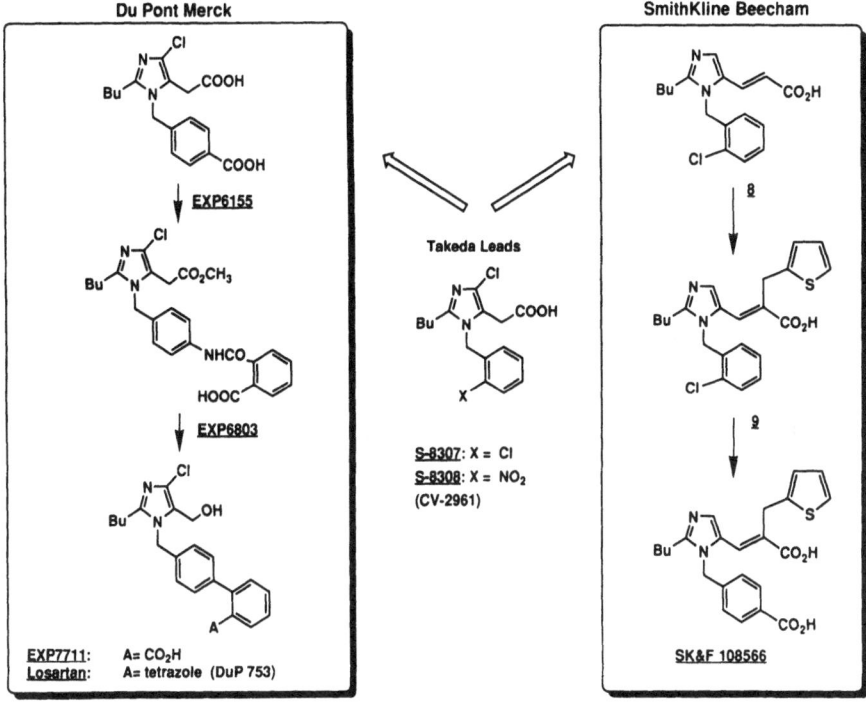

Figure 4. Discovery of losartan and SK&F 108566.

ANG II[66] by (1) aligning the carboxylic acid group on the benzylimidazole with the C-terminal carboxylic acid of ANG II (pointing both to a hypothetical positive charge[46] on the receptor); (2) overlapping the imidazole ring of the lead with the His[6] residue of ANG II; (3) pointing the *n*-butyl side chain of the Takeda compound into the area of the Ile[5] side chain; and (4) directing the benzyl group of the lead toward the N-terminus of ANG II. According to this model, the para position of the benzyl group had the most promise for extension or enlargement of the S-8307 structure. Since ANG II contains two acidic groups near the N-terminus, namely the Asp[1] β-carboxylic acid and the Tyr[4] phenolic OH, it was hypothesized that the benzyl group of the Takeda compound would optimally be substituted with a carboxylic acid.[23,55] Thus, substitution of a carboxyl moiety at the para position resulted in EXP6155, which had a binding affinity tenfold greater than that of S-8307.[23,67,68] Encouraged by this enhancement in binding affinity, further extensions toward the N-terminus produced a series of phthalamic acid analogues, including EXP6803, which afforded a second order of increased binding affinity.[23,68,69] Although both EXP6803 and EXP6155 were specific antagonists of ANG II and caused substantial decreases in mean arterial pressure of renal-hypertensive rats when

administered IV, both were devoid of oral activity.[68-70] Replacement of the
N-phenyl phthalamic acid moiety of EXP6803 with a biphenyl group bearing
a 2-carboxylic acid resulted in EXP7711,[71] which for the first time exhibited
good oral activity. Upon oral administration (10, 30, and 100 mg/kg) to conscious renal artery-ligated hypertensive rats (RHR) and furosemide-treated
dogs, EXP7711 caused a dose-dependent decrease in mean arterial pressure. A
dose of 100 mg/kg (RHR) lowered blood pressure to normotensive levels
without an effect on cardiac frequency.[72] Finally, while searching to replace
the polar carboxylic acid with a more lipophilic isostere, it was discovered that
the tetrazole group increased the intrinsic activity by an order of magnitude
(ED_{30} upon oral dosing in the RHR is 0.59 mg/kg) while maintaining good
oral bioavailability.[71,73] Hence, the Du Pont efforts culminated in the discovery
of losartan (DuP 753, MK-954). More detailed accounts of the discovery of
losartan have appeared.[23,73-77]

Using a modification of the Fermandjian model for the bioactive conformation of ANG II,[66] the SmithKline Beecham group hypothesized that the
N-benzyl and carboxyl moieties of the Takeda compounds corresponded to the
Tyr^4 aromatic side chain and the Phe^8 carboxyl group of ANG II.[28] Using this
model, chain extension at the imidazole 5-position appeared promising for
better mimicry of the Tyr^4 to C-terminal region of ANG II. An analogue **8**,
which incorporates a *trans* 5-acrylic acid group, was found to be more potent
than the acetic acid analogue S-8307. Addition of an α-benzyl group to the
acrylic acid moiety to better mimic the Phe^8 side chain of ANG II resulted in
further increase in affinity, with the 2-thienylmethyl analogue **9** demonstrating
fivefold enhanced binding affinity compared to the corresponding benzyl analogue. Addition of a 4-carboxyl group to the N-benzyl ring afforded SK&F
108566 and resulted in yet another order of magnitude increase in *in vitro* and
in vivo potencies and was suggested to be mimicking the Tyr^4 phenol.[28,29] The
SmithKline Beecham model suggests that the α-thienylacrylic acid moiety
aligns with Phe^8, the N-C-N imidazole region, and the acrylic acid double bond
serve as mimics of the peptide bonds and that the 2-butyl group lies in the
hydrophilic region near Ile^5. SK&F 108566 is a potent, AT_1-selective antagonist
that exhibits competitive inhibition of $[^{125}I]$-ANG II binding to rat mesenteric
artery and adrenal cortical membranes ($IC_{50} = 1.5$ and 9.2 nM, respectively).
Additional details of the discovery and the structure–activity relationships
leading to this antagonist are found in reference 28 and a summary of its
pharmacology is found in reference 29.

4.2. Development of New Nonpeptidic ANG II Antagonists

Losartan and its more potent carboxylic acid metabolite EXP3174 have
served as prototypes for numerous drug discovery programs in the pharmaceu-

tical industry, and an explosion of new and interesting receptor antagonist structures has emerged from this work. Some of these have entered clinical trials, and it is expected that new and valuable information about their properties will soon emerge. In the following sections, we have attempted to draw together new structural types explored in this recent work and to highlight antagonists that appear to have promise as *in vivo* candidates. In cases where literature citation to a new class of antagonists is not possible, reference has been made to the patent literature. However, due to limitations of space, many references (especially those in the patent literature) could not be included. Recent reviews should be consulted for a more complete account of the innovative contributions of the many research groups working in this area.[41,78]

4.3. Imidazole Biphenyltetrazole Antagonists Related to Losartan

Studies of the structure–activity relationships around the substituted imidazole ring of losartan have demonstrated the requirement for a 2-alkyl group or equivalent for high potency (Fig. 5). A variety of substituents at the imidazole C4 and C5 positions are acceptable, but the high *in vitro* potency discovered for EXP3174 (Table I)[23,77] demonstrated the superiority of a C5-carboxylic acid substituent (Fig. 6). Further refinement of the imidazole substituents in the losartan series led to the discovery of DuP 532 (Table I), in which a lipophilic and electron-withdrawing pentafluoroethyl group is positioned at the imidazole C4 position.[24,79] Recently, simple 4-alkyl substituents (e.g., 4-ethyl) have been found to yield antagonists with high *in vitro* potency.[80]

The tolerance of a large group at the imidazole C4 position is demonstrated by the high binding potency of imidazoles that incorporate bulky 4-aryl[81,82] or heteroaryl substituents.[83] Recently, new imidazole antagonists have been disclosed in which novel C4 substituents have been incorporated, including 4-alkylthio[84,85] or hydroxyalkyl.[86]

Interestingly, EXP3174 and DuP 532 differ from losartan in that they

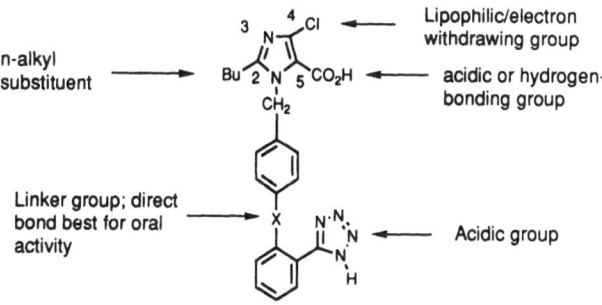

Figure 5. Structure–activity relationships of losartan analogues.

Table I. Pharmacological Characterization of AT₁-Selective Nonpeptide Angiotensin Receptor Antagonists

Compound	ANG II antagonism Rabbit aorta pA₂ or K_B (nM)	ANG II antagonism *in vivo* rat "effective" dose mg/kg IV	PO	Blood pressure lowering renal hypertensive rat "effective" dose mg/kg IV	PO	Reference
Losartan	pA₂ = 10	—	—	0.78	0.59	23, 73–77
EXP3174	K_B = 0.10	—	—	0.04	0.66	87
DuP 532	K_B = 0.11	0.03–1.0	0.03–10	0.02	0.21	79
L-158,809	IC₅₀ = 0.3	0.026	0.023	—	0.1	25–27, 94
SK&F 108566	K_B = 0.26	0.08	5.5 (ID)	—	—	28, 29
TCV-116	pA₂ = 9.97	0.01	0.03	—	0.03	95, 96, 163
SR 47436	IC₅₀ = 4	0.1–3	0.3–30	—	—	108, 109
SC-51895	pA₂ = 8.6	—	2.7 (IG)	—	—	110, 164
D8731	pA₂ = 8.3	1	5	—	—	130, 131
D6888	pA₂ = 10.3	—	3	—	—	130, 165
A-81988	pA₂ = 10.6	—	—	—	0.3	136, 137
CGP48933	IC₅₀ = 1.4	—	3–10	—	3–10	141, 143, 166
GR117289	K_B = 9.8	—	—	0.3–3 (IA)	0.3–10	143–147

Abbreviations: ID, intraduodenal; IG, intragastric; IA, intra-arterial.

produce nonparallel shifts of the ANG II concentration–functional (contractile) response curves and suppress the maximal response to ANG II in the rabbit aorta,[78,87] a result that has led to these antagonists being termed *noncompetitive*. More recently, the difference between these two types of antagonists has been described as *surmountable* (parallel shifts with no suppression of maximal response) versus *insurmountable* (nonparallel shifts with suppression of maximal response), and the latter phenomenon may be due to a slow off-rate of bound antagonist from the receptor.[88]

Diacidic antagonists such as DuP 532 have been found to bind tightly to plasma proteins, including the bovine serum albumin (BSA) commonly used to reduce nonspecific binding in ANG II receptor antagonist assays.[24,78] This

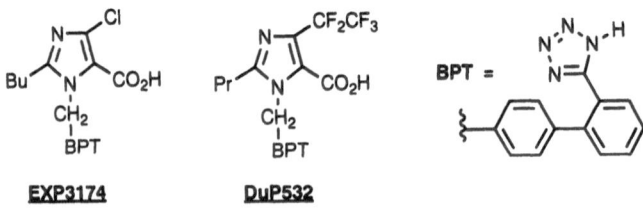

Figure 6. Imidazole antagonists related to losartan.

led to a discrepancy between *in vitro* binding to membrane preparations (micromolar) for DuP 532 and potency in functional (aortic contraction) assays (nanomolar). Removal of BSA from the binding assays enhanced the binding affinity for DuP 532 by 1500-fold and that of EXP3174 by 20-fold. Although antagonists lacking the 5-carboxylic acid group show less of an effect of BSA, a high degree of plasma binding may prove to be an important characteristic in other antagonist series also.

4.4. Fused Imidazole Antagonists

That a variety of substituents are acceptable at the imidazole C4 and C5 positions of antagonists in the losartan series suggested that these substituents could be joined to yield fused imidazoles. Benzimidazoles have been investigated by several groups,[89–91] and potent antagonists have been reported (Fig. 7), including BIBS 39 ($K_i = 29$ nM; rat lung).[92] Related tetrahydrobenzimidazoles with high potency have also been disclosed.[93] Also of great interest are imidazo[4,5-b]pyridines of which the potent antagonist L-158,809 (Table I)[25–27,94] is an example. Fused imidazoles TCV-116 (Table I)[95,96] and L-158,978 ($IC_{50} = 0.5$ nM; rabbit aorta)[97] incorporate a carboxylic acid substituent (or a prodrug ester) in a location similar to that of EXP3174. Other six-membered ring fused imidazoles that have been disclosed include potent purines,[98] xan-

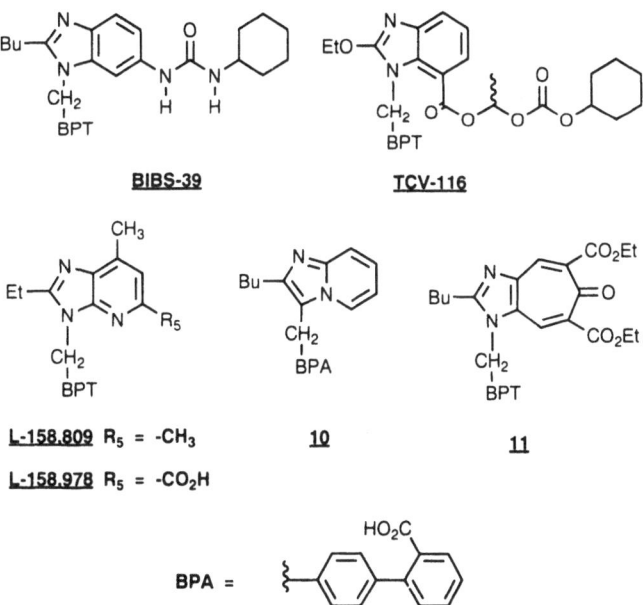

Figure 7. Fused imidazole ANG II antagonists.

thines,[98–100] pyrazolo[1,2-a]pyrimidines,[101] imidazo[1,2-b]pyridazines,[102] and imidazo[1,2-a]pyridines such as **10** ($IC_{50} = 7.4$ µM; guinea pig adrenal).[91] A seven-membered ring is fused to an imidazole in the modestly potent imidazocycloheptadienone **11** ($IC_{50} = 0.4$ µM; rat uterus).[103]

4.5. Other Five-Membered Ring Antagonists

A number of antagonist designs have been reported in which various five-membered ring heterocycles replace the substituted imidazole of losartan (Fig. 8), including pyrazoles such as **12** ($IC_{50} = 0.3$ µM; rat adrenal)[104,105] and pyrroles such as **13** ($IC_{50} = 1.6$ µM; rat adrenal).[104,105] Potent 1-substituted triazoles including SC-50560 ($IC_{50} = 5.6$ µM; rat uterus)[106] and 4-substituted triazoles[104,107] have also been reported. The potent antagonist SR 47436 (Table I) incorporates an imidazolinone ring in place of imidazole,[108,109] as does the isomeric imidazolone SC-51895 (Table I).[110] Other five-membered ring antagonists disclosed include triazolinones[111,112] such as SC-51316 ($IC_{50} = 3.6$ nM; rat adrenal).[113] Five-membered ring heterocycles linked through a carbon atom to a (methyl)biphenyl tetrazole element have also been reported, including pyrazoles[114] such as **14** ($pK_B = 10.5$; rabbit aorta)[115] and (3-substituted) triazoles.[116]

4.6. Antagonists Incorporating Six-Membered Heterocyclic Rings

Potent antagonists have also been obtained by replacing the imidazole of losartan with a six-membered ring heterocycle (Fig. 9). Antagonists of this type that are linked through a nitrogen atom to the biphenyltetrazole element include

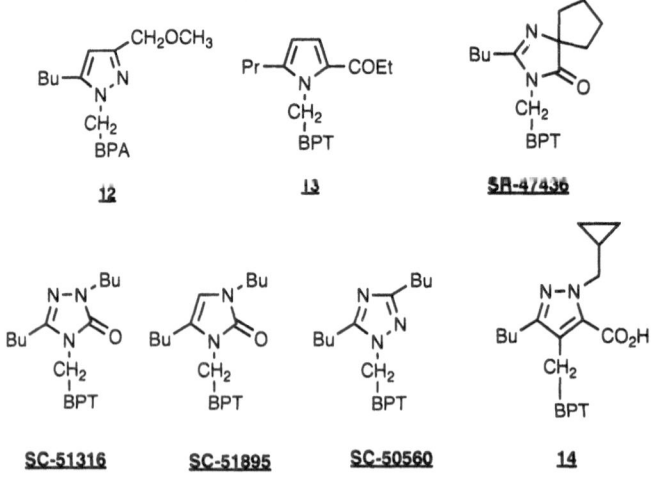

Figure 8. Five-membered ring ANG II antagonists.

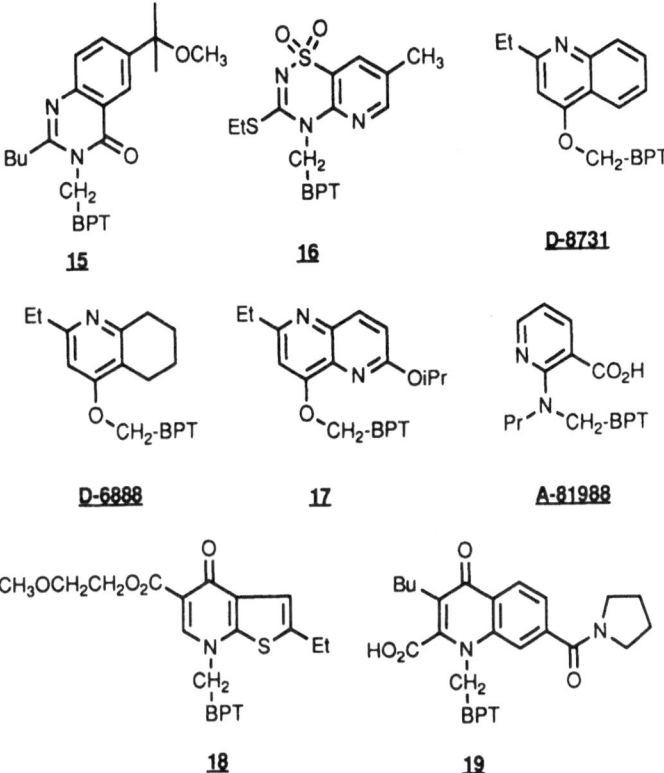

Figure 9. Six-membered ring ANG II antagonists.

pyrimidinone,[117–119] pyrimidinedione,[120] pyridinone,[121] and isoquinolinone[122] and quinazolinone antagonists,[117–120,123] the potent antagonist **15** (IC$_{50}$ = 6.4 nM; rat adrenal) being an example of the latter.[123] A related class of antagonists is exemplified by the pyridylthiadiazine dioxide **16** (K$_i$ = 0.6 nM; rat adrenal).[124]

Antagonists based on a carbon-linked six-membered ring heterocycle include pyrimidines,[125,126] pyrimidinones,[126–129] and pyridopyrimidinones.[126] The potent quinoline D8731[130,131] and the related tetrahydroquinoline D-6888 (Table I) make use of a novel oxymethylene linker connecting the heterocycle to the biphenyltetrazole, as do related pyridines[132,133] and naphthyridines such as **17** (IC$_{50}$ = 7.9 nM; guinea pig adrenal).[134,135] In the novel pyridine A-81988 (Table I), it appears that the *N*-propyl substituent plays the role of the imidazole 2-substituent of losartan; the carboxylic acid substituent may mimic that of EXP3174.[136,137] Thienopyridinone antagonists such as **18** (IC$_{50}$ = 10 nM; bovine adrenal) have been disclosed which depart from the standard heterocyclic designs,[138] as do quinolinone antagonists such as **19** (IC$_{50}$ = 8.7 nM; rat liver).[139]

4.7. Nonheterocyclic Antagonist Designs

In the nonheterocyclic antagonist valsartan (CGP48933; $IC_{50} = 8.9$ nM; rat aorta), an acylated amino acid element replaces the heterocycle of losartan (Fig. 10).[140,141]

4.8. Non-Biphenyltetrazole Antagonist Designs

In work leading to the development of biphenyltetrazole as the preferred (acidic) imidazole substituent in losartan, a number of designs were explored in which the two phenyl rings were connected by linker groups consisting of one-atom (-O-, -S-, -CO-), two-atom (-NHCO-, -CONH-, -OCH$_2$-), and three-atom (-NHCONH-) elements.[55,142] Although the high potency obtained with many of these designs demonstrated the flexibility allowed in the placement of the acidic group relative to the heterocycle, the biphenyl design was unique in providing analogues (EXP7711, losartan) with oral activity. Recently, antagonist designs based on Du Pont Merck amide-linked structures such as **20** (IC_{50} = 0.14 μM; rat adrenal)[55] have been reported (Fig. 11). These include series of indole, benzothiophene, and benzofuran antagonists disclosed by Glaxo, from which the potent benzofuran antagonist GR117289 (Table I)[143–147] has emerged as a clinical candidate. Interestingly, although losartan is more potent than its carboxylic acid analogue EXP-7711, GR117289 and its corresponding carboxylic acid analogue are equipotent. Related antagonist designs in which the central phenyl ring is replaced by various substituted indoles[148,149] include **21** (IC_{50} = 5 nM; rabbit aorta).[149]

In the antagonists **22** (IC_{50} = 43 nM; rabbit aorta).[150,151] and **23** (IC_{50} = 0.14 μM; rat liver),[152] novel acidic elements are introduced in place of biphenyltetrazole. Potent antagonists have been obtained by replacing the tetrazole-bearing phenyl ring of biphenyltetrazole with a heterocycle such as furan,[71] (N-linked) pyrrole,[153,154] thiophene,[155] pyridine,[156] and indole (or benzimidazole).[157] Surprisingly, replacement of this ring with a partially saturated carbocyclic ring yielded a loss of potency in **24** (IC_{50} = 0.31 uM; rat adrenal).[158]

Conformationally restrained biphenyltetrazole replacements have been reported. Introduction of methoxy groups on the upper ring of the biphenyl results

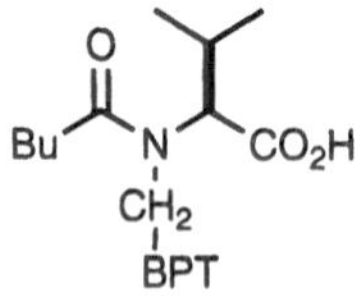

Valsartan

Figure 10. Nonheterocyclic ANG II antagonist.

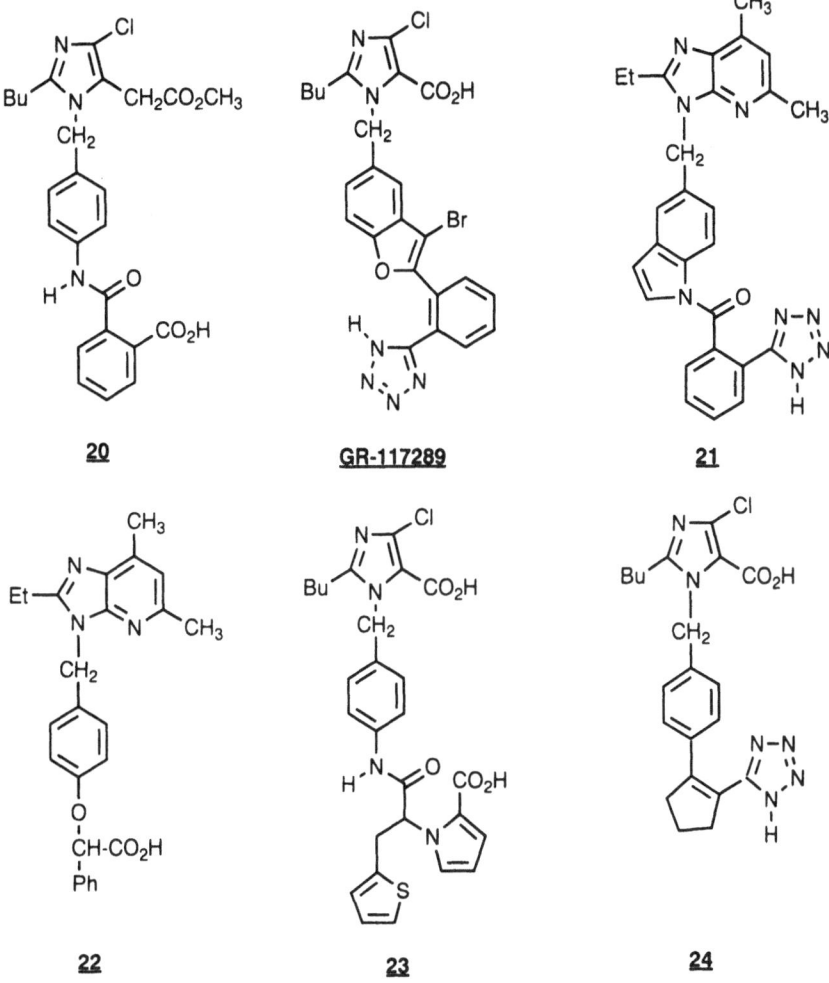

Figure 11. Nonbiphenyl ANG II antagonists.

in an analogue **25** ($IC_{50} = 1.9$ uM; rat uterus) with reduced antagonist potency,[159] possibly due to conformational effects (Fig. 12). Another conformationally restrained structure is the tetrahydronaphthyl antagonist **26** ($IC_{50} = 0.2$ µM; rabbit aorta), which retains significant antagonist potency.[160]

The substituted imidazole antagonist **27** ($IC_{50} = 3.8$ µM; rat adrenal) represents a novel class of antagonists that are structurally unrelated to losartan.[161,162]

4.9. "Balanced" (AT$_1$/AT$_2$) Antagonists

The anticipated increase in ANG II levels resulting from *in vivo* AT$_1$ blockade could influence the efficacy of a competitive AT$_1$ receptor antagonist,

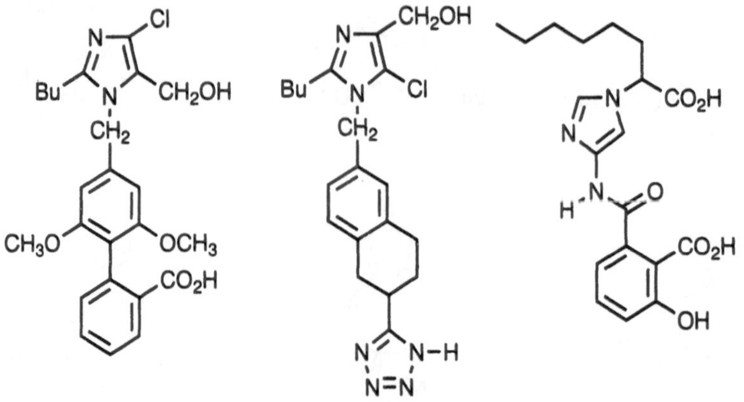

Figure 12. ANG II antagonists.

and more importantly might lead to activation of unblocked AT_2 receptors. Thus, although specific roles for the AT_2 receptor are not yet clearly defined, there is current interest in the development of "balanced" ANG II antagonists that will effectively block both receptors *in vivo*. The current lack of a functional response to ANG II that is mediated by the AT_2 receptor complicates the task of evaluating the *in vivo* properties of a balanced antagonist. Recently, benzimidazoles that are potent AT_1 antagonists and also have substantial AT_2 potency[92] have been reported, including BIBS39 (AT_1 $K_i = 29$ nM; rat lung; $AT_2 K_i = 0.48$ uM; rat adrenal; see Fig. 7). The availability of potent selective antagonists that bind to either AT_1 or AT_2 receptors presents a starting point for medicinal chemists interested in designing "balanced" ANG II antagonists.

5. IN VIVO PHARMACOLOGY OF ANG II ANTAGONISTS

The *in vivo* pharmacology of losartan and its metabolite EXP3174 has been reviewed recently and compared to other nonpeptide ANG II receptor antagonists in various animal models.[64,65] Pharmacological data reported for several other AT_1 selective antagonists including L-158,809,[25–27,94] SK&F 108566,[28,29] TCV-116,[95,96,163] SR 47436,[108,109] SC 51895,[110,164] D8731,[130,131] D6888,[130,165] A-81988,[136,137] CGP48933,[140,141,166] and GR117289,[143–147] are summarized in Table I. A quantitative comparison of these data is not possible since the pharmacological models differ among research groups. The Du Pont Merck group has studied L-158,809, SK&F 108566, GR117289, and SR 47436 in direct comparison to losartan and EXP3174.[64,87] Table II and Fig. 13 show comparative data for AT_1 receptor binding, antagonism of ANG II-induced

Table II. Comparison of Pharmacological Properties
of AT_1-Selective Nonpeptide ANG II Receptor Antagonists[a]

Compound	IC_{50} (nM)[b]	K_B (nM)[c]	ED_{30} (mg/kg)[d]	
			IV	PO
Losartan	5.5	3.3[e]	0.78	0.59
EXP3174	1.3	0.10[f]	0.04	0.66
DuP 532	3.1	0.11[f]	0.02	0.21
L-158,809	0.6	0.40[f]	0.03	0.03
SK&F 108566	3.0	2.6[e]	0.03	0.75
GR 117289[g]	2.5	0.24[f]	0.03	1.26
SF 47436	0.9	0.19[e]	0.16	> 10

[a]Data obtained at the Du Pont Merck Pharmaceutical Company.
[b]Inhibition of specific binding of [^{125}I]-ANG II (0.05 nM) to rat adrenal cortical microsomes in the absence of serum albumin; IC_{50} is the concentration to reduce specifid binding by 50%.
[c]Antagonism of ANG II induced constriction of rabbit aorta.
[d]Antihypertensive potency in renal hypertensive rats following intravenous (IV) or oral (PO) administration. ED_{30} is the dose to decrease mean arterial pressure by 30 mm Hg.
[e]Competitive antagonism.
[f]Noncompetitive antagonism.
[g]GR 117289 was a generous gift of Glaxo Research Laboratories.

constriction in isolated rabbit aorta, and oral antihypertensive potency and duration of action. In summary, losartan, SK&F 108566, and SR 47436 are competitive antagonists (surmountable antagonists) in the rabbit aorta, while the antagonists EXP3174, L-158,809, and GR117289 showed varying degrees of reduction in the maximal response to ANG II, and hence are considered noncompetitive (insurmountable) antagonists. Losartan, EXP3174, SK&F 108566, and GR117289 showed similar antihypertensive effect upon oral dosing in the renal hypertensive rats, while L-158,809 was approximately 20-fold more potent in this model. Losartan, EXP3174, L-158,809, and GR117289, following a single oral dose, resulted in a sustained decrease in blood pressure for greater than 24 hr, while SK&F 108566 produced a rapid lowering of blood pressure that returned essentially to control levels within 3 hr.

Many of the antagonists cited above, including losartan, DuP 532, SK&F 108566, GR117289, D8731, D6888, CGP 48933, TCV-116, and SR 47436, have now entered clinical trials. The clinical experience to date for losartan is described elsewhere in this book (Chapter 20), and results on other antagonists are awaited with interest.

6. PROPERTIES OF AT_1-SELECTIVE ANTAGONISTS

It is likely that in order to compete with the current angiotensin-converting enzyme inhibitors, ANG II antagonists will have to be effective during once-a-day administration for hypertension. Losartan is unique among reported

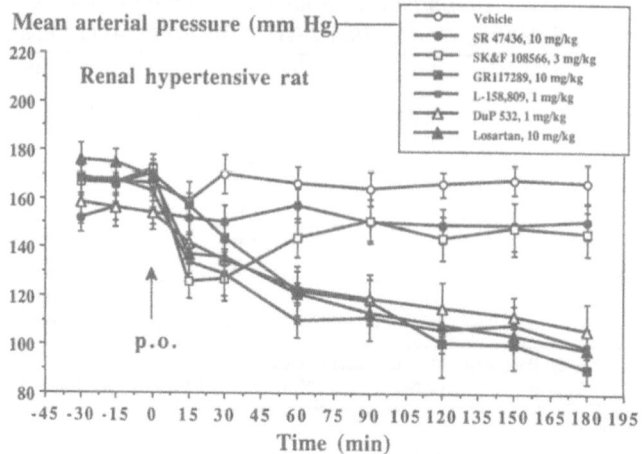

Figure 13. Effects of SR 47436 (10 mg/kg, $n = 5$); SKF 108566 (3 mg/kg, $n = 5$); GR117289 (10 mg/kg, $n = 5$); L-158,809 (1 mg/kg, $n = 6$); DuP 532 (1 mg/kg, $n = 5$; and losartan (10 mg/kg, $n = 6$) on mean arterial pressure of renal artery-ligated rats. Values represent the mean ± SEM.

ANG II antagonists currently in development in that after oral administration a monoacidic parent antagonist is absorbed and subsequently (in rats and in man) converted by metabolism to the more potent diacidic analogue EXP3174. It is now recognized that losartan is cleared more rapidly than EXP3174 and has short duration of action in dogs, where metabolite is not formed to a great extent. Losartan has been found to undergo irreversible *in vivo* metabolism to an (inactive) tetrazole N2 glucuronide,[167] a mode of metabolism that other tetrazole-containing antagonists (especially monoacidic ones) may share.

The discovery that the carboxylic acid metabolite (EXP3174) of losartan has greater affinity for AT_1 receptors and prolonged duration of action *in vivo* has resulted in many new diacidic antagonist designs (DuP 532, GR117289, A-81988). Since EXP3174 has lower oral bioavailability than losartan,[87] it will be interesting to learn whether diacidic antagonists will achieve good oral bioavailability in man (as they appear to do in animal models). The potent antagonists SK&F 108566 and CGP-48933 are also diacidic, but appear to be well absorbed (Section 5). In the case of TCV-116, a prodrug ester is used to deliver a more polar diacidic antagonist.[95] The phenomenon of "insurmountable" antagonism for some diacidic (and other) antagonists, and the possibility of (related) slow dissociation of antagonist from receptor, could add another dimension to their *in vivo* profiles.

Monoacidic antagonists (L-158,809, D8731, D6888, SR47436, SC-51895) have been reported that have high binding potency and appear to be interesting *in vivo* candidates (Section 5). Although these antagonists do not form diacidic metabolites *in vivo*, each has demonstrated good duration of action in at least one species. Since these antagonists are (in general) more lipophilic than the diacidic ones, it is possible that they could achieve more extensive tissue (and/or CNS) penetration, resulting in a different *in vivo* profile. Neutral or (overall) basic ANG II antagonists have not yet been reported; their *in vivo* properties could differ substantially from current antagonists.

Of the many potent AT_1-selective antagonist designs reported to date, only the Lilly series, exemplified by structure **27** (Fig. 12), appears not to have been derived from the original Takeda benzylimidazole lead. It will be interesting to see whether more potent antagonists can be developed from this series. That continued screening will result in new ANG II antagonist leads is suggested by the diversity of structures that bind potently at other G-protein-linked peptide receptors (CCK, PAF, substance P).

7. CONCLUSIONS

A number of potent, AT_1-selective ANG II antagonists have been developed that have entered clinical trials. Losartan, the most advanced of these, has shown efficacy similar to that of enalapril in mildly hypertensive patients,[168–170] and further results with this and other antagonists are awaited with interest. Potential advantages for ANG II antagonists relative to ACE inhibitors,[171,172] if realized, could make them particularly exciting as new drug candidates. The possible clinical significance of elevated levels of ANG II resulting from AT_1 receptor blockade will determine whether "balanced" (nonselective AT_1/AT_2) receptor antagonists may be of interest. The possible roles of the AT_2 receptor, expected to emerge from ongoing work to clone and express this receptor, and the possible existence of other clinically relevant ANG II receptors will continue to enliven drug discovery in the ANG II area.

ACKNOWLEDGMENTS. The authors wish to thank the many individuals in our organizations (MRL and DMPC) for their contributions to this manuscript.

REFERENCES

1. Sealey JE, Laragh JH: The renin–angiotensin–aldosterone system for normal regulation of blood pressure and sodium and potassium homeostasis. In Laragh JH, Brenner BM, (eds): *Hypertension: Pathophysiology, Diagnosis and Management*. Raven Press, New York, 1990, p. 1287.
2. Ondetti MA, Cushman DW: Inhibition of the renin–angiotensin system. A new approach to the therapy of hypertension. *J Med Chem* 24:355, 1981.
3. Streeten DHP, Anderson GH: Angiotensin-receptor blocking drugs. In Doyle A (ed): *Hand-

book of Hypertension, Vol. 5. *Clinical Pharmacology of Antihypertensive Drugs*. Elsevier, Amsterdam, 1984, p. 246.

4. Wyvratt MJ, Patchett AA: Recent developments in the design of angiotensin-converting enzyme inhibitors. *Med Res Rev* 5:483, 1985.

5. Pals DT, Masucci FD, Sipos F, et al: A specific competitive antagonist of the vascular action of angiotensin II. *Circ Res* 29:664, 1971.

6. Streeten DHP, Anderson GH, Freiberg JM, et al: Use of an angiotensin II antagonist (saralasin) in the recognition of "angiotensinogenic" hypertension. *N Engl J Med* 292:657, 1975.

7. Garrison JC, Peach MJ: Renin and angiotensin. In Gilman AG, Rall TW, Nies AS, Taylor P (eds): *The Pharmacological Basis of Therapeutics*, 8th ed. Pergamon Press, New York, 1990, p. 749.

8. Furukawa Y, Kishimoto S, Nishikawa K: Hypotensive imidazole derivatives. U.S. Patent 4,340,598. Issued to Takeda Chemical Industries, Ltd. (Osaka, Japan), 1982.

9. Furukawa Y, Kishimoto S, Nishikawa K: Hypotensive imidazole-5-acetic acid derivatives. U.S. Patent 4,355,040. Issued to Takeda Chemical Industries, Ltd. (Osaka, Japan), 1982.

10. Wong PC, Chiu AT, Price WA, et al: Nonpeptide angiotensin II receptor antagonists I. Pharmacological characterization of 2-n-butyl-4-chloro-1-(2-chlorobenzyl) imidazole-5-acetic acid, sodium salt (S-8307). *J Pharmacol Exp Ther* 247:1, 1988.

11. Chiu AT, Carini DJ, Johnson AL, et al: Nonpeptide angiotensin II receptor antagonists II. Pharmacology of S-8308. *Eur J Pharmacol* 157:13, 1988.

12. Whitebread S, Mele M, Kamber B, et al: Preliminary biochemical characterization of two angiotensin II receptor subtypes. *Biochem Biophys Res Commun* 163:284, 1989.

13. Chiu AT, Herblin WF, Ardecky RJ, et al: Identification of angiotensin II receptor subtypes. *Biochem Biophys Res Commun* 165:196, 1989.

14. Chiu AT, McCall DE, Ardecky RJ, et al: Angiotensin II receptor subtypes and their selective nonpeptide ligands. *Receptor* 1:33, 1990.

15. Timmermans PBMWM, Wong PC, Chiu AT, et al: Nonpeptide angiotensin II receptor antagonists. *Trends Pharmacol Sci* 12:55, 1991.

16. Wong PC, Chiu AT, Duncia JV, et al: Angiotensin II receptor antagonists and receptor subtypes. *Trends Endrocrinol Metab* 3:211, 1992.

17. Timmermans PBMWM, Chiu AT, Herblin WF, et al: Angiotensin II receptor subtypes. *Am J Hypertens* 5:1, 1992.

18. Chang RSL, Lotti VJ, Chen TB, et al: Two angiotensin II binding sites in rat brain revealed using ^{125}I-Sar[1]-Ile[8]-angiotensin II and selective nonpeptide antagonists. *Biochem Biophys Res Commun* 171:813, 1990.

19. Chang RSL, Lotti VJ: Two distinct angiotensin II receptor binding sites in rat adrenal revealed by new selective nonpeptide ligands. *Mol Pharmacol* 29:347, 1990.

20. Rowe BP, Grove KL, Saylor DL, et al: Angiotensin II receptor subtypes in rat brain *Eur J Pharmacol* 186:339, 1990.

21. Speth RC, Kim KH: Discrimination of two angiotensin II receptor subtypes with a selective agonist analogue of angiotensin II, *p*-aminophenylalanine angiotensin II. *Biochem Biophys Res Commun* 169:997, 1990.

22. Bumpus FM, Catt KJ, Chiu AT, et al: Nomenclature for angiotensin receptors. *Hypertension* 17:720, 1991.

23. Duncia JV, Carini DJ, Chiu AT, et al: The discovery of DuP 753, a potent, orally active nonpeptide angiotensin II receptor antagonist. *Med Res Rev* 12:149, 1992.

24. Chiu AT, Carini DJ, Duncia JV, et al: DuP 532: A second generation of nonpeptide angiotensin II receptor antagonists. *Biochem Biophys Res Commun* 177:209, 1991.

25. Mantlo NB, Chakravarty PK, Ondeyka D, et al: Potent, orally active imidazo[4,5-b]pyridine angiotensin II receptor antagonists. *J Med Chem* 34:2919, 1991.

26. Chang RSL, Siegl PKS, Clineschmidt BV, et al: *In vitro* pharmacology of L-158,809, a new highly potent and selective angiotensin II receptor antagonist. *J Pharmacol Exp Ther* 262:133, 1992.

27. Siegl PKS, Chang RSL, Mantlo NB, et al: *In vivo* pharmacology of L-158,809, a new highly potent and selective nonpeptide angiotensin II receptor antagonist. *J Pharmacol Exp Ther* 262:139, 1992.

28. Weinstock J, Keenan RM, Samanen J, et al: 1-(Carboxybenzyl)imidazole-5-acrylic acids: Potent and selective angiotensin II receptor antagonists. *J Med Chem* 34:1514, 1991.

29. Edwards RM, Aiyar N, Ohlstein EH, et al: Pharmacological characterization of the nonpeptide angiotensin II receptor antagonist, SK&F108566. *J Pharmacol Exp Ther* 260(1):175, 1992.

30. Chiu AT, McCall DE, Nguyen TT, et al: Discrimination of angiotensin II receptor subtypes by dithiotheitol. *Eur J Pharmacol* 170:117, 1989.

31. Dudley DT, Panek RL, Major TC, et al: Subclasses of angiotensin II binding sizes and their functional significance. *Mol Pharmacol* 38:370, 1990.

32. Blankley CJ, Hodges JC, Klutchko SR, et al: Synthesis and structure–activity relationships of a novel series of non-peptide angiotensin II receptor binding inhibitors specific for the AT_2 subtype. *J Med Chem* 34:3248, 1991.

33. Chang RSL, Lotti VJ: Selective ligands reveal subtypes of angiotensin receptors in rat vasculature and brain (Abstract). *Pharmacologist* 31(3):183, 1989.

34. Timmermans PBMWM, Benfield P, Chiu AT, et al: Angiotensin II receptors and functional correlates. *Am J Hypertens* 5:221S–235S, 1992.

35. Timmermans PBMWM, Wong PC, Chiu AT, et al: Angiotensin II receptors and angiotensin II receptor antagonists. *Pharmacol Rev* 45:205–247, 1993.

36. Khosla MC, Smeby RR, Bumpus FM: Structure–activity relationship in angiotensin II analogs. *Handbook Exp Pharmacol* 37:126, 1974.

37. Peach MJ: Renin–angiotensin system: Biochemistry and mechanisms of action. *Physiol Rev* 57:313, 1977.

38. Regoli D, Park WK, Rioux F: Pharmacology of angiotensin. *Physiol Rev* 26:69, 1974.

39. Moore AF, Fulton RW: Angiotensin II antagonists—saralasin. *Drug Dev Res* 4:331, 1984.

40. Bovy PR, Blaine EH: Peptidic and non-peptidic angiotensin II competitive antagonists. Current Cardiovascular Patents 2044, 1989.

41. Hodges JC, Hamby JM, Blankeley CJ: Angiotensin II receptor binding inhibitors. *Drugs Future* 17(7):575, 1992.

42. Hughes J, Boden P, Costall B, et al: Development of a class of selective cholecystokinin type B receptor antagonists having potent anxiolytic activity. *Proc Natl Acad Sci USA* 87:6728, 1990.

43. Hemmi K, Neya M, Fukami N, et al: European Patent Application 457,195, 1991.

44. Ishikawa K, Fukami T, Hayama T, et al: European Patent Application 460,679, 1991.

45. Kleinert HD, Rosenberg SH, Baker WR, et al: Discovery of a peptide-based renin inhibitor with oral bioavailability and efficacy. *Science* 257:1940, 1992.

46. Hsieh K, Marshall GR: Role of the C-terminal carboxylate in angiotensin II activity: Alcohol, ketone and ester analogues of angiotensin II. *J Med Chem* 29:1968, 1986.

47. De Gasparo M, Whitebread S, Kamber B, et al: Effect of covalent dimer conjugates of angiotensin II on receptor affinity and activity in vitro. *J Receptor Research* 11:247, 1991.

48. Bovy PR, Trapani AJ, McMahon EG, et al: A carboxy-terminus truncated analogue of angiotensin II, [sar1]angiotensin II-(1-7)-amide, provides an entry to a new class of angiotensin II antagonists. *J Med Chem* 32:520, 1989.

49. Bovy PR, O'Neal JM, Olins GM, et al: Structure–activity relationships for the carboxy-

terminus truncated analogues of angiotensin II, a new class of angiotensin II antagonists. *J Med Chem* 33:1477, 1990.

50. Holck M, Bosse R, Fischli W, et al: An angiotensin II antagonist with strongly prolonged action. *Biochem Biophys Res Commun* 160:1350, 1989.

51. Bosse R, Gerold M, Fischli W, et al: An angiotensin with prolonged action and blood pressure-lowering properties. *J Cardiovasc Pharmacol* 16(Suppl. 4):S50, 1990.

52. Hsieh K, LaHann TR, Speth RC: Topographic probes of angiotensin and receptor: Potent angiotensin II agonist containing diphenylalanine and long-acting antagonists containing biphenylalanine and 2-indan amino acid in position 8. *J Med Chem* 32:898, 1989.

53. Bovy PR, Getman DP, Olins GM, et al: Poster Abstracts, American Chemical Society, Medicinal Chemistry Division, 22nd National Medicinal Chemistry Symposium, July 29–August 2, 1990, Austin TX, p. 46.

54. Bovy PR, O'Neal JM, McMahon EM, et al: Synthesis and biological activity of N-terminus modified [Ile^8] angiotensin II analogues. *Eur J Med Chem* 25:589, 1990.

55. Duncia JV, Chiu AT, Carini DJ, et al: The discovery of potent nonpeptide angiotensin II receptor antagonists: A new class of potent antihypertensives. *J Med Chem* 33:1312, 1990.

56. Moore, GJ: Kinetics of acetylation–deacetylation of angiotensin II; intramolecular interactions of the tyrosine and histidine side-chains. *Int J Peptide Protein Res* 26:469, 1985.

57. Samanen J, Cash T, Narindray D, et al: An investigation of angiotensin II agonist and antagonist analogues with 5,5-dimethylthiazolidine-4-carboxylic acid and other constrained amino acids. *J Med Chem* 34:3036, 1991.

58. Spear KL, Brown MS, Reinhard EJ, et al: Conformational restriction of angiotensin II: Cyclic analogues having high potency. *J Med Chem* 33:1935, 1990.

59. Sugg EE, Dolan CA, Patchett AA, et al: Cyclic disulfide analogs of [Sar^1,Ile^8]-angiotensin II. In Rivier JE, Marshall GR, (eds): *Peptides: Chemistry, Structure and Biology*. ESCOM Science Publishers, Leiden, The Netherlands, 1990, p. 305.

60. Marshall GR, Kaczmarek K, Kataoka T, et al: Optimization of constraints forcing receptor-bound turn conformations of angiotensin. In Girald E, Andreu E (eds): *Peptides 1990, Proceedings of the 21st European Peptide Symposium*. ESCOM Science Publishers, BV, Leiden, 1991, p. 594.

61. Pierson ME, Freer RJ: Analysis of the active conformation of angiotensin II: A comparison of AII and non-peptide AII antagonists. *Peptide Res* 5:102, 1992.

62. Keenan RM, Weinstock J, Finkelstein JA, et al: Imidazole-5-acrylic acids: Potent, nonpeptide angiotensin II receptor antagonists designed using a novel peptide pharmacophore model. *J Med Chem* 35:3858, 1992.

63. Samanen JM, Weinstock J, Hempel JC, et al: A molecular model of angiotensin II (Ang II) for rational design of small Ang II antagonists with increased potency. In Smith JA, Rivier JE, eds: *Peptides, Chemistry and Biology*. Proceedings of the Twelfth American Peptide Symposium, June 16–21, 1991, Cambridge, MA, ESCOM Science Publishers, BV, Leiden, 1992, p. 386.

64. Timmermans PBMWM, Chiu AT, Smith RD, et al: Novel nonpeptide angiotensin II receptor antagonists. *Proc Western Pharmacol Soc* 35:65, 1992.

65. Wexler RR, Carini DJ, Duncia JV, et al: The rationale for the chemical development of angiotensin II receptor antagonists. *Am J Hypertens* 5:209S–220S, 1992.

66. The conformation of Ang II was taken from Smeby RR, Fermandjian S: Conformation of angiotensin II. In Weinstein B (ed): *Chemistry and Biochemistry of Amino Acids, Peptides and Proteins*. Marcel Dekker, New York, 1978, p. 117.

67. Chiu AT, Duncia JV, McCall DE, et al: Nonpeptide angiotensin II receptor antagonists. III. Structure–function studies. *J Pharmacol Exp Ther* 250:867, 1989.

68. Wong PC, Price WA, Chiu AT, et al: Nonpeptide angiotensin II receptor antagonists. IV. EXP6155 and EXP6803. *Hypertension* 13:489, 1989.

69. Wong PC, Price WA, Chiu AT, et al: EXP6803 (methyl 2-n-butyl-1-[4-(2-carboxybenzamido) benzyl]-4-chloroimidazole-5-acetate, sodium salt): A novel nonpeptide angiotensin II receptor antagonist. *Hypertension* 12:340, 1988.

70. Wong PC, Price WA, Chiu AT, et al: EXP6803, a nonpeptide angiotensin II receptor antagonist. *Cardiovasc Drug Rev* 7(4):285, 1989.

71. Carini DJ, Duncia JV, Aldrich PE, et al: Nonpeptide angiotensin II receptor antagonists: the discovery of a series of *N*-(biphenylmethyl)imidazoles as potent, orally-active antihypertensives. *J Med Chem* 34:2525, 1991.

72. Timmermans PBMWM, Carini DJ, Chiu AT, et al: Nonpeptide angiotensin II receptor antagonists: A novel class of antihypertensive agents. *Blood Vessels* 27:295, 1990.

73. Timmermans PBMWM, Carini DJ, Chiu AT, et al: Nonpeptide angiotensin II receptor antagonists. *Am J Hypertens* 3:599, 1990.

74. Timmermans PBMWM, Carini DJ, Chiu AT, et al: The discovery and physiological effects of a new class of highly specific angiotensin II receptor antagonists. In Laragh JH, Brenner BM (eds): *Hypertension: Pathophysiology, Diagnosis and Management.* Raven Press, New York, 1990, p. 2351.

75. Johnson AL, Carini DJ, Chiu AT, et al: Nonpeptide angiotensin II receptor antagonists. *Drug News Perspectives* 3:337, 1990.

76. Duncia JV, Carini DJ, Chiu AT, et al: DuP 753. *Drugs Future* 16(4):305, 1991.

77. Wong PC, Barnes TB, Chiu AT, et al: Losartan (DuP 753), an orally active nonpeptide angiotensin II receptor antagonist. *Cardiovasc Drug Rev* 9(4):317, 1991.

78. Buhlmayer P: Angiotensin-II antagonists: Patent activity since the discovery of DuP 753. *Curr Opin Ther Patents* 2(10):1693, 1992.

79. Wong PC, Hard SD, Chiu AT, et al: Pharmacology of DuP 532, a selective and noncompetitive AT₁ receptor antagonist. *J Pharmacol Exp Ther* 259(2):861, 1991.

80. Carini, DJ: 4-Alkylimidazole derivatives. U.S. Patent 5,137,902. Issued to E. I. du Pont de Nemours and Company (Wilmington, DE), 1992.

81. Carini DJ, Duncia JV, Wong PC: Substituted imidazoles. PCT Patent Application WO 91/00277, 1991.

82. Ardecky RF, Carini DJ, Duncia JV, et al: Angiotensin II receptor blocking imidazoles. PCT Patent Application WO91/00281, 1991.

83. Hodges JC, Sircar I, Edmunds JJ, et al: 4-(1H-pyrrol-1-yl)imidazoles with angiotensin II receptor (AT₁) antagonist activity (Abstract). American Chemical Society, 204th National Meeting, Washington, DC, August 23–28, 1992, MEDI No. 33.

84. Wagner A, Englert H, Kleeman HW, et al: European Patent Application 503,162, 1992.

85. Caille J, Corbier A, Fortin M, et al: European Patent Application 465,368, 1992.

86. Yanagisawa H, Shimoji Y, Fujimoto K, et al: European Patent Application 503,785, 1992.

87. Wong PC, Price WA, Chiu AT, et al: Nonpeptide angiotensin II receptor antagonists. XI. Pharmacology of EXP3174, an active metabolite of DuP 753—an orally active antihypertensive agent. *J Pharmacol Exp Ther* 255(1):211, 1990.

88. For references on surmountable and insurmountable antagonism see Wong PC, Timmermans PBMWM: Nonpeptide angiotensin II receptor antagonists: Insurmountable angiotensin II antagonism of EXP 3892 is reversed by the surmountable antagonist DuP 753. *J Pharmacol Exp Ther* 252:49, 1991.

89. Carini DJ, Duncia JV, Yoo SE: Angiotensin II blocking imidazoles. U.S. Patent 4,880,804. Issued to E.I. du Pont de Nemours and Company (Wilmington, DE), 1989.

90. Chakravarty PK, Camara VJ, Chen A, et al: Substituted benzimidazoles as novel non-peptide

angiotensin II receptor antagonists (Abstract). American Chemical Society, 200th National Meeting, Washington DC, August 26–31, 1990, MEDI No. 90.

91. Thomas AP, Allott CP, Gibson KH, et al: New nonpeptide angiotensin II antagonists. 1. Synthesis, biological properties, and structure-activity relationships of 2-alkyl benzimidazole derivatives. *J Med Chem* 35:877, 1992.

92. Zhang J, Entzeroth M, Wienen W, et al: Characterization of BIBS 39 and BIBS 222: Two new nonpeptide angiotensin II receptor antagonists. *Eur J Pharmacol* 218:35, 1992.

93. Yanagisawa I, Watanabe T, Kikuchi K, et al: PCT Patent Application WO 92/04343, 1992.

94. Kivlighn SD, Gabel RA, Siegl PKS: L-158, 809: Antihypertensive efficacy and effects on renal function in the conscious spontaneously hypertensive rat (SHR) [Abstract]. *FASEB J* 5(6, part III):A1576, 1991.

95. Shibouta Y, Inada Y, Ojima M, et al: Pharmacological profiles of TCV-116, a highly potent and long acting angiotensin II (AII) receptor antagonist (Abstract). *J Hypertens* 10(Suppl 4):S143, 1992.

96. Mizuno K, Niimura S, Tani M, et al: Hypotensive activity of TCV-116, a newly developed angiotensin II receptor antagonist, in spontaneously hypertensive rats. *Life Sci* 51:183, 1992.

97. Mantlo NB, Ondeyka D, Chang RSL, et al: Diacidic nonpeptide angiotensin II antagonists (Abstract). American Chemical Society, Fourth Congress of North America, New York, 1991, MEDI No. 103.

98. Chakravarty PK, Greenlee WJ, Mantlo NB, et al: European Patent Application 400,974, 1990.

99. Jaunin R: European Patent Application 467,207, 1992.

100. Morimoto A, Nishikawa K: European Patent Application 430,300, 1991.

101. Allen EE, Huang SX, Chang RSL, et al: Substituted pyrazolo[1,5-a]pyrimidines as potent orally active angiotensin II receptor antagonists (Abstract). American Chemical Society, 202nd National Meeting, New York, 1991, MEDI No. 104.

102. Walsh TF, Fitch KJ, MacCoss M, et al: Synthesis of new imidazo[1,2-b]pyridazine isosteres of potent imidazo[4,5-*b*]pyridine angiotensin II antagonists. *Bioorg Med Chem Lett* 4:219, 1994.

103. Bovy PR, O'Neal J, Collins JT, et al: New cycloheptimidazolones are nonpeptide antagonists for angiotensin II receptors. *Med Chem Res* 1:86, 1991.

104. Carini DJ, Wells GJ, Duncia JV: European Patent Application 323,841, 1989.

105. Wells GJ, Carini DJ, Chiu AT, et al: Substituted pyrrole, pyrazole and triazole AII receptor antagonists (Abstract). American Chemical Society, 200th National Meeting, Washington, DC, August 26–31, 1990, MEDI No. 101.

106. Reitz DB, Penick MA, Brown MS, et al: 1H-1,2,4-Triazoles as potent orally active angiotensin II antagonists (Abstract). American Chemical Society, 203rd National Meeting, San Francisco, CA, April 5–10, 1992, MEDI No. 189.

107. Ashton WA, Cantone CL, Chang LL, et al: Synthesis and evaluation of 3,4,5-trisubstituted-1,2,4-triazoles as angiotensin II antagonists (Abstract). American Chemical Society, 202nd National Meeting, New York, 1991, MEDI No. 105.

108. Nisato D, Cazaubon C, Lacour C, et al: Pharmacological properties of SR 47436, a nonpeptidic angiotensin II receptor antagonist (Abstract). *Br Pharmacol Soc Proc* 105,84P, 1992.

109. Cristophe B, Libon R, Chatelain P, et al: Effect of SR 47436 on angiotensin II induced pressor responses in pithed rats. *Br J Pharmacol Proc* (Suppl) 105,259P, 1992.

110. Reitz DB, Garland DJ, Norton MS, et al: 2H-Imidazol-2-ones as potent orally active angiotensin II receptor antagonists (Abstract). American Chemical Society, 204th National Meeting, Washington, DC, August 23–28, 1992, MEDI No. 31.

111. Chang LL, Ashton W, MacCoss M, et al: European Patent Application 412,594, 1991.

112. Chang LL, Ashton WT, Strelitz RA, et al: Synthesis and evaluation of substituted triazolinones as angiotensin II receptor antagonists. Poster Abstracts, American Chemical Soci-

ety, 23rd National Medicinal Chemistry Symposium, Poster No. 32, Buffalo, NY, June 14–18, 1992.

113. Olins GM, Corpus VM, McMahon EG, et al: In vitro pharmacology of a nonpeptidic angiotensin II receptor antagonist, SC-51316, *J Pharmacol Exp Ther* 261:1037, 1992.

114. Ashton WT, Hutchins SM, Greenlee WJ, et al: 1-Substituted-3-alkyl-1H-pyrazole-5-carboxylic acids as potent angiotensin II antagonists (Abstract). American Chemical Society, 203rd National Meeting, San Francisco, CA, April 5–10, 1992, MEDI No. 168.

115. Middlemiss D, Ross BC, Eldred C, et al: C-linked pyrazole biaryl tetrazoles as antagonists of angiotensin II. *Bioorg Med Chem Lett* 2:1243, 1992.

116. Reitz DB: 1,3,5-Trisubstituted-1,2,4-triazole compounds for treatment of circulatory disorders. U.S. Patent 5,140,036, 1992.

117. De Laszlo S, Allen EE, Chakravarty PK, et al: European Patent Application 411,766,1991.

118. Herold P, Buhlmayer P: European Patent Application 407,342, 1991.

119. Morimoto A, Nishikawa K: European Patent Application 445,811, 1991.

120. Naka T, Nishikawa K: European Patent Application 442,473, 1991.

121. Bantick JR, McInally T, Tinker AC, et al: European Patent Application 500,297, 1992.

122. Patchett AA, de Laszlo SE, Greenlee WJ: European Patent Application 502,575, 1992.

123. Venkatesan AM, Levin JI: European Patent Application 497,150, 1992.

124. Weller HN, Miller AV, Moquin RV, et al: Benzothiadiazine dioxides: A new class of potent angiotensin-II (AT$_1$) receptor antagonists. *Bioorg Med Chem Lett* 2:1115, 1992.

125. Herold P, Buhlmayer P: European Patent Application 424,317, 1991.

126. Allen EE, Greenlee WJ, MacCoss M, et al: PCT Patent Application WO 91/15209, 1991.

127. Hoornaert C, Daumas C, Aletru M, et al: European Patent Application 500,409, 1992.

128. Herold P, Buhlmayer P: European Patent Application 435,827, 1991.

129. Bru-Magniez N, Teulon J, Nicolai E: European Patent Application 465,323, 1992.

130. Oldham AA, Allott CP, Major JS, et al: ICI D8731: A novel, potent and orally-effective angiotensin II antagonist (Abstract). *Br Pharmacol Soc Proc* 105,83P, 1992.

131. Bradbury RH, Allott CP, Dennis M, et al: New nonpeptide angiotensin II receptor antagonists. 2. Synthesis, biological properties, and structure-activity relationships of 2-alkyl-4-(biphenylylmethoxy)quinoline derivatives. *J Med Chem* 35:4027, 1992.

132. Roberts DA, Bradbury RH, Ratcliffe AH: European Patent Application 453,210, 1991.

133. Katano K, Ogino H, Shitara E, et al: PCT Patent Application WO91/19697, 1991.

134. Roberts DA, Pearce RJ, Bradbury RH: PCT Patent Application WO91/07407, 1991.

135. Oldham AA, Allott CP, Major JS, et al: Substituted 4-benzyloxy-naphthyridine derivatives: A novel, potent and orally-effective angiotensin II receptor antagonist (Abstract). *Br Pharmacol Soc Proc* 105, 34P, 1992.

136. Winn M, De B, Zydowsky TM, et al: European Patent Application 475,206, 1992.

137. De B, Winn M, Zydowsky TM, et al: Discovery of a novel class of orally active, non-peptide angiotensin II antagonists. *J Med Chem* 35:3714, 1992.

138. Morimoto A, Nishikawa K, Naka T: European Patent Application 443,568, 1991.

139. Clemence F, Fortin M, Haesslein J: European Patent Application 498,721, 1992.

140. Criscione L, Degasparo M, Buhlmayer P, et al: Pharmacological profile of CGP48933, a novel, nonpeptide antagonist of AT$_1$ angiotensin II receptor (Abstract). *J Hypertension* 10(Suppl 4):196, 1992.

141. Buhlmayer P, Ostermayer F, Schmidlin T: European Patent Application 443,983, 1991.

142. Carini, DJ, Duncia JV, Johnson AL, et al: Nonpeptide angiotensin II receptor antagonists: N-[(benzyloxy)benzyl]imidazoles and related compounds as potent antihypertensives. *J Med Chem* 33:1330, 1990.

143. Marshall FH, Barnes JC, Brown JD, et al: The interaction of GR117289 with the angiotensin AT$_1$ and AT$_2$ binding sites (Abstract). *Br J Pharmacol* 104(Suppl):425P, 1991.

144. Middlemiss D, Drew GM, Ross BC, et al: Bromobenzofurans: A new class of potent, non-peptide antagonists of angiotensin II. *Bioorg Med Chem Lett* 1:711, 1991.
145. Robertson MJ, Middlemiss D, Ross BC, et al: GR117289: A novel, potent and specific nonpeptide angiotensin receptor antagonist (Abstract). *Br J Pharmacol* 104(Suppl):300P, 1991.
146. Hilditch A, Akers JS, Travers A, et al: Cardiovascular effects of the angiotensin receptor antagonist, GR117289, in conscious renal hypertensive and normotensive rats (Abstract). *Br J Pharmacol* 104:423P, 1991.
147. Hunt AAE, Hilditch A, Robertson MJ, et al: Cardiovascular effects of the angiotensin receptor antagonist GR117289, in conscious dogs and marmosets. *Br J Pharmacol* (Proceedings Suppl) 104:424P, 1991.
148. Poss MA: European Patent Application 488,532, 1992.
149. Dhanoa DS, Bagley SW, Chang RSL, et al: Potent AT_1 selective angiotensin II antagonists (Abstract). American Chemical Society, 204th National Meeting, Washington, DC, August 23–28, 1992, MEDI No. 34.
150. Greenlee WJ, Patchett AA, Hangauer DG, et al: PCT Patent Application WO91/11909, 1991.
151. Walsh T, Fitch KJ, Greenlee WJ, et al: Merck Research Laboratories, unpublished observations.
152. Sircar I: PCT Patent Application WO 92/06081, 1992.
153. Bovy PR, Collins JT, Manning RE: PCT Patent Application WO 92/15577, 1992.
154. Bovy PR, Reitz DB, Collins JT, et al: Toluyl-1H-pyrrolyl-tetrazoles are angiotensin II receptor antagonists. American Chemical Society, 203rd National Meeting, San Francisco, CA, April 5–10, 1992, MEDI No. 170.
155. Oku T, Kayakiri H, Satoh S, et al: European Patent Application 480,204, 1991.
156. Chakravarty PK, MacCoss M, Mantlo N, et al: European Patent Application 510,813, 1992.
157. Poss MA: European Patent Application 501,269, 1992.
158. Lin H-S, Rampersaud AA, Zimmerman K, et al: Nonpeptide angiotensin II receptor antagonists: Synthetic and computational chemistry of N-[[4-[2-(1H-tetrazol-5-yl)-1-cycloalken-1-yl]phenyl]methyl]imidazole derivatives and their in vitro activity. *J Med Chem* 35:2658, 1992.
159. Bovy PR, Collins JT, Olins GM, et al: Conformationally restricted polysubstituted biphenyl derivatives with angiotensin II receptors antagonist properties. *J Med Chem* 34:2410, 1991.
160. Buhlmayer P, Criscione L, Fuhrer W, et al: Nonpeptidic angiotensin II antagonists: Synthesis and in vitro activity of a series of novel naphthalene and tetrahydronaphthalene derivatives. *J Med Chem* 34:3105, 1991.
161. Lifer SL, Marshall WS, Mohamadi F, et al: European Patent Application 438,869, 1991.
162. Marshall WS, Whitesitt CA, Simon RL, et al: The synthesis and evaluation of a novel series of imidazole angiotensin II (AII) receptor antagonists. American Chemical Society, 203rd National Meeting, San Francisco, CA, April 5–10, 1992, MEDI No. 173.
163. Wada T, Inada Y, Shibouta Y, et al: Antihypertensive action of a nonpeptide angiotensin II (AII) antagonist, TCV-116, in various hypertensive rats (Abstract). *J Hypertens* 10(Suppl 4):S144, 1992.
164. Olins GM, Corpus VM, McMahon EG, et al: Pharmacology of SC-51895, a potent non-peptidic angiotensin II (AII) receptor antagonist (Abstract). *FASEB J* 6(5):A1775, 1992.
165. Edwards MP, Allott CP, Bradbury RH, et al: Potent, nonpeptidic angiotensin II antagonists (Abstract). American Chemical Society, 203rd National Meeting, San Francisco, April 5–10, MEDI No. 177, 1992.
166. Wood JM, Schnell CR, Levens NR: The kidney is an important target for the antihypertensive action of an angiotensin II (Ang II) receptor antagonist in spontaneously hypertensive rats (SHR) (Abstract). *Hypertension* 20(3):436, 1992.
167. Stearns RA, Miller RR, Doss GA, et al: The metabolism of DuP 753, a nonpeptide angio-

tensin II receptor antagonist, by rat, monkey and human liver slices. *Drug Metab Disp* 20:281, 1992.

168. Nelson E, Merrill D, Sweet CS, et al: Efficacy and safety of oral MK-954 (DuP 753), an angiotensin receptor antagonist, in essential hypertension (Abstract). *J Hypertens* 9(Suppl 6):S468, 1991.

169. Hagino T, Abe K, Tsunoda K, et al: Chronic effects of MK-954, a nonpeptide angiotensin II receptor antagonist, on 24-hour ambulatory blood pressure, renin angiotensin aldosterone system and renal function in essential hypertension. *J Hypertens* 10(Suppl 4):224, 1992.

170. Nelson E, Arcuri K, Ikeda L, et al: Efficacy and safety of losartan in patients with essential hypertension (Abstract). *J Hypertens* 10(Suppl 4):S122, 1992.

171. Bucknall CE, Neilly JB, Carter R, et al: Bronchial hyperreactivity in patients who cough after receiving angiotensin converting enzyme inhibitors. *Br Med J* 296:86, 1988.

172. Ferner RE, Simpson JM, Rawlins MD: Effects of intradermal bradykinin after receiving angiotensin converting enzyme inhibitors. *Br Med J* 294:1119, 1987.

5

The Angiotensin II AT$_2$ Receptor Subtype

Marc de Gasparo, Nigel R. Levens, Bruno Kamber, Pascal Furet, Steven Whitebread, Véronique Brechler, and Serge P. Bottari

During the past 30 years, evidence has accumulated from pharmacological studies indicating that there are multiple subtypes of the angiotensin II (ANG II) receptor. A great many peptide analogues of ANG II have been synthesized to obtain final proof of the existence of such subtypes.[1]

By the end of the 1970s, all that was known about the structure–activity relationship of ANG II could be summed up as follows: (1) the constituents essential for its biological activity and receptor binding are the aromatic amino acids Tyr[4], His[6], and Phe[8] and the free carboxylic acid group of Phe[8]; (2) a sarcosine residue instead of the Arg in position 1 increases biological activity and makes the compound more stable; (3) the sequence 4–8 is the shortest chain fragment with biological activity; and (4) the substitution of aliphatic amino acids for Phe in position 8 yields ANG II antagonists.[2]

It was only in the late 1980s that selective compounds became available with the help of which it was possible to demonstrate unequivocally the existence of at least two receptor subtypes for ANG II.[3,4] The prototypical antagonist of the AT$_1$ receptor is losartan (2-*n*-butyl-4-chloro-5-(hydroxymethyl)-

Marc de Gasparo, Nigel R. Levens, Bruno Kamber, Pascal Furet, Steven Whitebread, Véronique Brechler, and Serge P. Bottari • Cardiovascular Research Laboratories, CIBA-GEIGY Limited, CH-4002 Basel, Switzerland. *Present address for VB:* Laboratoire de Biochimie Moléculaire de l'Hypertension, Institut de Recherche Clinique de Montréal, H3W 1R7 Montréal, Quebec, Canada. *Present address for SPB:* Centre d'Etudes Nucléaires de Grenoble DBMS/BRCE, INSERM U 244, 38041 Grenoble Cedex, France.

Angiotensin Receptors, edited by Juan M. Saavedra and Pieter B.M.W.M. Timmermans. Plenum Press, New York, 1994.

1-[[2'-(1H-tetrazol-5-yl)biphenyl-4yl]methyl]imidazole). Examples of specific AT$_2$ receptor ligands are CGP 42 112, a markedly modified pentapeptide analogue of ANG II (nicotinyl-Tyr-(N^α-benzyloxycarbonyl-Arg)-Lys-His-Pro-Ile-OH), and tetrahydroimidazo-pyridines (spinacine) typified by PD 123177 (1-[(4-amino-3-methylphenyl)methyl]-5-(diphenylacetyl)-4,5,6,7-tetrahydro-1H-imidazol[4,5-c]pyridine-6-carboxylic acid). These compounds (Fig. 1) have been extensively used to characterize both AT$_1$ and AT$_2$ receptors in many somatic tissues and in the central nervous system of various species (for review, see reference 5). The recent classifications of ANG II receptors into two distinct subpopulations is based on the binding characteristics of the above peptide and nonpeptide ligands.[6]

1. STRUCTURE–ACTIVITY STUDIES WITH ANG II

On the basis of the structure–activity relationships identified with the help of synthetic analogues, combined with spectroscopic and theoretical studies of the conformation of ANG II, various models have been proposed in the literature to account for the mode of action of the peptide at its receptor site.

ANG II is a small peptide with no significant conformational restrictions and is an extremely flexible molecule in solution. We have performed conformational analyses of the full octapeptide sequence using the buildup approach of Gibson and Sheraga.[7] This analysis has led to the identification of 75 different types of backbone conformation in the calculated low-energy conformational minima of ANG II. Most of these states are folded, and comparison with the possible conformations of a cyclic analogue [Hcy^{3,5}]-ANG II retaining full activity[8] has suggested that one in particular, in which the central part of ANG II is involved in a beta-turn of type III giving an overall U-shape to the backbone of the peptide, might correspond to the conformation adopted by ANG II at its receptor. Recently, the crystallographic structure of an anti-idiotypic antibody complexed with ANG II has been published.[9] This structure reveals a compact, bound conformation of ANG II in which two turns bring the amino- and carboxyl-terminal portions into close proximity. Although there is no *a priori* reason why this conformation should also be the one adopted by NG II when it binds to its receptor, this experimental information on structure makes the model of folded bioactive conformation derived from structure–activity studies and theoretical calculations appear reasonable, since it shows that such a conformation of the octapeptide can exist in a protein environment.

As the 4–8 sequence of ANG II is the shortest chain fragment with biological activity, various peptide analogues have been synthesized (Table I) with substitutions at the N- and C-terminal ends as well as in the 4–8 core of

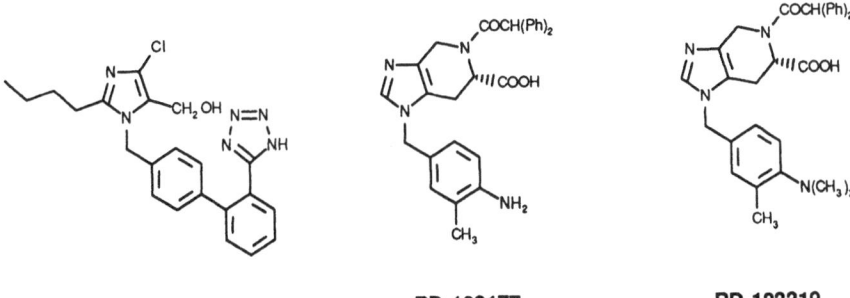

CGP42112

p-aminophenylalanine[6]- Angiotensin II

losartan **PD-123177** **PD-123319**

Figure 1. Angiotensin receptor ligands. Losartan is specific for the AT$_1$ subtype, whereas CGP 41112, p-aminophenylaline[6] ANG II, PD 123177, and PD 123319 are specific for the AT$_2$ subtype.

Table I. Affinity (IC$_{50}$) of Various Peptide Analogues
for the AT$_1$ (SMC) and the AT$_2$ (hUt) Subtypes

CGP		Human uterus (hUt)	Smooth muscle cell (SMC)
	Tyr - Ile - His - Pro - Phe	5.5 10^{-7}	
36734A	Tyr - Ile - Phe - Pro - Ile	3.0 10^{-5}	
36732A	Tyr - Ile - His - Pro - Ile	7.5 10^{-7}	30% at 10^{-4}
37013A	Gac-GABA - Tyr - Ile - His - Pro - Phe	2.0 10^{-7}	
37012A	Gac-GABA - Tyr - Ile - His - Pro - Ile	3.0 10^{-8}	
37837A	Gac-Ava - Tyr - Ile - His - Pro - Ile	1.6 10^{-9}	4.7 10^{-6}
39716	Z - Tyr - Ile - His - Pro - Phe	5.5 10^{-7}	
38108	Z - Tyr - Ile - His - Pro - Ile	1.0 10^{-8}	7.7 10^{-6}
41731	Z - Tyr - Ile - His - Pro - PheOMe	2.0 10^{-8}	
40982	Nic - Tyr - Ile - His - Pro - Ile	6.3 10^{-9}	3.0 10^{-6}
42099A	Nic-Glu(X) - Tyr - Ile - His - Pro - Ile	3.5 10^{-9}	
42103A	Nic-Lys(AVA) - Tyr - Ile - His - Pro - Ile	1.7 10^{-8}	
42105A	Nic-Tyr-Asp(Arg-OMe) - His - Pro - Ile	2.0 10^{-7}	
42104	Nic-Tyr-Asp(OBzl) - His - Pro - Ile	1.0 10^{-8}	
42112A	Nic-Tyr-Lys(Z-Arg) - His - Pro - Ile	1.9 10^{-10}	2.1 10^{-6}
54255A	Nic-Tyr-Lys(Z-Arg) - His - Pro - Phe	1.2 10^{-10}	2.4 10^{-7}

Abbreviations: Ava, *d*-aminovaleric acid; GABA, γ-aminobutyric acid; Gac, guanidinoacetic acid; Nic, nicotinic acid; Z, benzyloxycarbonyl; X, Gac-Ava-N(Me)-(Ch$_2$)$_3$-NH

the molecule. Tests on membranes isolated from human myometrium (AT$_2$ subtype) or rat adrenal cortex (AT$_1$ and AT$_2$ subtypes) and on rabbit aortic rings (AT$_1$ subtype) have confirmed that His^6 is important for binding to both receptor subtypes and cannot be substituted by Phe. However, Speth and Kim[10] have demonstrated that His^6 can be replaced by *p*-aminophenylalanine. This derivative molecule is indeed a strong ANG II ligand having an affinity in the nanomolar range for the AT$_2$ subtype. Its affinity for AT$_1$ is 1000-fold lower.

The N-terminal portion of ANG II has been substituted by various groups such as guanidinoacetic acid-γaminobutyric acid (Gac-GABA), guanidinoacetic acid-*d*-amino-valeric acid (Gac-Ava), benzyloxycarbonyl (Z), or nicotinic acid. The results clearly indicate the importance of this part of the molecule for binding to the AT$_2$ receptor expressed in human uterus.

2. CGP 42112: A SELECTIVE LIGAND FOR THE AT$_2$ RECEPTOR

Changes in the 4–8 core of ANG II have also been made, culminating in the synthesis of CGP 42112A, which has the highest affinity, in the sub-nanomolar range, for the AT$_2$ receptor.[3] However, this molecule does not affect contractions of rabbit aortic rings induced by ANG II, except at very high concentrations in the micromolar range, at which it proved to be an antagonist (IC_{50} 2 μM). This biological activity corresponds to a low binding affinity for the AT_1 subtype (K_i 1.7 μM) as measured in vascular smooth muscle cell membranes. The absence of a correlation between receptor binding in uterus and the functional assay in aortic rings suggested that these two tissues are not expressing the same receptor subtype.[11]

The first batch of CGP 42112 to be synthesized was the HCl salt (CGP 42112A). Subsequently, the trifluoroacetate salt (CGP 42112B) was produced, which is now commercially available (Neosystem, Strasbourg, France).

Replacement of Ile[8] of CGP 42112A by Phe turns the molecule into an agonist of the AT_1 subtype (EC_{50} in aortic rings 0.3 μM).

Novel substituted nonpeptide molecules published in a major patent by DuPont de Nemours and Company in 1988 and derived from blood pressure lowering imidazoles previously reported by Furakawa *et al.* in 1982 allowed us to clearly demonstrate the presence of two ANG II receptor subtypes. Losartan is characteristic of the AT_1 subtype, whereas CGP 42112A binds specifically to the AT_2 subtype.[3,4] Other imidazole derivatives, among them PD 123177 and PD 123319 ((S)-1-[[4-(di-methylamino)-3-methylphenyl]methyl]-5-(diphenylacetyl)-4,5,6,7-tetrahydro1H-imidazol[4,5-C]pyridine-6-carboxylic acid), were published in 1988 by Warner-Lambert. These compounds appear to have a greater affinity for the receptor recognized by CGP 42112 (AT_2) than for the receptor specific for losartan (AT_1).[4,12,13] However, CGP 42112 in plasma membranes from human uterus has an affinity for AT_2 roughly 200- and 30-fold greater than that of PD 123177 and PD 123319, respectively.[3]

ANG II and most large peptide derivatives (Sar[1]Ile[8] Ang II, Sar[1] Ang II, Ang III, Ang 1-7) bind to AT_1 and AT_2 receptors equally well and with similarly high affinities, but shorter peptides display rather weak binding.[14] It has been hypothesized, therefore, that CGP 42112, with its Arg-containing side chain, is superposable on a portion of the whole ANG II molecule that is essential for specificity for the AT_2 receptor subtype.

Since the nonpeptidic Parke-Davis compounds and the peptidic CGP 42112 despite its more bulky structure, recognize the same receptor subtype with high affinity, it has been of interest to superimpose one upon the other in an attempt to obtain information concerning the requirement of the AT_2 receptor for binding.

The structure–activity relationship recently published by Blankley *et al.*[15] shows that in their nonpeptidic ligands (PD 123177, PD 123319) the carboxylic acid function is crucial to obtain high affinity for the AT_2 receptor subtype. Similar behavior is observed in the peptidic analogues of ANG II concerning the C-terminus acid function, which suggests that the Parke-Davis compounds might mimic the C-terminal part of ANG II and the other peptidic ligands, in particular CGP 42112A. This hypothesis is illustrated by the superposition of

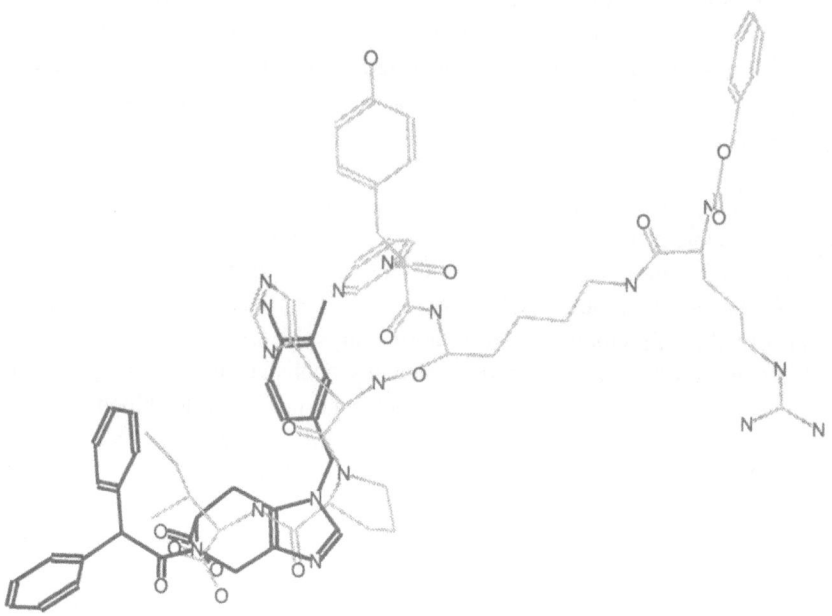

Figure 2. Superposition of CGP 42112 (gray) and PD 123177 (black). The diphenyl system of PD 123177 superimposes on the lipophylic pocket of isoleucine in CGP 42112.

CGP 42112 and PD 123177 (Fig. 2). By matching the positions of the carboxylic acid moieties, it is possible to bring the diphenyl system of PD 123177 into the region corresponding to the lipophilic pocket of residue 8 of ANG II (isoleucine in CGP 42112) to obtain an overlap of the imidazole ring with the peptide bond located between the last two C-terminal residues, Pro and Ile. In such a superposition, the *para*-aminophenylalanine group of the Parke-Davis structure is in proximity with the histidine residue corresponding to position 6 of ANG II, a residue that, interestingly, can be replaced by *para*-aminophenylalanine without loss of affinity for the AT_2 receptor subtype, as shown by Speth and Kim.[10]

3. THE AT_2 RECEPTOR: DISTRIBUTION AND BINDING CHARACTERISTICS

Synthetic peptides such as CGP 42112 and nonpeptidic agents like PD 123177 and PD 123319 have been extensively applied to characterize the AT_2 subtype, using radioactive ANG II or Sar^1Ile^8 ANG II (3-4,12-13,16-18).

These agents show at least a 2000-fold greater affinity for the AT_2 than for the AT_1 receptor.[6]

It has been clearly demonstrated that some tissues express only, or mainly, the AT_2 receptors. These include somatic tissues such as marmoset and human myometrium, pancreas, adrenal medulla, and ovarian granulosa, as well as brain tissues such as cerebellar cortex, locus coeruleus, thalamic nuclei, inferior olive, inferior colliculus, and medial amygdala.[5,16,17]

Other tissues have a nearly homogeneous population of AT_1 receptors (e.g., liver, lung, kidney, vascular smooth muscle cells in culture, subfornical organ) or are characterized by a mixture of both receptor subtypes in various proportions (e.g., adrenal cortex, heart, midbrain, brainstem).[5] In contrast to the distribution in adult tissues, the AT_2 subtype is transiently expressed in high density in fetal and neonatal tissues, some of which are not generally considered targets for ANG II. These tissues include skin, skeletal muscles, connective tissue, the gastrointestinal tract, and certain brain areas.[5,18]

The two major characteristics of the AT_2 receptor observed in binding studies are its interaction with sulfhydryl reducing agents such as dithiothreitol (DTT)[5,13,19] and its apparent absence of coupling to G-proteins.[20]

DTT almost totally suppresses the binding of radioiodinated ANG II to the AT_1 receptor with an IC_{50} of 2 nM, whereas either it increases the binding of the radioligand to the AT_2 subtype, to a maximum of 2.5 mM DTT,[3,12,21,22] or has no effect on ligand binding.[13,23]

The absence of interaction of the AT_2 receptors with G-protein was concluded from studies on the effect of the nonhydrolyzable analogue of GTP, GTPγS, on the affinity and binding kinetics of ANG II in plasma membrane particulates.[20] In contrast to the effect of GTP observed in vascular smooth muscle cells, expressing only the AT_1 receptor subtype, GTPγS does not affect binding parameters in myometrium and bovine cerebellar cortex. Neither the subnanomolar affinity of ANG II nor the capacity of these sites is affected by GTPγS. Furthermore, GTPγS did not affect the dissociation rate of ANG II from the AT_2 receptor, which excludes an exchange of GDP for GTP in the dissociation of the hormone-receptor–G-protein complex. Affinity and binding kinetics were not significantly affected when determined with or without sodium chloride, except in bovine cerebellum, where NaCl lowered the K_i two- to threefold, suggesting some heterogeneity within this subclass. These two parameters were also not affected when competition experiments were performed in the presence of the cytosolic fraction, indicating that the lack of effect of GTP analogues is not due to the loss of soluble G-protein subunits. Furthermore, as opposed to tissues expressing the AT_1 receptor, ANG II did not stimulate GTPγ-[^{35}S] incorporation into membranes prepared from tissues expressing only the AT_2 receptor.[20] These data strongly suggest that the

AT_2 receptor mediates ANG II effects through mechanisms independent of G-proteins. Similar results have been obtained by others.[12,23-27]

These binding data have suggested that the AT_2 receptor subtype probably does not belong to the superfamily of seven-transmembrane-domain receptors and should therefore differ significantly from the AT_1 receptor in signal transduction and molecular structure. Indeed, Northern blot analysis of PC12W mRNA with an AT_1 receptor gene probe failed to produce an RNA : DNA hybrid at low stringency.[27] Recently, however, a cDNA coding a unique 363-amino acid protein with pharmacological specificity, tissue distribution, and developmental pattern of the AT_2 receptor has been cloned from a rat fetus expression library. It is 34% identical in sequence to the AT_1 receptor, sharing a seven-transmembrane domain topology.[39b,42a] This receptor may belong to a new class of receptors including somatostatin SSTR1 in dopamine D_3, for which G-protein coupling has not been demonstrated.[42a]

The AT_1 : AT_2 ratio may change when it is calculated in competition assays using radioactive ANG II or Sar^1 Ile^8 ANG II as ligands and single blocking concentrations of specific AT_1 or AT_2 ligands as competitors. A correction for the different degree of occupancy of the receptor subtypes by the radioligand has been proposed.[26] This may not be necessary if the radioligands have the same affinity for both receptors.

In order to quantify specifically the AT_2 receptor subtype, CGP 42112 was radioiodinated and its binding characteristics have been compared with those of $[^{125}I]$-ANG II.[28]

In human myometrium, binding was saturable and reversible. The order of potency of a number of peptides and nonpeptides was the same as when radioiodinated ANG II was used as tracer and was typical for the AT_2 receptor: CGP 42112 > Ang II > PD 123319 >> losartan.[3,28] No specific binding of labeled CGP 42112 could be detected on membranes from vascular smooth muscle cells, confirming the absence of AT_2 receptor in this cell type. These data clearly indicate that CGP 42112 is a specific ligand for the AT_2 receptor.

In tissues expressing both receptor populations, such as rat adrenal cortex, competitive binding experiments using radioactive CGP 42112 as a tracer yielded monophasic competition curves characteristic of pure AT_2-type binding when using CGP 42112, PD 123319, and losartan as competitors. Biphasic curves were obtained with these ligands when $[^{125}I]$-ANG II was used as a tracer (Fig. 3).

Autoradiography was also performed using $[^{125}I]$-CGP 42112A. This is exemplified for the rat brain. There was no detectable binding of the radioligand in the cerebral cortex. However, specific binding was clearly detectable in the brainstem, below the cerebellum, and in certain areas of the

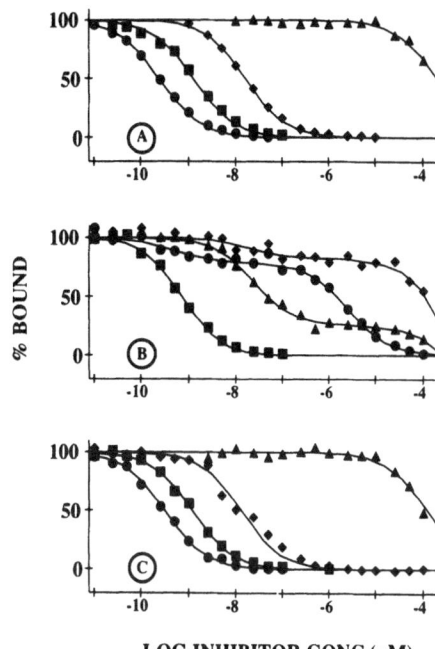

Figure 3. Competition curve with ANG II (■), CGP 42112A (●), PD 123319 (◆), and losartan (▲). (A) Radioiodinated CGP 42112 A binding to human myometrium membranes. (B) Radioiodinated ANG II binding to rat adrenal glomerulosa membranes. (C) Radioiodinated CGP 42112A binding to rat adrenal glomerulosa membranes. (From Whitebread et al.[28] Reprinted with permission.)

cerebellum that appear to be related to the white matter of the cerebellar cortex (Fig. 4).

Radioiodinated CGP 42112 is therefore especially useful to study AT_2 receptors in tissues where other ANG II receptor subtypes are also present.

4. SIGNAL TRANSDUCTION MECHANISM OF THE AT_2 RECEPTOR

4.1. Role of G-Proteins

By comparison with the AT_1 receptor, much less is known about the signaling mechanisms of the AT_2 receptor. However, as stated above, this receptor does not signal through "classical" G-proteins.[20]

Until very recently, there was no evidence to suggest which signaling pathway was linked to AT_2 receptors or the biological responses elicited by their stimulation. Indeed, several comprehensive reports, devoted to the functional aspects of the AT_2 receptor, ruled out a series of signaling mechanisms and possible biological responses.[23,24,27,29,30] A first clue came from a report from Sumners et al.,[31] suggesting that the ANG II-mediated decrease of intracellular cGMP in neuron cultures is mediated by AT_2 receptors.

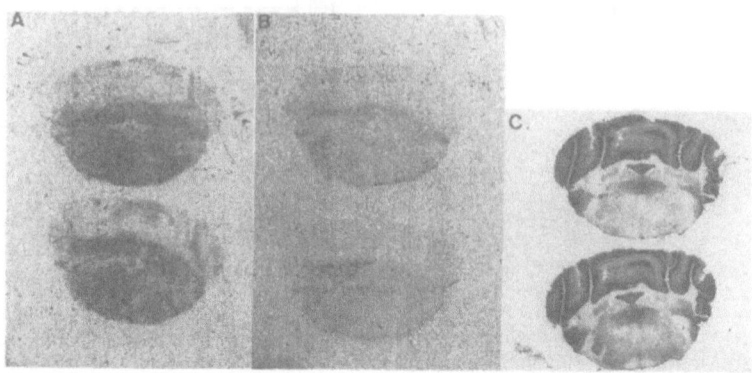

Figure 4. Autoradiogram of radioiodinated CGP 42112A binding in rat brain section at the levels of the cerebellum. (A) Total binding. (B) Nonspecific binding in the presence of 1 mM ANG II. (C) Hematoxylin-stained section showing the stained cerebellum and unstained brainstem.

To try to elucidate the mechanism through which the AT_2 receptor is linked to decreases in cellular cGMP, the effect of ANG II on PC12W cells, which express only the AT_2 subtype, was investigated. In these cells, ANG II induces a decrease in both basal and atrial natriuretic peptide (ANP)-stimulated intracellular cGMP levels.[32] This effect is not inhibited by losartan and is mimicked by CGP 42 112, which thus appears to be an agonist in this system. It is not affected by the phosphodiesterase inhibitor 3-isobutyl-1-methylxanthine, indicating that this AT_2-receptor-mediated effect does not involve a phosphodiesterase.[32] The lack of modulation of NO-stimulated guanylate cyclase (GC) activity by ANG II and the inability of NO-synthase inhibitors to modify the ANG II effect (Brechler *et al.*, unpublished data) indicate that ANG II most probably affects particulate guanylate cyclase (pGC).

In view of these observations, the effect of ANG II on pGC activity on plasma membrane prepared from tissues and cells expressing AT_1 and/or AT_2 receptors was investigated.[32] In these studies, ANG II was shown to inhibit basal and ANP stimulated pGC activity in membrane particulates prepared from cells or tissues expressing the AT_2 receptors. This mechanism was not observed in membrane particulates prepared from vascular smooth muscle cells (SMC), which express only the AT_1 subtype, and was not affected by the AT_1 antagonist losartan.

These findings suggest that the AT_2-receptor-mediated, ANG II-induced decrease in intracellular cGMP is due to inhibition of pGC activity.

Since it has been reported that the phosphorylation status of the ANP receptor (ANPR-A) is important for the regulation of its intrinsic GC activity,[33,34] it has been hypothesized that ANG II might inhibit ANPR-A GC activity by a mechanism involving the phosphorylation or dephosphorylation of this molecule. Thus the effect of inhibitors of selective tyrosine and serine/threonine phosphatase inhibitors on pGC activity was examined. Sodium orthovanadate, a selective inhibitor of protein tyrosine phosphatases (PTPases),[35] did not affect pGC stimulation by ANP, but completely abolished ANG II-induced inhibition of pGC.[32] This observation suggested the involvement of a PTPase in the AT_2-receptor-mediated inhibition of pGC. Okadaic acid, a selective inhibitor of protein phosphatase 2A and 1 (PP2A and PP1),[36] did not alter the inhibitory effect of ANG II. Similar findings were made on intact PC12W cells, suggesting that a PTPase is involved in the ANG II-mediated inhibition of pGC via the AT_2 receptor.

To further investigate this hypothesis, the effect of ANG II on the phosphotyrosine protein pattern was investigated. ANG II induces a rapid dephosphorylation of tyrosine-phosphorylated proteins in PC12W cells, which is inhibited by sodium orthovanadate.[32] This observation confirms the ability of ANG II to stimulate PTPase activity and suggests that, by analogy with the lymphocyte antigen receptor family,[37] the AT_2 receptor may use this pathway as a signal transduction mechanism. Early stimulation of a membrane PTPase has also been proposed as part of the signal transduction pathways mediated by somatostatin and could be involved in the inhibition of cell growth as observed with this peptide in a rat pancreatic acinar tumor cell line.[38,39]

Other authors have failed to reproduce these findings. Webb *et al.*,[39a] also using PC12W cells, have been unable to detect any tyrosine, serine, or threonine phosphorylation of the AT_2 receptor. In contrast, Kambayashi *et al.*[39b] have recently reported that ANG II, ANG III, as well as the AT_2-specific ligand CGP 42112 inhibit PTPase activity in plasma membranes of PC12W cells, rat fetal skin and COS-7 cells stably transfected with AT_2 cDNA, respectively. Thus there are reports indicating either a stimulation, an inhibition, or no effect induced by ANG II on PTPase activity. Collectively, it is therefore still unclear how the AT_2 receptor is coupled to PTPase activity.

4.2. Inhibition of T-Type Ca^{2+} Currents

Another cellular target of AT_2 receptors has recently been described in neurons cocultured from the hypothalamus and brainstem of 1-day-old rats[40] as well as in a hybridoma cell line NG108-15.[41] In their nondifferentiated state, NG108-15 cells exhibit only Ca^{2+} currents of the T-type. Using whole cell patch-clamping, both ANG II and CGP 42112 have been shown to inhibit T-type Ca^{2+} currents at membrane potentials higher than –40 mV and to shift the current voltage curve at lower potentials. These phenomena are not affected by the AT_1 receptor antagonist losartan.[5] This mechanism can also be selectively inhibited by Na orthovanadate and is not affected by alterations of the intracellular cGMP concentration, suggesting the involvement of a PTPase (Buisson *et al.*, unpublished observation).

Our current hypothesis is that AT_2 receptors may signal by altering PTPase activity, which modulates a series of cellular effector systems. Whether the AT_2 receptor molecule has intrinsic PTPase activity, such as the lymphocyte antigen receptor family,[37] would require studies with purifed or cloned receptors. Nevertheless, this hypothesis opens up new perspectives for the physiological role of ANG II.

Indeed, stimulation of PTPase activity together with observations of transient expression of the AT_2 receptor during development[18,25] suggests that this receptor subtype is involved in the control of proliferation and differentiation. Most growth factors are known to mediate their effects through receptors that have intrinsic tyrosine kinase activity and signal through tyrosine phosphorylation/dephosphorylation cascades. This is also true of many neurotransmitters and peptide hormones, which have been shown to stimulate tyrosine phosphorylation of various cellular substrates.[42] Since tyrosine dephosphorylation can lead to either activation or inhibition of a large variety of enzymes, the role of the AT_2 receptors in modulating cellular mechanisms can be expected to be complex.

5. HETEROGENEITY OF THE AT_2 RECEPTOR?

Estimates of the molecular weights of the AT_2 receptor show disparities. In nondifferentiated NG108-15, a cell line exclusively expressing the AT_2 subtype, the glycosylated and deglycosylated forms have molecular weights of 91.2 and 56.2 kDa, as determined by photoaffinity (Bonnafous *et al.*, unpublished observation). By contrast, cross-linking experiments performed in rat ovarian granulosa and R3T3 fibroblasts give estimates of 79 and 100 kDa, respectively.[23,24] The molecular weight deduced from the cDNA cloned is 41,300.[42a] These observations may suggest the existence of some heterogeneity among the AT_2 receptor subtype family, for example, several degrees of glycosylation in various tissues.

Tsutsumi and Saavedra[43,44] recently published arguments suggesting that AT_2 receptors in the brain constitute a heterogeneous population. Upon quantitative autoradiography of the brains of 2-week-old rats, they observed AT_2 receptors that differed in their sensitivity to DTT, GTPγS, and pertussis toxin. Whereas the inferior olive and hypoglossal nucleus express only the classic AT_2 subtype, ventral and mediodorsal thalamic, medial geniculate, oculomotor nuclei, superior colliculus, locus coeruleus, and the cerebellar cortex contain an AT_2 receptor that, although sensitive to CGP 42112A, is also sensitive to DTT, GTPγS, and pertussis toxin pretreatment, suggesting an association to specific

G-protein subtypes. These AT_2 receptors, although highly expressed early in development, do not reach a concentration comparable to that in the inferior olive. These data suggest that brain AT_2 receptors may be heterogeneous and may belong to two different pharmacological subtypes for which the designations AT_{2a} (coupled to G-protein) and AT_{2b} (not coupled to G-protein) receptor have been proposed.

Recently, Fluharty *et al.* (1992, personal communication), using N1E-115 neuroblastoma cell line,[45] proposed the existence of two populations of AT_2 receptor. Both exhibit a high affinity for CGP 42112 and no affinity for losartan. They differ, however, in their sensitivity to PD 123319 and to DTT. The data obtained with specific antisera suggest that one of these CGP 42112-sensitive receptor subpopulations is coupled to PLC_α through Gq.

Whether these receptors represent another subclass of AT_2 receptors or a third subtype of ANG II receptors is not yet clear. So far, there are no data available concerning their function.

6. HOW SELECTIVE ARE PD 123177 OR PD 123319 AND CGP 42112 FOR THE AT_2 THE RECEPTOR SUBTYPE?

Studies in various mammalian tissues indicate that CGP 42112 and the Parke-Davis compounds bind to the AT_2 receptor as defined above. However, in the literature there is some evidence that these compounds may not be entirely specific. It is known that CGP 42112 also has a low affinity (K_i 1.7 µM) for the mammalian AT_1 receptor.[3] Other reports suggest that these compounds can bind to other ANG II receptor subtypes with different affinities.[46–49]

Amphibian ANG II receptors differ from their mammalian counterparts in binding CGP 42112 (IC_{50} 400 nM) but not PD 123177 ($IC_{50} > 10$ µM).[46,47] Similar findings have been obtained in chicken chorioallantoic membranes.[48] The mammalian cytosolic ANG II-binding protein described by Kiron and Soffer[49] shows no affinity for either PD 123177 or losartan, whereas CGP 42112A binds to it with an IC_{50} of 300 nM (Whitebread *et al.*, unpublished observation). It is therefore likely that these binding sites are different from the known AT_1 and AT_2 receptors.

In renal mesangial cells, Ernsberger *et al.*[50] have described two AT_1 receptor subtypes that have different sensitivities to CGP 42112 and PD 123319. These authors termed these subtypes AT_{1a} and AT_{1b}. This nomenclature is, however, unrelated to the classification based on the cloning data,[51–53] which led to the discovery of highly homologous AT_1 receptor subpopulations, none having a higher-than-micromolar affinity for CGP 42112, PD 123177 or PD 123319. By contrast, the affinity (K_i) of PD 123319 for the renal AT_{1a}

receptor is 24 μM but 2.2 nM for the renal AT_{1b}. The affinity of CGP 42112 for renal AT_{1b} is 1000-fold lower (1.4 μM).[50] This nomenclature should be standardized in the future to avoid further confusion.

In functional assays, differences between CGP 42112 and the Parke-Davis compounds were reported by Jaiswal *et al.*[54,55] Using porcine endothelial and vascular smooth muscle cells, these authors compared the efficacy of CGP 42112, PD 123177, and losartan in inhibiting the prostaglandin synthesis induced by ANG I, ANG II, and ANG 1-7. In some cases, PD 123177 was able to inhibit, whereas CGP 42112 was not, and vice versa. Moreover, evidence was given for independent regulation of prostaglandins PGE_2 and PGI_2 even in the same cell type, CGP 42112A inhibiting PGE_2 production, but not PGI_2.

These pieces of circumstantial evidence may suggest that CGP 42112, because of its structural differences from PD 123177 and PD 123319, may recognize additional epitopes of the ANG II receptor. Alternatively, PD 123319 has been proposed to be an AT_2 antagonist, whereas CGP 42112 has been proposed to be an agonist.[5] Data on particulate guanylate cyclase activity in PC12W cells,[5,32,39b,56] as well as the results of patch-clamp experiments,[41] indicate that CGP 42112 is indeed a full agonist, whereas PD 123319 acts as an antagonist. Provided they are used at the right concentrations, CGP 42112 (<10 nM or 50-fold the K_i for AT_2) and PD 123319 (<0.5 μM or 100-fold the K_i for AT_2) constitute excellent complementary tools for the characterization of the physiological role of the AT_2 receptors.[56]

7. DOES THE AT$_2$ RECEPTOR PLAY A PHYSIOLOGICAL ROLE?

The use of subtype-selective ligands in recent years has furnished a wealth of information demonstrating both the intracellular mechanisms and the physiological processes mediated by ANG II acting through the AT_1 receptor subtype. On the other hand, intracellular mechanisms linked to the AT_2 receptor subtype have only recently been proposed.

The role of cGMP in cell metabolism is still poorly documented. Interestingly, cGMP has been suggested to affect the renin–angiotensin system through inhibition of renin release[57] and ANG II-stimulated aldosterone secretion.[58] Since the AT_2 receptor is expressed by adrenal glomerulosa cells,[3,4,12,13] one may speculate on its involvement in the modulation of aldosterone secretion in response to various stimulatory peptides. Similarly, this mechanism may also play a role in steroidogenesis in the zona fasciculata, which also expresses the AT_2 receptor. No proof, however, has so far been advanced for a role of the AT_2 subtype in the fine tuning of aldosterone secretion and steroidogenesis.[21]

cGMP is also generally considered to act as an inhibitor of cell proliferation. Modulation of its synthesis through the AT_2 receptor may also play a role in proliferation and differentiation. The transient expression of the AT_2 receptor during fetal development[18,25] may support this hypothesis.

Concerning the other cellular effect found to be mediated through AT_2 receptors, namely the inhibition and modulation of T-type Ca^{2+} currents, it is now well established that these currents support pacemaker activities in neurons.[59] Interestingly, the brain areas where AT_2 receptors have been found to be most abundant correspond to those where pacemaker activities resting on a low-threshold-activated Ca^{2+} conductance, corresponding to T-type currents, have been described previously.[60] One may thus speculate that at least one of the roles of the AT_2 receptor in the brain may be to modulate the frequency of membrane-potential oscillations, thereby affecting neurosecretion.

Recently, several studies have appeared in the literature showing that AT_2-receptor-selective ligands affect various function. These findings may suggest a physiological role for the AT_2 receptor.

7.1. The AT_2 Receptor and Renal Function

The kidney has been the focus of the majority of studies designed to investigate the physiological role of ANG II receptors, and, in dogs and rats, the AT_2-specific ligands PD 123177 and PD 123319 have been shown to influence renal function, although there is no convincing evidence for the presence of AT_2 receptors in the kidney of these species. In the anesthetized dog, intravenous administration of the AT_1-specific antagonist GR 117289 acts both to decrease blood pressure and to dilate the renal vasculature.[61] Intrarenal infusion of the AT_2-specific ligand PD 123177 at a rate of 20 µg/kg per min, unlike GR 117289, has no effect on basal renal function, but blocks the renal vascular constriction produced by intrarenal infusion of ANG II.[52] These apparently antagonistic actions of PD 123177 on ANG II-induced renal vasoconstriction are qualitatively the same as those produced by the AT_1 receptor antagonist GR 117289. In the anesthetized dog, intrarenal infusion of PD 123319 (30 and 300 µg/kg per min) has also been shown to induce an increase in urine volume and freewater formation.[62] This effect was not due to inhibition of vasopressin release, and a direct effect on the renal tubule could not be ruled out.[62]

In the anesthetized rat, intravenous administration of 120 mg/kg PD 123177 increases glomerular filtration rate and produces a chloridiuresis.[63] These actions of PD 123177 are, however, essentially identical to those produced by losartan in this animal model. Finally, PD 123177 (3 mg/kg bolus + 15-min intravenous infusion at 3 mg/kg per min) has, like losartan, been shown to increase renal blood flow in conscious rats after activation of the endogenous renin–angiotensin system by prior myocardial infarction.[64]

The results of all of these studies should, however, be interpreted with caution as the doses of AT_2 ligands that were administered are compatible with interaction with the AT_1 receptor as well.

In a recent series of experiments, we have further explored the possibility that ANG II acting through AT_2 receptors influences renal function. In these experiments, we have shown that the selective AT_1 antagonist losartan decreases blood pressure and renal vascular resistance while increasing renal blood flow in the anesthetized sodium-depleted rat. Administration of the AT_2-selective ligand PD 123319, to produce plasma levels compatible with an interaction with AT_2 receptors, had no effect on renal function. Another AT_2-selective ligand, CGP 42112B, also had no effect on renal function except at very high doses where effective renal blood flow was increased and renal vascular resistance decreased. Since the renal effects of CGP 42112B occurred at doses where an interaction of this compound with AT_1 receptors would be expected and exerted effects similar to losartan, our data support only a role for AT_1 receptors and not AT_2 receptors in the control of renal function in the rat.[64a]

7.2. The AT_2 Receptor and Cerebral Blood Flow

Within the systemic circulation, ANG II is an arterial constrictor. By contrast, recent studies have shown ANG II to induce relaxation of cerebral arteries. Since large cerebral arteries appear to contain an exclusive population of AT_2 receptors,[25,65] ANG II-induced relaxation is probably mediated through the AT_2 receptor. Increased systemic blood pressure is associated usually with an increase in cerebral blood flow, which is secondary to dilation of the cerebral vessels. Recent studies have shown that PD 123319 (10 mg/kg) is able to shift the upper limit of cerebral blood flow autoregulation toward higher pressures and therefore prevents the increase in cerebral blood flow produced by a pressor dose of norepinephrine. Since in these studies PD 123319 did not prevent ANG II-mediated peripheral vasoconstriction, the results have been interpreted to imply that ANG II acting through AT_2 receptors is responsible for the dilation of cerebral blood vessels.[66]

7.3. The AT_2 Receptor and Cognitive Function

In conscious rats, doses of PD 123177 as low as 10 ng/kg intraperitoneally (IP) enhanced the performance of mice in a light–dark aversion and habituation test. These observations would suggest that ANG II acting through the AT_2

receptor has anxiolyticlike and cognition-enhancing properties.[67] Losartan has been reported to have similar effects.[68]

7.4. The AT$_2$ Receptor and Balloon Injury

While in normal rat carotid artery the majority of ANG II receptors are of the AT$_1$ subtype, 2 weeks after balloon injury, AT$_2$ receptors become dominant.[69] By contrast, neointimal cells of injured rat thoracic aorta express almost exclusively the AT$_1$ receptor.[70] Treatment of rats after carotid ballooning with PD 123319 (20 mg/kg per day IP), which produced plasma levels of 248 nM, results in a 73% decrease in intimal area, while the AT$_1$ receptor ligand losartan at the same dosage produced a 95% decrease in intimal area.[69] In a similar model, CGP 42112 continuously infused perivascularly for 14 days (1 mg/kg per day) in the vicinity of the injured carotid artery was as effective as an angiotensin-converting enzyme inhibitor in preventing neointima formation. Low doses of losartan (5 mg/kg per day) were ineffective.[71] This discrepancy may be due to the low dose of losartan used in the latter study compared to other reports.[72,73] Both AT$_1$ and AT$_2$ receptors may therefore play a role in the processes of excessive repair that lead to lesion formation after balloon injury.

7.5. The AT$_2$ Receptor and Reproduction

In estrogen- and progesterone-treated ovariectomized rats, intracerebroventricular injection of ANG II stimulates luteinizing hormone and inhibits prolactin release through a mechanism partially mediated by both AT$_1$ and AT$_2$ receptor subtypes.[74] Granulosa cells in Graafian follicles express only AT$_2$ receptors.[23,75] Since the ovary possesses a complete renin–angiotensin system, it would therefore be surprising if this system were not found to play a role in the process of follicular maturation through the AT$_2$ receptors. Indeed, ANG II induces ovulation in perfused rabbit ovaries *in vitro* in the absence of gonadotropin,[76] and saralasin totally blocks this effect. The AT$_2$ receptor being the only subtype detectable in granulosa cells, it is perfectly conceivable that it might somehow be involved in the maturation of the follicle and in ovulation.

7.6. The AT$_2$ Receptor and Water Intake

PD 123319 and CGP 42112 administered intracerebroventricularly to rats inhibits sodium intake after sodium deprivation or converting-enzyme inhibition. PD 123319 also inhibits ANG II-induced water intake but at doses about ten times higher than those of losartan.[77]

Similarly, Beresford and Fitzsimons[78] as well as Hogarty *et al.*[79] reported blockade of water intake by losartan. However, in their hands, CGP 42112 and

PD 123177 were ineffective. Stephenson and Steele[74] also demonstrated that the AT_2 receptor does not play a role in water intake in estrogen- and progesterone-treated ovariectomized rats.

These conflicting data may suggest that both AT_1 and AT_2 receptor subtypes in the brain are involved in water intake, but this possibility still requires further investigation.

8. CONCLUSIONS

While the above results demonstrate that AT_2-selective ligands influence several physiological functions, it should be clearly stated that these observations do not necessarily mean that ANG II acting through the AT_2 receptor plays an important regulatory role in these functions. For example, the current generation of AT_2 selective ligands, such as PD 123177, PD 123319, and CGP 42112, are not totally specific for the AT_2 receptor, and in the majority of the above experiments they were given to the experimental animals in high doses. Our interpretation of published studies brings us to the conclusion that not enough care has been taken to ensure that the available ligands are given at doses compatible with an exclusive interaction with the AT_2 receptor subtype, and so far as concerns the majority of experiments conducted to date, it is equally possible to argue that the results obtained are due to an interaction of the ligands with the AT_1 receptor. Indeed PD 123177 and PD 123319 are valuable selective antagonists provided they are used at selective concentrations, ie., not higher than 0.5 μM. CGP 42112, on the other hand, is a full and selective AT_2 agonist if its concentration does not exceed 10 nM.[56] Possible exceptions to this criticism are the studies of Stromberg *et al.*[66] and those of Barnes *et al.*,[67,68] who either used doses of the compound too low to affect AT_1 receptors or demonstrated that the effects of AT_2 ligands were not influenced by AT_1 receptor blockade.

The distribution of AT_2 receptors in the growing and differentiating fetus, in the brain, in the ovary, and in some pathological situations[80,81] indicates that these receptors may be of relevance. Moreover, the finding that the AT_2 receptor may be linked to a tyrosine phosphatase activity suggests that it may be involved in various mechanisms in which tyrosine phosphorylation and dephosphorylation play a central role. The discovery of other members of the AT_2 subtype family may also broaden our field of investigation. Therefore, a more detailed knowledge of the pathophysiology of the AT_2 receptor based on studies performed with specific agonists and antagonists at selective concentrations could well open a new chapter in ANG II research and possibly also disclose new therapeutic applications.

REFERENCES

1. Peach MJ: Renin–angiotensin system: Biochemistry and mechanisms of action. *Physiol Rev* 57:313–370, 1977.

2. Khosla MC, Smeby RR, Bumpus FM: Structure–activity relationship in angiotensin II analogs. In Page IH, Bumpus FM (eds): *Angiotensin Handbook of Pharmacology.* Springer-Verlag, Berlin, Heidelberg, 1974, pp. 126–161.

3. Whitebread S, Mele M, Kamber B, et al: Preliminary biochemical characterization of two angiotensin II receptor subtypes. *Biochem Biophys Res Commun* 163:284–291, 1989.

4. Chiu AT, Herblin WF, McCall DE, et al: Identification of angiotensin II receptor subtypes. *Biochem Biophys Res Commun* 165:196–203, 1989.

5. Bottari SP, de Gasparo M, Steckelings UM, et al: Angiotensin II receptor subtypes. Characterization, signaling mechanisms and possible physiological implications. *Front Neuroendocrinol* 14:123–171, 1993.

6. Bumpus FM, Catt KJ, Chiu AT, et al: Nomenclature for angiotensin receptors. A report of the nomenclature committee of the council for High Blood Pressure research. *Hypertension* 17:720–721, 1991.

7. Gibson KD, Scheraga HA: Revised algorithms for the build-up procedure for predicting protein conformations by energy minimization. *J Comp Chem* 8:826–834, 1987.

8. Spear KL, Brown MS, Reinhard EJ, et al: Conformation restriction of angiotensin II: Cyclic analogues having high potency. *J Med Chem* 33:1935–1940, 1990.

9. Garcia KC, Ronco PM, Verroust PJ, et al: 3-Dimensional structure of an angiotensin-II Fab complex at 3 Angstrom—Hormone recognition by an anti-idiotypic antibody. *Science* 257:502–507, 1992.

10. Speth RC, Kim KH: Discrimination of two angiotensin II receptor subtypes with a selective agonist analogue of angiotensin II, p-aminophenylalanine[6] angiotensin II. *Biochem Biophys Res Commun* 169:997–1006, 1990.

11. Criscione L, Thomann H, Whitebread S, et al: Binding characteristics and vascular effects of various angiotensin II antagonists. *J Cardiovasc Pharmacol* 16(Suppl 4):S56–S59, 1990.

12. Dudley DT, Panek RL, Major TC, et al: Subclasses of angiotensin II binding sites and their functional significance. *Mol Pharmacol* 38:370–377, 1990.

13. Chang RS, Lotti VJ: Two distinct angiotensin II receptor binding sites in rat adrenal revealed by new selective nonpeptide ligands. *Mol Pharmacol* 37:347–351, 1990.

14. de Gasparo M, Whitebread S, Bottari SP, et al: Angiotensin II receptor heterogeneity. In Timmermans PB, Wexler RR (eds): *Medicinal Chemistry of the Renin Angiotensin System.* Elsevier, Amsterdam, 1994 (in press).

15. Blankley CJ, Hodges JC, Klutchko SR, et al: Synthesis and structure–activity relationships of a novel series of non-peptide angiotensin II receptor binding inhibitors specific for the AT_2 subtype. *J Med Chem* 34:3248–3260, 1991.

16. Hodges JC, Hamby JM, Blankley CJ: Angiotensin II binding inhibitors. *Drugs Future* 17:575–593, 1992.

17. Smith RD, Chiu AT, Wong PC, et al: Pharmacology of nonpeptide angiotensin II receptor antagonists. *Annu Rev Pharmacol Toxicol* 32:135–165, 1992.

18. Grady EF, Sechi LA, Griffin CA, et al: Expression of AT_2 receptors in the developing rat fetus. *J Clin Invest* 88:921–933, 1991.

19. Speth RC, Rowe BP, Grove KL, et al: Sulfhydryl reducing agents distinguish two subtypes of angiotensin II receptors in the rat brain. *Brain Res* 548:1–8, 1991.

20. Bottari SP, Taylor V, King IN, et al: Angiotensin II AT_2 receptors do not interact with guanine nucleotide binding proteins. *Eur J Pharmacol* 207:157–163, 1991.

21. Ouali R, Poulette S, Penhoat A, et al: Characterization and coupling of angiotensin-II

receptor subtypes in cultured bovine adrenal fasciculata cells. *J Steroid Biochem Mol Biol* 43:271–280, 1992.

22. Chiu AT, McCall DE, Nguyen TT, et al: Discrimination of angiotensin II receptor subtypes by dithiothreitol. *Eur J Pharmacol* 170:117–118, 1989.

23. Pucell AG, Hodges JC, Sen I, et al: Biochemical properties of the ovarian granulosa cell type 2-angiotensin II receptor. *Endocrinology* 128:1947–1959, 1991.

24. Dudley DT, Hubbell SE, Summerfelt RM: Characterization of angiotensin II (AT$_2$) binding sites in R3T3 cells. *Mol Pharmacol* 40: 360–367, 1991.

25. Tsutsumi K, Stromberg C, Viswanathan M, et al: Angiotensin-II receptor subtypes in fetal tissue of the rat: Autoradiography, guanine nucleotide sensitivity, and association with phosphoinositide hydrolysis. *Endocrinology* 129:1075–1082, 1991.

26. Rowe BP, Saylor DL, Speth RC: Analysis of angiotensin II receptor subtypes in individual rat brain nuclei. *Neuroendocrinology* 55:563–573, 1992.

27. Webb ML, Liu ECK, Cohen RB, et al: Molecular characterization of angiotensin-II type-II receptors in rat pheochromocytoma cells. *Peptides* 13:499–508, 1992.

28. Whitebread SE, Taylor V, Bottari SP, et al: Radioiodinated CGP 42112A: A novel high affinity and highly selective ligand for the characterization of angiotensin AT$_2$ receptors. *Biochem Biophys Res Commun* 181:1365–1371, 1991.

29. Balla T, Baukal AJ, Eng S, et al: Angiotensin II receptor subtypes and biological responses in the adrenal cortex and medulla. *Mol Pharmacol* 40:401–406, 1991.

30. Leung KH, Roscoe WA, Smith RD, et al: Characterization of biochemical response of angiotensin II (AT$_2$) binding sites in the rat pheochromocytoma PC12W cells. *Eur J Pharmacol,* 227:63–70, 1992.

31. Sumners C, Myers LM: Angiotensin II decreases cGMP levels in neuronal cultures from rat brain. *Am J Physiol* 260:C79–C87, 1991.

32. Bottari SP, King IN, Reichlin S, et al: The angiotensin AT$_2$ receptor stimulates protein tyrosine phosphatase activity and mediates inhibition of particulate guanylate cyclase. *Biochem Biophys Res Commun* 183:206–211, 1992.

33. Ballermann BJ, Marala RB, Sharma RK: Characterization and regulation by protein kinase C of renal glomerular atrial natriuretic peptide receptor-coupled guanylate cyclase. *Biochem Biophys Res Commun* 157:755–761, 1988.

34. Chinkers M, Garbers DL: The protein kinase domain of the ANP receptor is required for signaling. *Science* 245:1392–1394, 1989.

35. Gordon JA: Use of vanadate as a protein phosphotyrosine phosphatase inhibitor. In Hunter T, Sefton BM (eds): *Methods in Enzymology,* Vol. 201. Academic Press, San Diego, CA, 1991, pp. 477–482.

36. Cohen P, Holmes CFB, Tsukitani V: Okadaic acid: A new probe for the study of cellular regulation. *Trends Biochem Sci* 15:98–102, 1990.

37. Fisher EH, Charbonneau H, Tonks NK: Protein tyrosine phosphatase: A diverse family of intracellular and transmembrane enzymes. *Science* 253:401–406, 1991.

38. Tahiri-Jouti N, Cambillau C, Viguerie N, et al: Characterization of a membrane tyrosine phosphatase in AR42J cells: Regulation by somatostatin. *Am J Physiol* 262:G1007–G1014, 1992.

39. Colas B, Cambillau C, Buscail L, et al: Stimulation of a membrane tyrosine phosphatase activity by somatostatin analogues in rat pancreatic acinar cells. *Eur J Biochem* 207:1017–1024, 1992.

39a. Webb ML, Liu ECK, Cohen RB, et al: Molecular characterization of angiotensin II type II receptors in rat pheochromocytoma cells. *Peptides* 13:499–508, 1992.

39b. Kambayashi Y, Bardhan S, Takahashi K, et al: Molecular cloning of a novel angiotensin II type II receptor isoform involved in phosphotyrosine phosphatase inhibition. *J Biol Chem* 268:2453–24546, 1993.

40. Kang J, Sumners C, Posner P: Modulation of net outward current in cultured neurons by angiotensin II-Involvement of AT(1)-receptor and AT(2)-receptors. *Brain Res* 580:317–324, 1992.

41. Buisson B, Bottari SP, de Gasparo M, et al: The angiotensin II AT₂ receptor modulates T-type calcium currents in non-differentiated NG108-15 cells. *FEBS Lett* 309:161–164, 1992.

42. Hunter T: Protein modification: Phosphorylation on tyrosine residues. *Curr Opin Cell Biol* 1:1168–1181, 1989.

42a. Mukoyama M, Nakajima M, Horiuchi M, et al: Expression cloning of type-2 angiotensin II receptor reveals a unique class of seven-transmembrane receptors. *J Biol Chem* 268:24539–24542, 1993.

43. Tsutsumi K, Saavedra JM: Heterogeneity of angiotensin II AT₂ receptors in the rat brain. *Mol Pharmacol* 41:290–297, 1992.

44. Tsutsumi K, Zorad S, Saavedra JM: The AT₂ subtype of the angiotensin II receptors has differential sensitivity to dithiothreitol in specific brain nuclei of young rats. *Eur J Pharmacol* 226:169–173, 1992.

45. Mah SJ, Ades AM, Mir R, et al: Association of solubilized angiotensin-II receptors with phospholipase C-alpha in murine neuroblastoma NIE-115 cells. *Mol Pharmacol* 42:217–226, 1992.

46. Ji H, Sandberg K, Catt KJ: Novel angiotensin II antagonists distinguish amphibian from mammalian angiotensin II receptors express in *Xenopus laevis* oocytes. *Mol Pharmacol* 39:120–123, 1991.

47. Sandberg K, Ji H, Millan MA, et al: Amphibian myocardial angiotensin II receptors are distinct from mammalian AT₁ and AT₂ receptor subtypes. *FEBS Lett* 284:281–284, 1991.

48. Le Noble FAC, Schreurs NHJS, van Straaten HWM, et al: Evidence for a novel angiotensin II receptor involved in angiogenesis in chick embryo chorio-allantoic membrane. *Am J Physiol* 264:R460–R465, 1993.

49. Kiron MA, Soffer RL: Purification and properties of a soluble angiotensin II-binding protein from rabbit liver. *J Biol Chem* 264:4138–4142, 1989.

50. Ernsberger P, Zhou J, Damon TH, et al: Angiotensin-II receptor subtypes in cultured rat renal mesangial cells. *Am J Physiol* 263:F411–F416, 1992.

51. Elton TS, Stephan CC, Taylor GR, et al: Isolation of two distinct type I angiotensin II receptor genes. *Biochem Biophys Res Commun* 184:1067–1073, 1992.

52. Iwai N, Inagami T: Identification of two subtypes in the rat type I angiotensin II receptor. *FEBS Lett* 298:257–260, 1992.

53. Sasamura H, Hein L, Krieger JE, et al: Cloning, characterization, and expression of two angiotensin receptor (AT-1) isoforms from the mouse genome. *Biochem Biophys Res Commun* 185:253–259, 1992.

54. Jaiswal N, Diz DI, Chappell MC, et al: Stimulation of endothelial cell prostaglandin production by angiotensin peptides. Characterization of receptors. *Hypertension* 19:II49–II55, 1992.

55. Jaiswal N, Jaiswal RK, Tallant EA, et al: Alterations in prostaglandin production in SHR smooth muscle cells. (Abstract). *Hypertension* 20:432–P51, 1992.

56. Brechler V, Jones PW, de Gasparo M, et al: Agonistic and antagonistic properties of angiotensin analogs at the AT₂ receptor in PC12W cells. *Regul Pept* 44:207–213, 1993.

57. Henrich WL, Levi M: Ontogeny of renal renin release in spontaneously hypertensive rat and Wistar-Kyoto rat. *Am J Physiol* 260:F530–F535, 1991.

58. Barrett PQ, Bollag WB, Isales CM, et al: Role of calcium in angiotensin II-mediated aldosterone secretion. *Endocrinol Rev* 10:496–518, 1989.

59. Tsien RW, Ellinor PT, Horne WA: Molecular diversity of voltage-dependent calcium channels. *Trends Pharmacol Sci* 12:349–354, 1991.

60. Steriade M, Llinas RR: The functional status of the thalamus and the associated neuronal interplay. *Physiol Rev* 68:649–742, 1991.

61. Clark KL, Robertson MJ, Drew GM: Effects of angiotensin AT1 or AT2 receptor blockade on basal renal function and on the renal effects of angiotensin II (Ang II) in the anaesthetised dog (Abstract). *Br J Pharmacol* 106:19p, 1992.

62. Keiser JA, Bjork FA, Hodges JC, et al: Renal hemodynamic and excretory responses to PD 123319 and losartan, nonpeptide AT1 and AT2 subtype-specific angiotensin II ligands. *J Pharmacol Exp Ther* 262:1154–1160, 1992.

63. Cogan MG, Liu FY, Wong PC, et al: Comparison of inhibitory potency by nonpeptide angiotensin II receptor antagonists PD 123177 and DuP 753 on proximal nephron and renal transport. *J Pharmacol Exp Ther* 259:687–691, 1991.

64. Mento PF, Maita ME, Wilkes BM: Renal hemodynamics in rats with myocardial infarction: Selective antagonism of angiotensin receptor subtypes (Abstract). *Hypertension* 20:426–P66, 1992.

64a. Macari D, Bottari S, Whitebread S, et al: Renal actions of the selective angiotensin AT2 ligands CGP 42112B and PD 123313 in the sodium-depleted rat. *Eur J Pharmacol* 249:85–93, 1993.

65. Tsutsumi K, Saavedra JM: Characterization of AT2 angiotensin II receptors in rat anterior cerebral arteries. *Am J Physiol* 261:H667–H670, 1991.

66. Stromberg C, Naveri L, Saavedra JM: Angiotensin AT2 receptors regulate cerebral blood flow in rats. *Neuroreport* 3:703–704, 1992.

67. Barnes NM, Costall B, Kelly ME, et al: Cognitive enhancing actions of PD 123177 detected in a mouse habituation paradigm. *Neuroreport* 2:351–353, 1991.

68. Barnes NM, Champaneria S, Costall B, et al: Cognitive enhancing actions of DuP 753 detected in a mouse habituation paradigm. *Neuroreport* 1:239–242, 1990.

69. Pratt RE, Wang D, Hein L, et al: The AT2 isoforms of the angiotensin receptor mediates myointimal hyperplasia following vascular injury (Abstract). *Hypertension* 20:432–P52, 1992.

70. Viswanathan M, Stromberg C, Seltzer A, et al: Balloon angioplasty enhances the expression of angiotensin II subtype AT1 receptors in neointima of rat aorta. *J Clin Invest* 90:1707–1712, 1992.

71. Janiak P, Pillon A, Prost JF, et al: Role of angiotensin subtype 2 receptor in neointima formation after vascular injury. *Hypertension* 20:737–745, 1992.

72. Forney Prescott M, Webb RL, Reidy MA: Angiotensin-converting enzyme inhibitor versus angiotensin II, AT1 receptor antagonist. Effect on smooth muscle cell migration and proliferation after balloon catheter injury. *Am J Pathol* 139:1291–1296, 1991.

73. Kauffman RF, Bean JS, Zimmerman KM, et al: Lasartan, a nonpeptide angiotensin II (Ang II) receptor antagonist, inhibits neointima formation following balloon injury to rat carotid arteries. *Life Sci* 49:223–228, 1991.

74. Stephenson KN, Steele MK: Brain angiotensin-II receptor subtypes and the control of luteinizing hormone and prolactin secretion in female rats. *J Neuroendocrinol* 4:441–447, 1992.

75. Brunswigspickenheier B, Mukhopadhyay AK: Characterization of angiotensin II receptor subtype on bovine thecal cells and its regulation by luteinizing hormone. *Endocrinology* 131:1445–1452, 1992.

76. Yoshimura Y, Karube M, Koyama N, et al: Angiotensin-II directly induces follicle rupture and oocyte maturation in the rabbit. *FEBS Lett* 307:305–308, 1992.

77. Rowland NE, Rozelle A, Riley PJ, et al: Effect of nonpeptide angiotensin receptor antagonists on water intake and salt appetite in rats. *Brain Res Bull* 29:3–4, 1992.

78. Beresford MJ, Fitzsimons JT: Intracerebroventricular angiotensin II-induced thirst and sodium

appetite in rat are blocked by the AT_1 receptor antagonist, losartan (DuP 753) but not by the AT_2 antagonist CGP 42 112B. *Exp Physiol* 77:761–764, 1992.

79. Hogarty DC, Speakman EA, Puig V, et al: The role of angiotensin, AT_1-receptor and AT_2-receptor in the pressor, drinking and vasopressin responses to central angiotensin. *Brain Res* 586:289–294, 1992.

80. de Gasparo M, Rogg H, Brink M, et al: Angiotensin II receptor subtypes and cardiac function. *Eur Heart J* 1994 (in press).

81. de Gasparo M, Whitebread S, Levens N, et al: Pharmacology of angiotensin II receptor subtypes. In Saez Jm *et al.* (eds): *Cellular and Molecular Biology of the Adrenal Cortex.* Linbbey, London, 1992, pp. 3–17.

6

Medicinal Chemistry of AT$_2$ Receptors

Jeremy J. Edmunds and John C. Hodges

1. INTRODUCTION

Angiotensin II (ANG II) produces its physiological effects via interaction with its specific receptors that are distributed among a number of different tissues.[1-4] The wide variety of receptor locations as well as their varying affinity for angiotensin peptidic fragments[5,6] in the presence or absence of dithiothreitol (DTT) allowed some authors to speculate about ANG II receptor subtypes.[7] Indeed, some authors referred to different affinity states,[8-10] the ability of ligands to bind competitively versus noncompetitively,[11-13] and to different signal transduction pathways[14-17] to validate their case. However, it was not until the recent discovery of selective ligands that this receptor heterogeneity was clearly established.[18-28] Since the AT$_1$ receptor subtype[29] mediates a variety of therapeutically important events, in particular, vascular smooth muscle contraction, there has been little effort directed toward the design and synthesis of AT$_2$-specific ligands. However, as losartan (DuP 753)[30-33] and numerous other AT$_1$-selective antagonists[34-43] progress through the clinical departments[44-46] of pharmaceutical companies, one is left to consider the clinical consequence of AT$_2$ receptor activation. Although at present there have been very few functional correlates identified with the AT$_2$ receptor, it would seem reasonable to expect the role of the AT$_2$ receptor to be revealed during the long-term administration of AT$_1$ antagonists, as a consequence of elevated

Jeremy J. Edmunds and John C. Hodges • Department of Medicinal Chemistry, Parke-Davis Pharmaceutical Research, Division of Warner-Lambert Company, Ann Arbor, Michigan 48105.

Angiotensin Receptors, edited by Juan M. Saavedra and Pieter B.M.W.M. Timmermans. Plenum Press, New York, 1994.

ANG II levels. However, a more direct approach to addressing the function of the AT$_2$ receptor involves the use of AT$_2$ selective ligands. In this chapter we will review the medicinal chemistry associated with AT$_2$ receptor ligands.

2. AT$_2$ SELECTIVE PEPTIDES

As with any quest for an agonist or antagonist to a receptor, initial experimentation begins with the native ligand. In fact, peptide analogues of ANG II were heavily explored before investigators even considered receptor heterogeneity, and consequently many of these initial ligands were not screened for subtype specificity. Initial modification of the constituent amino acids of ANG II first defined the residues critical for binding. With this firmly established, further experimentation revealed that even subtle modification afforded receptor subtype specific ligands, an example of which is (4-aminophenyl)ala-nine[6] ANG II[47] (see Fig. 1).

This ligand exhibits appreciable binding to AT$_2$ binding sites in PC12 cell membranes (IC$_{50}$ 12 nM) and negligible binding in the rat liver (9 μM), predominantly AT$_1$. Furthermore, the fact that this binding affinity at the AT$_1$ site is decreased by the addition of guanosine 5'-O-(3-thiotriphosphate) may indicate that this compound is an agonist, at least at the AT$_1$ receptor.

Working from the premise that dimerized hormones often exhibit enhanced receptor binding affinity, a number of ANG II analogues have been prepared that incorporate a covalent link between the amino- or carboxy-terminus. These ligands were assessed for their binding affinity for both the AT$_1$ and AT$_2$ receptor subtypes, which revealed that although AT$_1$ affinity remained in the micromolar range, the AT$_2$ receptor affinity increased and was dependent on the linker length between the N-termini[48] (Fig. 2).

The most prevalent peptide analogue of ANG II that shows good selectiv-

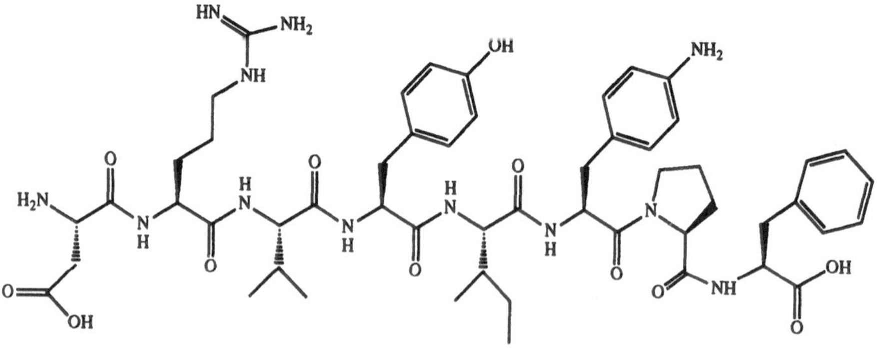

Figure 1. [*p*-NH$_2$Phe]^6-ANG II.

IC_{50} (M)

CGP 40937

$$
\begin{array}{l}
\text{[CO-TYR-VAL-HIS-PRO-PHE-OH} \\
\text{(CH}_2\text{)}_6 \\
\text{[CO-TYR-VAL-HIS-PRO-PHE-OH}
\end{array}
$$

AT_1	AT_2
2.00×10^{-5}	3.75×10^{-9}

CGP 42112A

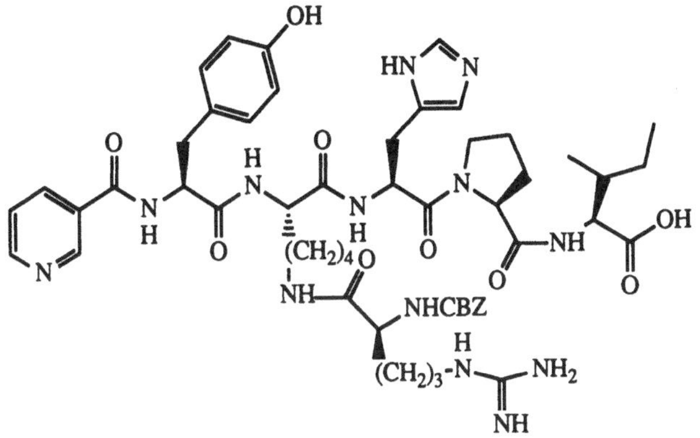

Figure 2. AT₂-selective peptides.

ity for the AT_2 receptor is the ligand CGP 42112A.[49-51] This latter ligand, being a branded peptide with nonpeptide modifications, is quite unlike the native ligand, ANG II, but is regarded as a primary pharmacological tool for determining AT_2 receptor location and function.

In concert with discovery efforts directed toward the determination of ANG II amino acid residues critical for binding and activation, studies proceeded to determine the receptor-bound conformation of ANG II.[52-54] Although initial efforts were not directed toward investigating the possible different binding conformations of ANG II at the AT_1 or AT_2 receptor, several ANG II analogues[55-58] that incorporate disulfide bridges as well as other ANG II analogues containing conformationally constrained lipophilic amino acids, at position 8, are reasonably selective AT_2 ligands[59] (Fig. 3).

3. AT₂ SELECTIVE NONPEPTIDES

As mentioned previously the design and synthesis of ligands for the AT_2 receptor is in its relative infancy compared to ligands prepared for the AT_1 receptor. However, in 1989, a U.S. patent appeared[60] that described a series of 4,5,6,7-tetrahydroimidazo[4,5-c]pyridine-6-carboxylic acids (spinacine) as potent ANG II-receptor-binding inhibitors.[23,61] These ligands were subsequently shown to be highly specific for the AT_2 receptor[19-22,62] (Fig. 4). It should be

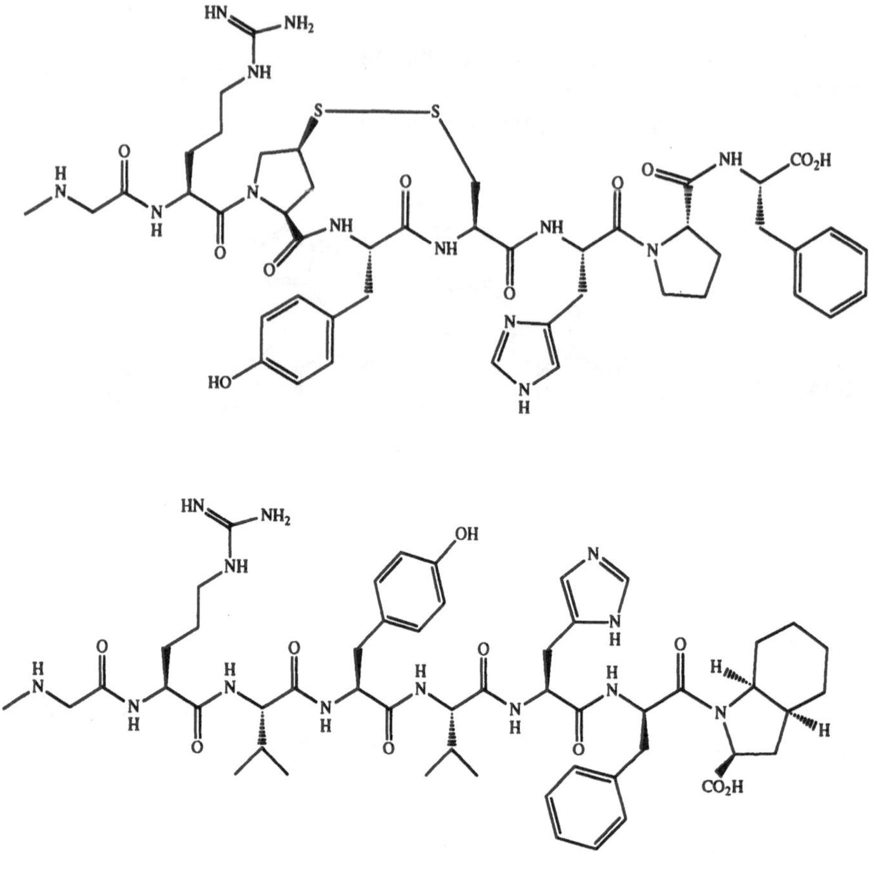

Figure 3. Conformationally constrained ANG II analogues.

noted at this point, however, that there exists some disparity between the effect of receptor blockade with CGP 42112A and PD 123177 (and presumably the related ligand PD 123319).[63] This discrepancy may reflect the possibility of AT_2 receptor subtypes or the agonist/antagonist nature of these ligands.[64]

Historically the discovery of these AT_2-selective agents was based on a strategy to discover new antihypertensive agents through the screening of numerous nonpeptide agents in an ANG II-receptor-binding assay that included DTT as an additive.[65–67] The fact that DTT inactivates the AT_1 receptor,[68] which was not appreciated at that time, precluded the possibility of the discovery of antihypertensive AT_1 antagonists and instead led to the identification of AT_2-specific agents. Compounds were selected for screening based on the inclusion of functional groups resembling the side chains of the C-terminal pentapeptide sequence of ANG II and the incorporation of functionally

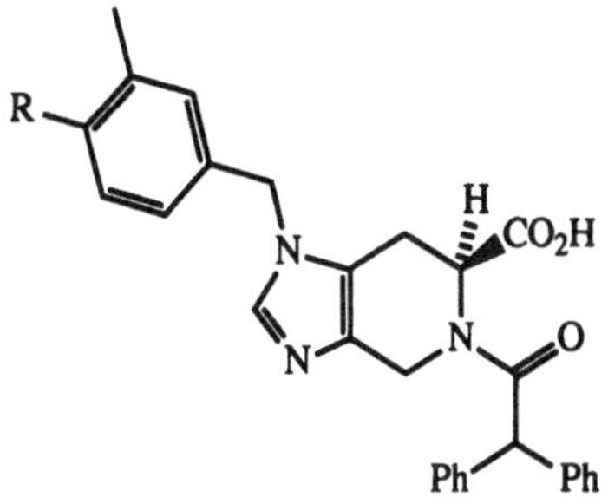

R = -OCH$_3$ PD 121981   (WL-19)

R = -NH$_2$ PD 123177   (EXP 655)

R = -NMe$_2$ PD 123319

Figure 4. AT$_2$-selective nonpeptide ligands.

resembling the 1-benzylimidazoles described in the first patents of nonpeptide ANG II antagonists, disclosed by Takeda.[69-71]

One such class of compounds, the *N*(im)-substituted-5-acyl-spinacines showed micromolar potency as inhibitors of [^{125}I]-Sar binding (Fig. 5). The N-1 regioisomer displayed higher affinity than the corresponding 3-isomer and a lipophilic acyl group at N-5 increased potency. For study of initial structure–activity relationships, the N-1 substituent was maintained as *N*-benzyl, while the N-5 substituent was varied. This resulted in the diphenyl acetyl group being selected as the N-5 substituent based on its potency and chemical simplicity. It should be noted that although the 2-cycloalkyl-2-phenacetyl groups at N-5 displayed comparable potency, the incorporation of an unresolved chiral center was considered undesirable, at least in early synthetic efforts.

Thus having selected the N-5 substituent, the N-1 substituent was system-

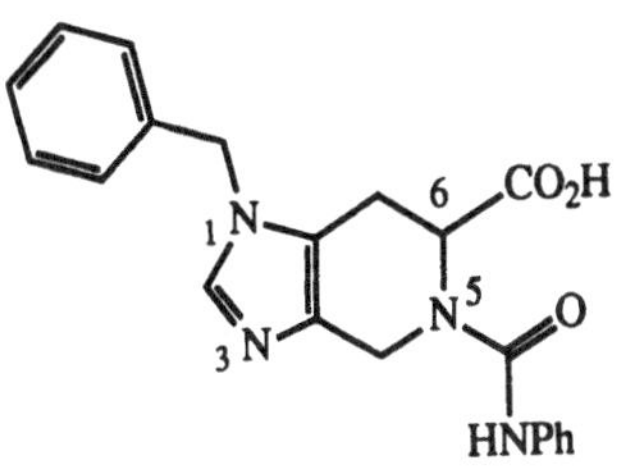

Figure 5. Spinacine lead.

atically varied to incorporate a number of alkyl and arylalkyl groups bearing a spectrum of electron-withdrawing and electron-donating functional groups. From this exploration resulted the electron-donating 4-dimethylamino benzyl as the favored N-1 substituent. The incorporation of a methly substituent ortho to the dimethylamino group further increased binding.

Investigation of substituent effects at C-6 showed that a carboxylic acid, rather than an ester, amide, or hydroxymethyl group, was required for potent binding. Additionally, the S-isomer displays greater affinity for the AT_2 receptor than the corresponding R-isomer.

From the extensive number of analogues prepared in this series of compounds, it should be noted that the effect of various substituents at the positions described are only partly additive. Rather, it appears that the spatial orientation of the whole molecule is important and it is this fact that resulted in the identification of PD 123319 as the most potent and AT_2-selective of the series of compounds prepared. PD 123319 is now prepared on a multigram scale, via the route outlined in Fig. 6, for the *in vivo* pharmacological study of the AT_2 receptor.

Recently, a number of AT_2-specific ligands have been disclosed.[72–78] These compounds incorporate a number of features common to PD 123319 and as such exhibit similar structure–activity relationships. Generally, a carboxylic acid and lipophilic amide in close proximity are crucial for activity. Unlike the PD 123319 series, however, the addition of para-electron-donating substituents to the aryl ring, substituting the tetrahydroisoquinoline, did not markedly improve AT_2 affinity (Fig. 7).

4. TOOLS FOR THE STUDY OF THE ANGIOTENSIN AT_2 RECEPTOR

In order to adequately characterize and locate the AT_2 receptor,[79,80] radio-labeled [^{125}I]-ANG II analogues are frequently displaced with subtype-specific ANG II receptor ligands. A more direct approach to identifying the locations of the AT_2 receptor, however, is possible through the use of ^{125}I-labeled AT_2 ligands. Both [^{125}I]-CGP 42112A and [^{125}I]-PD 123319 analogues have been reported. In fact, the ligand [^{125}I]-CGP 42112A displays impressive binding affinity in human myometrium membranes (IC_{50} 0.2 nM), whereas PD 123319 binds with lower affinity (IC_{50} 14 nM).[81] This degree of binding affinity of [^{125}I]-CGP makes it the ligand of choice for *in vitro* study. For *in vivo* use, however, the need for a metabolically stable AT_2 radioligand was addressed by the preparation of an iodinated analogue of PD 123319.[82] While PD 123319 could not be iodinated under typical iodination conditions in a satisfactory manner, it was found that the phenolic analogue, **4,** was a better substrate and

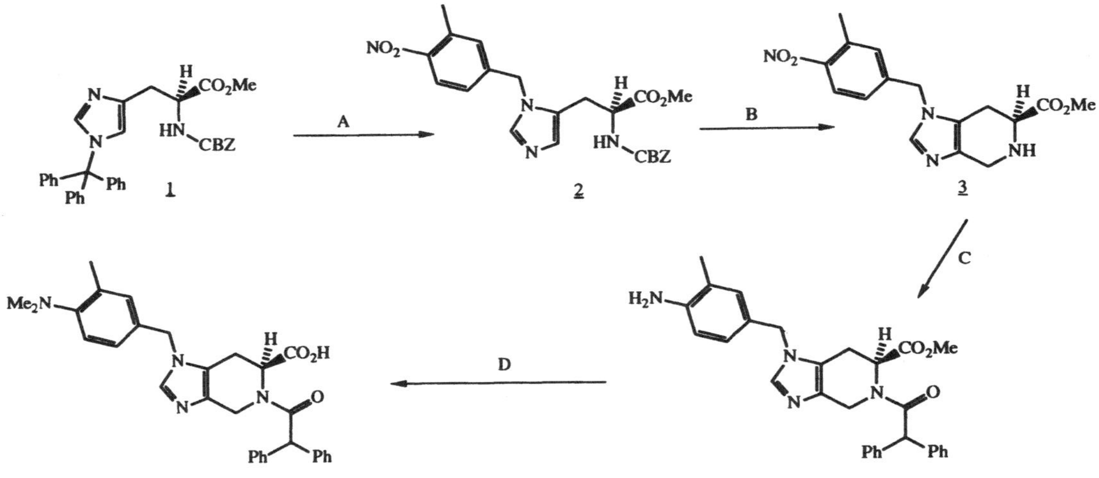

PD 123319 **PD 123177** methyl ester

A i) 3-Me,4-NO_2PhCH_2OH, Tf_2O, $EtN(i\text{-}Pr)_2$, CH_2Cl_2, -78°C. ii) AcOH, 85 °C, 1h.
B i) 6N HCl, reflux, 2h. ii) HCHO, HCl aq, reflux 2h. iii) MeOH, HCl, reflux 16h.
C i) Ph_2CHCO_2H, DCC, HOBT, MeCN, 48h. ii) H_2, Ra / Ni, THF, RT, 1h.
D i) HCHO, $NaCNBH_3$, HCl aq, EtOH, 16h. ii) NaOH, THF/ MeOH / H_2O, 3h.

Figure 6. PD 123319 synthesis.

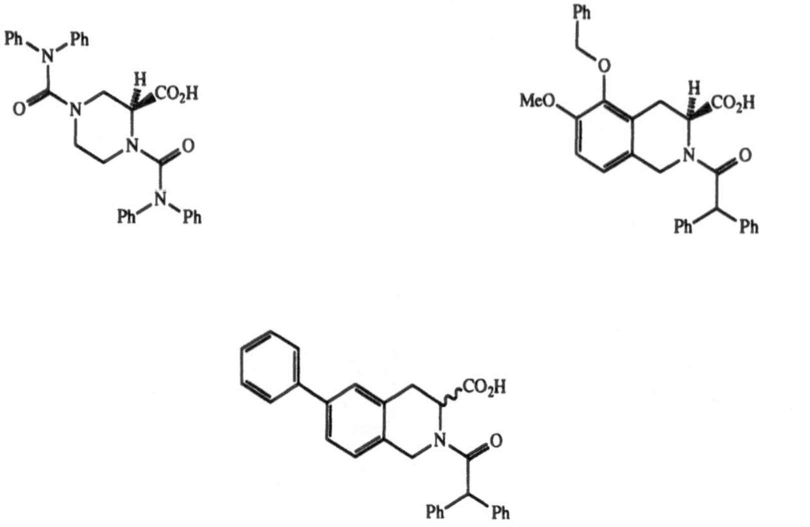

Figure 7. Nonpeptide AT$_2$-specific ligands.

could be rapidly iodinated to afford **5**. The ortho methyl substituent played a rather fortuitous role during this iodination as it prevented the di-iodination of the aromatic ring. This considerably aided the purification of **5** by reverse-phase high-pressure liquid chromotography. Gratifyingly, **5** (IC$_{50}$ 5 nM) displays comparable finding affinity to PD 123319 (IC$_{50}$ 8 nM) (Fig. 8).

The prototypical AT$_2$ receptor antagonist PD 123319 has also been modified to incorporate fluorescent and biotin residues.[82] The biotin analogue was prepared by using the known broad tolerance of the AT2 receptor for $N(1)$-substituents in the spinacine series. Thus, addition of a *n*-hexyl amino group to the aniline of PD 123177 methyl ester gave **6**, which was further extended with an ε-lysyl residue to provide a 14-atom spacer between biotin

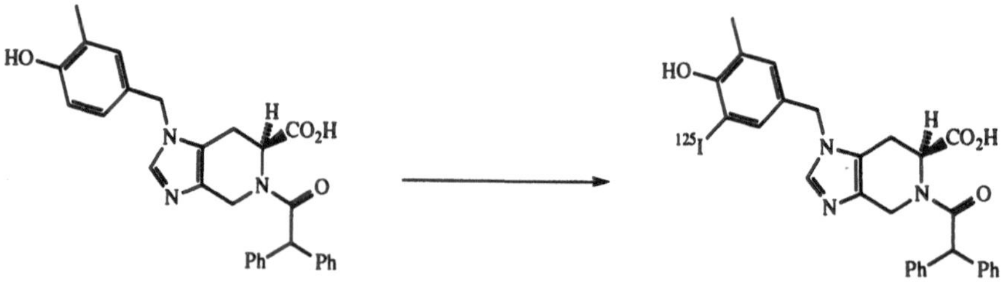

Figure 8. ^{125}I-labeled AT$_2$-specific nonpeptide.

and the AT$_2$ ligand. Remarkably, this large addition of additional functionality is well tolerated by the AT$_2$ receptor since **7** and PD 123319 displace [^{3}H]-ANG II with equal potency (IC$_{50}$ = 8 and 7 nM, respectively). Prior complexation of streptavidin to **7**, however, destroys binding affinity at the AT$_2$ receptor, limiting the utility of this biotinyl-AT$_2$ probe.[83]

A similar strategy, albeit with a shorter linker, was also employed to prepare a fluorescent derivative of PD 123319. This compound (**8**) again displays excellent affinity for the AT$_2$ receptor (IC$_{50}$ = 4 nM) (Fig. 9).

With the knowledge that the 8 position of ANG II is somewhat tolerant of large lipophilic amino acid residues,[84] it is reasonable to incorporate photoaffinity labels at this position.[85–88] These agents have utility in the isolation and characterization of the AT$_2$ receptor. Initial experiments with [Sar^{1}Tyr(^{125}I)^{4}Val^{5}D-Phe(N$_3$)^8]Ang II, which is a nonselective AT$_1$/AT$_2$ ligand, revealed that surprisingly only the AT$_1$ receptor subtype, on bovine adrenal cortex membranes, were adequately labeled.[89] A new photoaffinity ligand was developed to replace the DTT-sensitive phenylazide amino acid residue, at position 8, with (D/L)benzoylphenylalanine (Bpa). This ligand, which although a mixed AT$_1$/AT$_2$ ligand, was successfully used in the presence of the AT$_1$ ligand L-158809 (presumably this is preferable to DTT for inactivation of AT$_1$ binding by the affinity label) to label AT$_2$ receptors on human myometrium. Gel electrophoresis of the photolabeled [Sar^{1}Tyr(^{125}I)^{4}Val^{5}Bpa^8]ANG II–AT$_2$ receptor complex then revealed a molecular weight of 70.7 ± 2.9 kDa (Fig. 10).

5. DUAL ACTION ANG II ANTAGONISTS

Recently, a number of disclosures of nonpeptide ANG II-binding inhibitors with mixed AT$_1$/AT$_2$ affinity have appeared.[90–92] The design of these inhibitors follows the known structure–activity relationship of both AT$_1$- and AT$_2$-specific inhibitors. Representative structures are depicted in Fig. 11.

6. SUMMARY

The array of AT$_2$-selective ligands are presently enabling the characterization and localization of the AT$_2$ receptor, as well as beginning to indicate possible roles in the CNS, the cardiovascular system, kidney, and reproductive tissues (see Chapter 7).

However, the question concerning the utility of an AT$_2$ receptor agonist/antagonist, or even that of a dual-acting AT$_1$/AT$_2$ ANG II ligand, still remains largely unanswered but will be addressed in the near future.

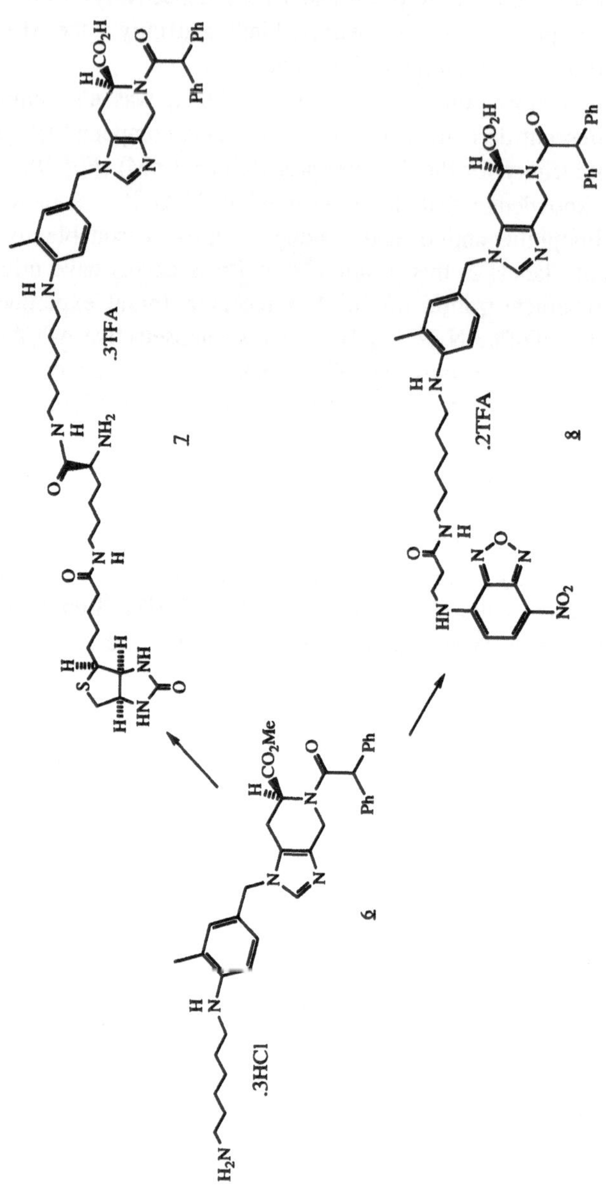

Figure 9. Biotinyl and fluorescent AT$_2$ ligands.

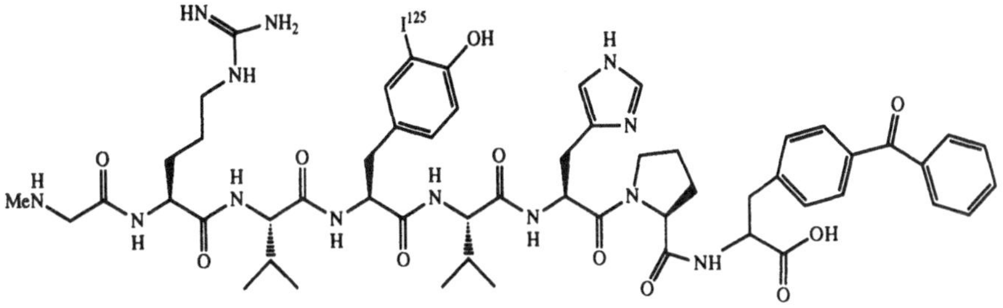

Figure 10. AT₂ photoaffinity ligand.

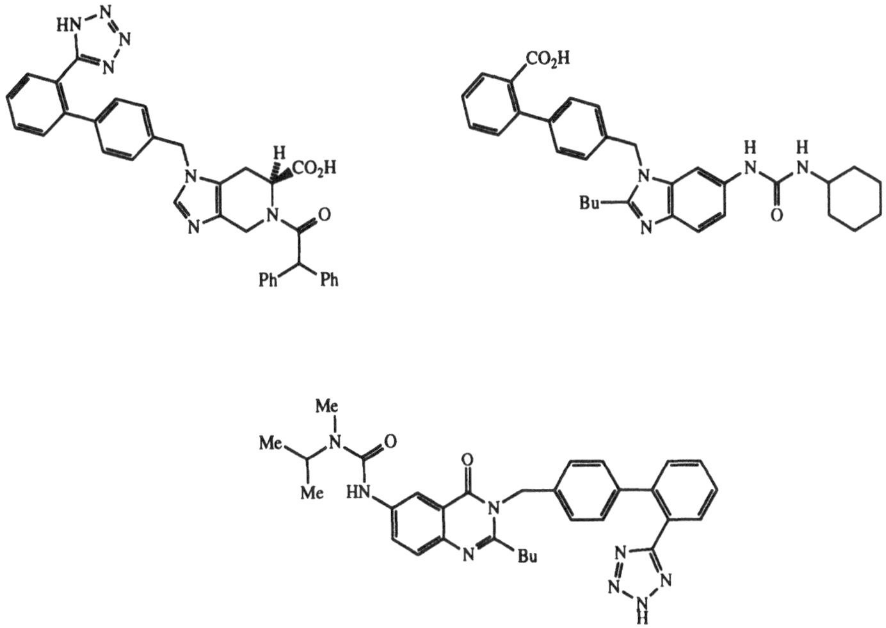

Figure 11. Mixed affinity AT₁/AT₂ ligands.

REFERENCES

1. Hodges JC, Hamby JH, Blankley CJ: Angiotensin II receptor binding inhibitors. *Drugs Future* 17(7):575–593, 1992.
2. Greenlee WJ, Siegl PKS: Angiotensin/renin modulators. *Ann Rep Med Chem* 26:63–72, 1991.
3. Greenlee WJ, Siegl PKS: Angiotensin/renin modulators. *Ann Rep Med Chem* 27:59–69, 1992.
4. Bovy PR, Blaine EH: Peptidic and non-peptidic angiotensin II competitive antagonists. *Curr Cardiovasc Patents* 1:2044–2056, 1989.

5. Douglas JG, Michailov M, Khosla MC, et al: Comparative receptor-binding properties of heptapeptide and octapeptide antagonists of angiotensin II in rat adrenal glomerulosa and uterine smooth muscle. *Endocrinology* 106:120–124, 1980

6. Bennett JP, Snyder SH: Receptor binding interactions of the angiotensin II antagonist [125]I-[sarcosine[1]leucine[8]] angiotensin II with mammalian brain and peripheral tissues. *Eur J Pharmacol* 67:11–25, 1980.

7. Catt K, Abbott A: Molecular cloning of angiotensin II receptors may presage further receptor subtypes. *Trends Pharmacol Sci* 12:279–281, 1991.

8. Gunther S: Characterizations of angiotensin II receptor subtypes in rat liver. *J Biol Chem* 259:7622–7629, 1984.

9. Wright GB, Alexander RW, Ekstein LS, et al: Characterization of the rabbit ventricular myocardial receptor for angiotensin II: Evidence for two sites of different affinities and specificities. *Mol Pharmacol* 24:213–221, 1983.

10. Moore GJ, Kwok YC: Angiotensin receptors in resting smooth muscle are the low affinity sites observed in binding studies. *Life Sci* 41:505–511, 1987.

11. Saltman S, Baukal A, Waters S, et al: Competitive binding activity of angiotensin II analogues in an adrenal cortex radioligand-receptor assay. *Endocrinoloy* 97:275–282, 1975.

12. Goghari MH, Franklin KJ, Moore GJ: Structure–activity relationships for the competitive angiotensin antagonist [sarconsine[1]O-methyltyrosine[4]] angiotensin II (Sarmesin). *J Med Chem* 29:1121–1124, 1975.

13. Trachte GJ, Peach MJA: Potent noncompetitive angiotensin II antagonist induces only competitive inhibition of angiotensin III responses. *J Cardiovasc Res* 5:1025–1033, 1983.

14. Douglas JG, Romero M, Hopfer U: Signaling mechanisms coupled to the angiotensin receptor of proximal tubular epithelium. *Kidney Int.* 38:S-43–S-47, 1990.

15. Summers C, Tang W, Zelenza B, et al: Angiotensin II receptor subtypes are coupled with distinct signal-transduction mechanisms in neurons and astrocytes from rat brain. *Proc Natl Acad Sci USA* 88:7567–7571, 1991.

16. Botarri SP, Taylor V, King IN, et al: Angiotensin AT$_2$ receptors do not interact with guanine nucleotide binding proteins. *Eur J Pharmacol Mol Pharmacol* 207:157–163, 1991.

17. Johnson MC, Aguilera G: Angiotensin-II receptor subtypes and coupling to signaling systems in cultured fetal fibroblasts. *Endocrinology* 129:1266–1274, 1991.

18. Herblin WF, Chiu AT, McCall DE, et al: Angiotensin II receptor heterogeneity. *Am J Hypertens* 4:299S–302S, 1991.

19. Chiu AT, Herblin WF, McCall DE, et al: Identification of angiotensin II receptor subtypes. *Biochem Biophys Res Commun* 165:196–203, 1989.

20. Chang RSL, Lotti VJ: Selective ligands reveal subtypes of angiotensin receptors in rat vasculature and brain. *Pharmacologist* 31:150, 1989.

21. Chang RSL, Lotti VJ: Two distinct angiotensin II receptor binding sites in rat adrenal revealed by new selective nonpeptide ligands. *Mol Pharmacol* 37:347–351, 1990.

22. Dudley DT, Hubbell SE, Summerfelt RM: Characterization of angiotensin II (AT$_2$) binding sites in R3T3 cells. *Mol Pharmacol* 40:360–367, 1991.

23. Dudley DT, Panek RL, Major TC, et al: Subclasses of angiotensin II binding sites and their functional significance. *Mol Pharmacol* 38:370–377, 1990.

24. Gehlert DR, Gackenheimer SL, Reel JK, et al: Non-peptide angiotensin II receptor antagonists discriminate subtypes of ^{125}I-angiotensin II binding sites in the rat brain. *Eur J Pharmacol* 187:123–126, 1990.

25. Ji H, Sandberg K, Catt KJ: Novel angiotensin II antagonists distinguish amphibian from mammalian angiotensin II receptors expressed in *Xenopus laevis* oocytes. *Mol Pharmacol* 39:120–123, 1991.

26. Sandberg K, Ji H, Millan MA, et al: Amphibian myocardial angiotensin II receptors are distinct from mammalian AT₁ and AT₂ receptor subtypes. *FEBS Lett* 284:281–284, 1991.

27. Chang RSL, Lotti VJ, Chen TB, et al: Two angiotensin II binding sites in rat brain revealed using [^{125}I]Sar^{1}Ile^8-angiotensin II and selective nonpeptide antagonists. *Biochem Biophys Res Commun* 171:813–817, 1990.

28. Rowe BP, Grove KL, Saylor DL, et al: Discrimination of angiotensin II receptor subtype distribution in the rat brain using non-peptidic receptor antagonists. *Regul Pept* 33:45–53, 1991.

29. Bumpus FM, Catt KJ, Chiu A, et al: Nomenclature for angiotensin receptors. *Hypertension* 17:720–721, 1991.

30. Timmermans PBMWM, Wong PC, Chiu AT, et al: Nonpeptide angiotensin II receptor antagonists. *Trends Pharmacol Sci* 12:55–62, 1991.

31. Carini DJ, Duncia JV, Aldrich PE, et al: Nonpeptide angiotensin II receptor antagonists: The discovery of a series of *N*-(biphenylylmethyl)-imidazoles as potent orally active antihypertensives. *J Med Chem* 34:2525–2547, 1991.

32. Wong PC, Tam SW, Herblin WF, et al: Further studies on the selectivity of DuP 753. A nonpeptide angiotensin II receptor antagonist. *Eur J Pharmacol* 196:201–203, 1991.

33. Rhaleb NE, Rouissi N, Nantel F, et al: DuP 753 is a specific antagonist for the angiotensin receptor. *Hypertension* 17:480–484, 1991.

34. Chiu AT, Carini DJ, Duncia JV, et al: DuP 532: A second generation of nonpeptide angiotensin II receptor antagonists. *Biochem Biophys Res Commun* 177:209–217, 1991.

35. Weinstock J, Keenan RM, Samanen J, et al: 1-(Carboxybenzyl)imidazole-5-acrylic acids: Potent and selective angiotensin II receptor antagonists. *J Med Chem* 34:1514–1517, 1991.

36. Mantlo NB, Chakravarty PK, Ondeyka DL, et al: Potent orally active imidazo[4,5-b]pyridine-based angiotensin II receptor antagonists. *J Med Chem* 34:2919–2922, 1991.

37. Oku T, Setoi H, Kayakiri H, et al: European Patent Application 0426021. Fujisawa Pharmaceutical Co Filed October 26, 1990.

38. Narr B, Bomhard A, Hauel N, et al: European Patent Application 0392317. Dr. Karl Thomae Gmbh. Filed April 3, 1990.

39. Oldham AA, Allott CP, Major JS, et al: ICI D8731: A novel potent and orally-effective angiotensin II antagonist. *Br J Pharmacol* 105(Suppl):83P, 1992.

40. Middlemiss D, Drew GM, Ross BC: Bromobenzofurans: A new class of potent non-peptide antagonists of angiotensin II. *BioMed Chem Lett* 1:711–716, 1991.

41. Naka T, Nishikawa K: European Patent Application 0425921. Takeda Chemical Industries Ltd Filed October 19, 1990.

42. Dower M (ed): SCRIP. PJB Publications Ltd, New York, 1991(1648), p. 21.

43. Nisato D, Cazaubon C, Lacour C, et al: Pharmacological properties of SR 47436 a nonpeptidic angiotensin II receptor antagonist. *Br J Pharmacol* 105(Proc Suppl April):84P, 1992.

44. Nakashima M, Uematsu T, Kosuge K, et al: Pilot study of the uricosuric effect of DuP 753, a new angiotensin II receptor antagonist, in healthy subjects. *Eur J Clin Pharmacol* 42(3):333–335, 1992.

45. Nelson E, Merrill D, Sweet C, et al: Efficacy and safety of oral MK-954 DUP753. An angiotensin receptor antagonist in essential hypertension. *J Hypertens* 9(Suppl 6):S468–S469, 1991.

46. Munafo A, Christen Y, Nussberger J, et al: Drug concentration response relationship in normal volunteers after oral administration of losartan, an angiotensin II receptor antagonist. *Clin Pharmacol Ther* 51(5):513–521, 1992.

47. Speth RC, Kim KH: Discrimination of two angiotensin II receptor subtypes with a selective

agonist analogue of angiotensin II p-aminophenylalanine[6] angiotensin II. *Biochem Biophys Res Commun* 169:997–1006, 1990.

48. deGasparo M, Whitebread S, Kamber B, et al: Effect of covalent dimer conjugates of angiotensin II on receptor affinity and activity in vitro. *J Recep Res* 11:247–257, 1991.

49. Whitebread S, Mele M, Kamber B, et al: Preliminary biochemical characterization of two angiotensin II receptor subtypes. *Biochem Biophys Res Commun* 163:284–291, 1989.

50. Balla T, Baukal AJ, Eng S, et al: Angiotensin II receptor subtypes and biological responses in the adrenal cortex and medulla. *Mol Pharmacol* 40:401–406, 1991.

51. deGasparo M, Whitebread S, Mele M, et al: Biochemical characterization of two angiotensin II receptor subtypes in the rat. *J Cardiovasc Pharmacol* 16(Suppl 4):S31–S35, 1990.

52. Turner RJ, Matsoukas JM, Moore GJ: Fluorescence properties of angiotensin II analogs in receptor-simulating environments: Relationship between tyrosinate fluorescence lifetime and biological activity. *Biochim Biophys Acta* 1065(1):P21–28, 1991.

53. Garcia KC, Ronco PM, Verroust PJ: Three-dimensional structure of an angiotensin II–fab complex at 3 A: Hormone recognition by an anti-idiotypic antibody. *Science* 257:502–507, 1992.

54. Pierson ME, Freer RJ: Analysis of the active conformation of angiotensin II: A comparison of AII and non-peptide AII antagonists. *Pept Res* 5(2):102–105, 1992.

55. Spear KL, Brown MS, Reinhard EJ, et al: Conformational restriction of angiotensin II: Cyclic analogs having high potency. *J Med Chem* 33:1935–1940, 1990.

56. Marshall GR, Kaczmarek K, Kataoka T, et al: Evidence for receptor-bound turn conformations of bradykinin and angiotensin II. In: Giralt E, Andreu D (eds): *Peptides 1990*. ESCOM, Leiden, 1991, 594–596.

57. Plucinska K, Kataoka T, Yodo M, et al: Multiple binding modes for the receptor-bound conformations of cyclic AII agonists. *J Med Chem* 36:1902–1913, 1993.

58. Padmaja J, Cody W, Dooley D, et al: Bradykinin and angiotensin II analogs containing a conformationally constrained proline analog. Abstracts from Twelfth American Peptide Symposium, Cambridge, MA, June 16–21, 1991, No. P-474.

59. Cody WL, He JX, Lunney EA, et al: Modification of the C-terminus of angiotensin II peptides lead to type 2 (AT2) receptor selectivity. Thirteenth American Peptide Symposium, Edmonton, Alberta, Canada, June 20–25, 1993, No. 607.

60. Blankley CJ, Hodges JC, Kiely JS, et al: 4,5,6,7-Tetrahydro-1H-imidazo(4,5-C)pyridine-6-carboxylic acid analogs having antihypertensive activity. US Patent 4812462, March 14, 1989.

61. Dudley DT, Hodges JC, Pugsley TA, et al: 4,5,6,7-Tetrahydro-1H-imidazo(4,5-c)pyridine derivatives and analogues as angiotensin II receptor antagonists. WO Patent 9205784, April 16, 1992.

62. Blankley CJ, Hodges JC, Klutchko SR, et al: Synthesis and structure–activity relationships of a novel series of non-peptide angiotensin II receptor binding inhibitors specific for the AT2 subtype. *J Med Chem* 34:3248–3260, 1991.

63. Jaiswal N, Diz DI, Chappell MC: Stimulation of endothelial cell prostaglandin production by angiotensin peptides. Characterization of receptors. *Hypertension* 19(2 Suppl):49–55, 1992.

64. Tsutsumi K, Saavedra JM: Heterogeneity of angiotensin AT2 receptors in the rat brain. *Mol Pharmacol* 41(2):290–297, 1992.

65. Douglas J, Aguilera G, Kondo T, et al: Angiotensin II receptors and aldosterone production in rat adrenal glomerulosa cells. *Endocrinology* 102:685–696, 1978.

66. Glossmann H, Baukal A, Aguilera G, et al: Radioligand assay for angiotensin II receptors. *Methods Enzymol* 109:110–126, 1985.

67. Moore GJ, Kwok YCA: Comparison of binding assay and bioassay data for angiotensin analogues in uterine smooth muscle. *Biochem Arch* 4:145–149, 1988.

68. Chiu AT, McCall DE, Nguyen TT, et al: Discrimination of angiotensin II receptor subtypes by dithiothreitol. *Eur J Pharmacol* 170:117–118, 1989.

69. Furukawa Y, Kishimoto S, Nishikawa K: 4-Chloro-2-phenylimidazole-5-acetic acid derivatives. European Patent Application 103647, 1984.

70. Furukawa Y, Kishimoto S, Nishikawa K: Hypotensive imidazole derivatives. US Patent 4,340,598, 1982.

71. Furukawa Y, Kishimoto S, Nishikawa K: Hypotensive imidazole-5-acetic acid derivatives. US Patent 4,355,040, 1982.

72. Wu MT, Ikeler TJ, Ashton WT, et al: Synthesis and structure–activity relationships of a novel series of non-peptide AT₂ selective angiotensin II receptor antagonists. 205th American Chemical Society Meeting, Denver, CO, March 28–April 2, 1993, MEDI No. 100.

73. Ashton WT, Greenlee WJ, Wu MT, et al: PCT International Patent Application, Publication No. WO 92/20661.

74. Wu MT, Ikeler TJ, Ashton WT, et al: Synthesis and structure–activity relationships of a novel series of non-peptide AT₂ selective angiotensin II receptor antagonists. *BioMed Chem Lett* 3:2023–2028, 1993.

75. Klutchko S, Hamby JM, Hodges JC: Tetrahydroisoquinoline derivatives with AT₂-specific angiotensin II receptor binding inhibitory activity. *BioMed Chem Lett* 4:57–62, 1994.

76. Blankley CJ, Hodges JC, Klutchko S: Substituted 1,2,3,4-tetrahydroisoquinolines with angiotensin II receptor antagonist properties. US Patent No. 5,246,943, Sept. 21, 1993.

77. VanAtten MK, Ensinger CL, Wexler RR, et al: 1,2,3,4-tetrahydroisoquinoline-3-carboxylic acids as novel selective inhibitors of angiotensin II binding to the AT₂ site. 206th American Chemical Society Meeting, Chicago, IL, August 1993, MEDI No. 85.

78. VanAtten MK: 1,2,3,4-Tetrahydroisoquinolines useful in the treatment of CNS disorders. US Patent No. 5,236,934, Aug. 17, 1993.

79. Tsutsumi K, Saavedra JM: Characterization and development of angiotensin II receptor subtypes AT-1 and AT-2 in rat brain. *Am J Physiol* 261:R209–R216, 1991.

80. Zemel S, Millan MA, Feuillan P: Characterization and distribution of angiotensin-II receptors in the primate fetus. *J Clin Endocrinol Metab* 71:1003–1007, 1991.

81. Whitebread SE, Taylor V, Bottari SP, et al: Radioiodinated CGP 42112: A novel high affinity and highly selective ligand for the characterization of angiotensin AT₂ receptors. *Biochem Biophys Res Commun* 181:1365–1371, 1991.

82. Hodges JC, Edmunds JJ, Nordblom GD, et al: The syntheses and binding affinities of tools for the study of angiotensin AT₂ receptors. *BioMed Chem Lett* 3:905, 1993.

83. Dudley DT: Personal communication.

84. Hsieh K, LaHann TR, Speth RC: Topographic probes of angiotensin and receptor: Potent angiotensin II agonist containing diphenylalanine and long-acting antagonists containing biphenylalanine and 2-indan amino acid in position 8. *J Med Chem* 32:898–903, 1989.

85. Seyer R, Aumelas A: Synthesis of biotinylated and photoreactive probes for angiotensin receptors. *J Chem Soc Perkin Trans* 1:3289–3299, 1990.

86. Moore GJ: Photoaffinity labeling of angiotensin receptors: Functional studies on responding tissues. *Pharmacol Ther* 33:349–381, 1987.

87. Escher E: Photoaffinity labeling of angiotensin II and bradykinin receptors. *Pharmacol Ther* 37:37–55, 1988.

88. Eberle AN, deGraan PNE: General principles for photoaffinity labeling of peptide hormone receptors. *Methods Enzymol* 109:129–156, 1985.

89. Bosse R, Servant G, Zhou L-M, et al: Selective photo-affinity labeling of angiotensin II receptors. *Fed Am Soc Exp Biol* 6(4):A1577, 1992.
90. Ardecky RJ, Chiu AT, Duncia JJV, et al: Treatment of CNS disorders with 4,5,6,7-tetrahydro-1H-imidazo(4,5-)-pyridine derivatives and analogs. US Patent 5,091,390, February 25, 1992.
91. Zhang JS, vanZwieten PA: Characterization of two novel nonpeptide angiotensin II antagonists. *Br J Pharmacol* 105(Suppl):85P, 1992.
92. de Laszlo SE, Quagliato CS, Greenlee WJ, et al: A potent, orally active, balanced affinity angiotensin II AT$_1$ antagonist and AT$_2$ binding inhibitor. *J Med Chem* 36:3207–3210, 1993.

7

Pharmacology of AT_2 Receptors

Joan A. Keiser and Robert L. Panek

1. INTRODUCTION

The existence of angiotensin II (ANG II) receptor subtypes has been postulated for several years[1–6]; however, receptor subtypes were not convincingly demonstrated until the availability of subtype selective ligands.[7–10] Subsequent cloning of the AT_1 receptor[11,12] served to clearly differentiate the receptor subtypes. Two major classes of mammalian receptor subtypes were first described in late 1989 and early 1990 by several research groups[13–19] using the selective ligands. Several naming schemes were employed by investigators, resulting in a confusing body of literature. Conventions for the nomenclature of ANG II receptor subtypes, [20] published in 1990, are now gaining widespread acceptance in the literature. Using this nomenclature, ANG II receptors are classified as either AT_1 or AT_2 based on selective displacement of radiolabeled ligand from the receptor *in vitro* as well as their localization and signaling properties. The pharmacology of AT_2 receptors is the focus of this chapter; AT_1 receptors[21,22] as well as cytosolic and nuclear angiotensin-binding proteins[23–26] have been described elsewhere in this book and are not covered here.

Although angiotensin, saralasin, and several other peptides bind non-selectively to both AT_1 and AT_2 receptors, a number of selective peptide and nonpeptide ligands for the AT_2 receptor have become available in the last few years. The chemistry of these AT_2-selective agents has been reviewed elsewhere[8,9,18,27] and is detailed in Chapter 6. In this chapter, we will review the characterization of AT_2 receptors *in vitro*, as well as their localization and

Joan A. Keiser and Robert L. Panek • Cardiovascular Pharmacology, Parke-Davis Pharmaceutical Research, Division of Warner-Lambert Company, Ann Arbor, Michigan 48105.

Angiotensin Receptors, edited by Juan M. Saavedra and Pieter B.M.W.M. Timmermans. Plenum Press, New York, 1994.

pharmacology in the central nervous system, reproductive tissues, cardiovascular system, and kidney.

2. AT$_2$ RECEPTOR EXPRESSION IN VITRO

2.1. Biochemical Characterization

AT$_2$ receptors have been shown to selectively recognize modified peptides such as CGP 42112A,[10,27–29] *para*-aminophenylalanine ANG II,[10,18] [3,5]cyclic angiotensin analogues,[30–32] and the nonpeptide agents typified by PD 121981 (also referred to as WL-19), PD 123177, and PD 123319.[8,16] The selectivity of these agents has been demonstrated by a greater than 1000-fold potency difference to compete with ANG II binding at AT$_2$ versus AT$_1$ binding sites; a summary of their affinity for AT$_1$ versus AT$_2$ receptors is detailed in Table I. Unlike the AT$_1$ site, to date there has been no clear function associated with AT$_2$ sites, although several have been suggested.[33–39]

Dudley *et al.*[40] recently described a mouse fibroblast cell line (R3T3) that selectively expresses AT$_2$ sites. These binding sites are not coupled to guanine nucleotide-binding proteins (GTP-γ-S did not affect binding affinity) and did not internalize in the presence of ANG II. Affinity-labeling experiments revealed a specifically labeled protein with an apparent molecular weight of about 100 kDa. Addition of ANG II to the R3T3 cells revealed no effect on common signaling pathways, including stimulation of phosphotidylinositol turnover, cAMP, tyrosine kinase activity, and release of arachidonic acid. Furthermore, neither ANG II or PD 123319 affected cell growth, mitogenesis, or hypertrophy. In addition, PD 123319 did not affect mitogenesis or hypertrophy stimulated by several growth factors. These results show the AT$_2$ binding

Table I. Selectivity of Peptide and Nonpeptide Ligands for AT$_1$ and AT$_2$ Receptors

Compound	AT$_2$ (IC$_{50}$ nM)[a]	AT$_1$ (IC$_{50}$ nM)[b]
Angiotensin	0.84	1.01
Saralasin	1.62	1.7
CGP 42112A	0.05	2800
PD123319	8.3	> 10,000
PD123177	34.3	> 10,000
PD121981	17.9	> 10,000

[a]IC$_{50}$ values for binding to AT$_2$ receptors. Competition curves were guaranteed using membranes prepared from rabbit uterus (see ref. 16 for details).
[b]IC$_{50}$ values for binding to AT$_1$ receptors. Competition curves were generated using membranes prepared from rat liver (see ref. 16 for details).

site is quite distinct from the AT$_1$ receptor in terms of molecular weight, binding properties [dithiothreitol (DTT) slightly enhances binding to AT$_2$ sites but inactivates AT$_1$ sites], and coupling to second messengers.

In a similar series of studies, Pucell *et al.*[41] demonstrated that primary cultures of rat ovarian granulosa cells from atretic follicles exclusively express AT$_2$ sites. The molecular weight of the granulosa cell AT$_2$ receptor as estimated by affinity cross-linking studies was 79 kDa. However, like the R3T3 cell AT$_2$ receptor, the granulosa cell AT$_2$ receptor did not undergo agonist-induced internalization. Furthermore, ANG II did not affect basal or stimulated inositol phosphate production, intracellular calcium mobilization, or cAMP or cGMP levels. The granulosa cell AT$_2$ receptor was not coupled to guanine nucleotide-binding regulatory proteins since addition of GTP-γ-S had no effect on ANG II binding affinity and the receptor was not an ANG II-binding protease.

Sumners *et al.*[42] were the first to report a putative signaling pathway for the AT$_2$ receptor. They showed that ANG II binding to AT$_2$ receptors located on primary cultures of neonatal rat neurons mediated a decrease in basal cGMP levels, an effect that was antagonized by PD 123177 and CGP 42112A.[42] Using voltage clamp techniques, these same authors reported that ANG II altered net outward current in cultured rat neurons; this response was blocked by PD 123319 and PD 123177, but not by losartan.[43] The physiological significance of this angiotensin response remains to be determined; to date there are no reported electrophysiologic studies *in vivo*.

More recently, it was reported that ANG II inhibited basal and atrial natriuretic peptide-stimulated particulate guanylate cyclase activity through AT$_2$ receptors in rat adrenal glomerulosa and PC12W cells.[44] Moreover, ANG II induced a rapid, transient dephosphorylation of phosphotyrosine-containing proteins in PC12W cells, suggesting that AT$_2$ receptors signal through stimulation of a phosphotyrosine phosphatase and that this mechanism is implicated in the regulation of particulate guanylate cyclase activity. Finally, a recent report suggests that in human astrocytes the AT$_2$ receptor is coupled to prostaglandin release through a calcium-independent mechanism.[45,46] Thus these three reports describe putative signal transduction pathways for the AT$_2$ receptor. However, these signaling mechanisms have not been linked to physiological responses *in vivo*. These signaling pathways need to be examined in other cells and tissues expressing the AT$_2$ receptor.

It is noteworthy that cross-linking studies have revealed different molecular weights for the AT$_2$ receptor in several preparations (79 kDa in rat ovarian granulosa cells,[41] 100 kDa in R3T3 cells,[40] and 71 kDa[47] in bovine adrenal cortex membranes), suggesting the existence of AT$_2$ subunits or heterogeneity of the AT$_2$ receptor.

2.2. Species and Tissue Distribution of AT_2 Receptors

Both species- and age-related changes in AT_2 receptor expression have been described. Species examined to date that express an AT_2 receptor include the rat, rabbit, pig, sheep, cow, monkey, and human. Most AT_2 receptor expression in adult animals is localized to a few anatomic regions including brain, adrenal, kidney, heart, and reproductive tissues, although not all species have the same organ distribution of AT_2 receptor expression. For example, rat adrenal contains predominately AT_1 sites in the cortex and AT_2 sites in medulla[9,13,15]; rabbit adrenal has both AT_1 and AT_2 sites in cortex and few sites of either subtype in medulla[48,49]; and human adrenal contains both AT_1 and AT_2 receptors in cortex.[9,50] In addition, subtype-specific ligands were reported to have potency differences between species.[27] The significance of these species differences remains to be elucidated.

In marked contrast to the adult distribution pattern, studies in fetal and neonatal animals indicate a high density of AT_2 receptors in skin, connective tissues, skeletal muscle, diaphragm, gastrointestinal tract, and brain.[51–54] This abundance of AT_2 receptors rapidly declines to the adult pattern of expression following birth, suggesting a role for the AT_2 receptor in mediating growth and/or cell differentiation.[51]

3. ROLE OF AT_2 RECEPTORS IN THE CENTRAL NERVOUS SYSTEM

Numerous investigators have used autoradiography and/or binding assays to identify AT_2 receptors in the central nervous system (CNS). AT_2 receptor localization studies have been conducted in brain tissue or CNS-derived cell lines from rat,[17,33,55–61] rabbit,[62] cow,[63] monkey,[49] and human.[34,50,64,65] The majority of localization studies and most detailed experiments to date have been conducted in rats. The first description of AT_2 receptors in rat brain appeared in 1989.[14] In that report, Chang and Lotti detailed that PD 121981 (WL-19) displaced radiolabeled ANG II with a 27 nM IC_{50}, similar to its affinity for the AT_2 receptor in rat adrenals.[18] AT_2 sites have been detected in rat brain septum, medial amygdala, thalamus, subthalamus, coliculi, locus coeruleus, and inferior olive.[17,55,56] AT_2 binding in these brain regions of adult rats is not inhibited by sulfhydryl-reducing agents.[33] Thus AT_2 receptors are compartmentalized into discrete areas in brain and are not distributed homogeneously. The presence of AT_2 sites in the thalamus and areas of the brain that process sensory information have led scientists to propose a novel modulatory role for ANG II in information processing via the AT_2 receptor.[33]

The functional role of AT_2 receptors in CNS has remained elusive. Schiavone and co-workers[35] reported that ANG II-induced vasopressin secretion from rat hypothalamoneurohypophyseal explants were inhibited by PD 123177 and CGP 42 112A, whereas losartan was ineffective. In these studies the concentrations of both peptide and nonpeptide ligands was in the molar range. One could readily argue that selectivity for the AT$_2$ receptor is compromised at these high concentrations; however, it is worth noting that losartan failed to inhibit vasopressin secretion at similar concentrations. Steele[66] reported that ANG II-induced luteinizing hormone (LH) release and inhibition of prolactin secretion were blocked by intracerebral ventricular (ICV) administration of PD 123177. However, in Steele's conscious rat studies, losartan also inhibited ANG II-stimulated responses, suggesting multiple sites of action for ANG II in this neurohormonal pathway. It is important to note that autoradiographic studies have failed to document AT$_2$ receptors in hypothalamic or pituitary regions of the brain.[56,67] Several investigators have clearly shown that AT$_2$ receptors are not involved in the drinking behavior associated with ICV ANG II.[66,68,69]

The unique location of AT$_2$ receptors in brain regions unrelated to cardiovascular function are suggestive of novel ANG II actions. Interestingly, Costall and co-workers[36] reported that PD 123177 treatment at doses as low as 0.1 mg/kg intraperitoneally overcame the cognitive impairment induced by scopolamine in a mouse habituation paradigm. In addition, both PD 121981 (WL-19) and losartan are reported to block apomorphine-induced stereotypy in the rat,[70] although again the activity of both AT$_1$ and AT$_2$ ligands in this model makes interpretation of the data more difficult.

Although early investigations of central ANG II receptors suggested that the brain contained two receptor subtypes similar to those found in the periphery, recent studies detail a more complex picture. Tsutsumi and Saavedra[71] reported a group of AT$_2$ receptors in rat brain medial geniculate and locus coeruleus sensitive to GTP-γ-S and pertussis toxin. In contrast, AT$_2$ sites in the inferior olive showed no sensitivity to GTP-γ-S or pertussis toxin, suggesting AT$_2$ receptor heterogeneity. Further studies from these same investigators detailed a rat brain AT$_2$ receptor in ventral and medial thalamic, medial geniculate, and oculomotor nuclei and superior colliculus and the cerebellar cortex that was inhibited by dithiothreitol.[72] (In general, AT$_2$ receptors are reported to be insensitive to sulfhydryl-reducing agents.)

Thus, in summary, brain AT$_2$ receptors are present in all species described to date. Localization and functional studies clearly demonstrate the involvement of AT$_1$ receptors in central regulation of cardiovascular function.[56,58,59,68] A physiological role for AT$_2$ receptors remains to be determined.

More extensive characterization of AT$_2$ receptors has been performed in cultured cell of CNS origin. Several neuronal cell lines express a mixture of AT$_1$ and AT$_2$ sites; examples include primary neuronal cell cultures from 1-day-old rats,[42,43] N1E-115 murine neuroblastoma cells,[73,74] and human astrocytes.[45,46] Neuroblastoma X glioma hybrid cells (NG-108-15) primarily express AT_1 receptors in the undifferentiated state, but after 4–5 days of differentiation the receptor subtype pattern changes. Absolute ANG II receptor numbers are markedly upregulated and more importantly the majority of receptor sites are now AT_2 (80%); this represents a 40-fold increase over baseline AT_2 receptor density.[64,75] Upregulation of AT_2 sites has also been described in N1E-115 cells.[74] Interestingly, other investigators have described ANG II receptors in neuro-2A cells that lack affinity for either the AT_2 (PD 123319) or AT_1 (losartan) ligands,[76,77] giving further credence to a unique population of brain ANG II receptors.

Some of the most provocative data regarding CNS ANG II receptors details changes in ANG II receptor expression during development. Work from the laboratories of Aguilera,[78] Saavedra[59,60] and Speth[79,80] have described the changes in brain angiotensin receptors in the fetal and neonatal rat. In fetal rat brain, AT_2 receptors are abundant even in regions that are exclusively AT_1 in adult animals. The prevalence and location of brain AT_2 receptors in neonates and the age-related changes in relative expression of the receptor subtypes have led investigators to postulate the ANG II may have a role in CNS development.

4. AT₂ RECEPTORS IN REPRODUCTION

Proteins comprising the cascade leading to the formation of ANG II, and ANG II itself, have been identified in the reproductive tract.[81–85] Recently, the evidence for an active renin–angiotensin system in the ovary has raised the possibility of ANG II acting in follicle development and ovulation.[84] Although receptors for ANG II have been found on granulosa cells of ovarian follicles,[41,86,87] a physiological role of angiotensin in maturation of the follicle, and of granulosa cells in particular, has not been determined. Daud et al.[88] reported that ANG II receptors predominate on granulosa cells of atretic follicles, suggesting a possible role of angiotensin in atresia. In the bovine ovary, the follicular fluid content of prorenin–renin is severalfold higher in atretic versus mature follicles.[86] In a recent study[41] granulosa cells isolated from atretic follicles of immature hypophysectomized rats bearing diethylstilbesterol implants were shown to contain exclusively AT_2 receptors. Moreover, Pucell et al.[86] showed that follicle-stimulating hormone (FSH) treatment of rat granulosa cells resulted in a decrease in AT_2 receptor content, suggesting a regulation of AT_2 receptors by gonadotropins. The onset of atresia in rat ovarian follicles has been associated with a marked increase in ANG II receptor number.[86,87] However, these same authors[41] showed that FSH-stimulated progesterone and estrogen secretion from cultured rat ovarian granulosa cells was not affected by

ANG II. Thus, in the atretic follicle, AT$_2$ receptor expression may be regulated by gonadotropins, but hormone secretion is not modulated via ANG II. Thus, in the atretic follicle, AT$_2$ receptor expression may be regulated by gonadotropins, but hormone secretion is not modulated via ANG II. Whether a similar pattern of AT$_2$ receptor expression and gonodotropin regulation exists in preovulatory follicles remains to be determined.

ANG II receptors are also found in diverse ovarian structures such as theca cells, corpora lutea, blood vessels, and surface epithelium.[87-89] In the rat, ovarian ANG II receptors other than the granulosa cells are predominately AT$_1$. However, data obtained from a study by Brunswig-Spickenheier and Mukhopadhyay[90] provide evidence for the existence of high-affinity AT$_2$ binding sites on theca cells isolated from bovine ovarian follicles. Furthermore, the density of these AT$_2$ receptors was shown to be increased by exposure of the cultured theca cells to LH via a cAMP-dependent mechanism. Thus, both existing and newly induced ANG II receptors were of the AT$_2$ subtype. In contrast to distribution in the rat, neither granulosa nor luteal cells of the bovine ovary possess significant ANG II binding sites. Similarly, in monkey ovary abundant ANG II receptors were detected on theca interna cells but not on granulosa cells.[91] In the monkey study the receptor subtype was not characterized; however, DTT was used in the binding assay, leading one to infer that these were AT$_2$ receptors.

There is considerable interspecies variation in the distribution of ovarian ANG II receptors. The pattern of AT$_2$ receptors in monkey and bovine ovaries is distinct from that seen in rat ovary. The absence of ANG II receptors on bovine granulosa and luteal cells could be the result of receptor expression only during a particular phase of the ovarian cycle; however, this needs to be verified. Similarly, monkey ovarian ANG II receptors have not been profiled across the ovarian cycle. To date no reports could be found describing ANG II receptor subtypes in human ovaries.

AT$_2$ receptor distribution in the uterus also differs when profiled across species. For example, human uterus contains only AT$_2$ sites,[9] whereas rat and rabbit uteri express both AT$_1$ and AT$_2$ sites with the latter predominating.[16,49] During ovine pregnancy myometrial smooth muscle ANG II receptors change subtype.[92] In pregnant ewes, from gestational day 113 to term (145 days), myometrial smooth muscle receptors were primarily AT$_1$ (85%). At about 7 days postpartum both subtypes were observed (AT$_1$ = 60%, AT$_2$ = 40%), and at 2 weeks postpartum AT$_2$ receptors predominated (> 90%). This change in subtype expression is associated with the development of ANG II contractile responses measured in myometrial strips from pregnant but not postpartum ewes. In all species examined to date, ANG II-induced uterine vasoconstriction is mediated via an AT$_1$ receptor.

5. EXPRESSION AND FUNCTION OF AT$_2$ RECEPTORS IN HEART AND KIDNEY

With the first reports of ANG II receptor subtypes, Wong and collaborators[7,68] clearly demonstrated that the vascular smooth muscle vasoconstrictor effects of angiotensin were mediated via an AT$_1$ receptor. In fact, early receptor localization studies using both membrane binding and autoradiographic techniques failed to identify AT$_2$ receptors in vascular tissue. However, in a pivotal study published in late 1991, Viswanathan *et al.*[53] described the presence of AT$_2$ receptors in the aorta of fetal and neonatal rats. This study demonstrated for the first time the presence of AT$_2$ receptors in vascular tissue and raised the possibility that the AT$_2$ receptor may modulate cardiovascular function. However, despite the fact that neonatal aorta expresses AT$_2$ and AT$_1$ receptors in a ratio of 20:1, ANG II-induced contraction of aortic rings from neonatal rats is blocked with losartan, whereas PD 123319 is ineffective (T. Major, personal communication). AT$_2$ receptors have also been described in rat carotid arteries undergoing intimal proliferation after balloon angioplasty[93] and in rat anterior cerebral arteries.[94] The function of AT$_2$ receptors in these vascular sites remains to be determined.

Two interesting reports were recently published describing hemodynamic effects of nonpeptide AT$_2$ ligands in rodents. Scheuer and co-workers[37] reported that rats treated with losartan had ANG III-induced vasodilator responses that were blocked with PD 123319. In a similar study, Widdop et al.[38] pretreated rats with an AT$_1$ antagonist and observed at 24 hr later that PD 123177 attenuated the hemodynamic response to ANG II. Administration of PD 123177 alone had no hemodynamic activity in Widdop's model. Both reports suggest that AT$_2$ receptor ligands can interact and/or modulate AT$_1$-mediated responses in rat vascular tissue. These studies are difficult to reconcile with the apparent absence of AT$_2$ receptors in vascular tissue from normal adult rats. In addition, there are no reports of AT$_2$ ligands modulating ANG II responses in isolated tissue strips. However, the possibility of indirect actions has not been eliminated. For example, in Widdop's study,[38] a 24-hour pretreatment schedule altered ANG II responses and in other models receptor subtype expression has changed markedly in a 24-hr interval.[52] In addition, Wong *et al.*[95] reported that PD 123177 displaced losartan from plasma proteins, and thus appeared to potentiate losartan inhibition of ANG II pressor responses in pithed rats.

AT$_2$ receptors have been reported in rabbit cardiac tissue,[96,97] but the inotropic responses to ANG II in rat and rabbit heart are blocked with losartan. Thus, the function of AT$_2$ receptor in cardiac tissue remains unclear.

Angiotensin exerts a variety of actions in the kidney, including effects on blood flow, glomerular tubular filtration, and tubular reabsorption. Using autoradiographic techniques and nonselective peptide ligands, Gibson *et al.*[98] have carefully detailed multiple anatomic sites for ANG II binding in the kidney, including vascular, glomerular, tubular, and interstitial locations. Marked species heterogeneity for angiotensin receptor subtype expression in renal tissue exists. Adult rat kidneys do not express AT$_2$ receptors,[28,99] rabbit kidneys have AT$_2$ receptors in the capsule,[49] and primate kidneys express both AT$_1$ and AT$_2$ receptors.[62] Human kidneys have a pattern of ANG II receptor expression similar to that reported in monkeys; both AT$_1$ and AT$_2$ receptors have been described in human kidneys.[100–102] Human fetal kidneys have more AT$_2$ binding sites than adult kidneys.[100] Glomeruli isolated from adult kidneys are exclusively AT$_1$.[101]

Although several investigators have reported receptor localization studies, the function of AT$_2$ receptors in the kidney is not understood. Clearly, the renal blood flow effects of ANG II are mediated via an AT$_1$ receptor. Cogan *et al.*[103] measured the renal tubular effects of PD 123177 and losartan in anesthetized rats. They concluded that the effects of angiotensin on proximal tubular reabsorption were mediated via an AT$_1$ receptor since the response was attenuated with losartan. However, PD 123319 infused intrarenally produced diuresis and increased free water clearance in anesthetized dogs,[39] suggesting a late renal tubular effects of this agent. Consistent with Cogan's studies, AT$_2$ receptors are not detected in normal rat kidney. There are no published autoradiographic or membrane binding data on angiotensin receptor subtypes from canine kidneys. The expression of AT$_2$ receptors in renal and cardiac tissues from fetal animals is intriguing and the potential role in pathological states remains to be determined.

6. SUMMARY

A large body of literature has accumulated that details AT$_2$ receptor localization and changing patterns of receptor expression throughout development. However, data addressing the function of AT$_2$ receptors are incomplete. The identification of a signaling/second messenger pathway that could be quantitatively linked to receptor activation and physiological responses would allow investigators to solidify the observations made to date. Cloning of the AT$_2$ receptor has, to date, eluded investigators. Progress in this area would allow us to probe the regulation of receptor expression and more systematically explore potential AT$_2$ receptor subtypes. Research in the CNS, reproductive, and renal fields are promising areas for understanding AT$_2$ receptor function. Continuing availability of subtype-selective ligands and ongoing cloning efforts should provide investigators with the tools necessary to unravel the function(s) of the AT$_2$ receptor.

REFERENCES

1. Peach MJ: Renin–angiotensin system: Biochemistry and mechanisms of action. *Physiol Rev* 57:313–370, 1977.
2. Regoli D, Park WK, Rioux F: Pharmacology of angiotensin. *Physiol Rev* 26:69–123, 1974.
3. Douglas JG: Angiotensin receptor subtypes of the kidney cortex. *Am J Physiol* 253:F1–F7, 1987.
4. Douglas JG, Michailov M, Khosla MC, et al: Comparative receptor-binding properties of heptapeptide and octapeptide antagonists of angiotensin II in rat adrenal glomerulosa and uterine smooth muscle. *Endocrinology* 106:120–124, 1980.
5. Wright GB, Alexander RW, Ekstein LS, et al: Characterization of the rabbit ventricular myocardial receptor for angiotensin II: Evidence for two sites of different affinities and specificities. *Mol Pharmacol* 24:213–221, 1983.
6. Gunther S: Characterization of angiotensin II receptor subtypes in rat liver. *J Biol Chem* 259:7622–7629, 1984.
7. Wong P, Chiu AT, Price WA, et al: Nonpeptide angiotensin II receptor antagonists. I. Pharmacological characterization of 2-n-butyl-4-chloro-1-(2-chlorobenzyl)-imidazole-5-acetic acid, sodium salt (S-8307). *J Pharm Exp Ther* 247:1–7, 1988.
8. Blankley CJ, Hodges JC, Klutchko SR, et al: Synthesis and structure–activity relationships of a novel series of nonpeptide angiotensin II receptor binding inhibitors specific for the AT2 subtype. *J Med Chem* 34:3248–3260, 1991.
9. Whitebread S, Mele M, Kamber B, et al: Preliminary biochemical characterization of two angiotensin II receptor subtypes. *Biochem Biophys Res Commun* 163:284–291, 1989.
10. Speth RC, Kim KH: Discrimination of two angiotensin II receptor subtypes with a selective agonist analogue of angiotensin II, p-aminophenylalanine 6-angiotensin II. *Biochem Biophys Res Commun* 169:997–1006, 1990.
11. Murphy TJ, Alexander RW, Griendling KK, et al: Isolation of a cDNA encoding the vascular type-1 angiotensin II receptor. *Nature* 351:233–236, 1991.
12. Sasaki K, Yamono Y, Bardhan S, et al: Cloning and expression of a complementary DNA encoding a bovine adrenal angiotensin II type-1 receptor. *Nature* 351:230–236, 1991.
13. Chiu AT, Herblin WF, McCall DE, et al: Identification of angiotensin II receptor subtypes. *Biochem Biophys Res Commun* 165:196–203, 1989.
14. Chang RSL, Lotti VJ: Selective ligands reveal subtypes of angiotensin receptors in rat vasculature and brain. *Pharmacologist* 31:150, 1989.
15. Chang RSL, Lotti VJ: Two distinct angiotensin II receptor binding sites in rat adrenal revealed by new selective nonpeptide ligands. *Mol Pharmacol* 29:347–351, 1990.
16. Dudley D, Panek RL, Major TC, et al: Subclasses of angiotensin II binding sites and their functional significance. *Mol Pharmacol* 38:370–377, 1990.
17. Gehlert DR, Gackenheimer SL, Reel JK, et al: Non-peptide angiotensin II receptor antagonists discriminate subtypes of ^{125}I-angiotensin II binding sites in the rat brain. *Eur J Pharmacol* 187:123–126, 1990.
18. Rowe BP, Grove KL, Saylor DL, et al: Discrimination of angiotensin II receptor subtype distribution in the rat brain using non-peptidic receptor antagonists. *Regul Pept* 33:45–53, 1991.
19. Chang RSL, Lotti VJ, Chen TB, et al: Two angiotensin II binding sites in rat brain revealed using [125I]Sar[1],Ile[8]-angiotensin II and selective nonpeptide antagonists. *Biochem Biophys Res Commun* 171:813–817, 1990.

20. Bumpus FM, Catt KJ, Chiu A, et al: Nomenclature for angiotensin receptors. *Hypertension* 17:720–721, 1991.

21. Timmermans PBMWM, Wong PC, Chiu AT, et al: Nonpeptide angiotensin II receptor antagonists. *Trends Pharmacol Sci* 12:55–62, 1991.

22. Herbli WF, Chiu AT, McCall DE, et al: Angiotensin II receptor heterogeneity. *Am J Hypertension* 4:299S–302S, 1991.

23. Burkard MR, Rauch AL, Mangiapane ML, et al: Differentiation of the rat liver cytoplasmic AII binding protein from the membrane angiotensin II receptor. *Hypertension* 16:323, 1991.

24. Hagiwara H, Sugiura N, Wakita K, et al: Purification and characterization of angiotensin-binding protein from porcine liver cytosolic fraction. *Eur J Biochem* 185:405–410, 1989.

25. Kiron RA, Soffer RL: Purification and properties of a soluble angiotensin II binding protein from rabbit liver. *J Biol Chem* 264:4138–4142, 1989.

26. Tang SS, Rogg H, Schumacher R, et al: Evidence and characterization of distinct nuclear and plasma membrane angiotensin II binding sites in the rat liver. *Hypertension* 16:323, 1991.

27. Balla T, Baukal AJ, Eng S, et al: Angiotensin II receptor subtypes and biological responses in the adrenal cortex and medulla. *Mol Pharmacol* 40:401–406, 1991.

28. de Gasparo M, Whitebread S, Mele M, et al: Biochemical characterization of two angiotensin II receptor subtypes in the rat. *J Cardiovasc Pharmacol* 16(Suppl 4):S31–S35, 1990.

29. Carini DJ, Duncia JJV, Yoo SE: US Patent 4,880,804, DuPont de Nemours and Co., November 14, 1989.

30. Marshall GA, Kaczmarek K, Kataoka T, et al: Evidence for receptor-bound turn conformations of bradykinin and angiotensin II. In Giralt E, Andreu D (eds): *Peptides 1990.* ESCOM, Leiden, 1991, pp. 594–596.

31. Marshall GA, Kataoka T, Plucinska K, et al: Optimization of constraints forcing receptor-bound turn conformations of angiotensin. Abstracts: Twelfth American Peptide Symposium, Cambridge, MA, June 16–21, 1991, No. P126.

32. Padmaja J, Cody W, Dooley D, et al: Bradykinin and angiotensin II analogs containing a conformationally constrained proline analog. Abstracts: Twelfth American Peptide Symposium, Cambridge, MA, June 16–21, 1991, No. P-474.

33. Speth RC, Rowe BP, Grove KL, et al: Sulfhydryl reducing agents distinguish two subtypes of angiotensin II receptors in rat brain. *Brain Res* 548:1–8, 1991.

34. Barnes JM, Barber PC, Barnes NM: Identification of angiotensin II receptor subtypes in human brain. *Neuroreport* 2(10):605–608, 1991.

35. Schiavone MT, Brosnihan KB, Khosla MC, et al: Angiotensin II activation of vasopressin release in the rat hypothalamoneurohypophysial system is mediated by the type 2 angiotensin receptor. *Hypertension* 17(3):425, 1991.

36. Barnes, NM, Costall B, Kelly ME, et al: Cognitive enhancing actions of PD 123177 detected in a mouse habituation paradigm. *Neuroreport* 2:351–353, 1991.

37. Scheuer DA, Bochnowicz S, Perrone MH: AT1 receptors do not mediate depressor response to angiotensin in rats. *FASEB J* 6(4, part 1):A1173, 1992.

38. Widdop RE, Gardiner SM, Kemp PA, et al: PD 123177-evoked inhibition of the hemodynamic effects of angiotensin II in conscious rats after exposure to EXP 3174. *Br J Pharmacol* 105(Suppl):86P, 1992.

39. Keiser JA, Bjork FA, Hodges JC, et al: Renal hemodynamic and excretory responses to PD 123319 and losartan, nonpeptide AT1 and AT2 subtype-specific angiotensin II ligands. *J Pharmacol Exp Ther* 263(3):1154–1160, 1992.

40. Dudley DT, Hubbell SE, Summerfelt RM: Characterization of angiotensin II (AT2) binding sites in R3T3 Cells. *Mol Pharmacol* 40:360–367, 1991.

41. Pucell AG, Hodges JC, Sen I, et al: Biochemical properties of the ovarian granulosa cell type 2-angiotensin II receptor. *Endocrinology* 128:1947–1959, 1991.
42. Sumners C, Tang W, Zelezna B, et al: Angiotensin II receptor subtypes are coupled with distinct signal transduction mechanisms in neurons and astrocytes from rat brain. *Proc Natl Acad Sci USA* 88(17):7567–7571, 1991.
43. Kang J, Sumners C, Posner P: Modulation of net outward current in cultured neurons by angiotensin II—involvement of AT(1)-receptor and AT(2) receptors. *Brain Res* 580(1-2):317–324, 1992.
44. Bottari SP, King IN, Reichlin S, et al: The angiotensin AT2 receptor stimulates protein tyrosine phosphatase activity and mediates inhibition of particulate guanylate cyclase. *Biochem Biophys Res Commun* 183(1):206–211, 1992.
45. Jaiswal N, Tallant EA, Diz DI, et al: Identification of two distinct angiotensin receptors on human astrocytes using an angiotensin receptor antagonist. *Hypertension* 16(3):30, 1990.
46. Jaiswal N, Tallant EA, Diz DI, et al: Subtype 2 angiotensin receptors mediate prostaglandin synthesis in human astrocytes. *Hypertension* 17(6, Pt 2):1115–1120, 1991.
47. Bosse R, Servant G, Zhou L-M, et al: Selective photoaffinity labeling of angiotensin II receptors. *FASEB J* 6(4):A1577, 1992.
48. Chiu AT, McCall DE, Nguyen TT, et al: Discrimination of angiotensin II receptor subtypes by dithiothreitol. *Eur J Pharmacol* 170:117–118, 1989.
49. Chang RSL, Lotti VJ: Angiotensin subtypes in rat, rabbit and monkey tissues: Relative distribution and species dependency. *Life Sci* 49:1485–1490, 1991.
50. Bottari SP, Taylor V, King IN, et al: Angiotensin AT2 receptors do not interact with guanine nucleotide binding proteins. *Eur J Pharmacol Mol Pharmacol* 207:157–163, 1991.
51. Grady EF, Sechi LA, Griffin CA, et al: Expression of AT2 receptors in the developing fetus. *J Clin Invest* 88(3):921–933, 1991.
52. Johnson MC, Aguilera G: Angiotensin II receptor subtypes and coupling to signaling systems in cultured fetal fibroblasts. *Endocrinology* 129(3):1266–1274, 1991.
53. Viswanathan M, Tsutsumi K, Correa FM, et al: Changes in expression of angiotensin receptor subtypes in the rat aorta during development. *Biochem Biophys Res Commun* 179(3):1361–1367, 1991.
54. Lu GH, Overhiser RW, Keiser JA, et al: Characterization of angiotensin II receptors in the rat fetus following 7 day administration of PD 123319 or DuP 753. *FASEB J* 6(4):A1017, 1992.
55. Rowe BP, Grove KL, Saylor DL, et al: Angiotensin II receptor subtypes in the rat brain. *Eur J Pharmacol* 186:339–342, 1990.
56. Gelhart DR, Gackenheimer SL, Schober DA: Autoradiographic localization of subtypes of angiotensin II antagonist binding in the rat brain. *Neuroscience* 44(2):501–514, 1991.
57. Gelhart DR, Gackenheimer SL, Schober DA: Angiotensin II receptor subtypes in rat brain: Dithiothreitol inhibits ligand binding to AII-1 and enhances binding to AII-2. *Brain Res* 546:161–165, 1991.
58. Song K, Zhuo J, Allen AM, et al: Angiotensin II receptor subtypes in rat brain and peripheral tissues. *Cardiology* 79:45–54, 1991.
59. Tsutsumi K, Saavedra JM: Characterization and development of angiotensin II receptor subtypes AT-1 and AT-2 in rat brain. *Am J Physiol* 261:R209–R216, 1991.
60. Tsutsumi K, Saavedra JM: Differential development of angiotensin II receptor subtypes in the rat brain. *Endocrinology* 128:630–632, 1991.
61. Obermuller N, Unger T, Culman J, et al: Distribution of angiotensin II receptor subtypes in rat brain nuclei. *Neurosci Lett* 132:11–15, 1991.
62. Criscione L, Thomann H, Whitebread S, et al: Binding characteristics and vascular effects of various angiotensin II antagonists. *J Cardiovasc Pharmacol* 16(Suppl 4):S56–S59, 1990.

63. Wiest SA, Rampersaud A, Zimmerman K, et al: Characterization of distinct angiotensin II binding sites in rat adrenal gland and bovine cerebellum using selective nonpeptide antagonists. *J Cardiovasc Pharmacol* 17:177–184, 1991.

64. Tallant EA, Jaiswal N, Diz DI, et al: Human astrocytes contain two distinct angiotensin subtypes. *Hypertension* 18:32–39, 1991.

65. Tallant EA, Diz DI, Khosla MC, et al: Identification and regulation of angiotensin II receptor subtypes on NG108-15 cells. *Hypertension* 17:1135–1143, 1991.

66. Steele MK: Angiotensin receptor subtypes, water intake and LH and prolactin secretion in female rats. *Soc Neurosci Abtsr* 17(1):978, 1991.

67. Tsutsumi K, Saavedra JM: Angiotensin II receptor subtypes in median eminence and basal forebrain areas involved in regulation of pituitary function. *Endocrinology* 129(6):3001–3008, 1991.

68. Wong PC, Hart SD, Zaspel AM, et al: Functional studies of nonpeptide angiotensin II receptor subtype-specific ligands: DuP 753 (AII-1) and PD 123177 (AII-2). *J Pharmacol Exp Ther* 255:584–592, 1990.

69. Dourish CT, Duggan JA, Banks RJA: Drinking induced by subcutaneous injection of angiotensin II in the rat is blocked by the selective AT1 receptor antagonist DuP 753 but not by the selective AT2 receptor antagonist WL19. *Eur J Pharmacol* 211:113–116, 1992.

70. Banks RJA, Dourish CT: The angiotensin receptor antagonists DuP 753 and WL 19 block apomorphine-induced stereotypy in the rat. *Br J Pharmacol* 104(Suppl):63P, 1991.

71. Tsutsumi K, Saavedra JM: Heterogeneity of angiotensin II AT2 receptors in the rat brain. *Mol Pharmacol* 41(2):290–297, 1992.

72. Tsutsumi K, Zorad S, Saavedra JM: The AT2 subtype of the angiotensin II receptors has a differential sensitivity to dithiothreitol in specific brain nuclei of young rats. *Eur J Pharmacol Mol Pharmacol* 226(2):169–173, 1992.

73. Zaharn ED, Ye X, Ades AM, et al: Angiotensin-induced cyclic GMP production is mediated by multiple receptor subtypes and nitric oxide in N1E-115 neuroblastoma cells. *J Neurochem* 58(5):1960–1963, 1992.

74. Reagan LP, Ye X, Mir R, et al: Up-regulation of angiotensin II receptors by in vitro differentiation of murine N1E-115 neuroblastoma cells. *Mol Pharmacol* 38(6):878–886, 1990.

75. Bryson SE, Warburton P, Wintersgill HP, et al: Induction of the angiotensin AT2 receptor subtype expression by differentiation of the neuroblastoma X glioma hybrid, NG-108-15. *Eur J Pharmacol* 225(2):199–127, 1992.

76. Chaki S, Inagami T: A newly found angiotensin II receptor subtype mediates cyclic GMP formation in differentiated neuro-2A cells. *Eur J Pharmacol Mol Pharmacol* 225(4):355–356, 1992.

77. Chaki S, Inagami T: Identification and characterization of a new binding site for angiotensin II in mouse neuroblastoma neuro-2A cells. *Biochem Biophys Res Commun* 192(1):388–394, 1992.

78. Millan MA, Jacobowitz DM, Aguilera G, et al: Differential distribution of AT1 and AT2 angiotensin II receptor subtypes in the rat brain during development. *Proc Natl Acad Sci USA* 88(24):11440–11444, 1991.

79. Cook VI, Grove KL, McMenamin KM, et al: Localization of the AT2 angiotensin receptor subtype in the brain of the developing rat fetus. *Soc Neurosci Abstr* 17(1-2):809, 1991.

80. Cook VI, Grove KL, McMenamin KM, et al: The AT2 angiotensin receptor subtype predominates in the 18 day gestational fetal rat brain. *Brain Res* 560(1-2):334–336, 1991.

81. Deschepper CF, Mellon SH, Cumin F, et al: Analysis by immunocytochemistry and in situ hybridization of renin and its mRNA in the kidney, testis adrenal and pituitary of the rat. *Proc Natl Acad Sci USA* 83:7552–7556, 1986.

82. Naruse M, Naruse K, Kurimoto F, et al: Evidence for the existence of DES-ASP-angiotensin II in human uterine and adrenal tissues. *J Clin Endocrinol Metab* 61:480–483, 1985.

83. Jackson B, Cubela RB, Sakaguchi K, et al: Characterization of angiotensin converting enzyme (ACE) in the testis and assessment of the in vivo effects of the ACE inhibitor perindopril. *Endocrinology* 123:50–55, 1988.

84. Culler MD, Tarlatzis BC, Lightman A, et al: Angiotensin II immunoreactivity in human ovarian follicular fluid. *J Clin Endocrinol Metab* 62:613–615, 1986.

85. Schultze D, Brunswig-Spickenheier B, Mukhopadhyay AK: Renin and prorenin-like activities in bovine ovarian follicles. *Endocrinology* 124:1389–1398, 1989.

86. Pucell AG, Bumpus FM, Husain A: Regulation of angiotensin II receptors in cultured rat ovarian cells by follicle stimulating hormone and angiotensin II. *J Biol Chem* 263:11954–11961, 1988.

87. Husain A, Bumpus FM, DeSilva P, et al: Localization of angiotensin II receptors in ovarian follicles and the identification of angiotensin II in rat ovaries. *Proc Natl Acad Sci USA* 84:2489–2493, 1987.

88. Daud AI, Bumpus FM, Husain A: Evidence for a selective expression of angiotensin II receptors on atretic follicles in the rat ovary: An autoradiographic study. *Endocrinology* 122:2727–2734, 1988.

89. Speth RC, Husain A: Distribution of angiotensin-converting enzyme and angiotensin II-receptor binding sites in the rat ovary. *Biol Reprod* 38:695–702, 1988.

90. Brunswig-Spickenheier B, Mukhopadhyay AK: Characterization of angiotensin II receptor subtypes on bovine theca cells and its regulation by luteinizing hormone. *Endocrinology* 131(3):1445–1452, 1992.

91. Aguilera G, Millan MA, Harwood JP: Angiotensin II receptor in the gonads. *Am J Hypertension* 2:395–402, 1989.

92. Cox BE, Ipson MA, Kamm KE, et al: Angiotensin II myometrial smooth muscle receptors change subtype during ovine pregnancy. *FASEB J* 5:A869, 1992.

93. Panek R, Kaplan C, Overhiser R, et al: Expression of angiotensin II (Ang II) AT1 and AT2 receptor subtypes in the neointima of rat carotid arteries following balloon injury. *Circulation* 86(4):I-168, 1992.

94. Tsutsumi K, Saavedra JM: Characterization of AT2 angiotensin II receptors in rat anterior cerebral arteries. *Am J Physiol* 261(3, part 2):H667–H670, 1991.

95. Wong PC, Christ DD, Timmermans PBMWM: Nonpeptide angiotensin II (AII) receptor antagonists: Enhancement of losartan-induced AII antagonism by PD 123177 in the pithed rat. *J Hypertens* 10(4):S75, 1992.

96. Rogg H, Schmid A, de Gasparo M: Identification and characterization on angiotensin II receptor subtypes in rabbit ventricular myocardium. *Biochem Biophys Res Commun* 173(1):416–422, 1990.

97. Scott AL, Chang RS, Lotti VJ, et al: Cardiac angiotensin receptors: Effects of selective angiotensin II receptor antagonists, DuP 753 and PD 121981, in rabbit heart. *J Pharmacol Exp Ther* 261(3):931–935, 1992.

98. Gibson RE, Thorpe HH, Cartwright ME, et al: Angiotensin II receptor subtypes in renal cortex of rats and monkeys. *Am J Physiol* 261:F512–F518, 1991.

99. Edwards RM, Stack EJ, Wiedley EF, et al: Characterization of renal angiotensin II receptors using subtype selective antagonists. *J Pharmacol Exp Ther* 260(3):933–938, 1992.

100. Grone HJ, Simon M, Fuchs E: Autoradiographic characterization of angiotensin subtypes in fetal and adult human kidney. *Am J Physiol* 262:F326–F331, 1992.

101. Chansel D, Czekalski S, Pham P, et al: Characterization of angiotensin II receptor subtypes in human glomeruli and mesangial cells. *Am J Physiol* 262:F432–F441, 1992.

102. Simon M, Flugge G, Fuchs E, et al: Evidence for two angiotensin II receptor subtypes in human fetal and adult kidney demonstrated by the AII receptor antagonists DuP 753 and PD 123177. *FASEB J* 5(4):A870, 1991.

103. Cogan MG, Liu FY, Wong PC, et al: Comparison of inhibitory potency by nonpeptide angiotensin II receptor antagonists PD 123177 and DuP 753 on proximal nephron and renal transport. *J Pharmacol Exp Ther* 259(2):687–691, 1991.

8

Brain Angiotensin II Receptor Subtypes

Juan M. Saavedra

1. BRAIN ANGIOTENSIN II RECEPTORS

After injection in the circulation, angiotensin II (ANG II) produces effects in the brain.[1] However, circulating ANG II does not cross the blood–brain barrier,[2,3] suggesting that the brain contains ANG II receptors in the circumventricular organs, which are areas outside the blood–brain barrier. This hypothesis was confirmed with the use of autoradiographic and fluorescence methods, and ANG II binding sites were identified in the pituitary gland, area postrema, organon vasculosum of the lamina terminalis, subfornical organ, and median eminence.[4]

The brain ANG II receptors were later characterized *in vitro* using conventional binding methods with membrane preparations and $[^{125}I]$-ANG II.[5,6] These studies revealed that the brain ANG II receptors were not restricted to areas containing the circumventricular organs, but were present throughout the brain,[7–10] including brain structures inside the blood–brain barrier. The presence of brain ANG II receptors inside the blood–brain barrier was only conclusively demonstrated by autoradiography after *in vitro* incubation of brain slices with radiolabeled ANG II or ANG II derivatives.[11] The ANG II receptors inside the blood–brain barrier were connected to ANG II endogenously produced in the brain, and therefore to the "central" ANG II system.[12]

ANG II, when administered in minute amounts into selective brain structures, resulted in increased blood pressure, drinking behavior, salt appetite, and vasopressin release.[12] The brain structures that responded to ANG II injections

Juan M. Saavedra • Section on Pharmacology, Laboratory of Clinical Science, National Institute of Mental Health, National Institutes of Health, Bethesda, Maryland 20814.

Angiotensin Receptors, edited by Juan M. Saavedra and Pieter B.M.W.M. Timmermans. Plenum Press, New York, 1994.

were those in which autoradiography demonstrated dense ANG II binding. Thus, the concept emerged of specific, selective, anatomically discrete brain areas containing physiologically active ANG II receptors, some accessible to circulating ANG II and some connected to the brain ANG II system. Both groups of receptor sites are anatomically connected by the forebrain ANG II pathway.[13] This pathway extends in continuity from the subfornical organ to the median preoptic nucleus, lamina terminalis, and organon vasculosum of the lamina terminalis, and contains both ANG II-immunopositive nerve terminals and ANG II receptors (Fig. 1).[12,13] The ANG II receptor band extends laterally to the paraventricular nucleus, where ANG II receptors are concentrated in its parvocellular zone.[14-16] The ANG II forebrain pathway may be considered to be one of the most important neuroanatomical connections between the brain ANG II system and its peripheral counterpart.[12]

In the brain, the distribution of ANG II receptors in mammalian species, including human, shows a distribution in general agreement with that of the

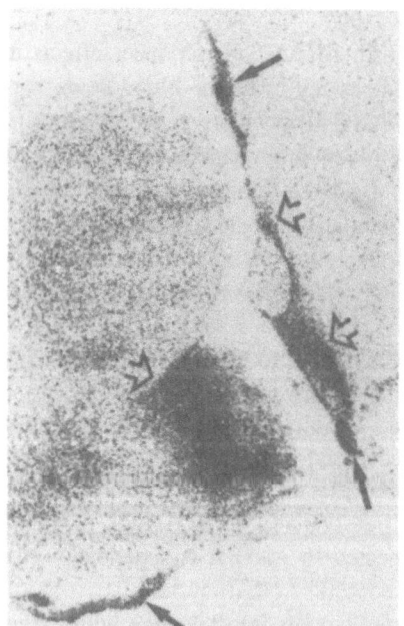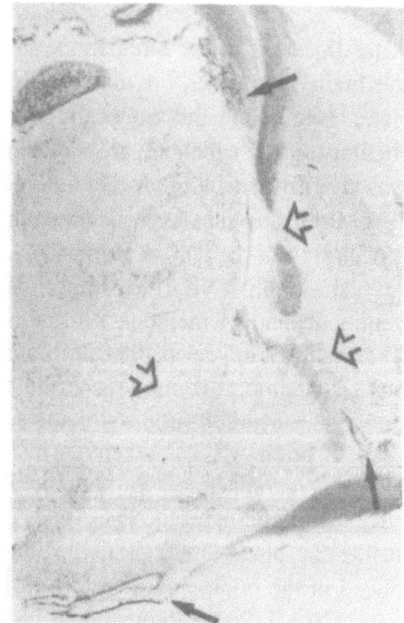

Figure 1. The forebrain ANG II receptor band. The figure represents parasagittal sections of the rat forebrain. (A) Autoradiograph after incubation with [^{125}I]-Sar[1]-ANG II. (B) Consecutive section stained with toluidine blue. Full arrows point to the subfornical organ (upper right), the organon vasculosum of the lamina terminalis (lower right), and the median eminence (lower left). All these areas are outside the blood–brain barrier. Open arrows point to the median preoptic nucleus (upper right), the lamina terminalis (lower right), and the paraventricular nucleus (lower left). (From Saavedra *et al.*[82] Reprinted with permission.)

rat.[17-22] The cellular localization of brain ANG II receptors was studied with a combination of autoradiographic methods and relatively specific lesions. In the intact brain, ANG II receptors seem to be associated with neuronal cell bodies. ANG II receptors are present in neurons of the inferior olivary complex[23] and in presynaptic vagal afferent terminals. These receptors originate in the nodose ganglion, are transported bidirectionally by the vagus nerve, and are localized in the nucleus of the solitary tract.[24,25] ANG II receptors are also present in catecholamine neurons in the locus coeruleus.[26]

The quantitative autoradiographic methods are the techniques of choice to study localization, characterization, and regulation of ANG II receptors in the brain.[27-29] Quantification is possible by comparison of optical densities with [^{125}I]-standard curves.[30] The ligand of choice is [^{125}I]-Sar^1-ANG II,[22,27-29] an ANG II agonist with lower degradation and higher affinity than ANG II itself. With these methods, it has been recently possible to study and characterize different types of brain ANG II receptor subtypes.

2. BRAIN ANG II RECEPTOR SUBTYPES

2.1. Discovery and Nomenclature

The existence of different ANG II receptor subtypes was first proposed in peripheral tissues,[31] where differences were noted in agonist and antagonist binding affinity and potency, and in sensitivity to disulfide-reducing agents.[32,33] In the brain, different groups of ANG II receptors were differentially regulated, and this also suggested receptor heterogeneity. For example, receptors in the paraventricular nucleus were upregulated during repeated stress but not by water deprivation,[14,15] whereas receptors in the subfornical organ were upregulated both during stress and water deprivation.[12]

Conclusive proof of the existence of different ANG II receptor subtypes in both peripheral organs and the brain, however, was obtained only after the development of more selective ligands.[34-39] A nomenclature for ANG II receptor subtypes has recently been proposed, based on the selective antagonism by nonpeptidic and peptidic ligands.[40] Two receptor subtypes were originally described, AT$_1$ and AT$_2$. Binding of ANG II to AT$_1$ receptors is selectively antagonized by the nonpeptidic ligand losartan potassium (DuP 753); binding to AT$_2$ receptors is selectively displaced by the nonpeptidic ligands PD 123177 and PD 123319, and by the peptide CGP 42112A.[35,39,40]

2.2. Localization and Development

Following the development of the selective receptor competitors, most of the initial studies were performed in peripheral tissues known as target sites for circulating ANG II. AT$_1$ receptors were present in high concentrations in vasculature, kidney, adrenal cortex, liver, and lung parenchyma.[12,41–45] AT_2 receptors in the periphery have been located so far in the adrenal cortex and medulla, uterus, ovaries, aorta, heart muscle, and in embryonic connective tissues.[35,46–50]

Soon after the demonstration of AT_1 and AT_2 receptors in the periphery, the presence of both these subtypes was reported in the brain.[51–58] As in peripheral tissues, binding to brain AT_1 receptors was blocked by losartan and binding to brain AT_2 receptors was selectively displaced by CGP 42112A and PD 123177.[51–59] Because of large differences in affinity, the proportion of AT_1 and AT_2 receptors in single brain nuclei and discrete areas could be determined using single concentrations of selective AT_1 and AT_2 competitors and quantitative autoradiography.[56]

We have performed a detailed study of the receptor subtype concentration in many areas of the brain of the young adult (8-weeks-old) Sprague-Dawley rat (Table I). Most brain areas contained a single type of ANG II receptor (Table I). In many areas (the suprachiasmatic, solitary tract, lateral olfactory tract, median preoptic, paraventricular, and basolateral amygdaloid nuclei; the choroid plexus, dentate gyrus, piriform, entorhinal, and retrosplenial granular cortex; the subiculum and parasubiculum; the area postrema and the subfornical organ) (groups A–C, Table I), only AT_1 receptors were present. Other areas, such as the superior colliculus, cingulate cortex, and cerebellar cortex, contained both AT_1 and AT_2 receptors (groups D and E, Table I). The localization

Table I. Classification of Brain ANG II Receptor Subtypes by Displacement of [^{125}]Sar^1-ANG II with Specific Antagonists[a]

	2-Week-old, percent of control		8-Week-old, percent of control	
	Losartan (10^{-5} M)	CGP 42112A (10^{-7} M)	Losartan (10^{-5} M)	CGP 42112A (10^{-7} M)
		AT_1		
Group A				
Suprachiasmatic nucleus	0	93 ± 5	0	97 ± 6
Choroid plexus	0	75 ± 25	0	100 ± 17
Nucleus of the solitary tract	0	98 ± 2	0	92 ± 15
Dentate gyrus	ND		0	80 ± 7
Group B				
Nucleus of the lateral olfactory tract	0	100 ± 15	0	117 ± 25
Piriform cortex	0	112 ± 8	0	100 ± 5
Median preoptic nucleus	0	103 ± 6	0	93 ± 12
Paraventricular nucleus	0	101 ± 5	0	100 ± 15
Subiculum	0	94 ± 18	0	115 ± 15
Area postrema	0	100 ± 13	0	112 ± 18
Parasubiculum	0	125 ± 25	ND	
Entorhinal cortex	0	150 ± 17	ND	

(continued)

Table I. (Continued)

	2-Week-old, percent of control		8-Week-old, percent of control	
	Losartan (10^{-5} M)	CGP 42112A (10^{-7} M)	Losartan (10^{-5} M)	CGP 42112A (10^{-7} M)
Group C				
Subfornical organ	0	107 ± 12	0	111 ± 13
Basolateral amygdaloid nucleus	0	100 ± 18	0	114 ± 14
Retrosplenial granular cortex	0	100 ± 11	ND	
		AT_1 and AT_2		
Group D				
Superior colliculus	70 ± 15	35 ± 10^b	$70 \pm 10^{b,c}$	30 ± 10
Cingulate cortex				
Superficial	$18 \pm 9^{b,c}$	73 ± 9	ND	
Deep	100 ± 15	$15 \pm 8^{b,c}$	ND	
Group E				
Cerebellar cortex (molecular layer)				
Superficial	31 ± 8^b	69 ± 8^b	ND	
Deep	53 ± 6^b	29 ± 6^b	ND	
		AT_2		
Group F				
Lateral septal nucleus	104 ± 11	0	100 ± 25	0
Ventral thalamic nucleus	108 ± 16	0	ND	
Mediodorsal thalamic nucleus	105 ± 10	0	113 ± 13	0
Locus coeruleus	109 ± 11	0	100 ± 6	0
Principal sensory trigeminal nucleus	108 ± 17	0	ND	
Parasolitary nucleus	100 ± 3	0	82 ± 9	0
Inferior olive	105 ± 7	0	104 ± 20	0
Medial amygdaloid nucleus	103 ± 10	0	100 ± 29	0
Medial geniculate nucleus	93 ± 10	0	87 ± 17	0
Group G				
Anterior pretectal nucleus	85 ± 8	0	ND	
Nucleus of the optic tract	87 ± 7	0	ND	
Ventral tegmental nucleus	100 ± 8	0	ND	
Posterodorsal tegmental nucleus	111 ± 11	0	ND	
Hypoglossal nucleus	109 ± 9	0	ND	
Central medial and paracentral thalamic nuclei	107 ± 11	0	ND	
Laterodorsal thalamic nucleus	104 ± 22	0	ND	
Oculomotor nucleus	92 ± 8	0	ND	

[a]Values are mean $\pm$ SE in fmole/mg protein; $n = 4$–5 individually measured animals/group. Concentration of [^{125}I]-Sar^1-ANG II was 0.5×10^{-9} M. 0: Binding was totally displaced at the concentration of displacer used (binding was not statistically significant when compared to nonspecific binding). ND: Not detected in control sections (nonstatistically significant over nonspecific binding).
[b]$p < 0.05$, vs. control (incubated in the absence of ANG II receptor displacer).
[c]$p < 0.05$, losartan vs. CGP 42112A.
From ref. 56.

of each receptor subtype in these structures, however, appeared different in the different cell layers, but because of limitations in the resolution of our method, the different subtypes could not be quantified separately. In still other areas (lateral septal, ventral thalamic, mediodorsal thalamic, principal sensory trigeminal, parasolitary, medial amygdaloid, and medial geniculate nuclei; locus coeruleus; inferior olive), only AT_2 receptors were present (group F, Table I).

Once the localization of the brain ANG II receptor subtypes was established, we analyzed the developmental pattern of receptor subtype expression.

Primary cultures from fetal rat brain express ANG II receptors,[60] and this indicated the possibility of the presence of these receptors early during normal brain development. However, early autoradiographic studies failed to demonstrate ANG II receptors in fetal brain.[61] More detailed autoradiographic analysis, however, clarified that both AT_1 and AT_2 receptors were indeed present in fetal brain[50,62] (Fig. 2). In the 18-day-old fetus, brain ANG II receptors were

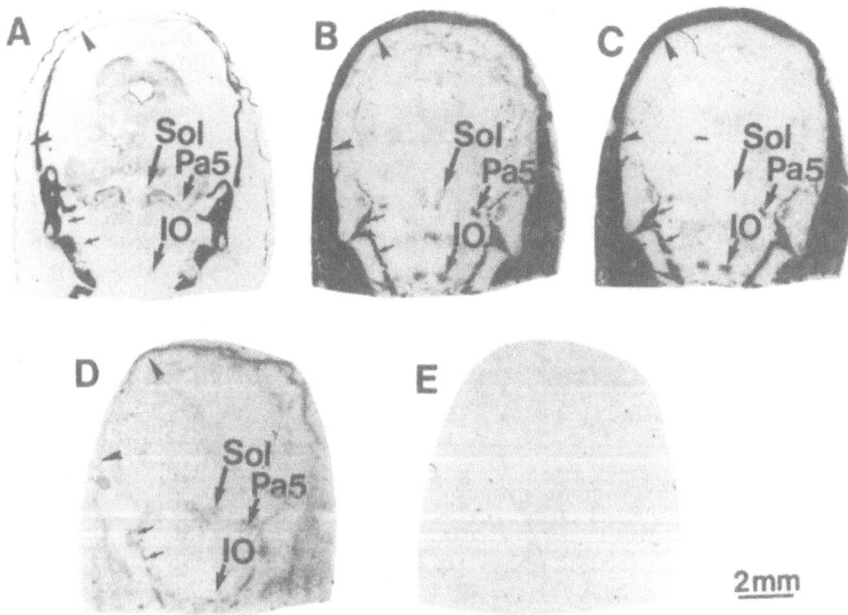

Figure 2. Autoradiography of ANG II receptors in rat fetal brain. (A) Coronal section of an 18-day-old rat fetus, stained with toluidine blue. B–E are adjacent sections to A incubated in the presence of 5×10^{-10} M [^{125}I]-Sar^1-ANG II. (B) Total binding. (C) Incubated in the presence of 10^{-5} M losartan. (D) Incubated in the presence of 10^{-7} M CGP 42112A. (E) Incubated in the presence of 5×10^{-6} M unlabeled ANG II (nonspecific binding). Sol, nucleus of the solitary tract; Pa5, paratrigeminal nucleus; IO, inferior olive. Arrowheads point to soft tissue surrounding the skull. Small arrows point to meninges. (From Tsutsumi *et al.*[62] Reprinted with permission.)

concentrated in a few areas: the solitary tract, paratrigeminal, and hypoglossal nuclei; the inferior olive; the choroid plexus; and the meninges (Fig. 2; Table II). As determined by selective displacement, the nucleus of the solitary tract and the choroid plexus expressed AT_1 receptors only, and the rest of the areas expressed only AT_2 (Fig. 2; Table II).

Earlier studies established that in the brain, the ANG II receptor number reached a maximum of ten times the adult level during the first 2 weeks after birth, and that the timing of development was different in different brain regions.[63] These studies, however, did not address the issue of development of receptor subtypes.

We have performed a comprehensive comparison of ANG II receptor subtype expression in the immature rat brain, at 2 weeks of postnatal life.[56,57] We have found that the receptor subtypes had different developmental patterns. In the case of AT_1 receptors, their numbers were similar, higher or lower between immature and mature animals, but the differences were never very large. AT_2 receptors, however, were remarkably expressed in immature animals, and in some areas absent in adults (Tables I and III; Fig. 3 and 4). For example, higher AT_1 receptor concentrations in adult rats occurred in the suprachiasmatic and solitary tract nuclei, choroid plexus, and dentate gyrus (group A in Tables I and III). No differences in AT_1 receptor expression were found in many other areas (group B in Tables I and III), and the AT_1 receptor number was reduced in mature when compared with immature rats in still other areas (groups C and D in Tables I and III).

The situation with regards to AT_2 receptors was very different. As a rule, the AT_2 receptor expression was lower in mature animals. In areas from group

Table II. ANG II Receptor Subtypes in Rat Fetal Brain and Soft Tissues[a]

Brain area	Specific binding fmole/mg protein	Losartan		CGP 42112A	
		fmole/mg protein	Percent	fmole/mg protein	Percent
Inferior olive	50 ± 5	51 ± 9	102 ± 2	ND	0
Paratrigeminal nucleus	40 ± 1	40 ± 1	100 ± 3	ND	0
Hypoglossal nucleus	151 ± 9	147 ± 7	97 ± 7	ND	0
Nucleus of the solitary tract	11 ± 0.8	ND	0	10 ± 1	91 ± 9
Choroid plexus	10 ± 0.8	ND	0	10 ± 0.8	100 ± 8
Meninges	143 ± 7	140 ± 9	98 ± 8	7 ± 0.6	5 ± 0.5
Cephalic soft tissues (striated muscle and skin)	218 ± 6	212 ± 20	97 ± 6	8 ± 0.8	4 ± 0.3

[a]Consecutive coronal sections were incubated in the presence of 5×10^{-10} M [^{125}I]-Sar^1-ANG II. Specific binding represents total binding after subtraction of nonspecific binding. Losartan: specific binding in the presence of 10^{-5} M losartan. CGP 42112A: specific binding in the presence of 10^{-7} M CGP 42112A. Results are expressed in apparent units and percent of specific binding, and are means ± SEM of groups of four fetuses studied individually.
From ref. 50.

Table III. Angiotensin II Receptor Concentrations
in Specific Brain Areas of 2-Week-old and 8-Week-old Rats[a]

	2-Week-old	8-Week-old
	AT_1	
Group A		
Suprachiasmatic nucleus	128 ± 13	220 ± 14[a]
Choroid plexus	8 ± 2	14 ± 2[a]
Nucleus of the solitary tract	266 ± 12	341 ± 18[a]
Dentate gyrus	18 ± 3	97 ± 7[a]
Group B		
Nucleus of the lateral olfactory tract	56 ± 5	60 ± 3
Piriform cortex	111 ± 5	94 ± 6
Median preoptic nucleus	206 ± 19	183 ± 36
Paraventricular nucleus	205 ± 7	172 ± 14
Subiculum	63 ± 11	66 ± 4
Area postrema	185 ± 13	223 ± 16
Parasubiculum	28 ± 4	22 ± 6
Entorhinal cortex	28 ± 4	20 ± 8
Group C		
Subfornical organ	507 ± 37	348 ± 45
Basolateral amygdaloid nucleus	62 ± 6	30 ± 7
Retrosplenial granular cortex	44 ± 4	ND
	AT_1 and AT_2	
Group D		
Superior colliculus	145 ± 5	65 ± 8
Cingulate cortex	19 ± 4	8 ± 4
Group E		
Cerebellar cortex	59 ± 6	ND
Group F		
Lateral septal nucleus	58 ± 8	18 ± 3
Ventral thalamic nuclei	101 ± 8	24 ± 3
Mediodorsal thalamic nucleus	165 ± 11	38 ± 13
Locus coeruleus	289 ± 19	98 ± 13
Principal sensory trigeminal nucleus	75 ± 6	15 ± 4
Parasolitary nucleus	220 ± 15	57 ± 14
Inferior olive	1328 ± 61	181 ± 32
Medial amygdaloid nucleus	159 ± 8	94 ± 9
Medial geniculate nucleus	338 ± 24	71 ± 6
Group G		
Anterior pretectal nucleus	53 ± 8	ND
Nucleus of the optic tract	101 ± 13	ND
Ventral tegmental area	101 ± 11	ND

(continued)

Table III. (Continued)

	2-Week-old	8-Week-old
Posterodorsal tegmental nucleus	110 ± 21	ND
Hypoglossal nucleus	141 ± 11	ND
Central medial and paracentral thalamic nuclei	202 ± 14	ND
Laterodorsal thalamic nucleus	110 ± 10	ND
Oculomotor nucleus	98 ± 13	ND

[a]Values are means ± SE in fmole/mg protein; $n = 6$ individually measured animals/group. The concentration of [^{125}I]-Sar^1-ANG II used was 3×10^{-9} M. ND: below the limit of sensitivity of the method (results not significantly different from nonspecific binding).
[b]$p < 0.05$, 8-week-old vs. 2-week-old.
From ref. 56.

F (Tables I and III), AT$_2$ binding was high in immature rat brain and much lower in adult rats. In areas from groups E and G (Tables I and III), binding was present in the immature brain but was no longer expressed in mature rats.

The selective distribution of AT$_1$ and AT$_2$ receptors indicate specific functions for each receptor subtype in the brain. Target sites for regulation of blood pressure, drinking, salt appetite, vasopressin formation and release, and anterior pituitary control by ANG II express only AT$_1$ receptors. The forebrain ANG II pathway, controlling the anterior and posterior pituitary function, autonomic responses, and the central response to peripheral ANG II, also expresses AT$_1$ only[64] (Fig. 5). Since in at least some of these areas, AT$_1$ receptors are expressed in the fetal, immature and adult brain, AT$_1$ receptors could have a role in cardiovascular function and fluid metabolism not only in adult animals but also during development. In addition, the presence of AT$_1$ receptors in some cortical, limbic system, and associated areas raise the possibility of behavioral roles for brain ANG II, as well as a functional role in the choroid plexus.[56]

The function of brain (and peripheral) AT$_2$ receptor is not presently understood. The AT$_2$ receptors are expressed in areas related to the control and learning of motor activity, sensory areas including the visual pathway, and in some limbic system areas. AT$_2$ receptors are already present during embryonic life, and in many areas their number is much higher in the immature brain. Therefore, a role of AT$_2$ receptors in the development of selected brain structures and the sensory–motor integration can be postulated.

2.3. Further Heterogeneity of Brain ANG II Receptors

There are indications that the mammalian brain does not only contain the two ANG II receptor subtypes, AT$_1$ and AT$_2$, but that these subtypes are in turn heterogeneous.

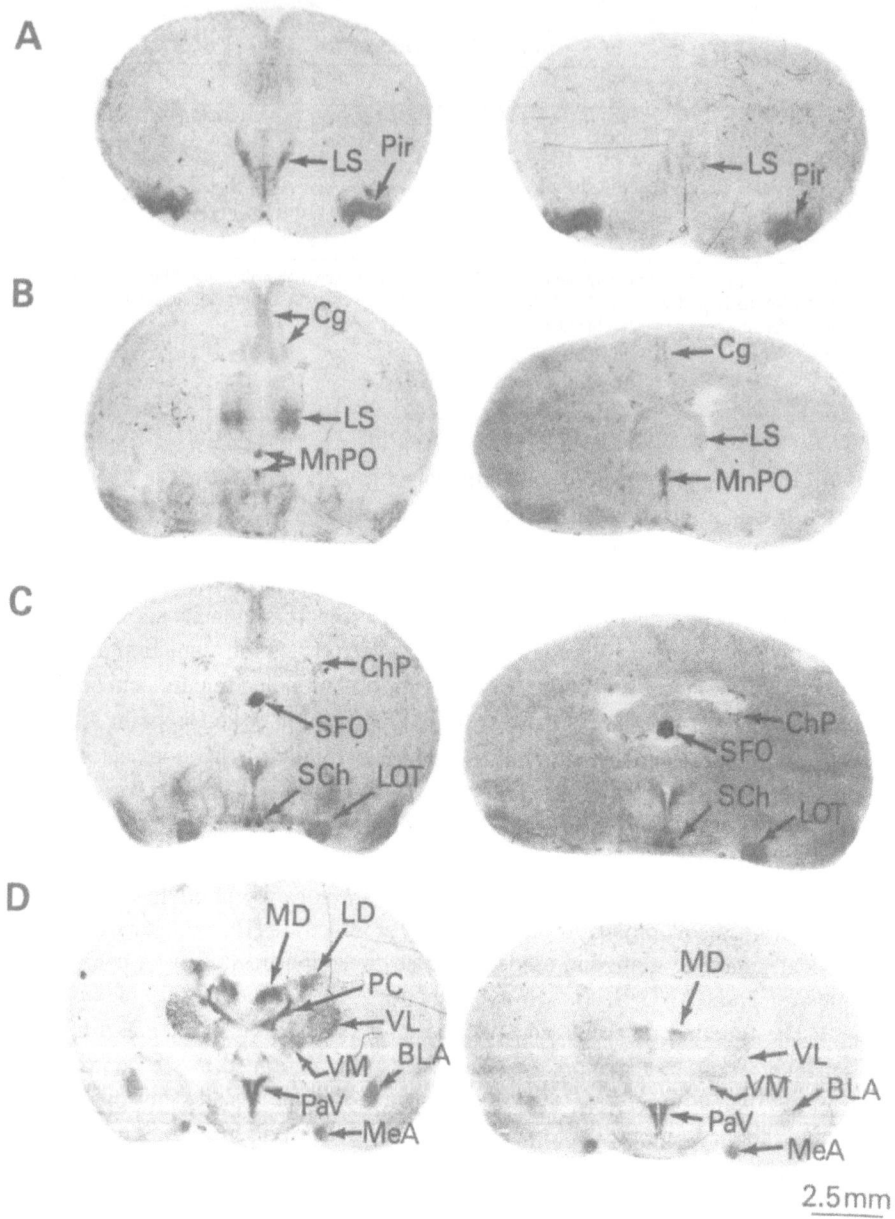

2.5mm

Figure 3. Autoradiography of ANG II receptors in brain of young (2-week-old) and adult (8-week-old) rats. Brain sections from 2- (A–D) and 8-week-old (E–I) rats were incubated in presence of 3×10^{-9} M [^{125}I]-Sar^1-ANG II. The bregma in A–I as follows: A: 1.6 mm; B: –0.26 mm; C: –1.30 mm; D: 1.80 mm; E: –5.30 mm; F: –7.04 mm; G: 10.04 mm; H: –10.04 mm, nonspecific binding of G, incubated as above in the presence 5×10^{-6} M unlabeled ANG II; I: 13.68 mm. Abbreviations: AP, area postrema; APT, anterior pretectal nucleus; BLA, basolateral amygdaloid nucleus, anterior part; Cg, cingulate cortex; Cx, cerebellar cortex (molecular layer); ChP, choroid plexus; DG, dentate gyrus; Ent, entorhinal cortex; IO, inferior olive; LC, locus coeruleus; LD, laterodorsal thalamic nuclei; LOT, nucleus of the lateral olfactory tract; LS, lateral septal nucleus; MD, mediodorsal

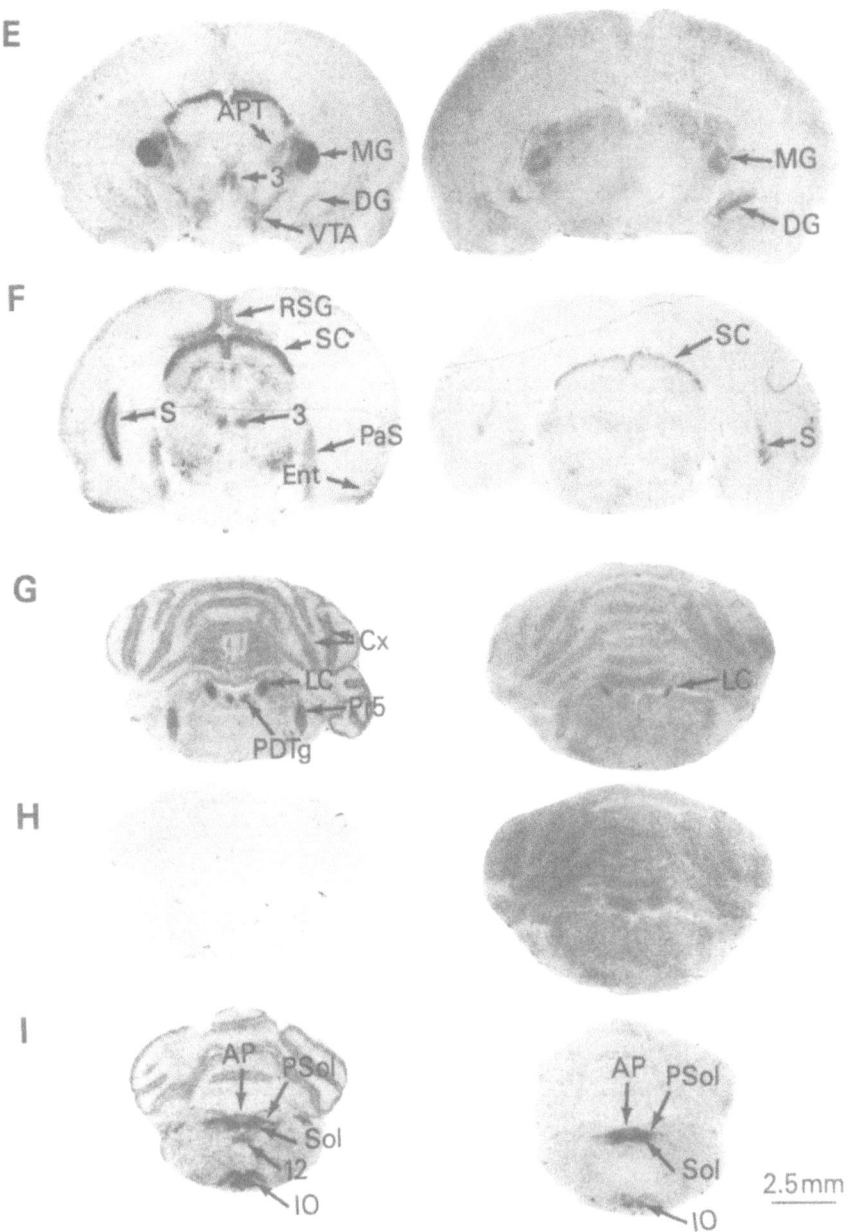

thalamic nucleus; MeA, medial amygdaloid nucleus, anterior part; MG, medial geniculate nucleus; MnPO, median preoptic nucleus; PaS, parasubiculum; PaV, hypothalamic paraventricular nucleus; PC, paracentral thalamic nucleus; PDTg, posterodorsal tegmental nucleus; Pir, piriform cortex; Pr5, principal sensory trigeminal nucleus; PSol, parasolitary nucleus; RSG, retrosplenial granular cortex; S, subiculum; SC, superior colliculus; SCh, suprachiasmatic nucleus; SFO, subfornical organ; Sol, nucleus of the solitary tract; VL, ventrolateral thalamic nucleus; VM, ventromedial thalamic nucleus; VTA, ventral tegmental area; 3, oculomotor nucleus; 12, hypoglossal nucleus. (From Tsutsumi *et al.*[56] Reprinted with permission.)

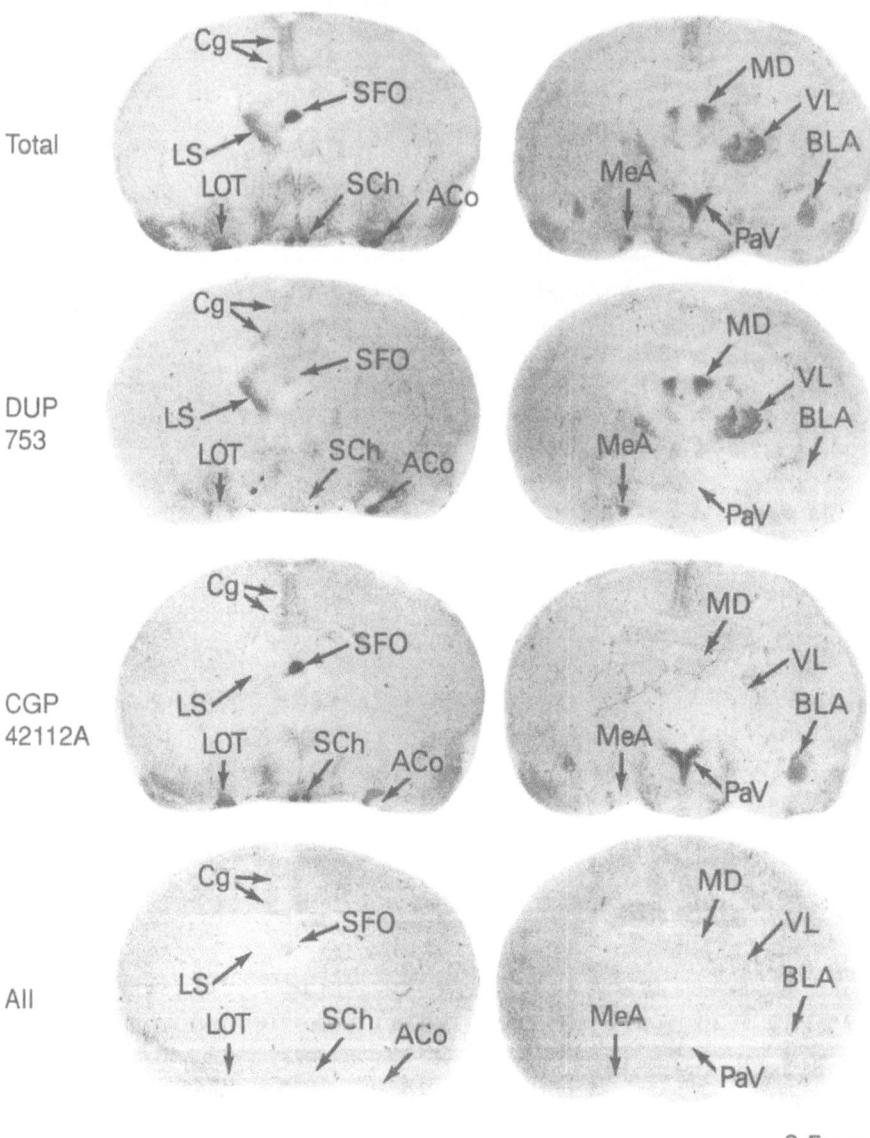

Figure 4. Selective displacement of ANG II receptors by losartan and CGP 42112A. Consecutive brain sections from 2-week-old male rats were incubated in presence of 0.5×10^{-9} M [^{125}I]-Sar^1-ANG II without (total) or with (displacement) single concentrations of losartan (DUP 753; 10^{-5} M), CGP 42112A (10^{-7} M), or unlabeled ANG II (5×10^{-6} M; nonspecific binding). OT, Nucleus of optic tract; see Fig. 3 for other abbreviations. (From Tsutsumi *et al.*[56] Reprinted with permission.)

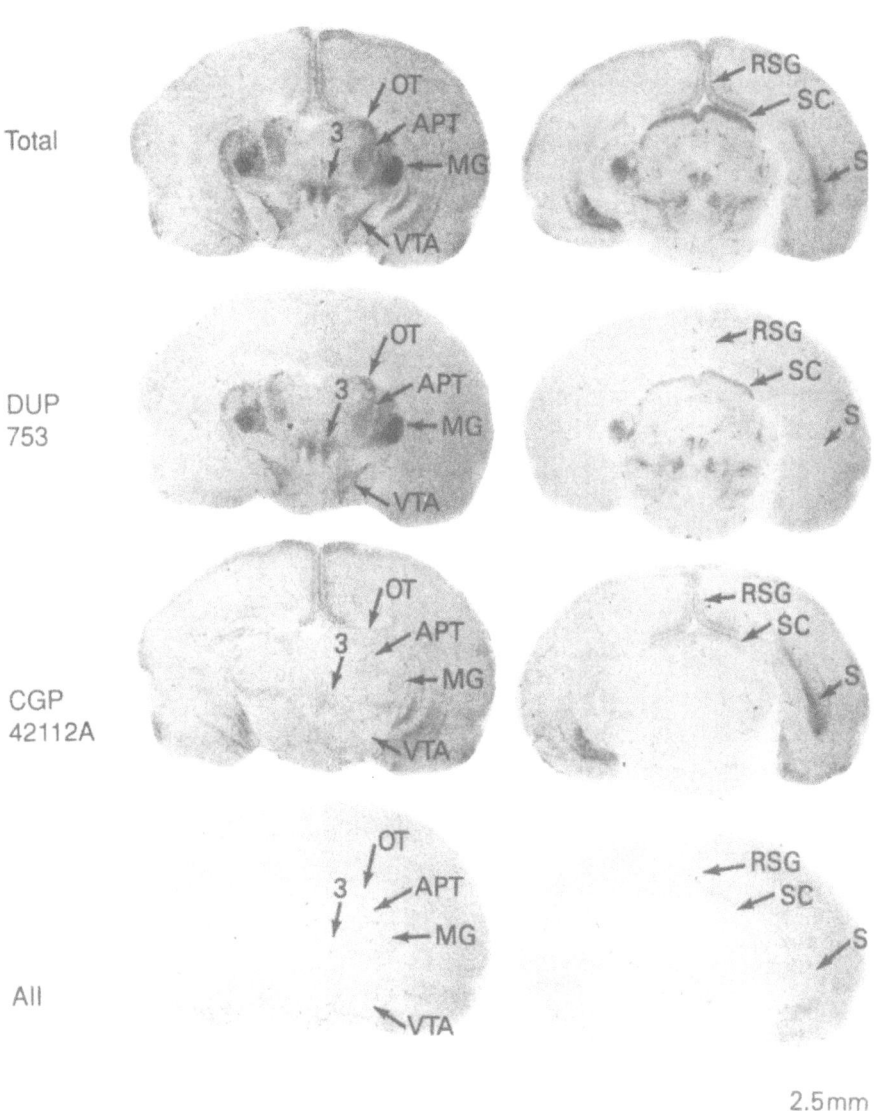

Figure 4. (Continued)

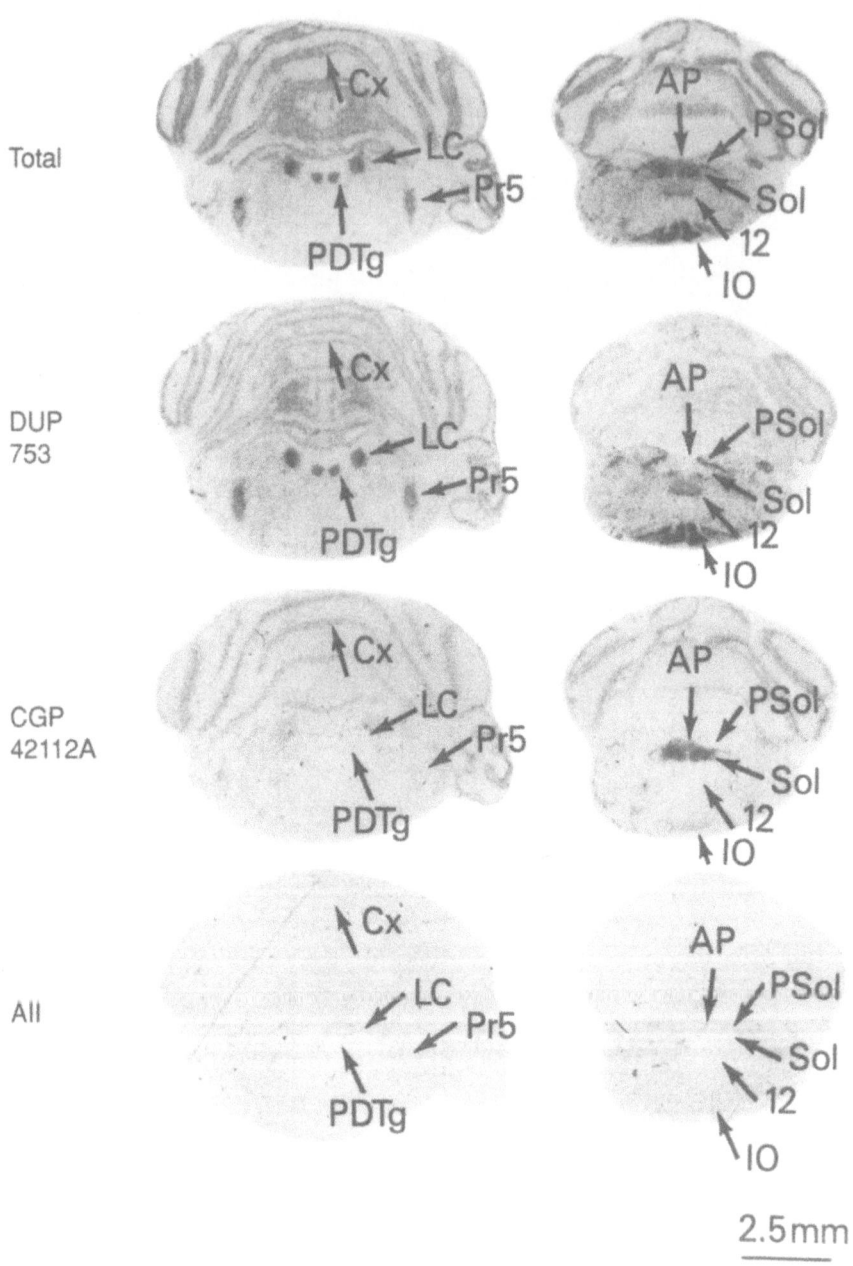

Total

DUP
753

CGP
42112A

All

2.5mm

Figure 4. (Continued)

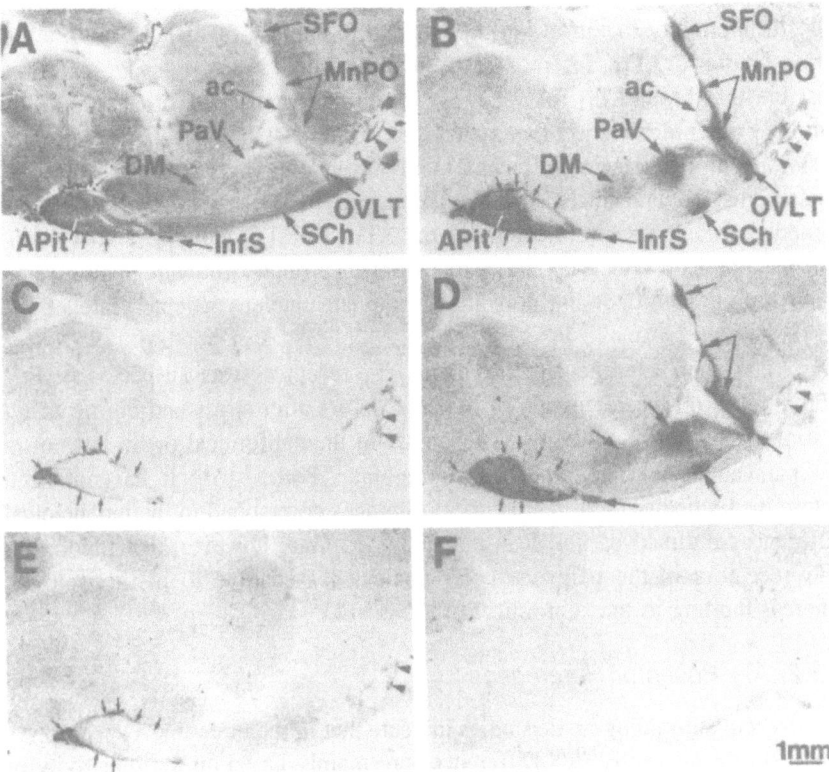

Figure 5. Autoradiography of ANG II binding in parasagittal sections of basal rat forebrain. (A) Stained with toluidine blue. (B) Total binding; section consecutive to A, incubated in the presence of 5×10^{-10} M $[^{125}\text{I}]$-Sar^1-ANG. (C) Consecutive section, incubated as in B, in the presence of 10^{-5} M losartan. (D) Consecutive section, incubated as in B, in the presence of 10^{-7} M CGP 42122A. (E) Consecutive section, incubated as in B, in the presence of 5×10^{-3} M dithiothreitol. (F) Consecutive section, incubated as in B, in the presence of 5×10^{-6} M unlabeled ANG. The arrowhead points to the anterior cerebral artery. Small arrows point to dura mater surrounding the pituitary gland. APit, anterior pituitary; DM, dorsomedial hypothalamic nucleus; PaV, paraventricular nucleus; ac, anterior commissure; SFO, subfornical organ; MnPO, median preoptic nucleus; OVLT, vascular organ of the lamina terminalis; SCh, suprachiasmatic nucleus; InfS, infundibulum stem. (From Tsutsumi *et al.*[64] Reprinted with permission.)

2.3.1. AT₁ Receptor Heterogeneity

In the rat brain, two very similar AT$_1$ receptor cDNAs have been recently described, and these are probably encoded by separate genes.[65] The AT$_{1A}$ subtype, initially found to be expressed in vascular smooth muscle, was found to be predominant in the rat hypothalamus. The AT$_{1B}$ subtype, expressed in the adrenal and pituitary, predominated in the circumventricular organs and in the rat cerebellum.[65] In peripheral tissues, the AT_{1A} and AT_{1B} receptors are not only differentially localized but probably differentially regulated.[66] In the anterior pituitary, AT_{1B} receptors are downregulated by estrogen,[67] and it is possible that brain AT_{1B} receptors in the circumventricular organs could be similarly downregulated.[65] Estrogen, when administered sequentially with progesterone, upregulates AT_1 receptors in a selective hypothalamic area, the dorsomedial arcuate nucleus of the female rat.[68] It is not known if the dorsomedial arcuate receptors belong to the AT_{1A} or AT_{1B} subtype. Either estrogen has opposite effects on AT_{1B} receptors in pituitary (downregulation) and brain (upregulation) or the dorsomedial arcuate nucleus receptors are of the AT_{1A} subtype.

Additional heterogeneity for brain AT_1 receptors was suspected earlier, from studies on the regulation of brain receptors after stress and during dehydration. Dehydration upregulates receptors in the subfornical organ but not in the paraventricular nucleus; stress upregulates both.[12,14,15] It has not been determined whether these regulatory differences correspond to the biochemical differences outlined earlier. It is interesting to note, however, that binding to AT_1 receptors in the paraventricular nucleus is sensitive to pertussis toxin, whereas binding in the subfornical organ is not.[69]

2.3.2. AT_2 Receptor Heterogeneity

Recent autoradiographic studies indicate that in the rat brain, AT_2 receptors are also heterogeneous.[69,70] Differences are mainly based on the sensitivity of the binding to guanine nucleotides and to incubation in the presence of dithiothreitol, a reducing agent.[69,70] In the rat brain, AT_{2A} receptors are those sensitive to guanine nucleotides, pertussis toxin, and dithiothreitol; AT_{2B} receptors are not. Examples of AT_{2A} receptors are those located in the ventral thalamic nuclei, medial geniculate nuclei, and in the locus coeruleus; AT_{2B} receptors are those in the inferior olive[69,70] (Fig. 6 and Table IV).

The structure of brain AT_2 receptors has not been yet determined. It is generally accepted, however, that AT_2 receptors in peripheral organs are not coupled to G-proteins, whereas AT_1 receptors are. Binding sensitivity to guanine nucleotides is considered to be a sign of G-protein coupling, and the findings outlined above suggest that the brain AT_{2A} receptors could indeed be considered as possibly coupled to G-proteins.[69] Further analysis of AT_2 receptor heterogeneity would require the biochemical characterization of the receptor subtypes and the elucidation of its transduction mechanisms.[12]

2.3.3 Other Brain Angiotensin Receptors

A previously unrecognized additional class of angiotensin receptors has been described recently in vascular smooth muscle and bovine adrenal cortical cells, a site that binds angiotensin (3-8) (ANG IV) with high affinity, saturably,

Table IV. Brain ANG II Receptor Subtypes

	AT₁ receptors		
	Subfornical organ	Nucleus of the solitary tract	Paraventricular nucleus
GTPγS sensitivity	Present	Present	Present
Pertussis toxin sensitivity	Absent	Absent	Present

	AT₂ receptors			
	AT₂A			AT₂B
	Ventral thalamic nuclei	Medial geniculate nucleus	Locus coeruleus	Inferior olive
GTPγS sensitivity	Present	Present	Present (higher)	Absent
Pertussis toxin sensitivity	Present	Present	Present	Present

From Tsutsumi *et al.*[69]

reversibly, and specifically.[71,72] These sites are present in a variety of peripheral tissues and in the guinea pig, but not the rat, brain.[71,72] The function of these sites is unknown; however, they have been proposed to play a role in vasodilation, perhaps through one endothelium-derived factor, nitric oxide.[71,72] Whether a role for ANG IV receptors exists in the brain, and the reason for the interspecies variation observed, is still a matter of speculation.

2.4. Selective Effects of Angiotensin Receptor Subtypes

In peripheral tissues, AT₁ receptors appear to mediate all the well-known effects of ANG II, such as vasoconstriction and aldosterone release. Brain AT₁ receptors may also mediate most, if not all, of the well-characterized central ANG II effects, such as drinking and salt appetite.[12] The function of AT₂ receptors, however, is still unknown. The developmental studies in brain AT₂ receptors suggest, as we indicated earlier, a possible function in the development of sensory and motor functions. In addition, we report here some preliminary studies that indicate a possible role of AT₂ receptors in the control of cerebral circulation.[12]

The selective localization of AT₂ receptors in large cerebral arteries of the rat[73] (Figs. 5, 7, and 8) was the first indication that ANG II could be involved in the control of cerebral blood flow, as proposed earlier through stimulation of AT₂ receptors.[12]

The effects of ANG II and ANG II receptor subtype competitors on cerebrovascular autoregulation were studied on anesthetized rats under phenylephrine infusion.[74] We found that the AT₂ receptor ligand PD 123319 increased the autoregulatory range and that the loss of autoregulation occurred at higher blood pressures than in control animals (Fig. 9). ANG II, in the presence of losartan to block AT₁ receptors, also shifted the upper limit of autoregulation

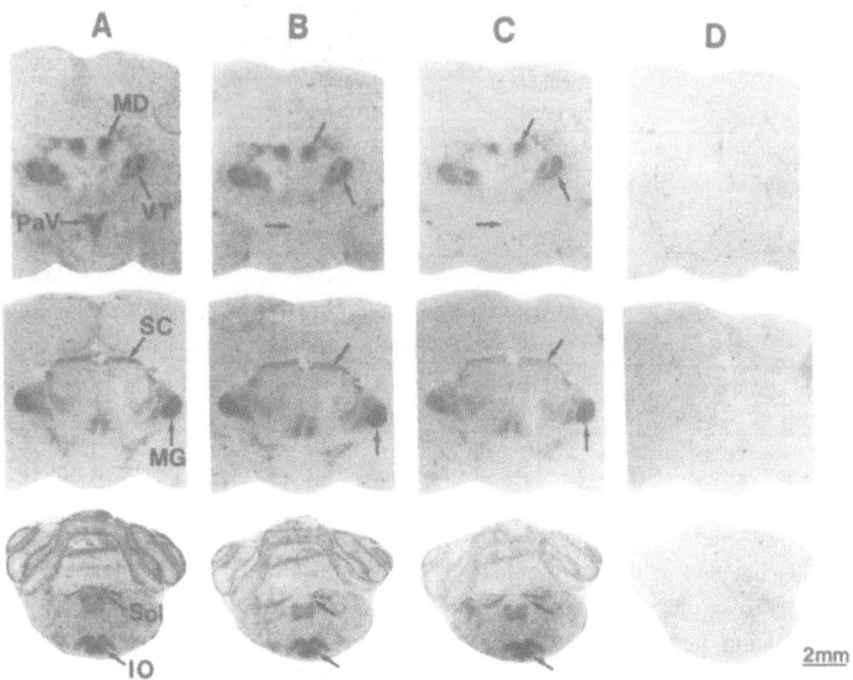

Figure 6. Autoradiography of ANG II receptors in selected brain nuclei of young rats. (A) Coronal sections of 2-week-old male rats, incubated in the presence of 0.5 nM [^{125}I]-Sar^1-ANG II (total binding). (B) Consecutive sections incubated as in A, in the presence of 10 μM losartan to displace binding to AT$_1$ receptors. (C) Consecutive sections incubated as in A, in the presence of 10 μM losartan and 5 mM DTT, showing decreased binding in selected areas containing AT$_2$ receptors. (D) Consecutive sections incubated as in A, in the presence of 5 μM unlabeled ANG II (nonspecific binding). MD, mediodorsal thalamic nucleus; VT, ventral thalamic nuclei; PaV, paraventricular nucleus; SC, superior colliculus; MG, medial geniculate nucleus; 3, oculomotor nucleus; Cb, cerebellar cortex; Sol, nucleus of the solitary tract; 12, hypoglossal nucleus; IO, inferior olive. (From Tsutsumi *et al.*[70] Reprinted with permission.)

to high pressures (Fig. 10). Thus, stimulation of cerebrovascular AT$_2$ receptors by ANG II administration in the presence of losartan produced a shift to the right in the autoregulation curve, similar to that obtained by administration of the AT$_2$ receptor ligand PD 123319.[74] Further experiments[75] demonstrated that the selective AT$_2$ ligand CGP 42112A dose-dependently shifted the upper limit of the cerebrovascular upregulation to the right, that is, to higher blood pressures. In addition, the effect of both CGP 42112A and PD 123319 could be blocked by simultaneous administration of the nonselective ANG II receptor blocker Sar^1, Ile^8-ANG II.[75]

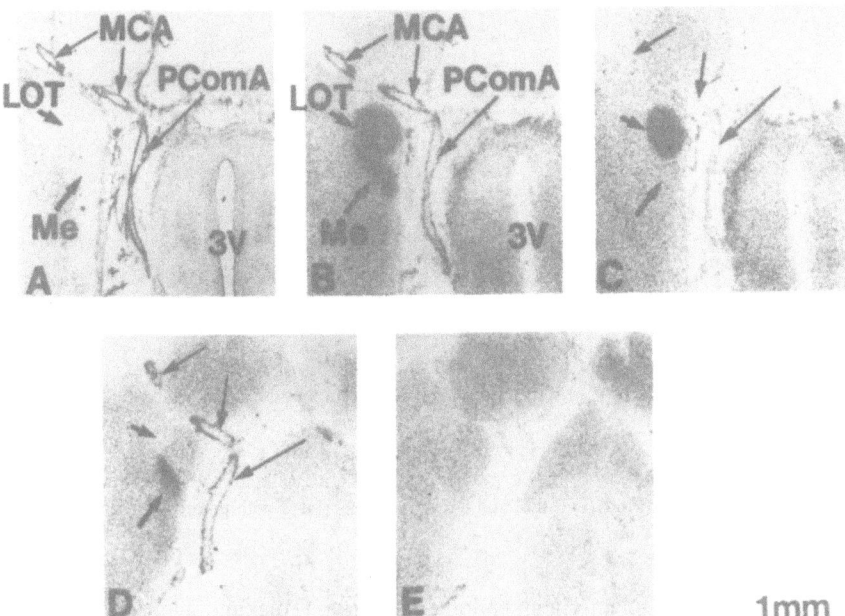

Figure 7. Autoradiography of cerebrovascular ANG II receptor subtypes. Figures represent horizontal sections of the basal forebrain. (A) Toluidine blue staining. (B) Total binding (0.5 nM [^{125}I]-Sar^1-ANG II). (C) AT$_1$ receptors, incubated as in B, in the presence of 10^{-5} M CGP 42112A. (D) AT$_2$ receptors, incubated as in B, in the presence of 10^{-5} M losartan. (E) Nonspecific binding, incubated as in B, in the presence of 10^{-6} M unlabeled ANG II. 3V, third ventricle; PComA, posterior communicating artery; MCA, medial communicating artery; LOT, nucleus of the lateral olfactory tract; Me, medial amygdaloid nucleus. (From Saavedra *et al.*[82] Reprinted with permission.)

The experiments outlined above indicate that both CGP 42112A and PD 123319 can be considered AT$_2$ receptor agonists and that they shift the autoregulatory range toward higher blood pressures probably by stimulation of AT$_2$ receptors located in large cerebral arteries.[12,73–75]

Our results support the hypothesis of a role for ANG II in the control of cerebral circulation[12] and indicate that the AT$_2$ agonists are useful tools to study the role of ANG II in this system, without the complication of AT$_1$ stimulation by ANG II and the resulting increase in blood pressure. The role of ANG II in the control of cerebrovascular function, however, is far from clear. ANG II is known to constrict large cerebral arteries,[76] and this is followed by compensatory autoregulatory dilation of intracerebral arterioles.[76] On the other hand, there are reports of direct vasodilating effects of ANG II in pial arteries, perhaps through the active ANG II derivative ANG IV [ANG (3–8)],[77] which

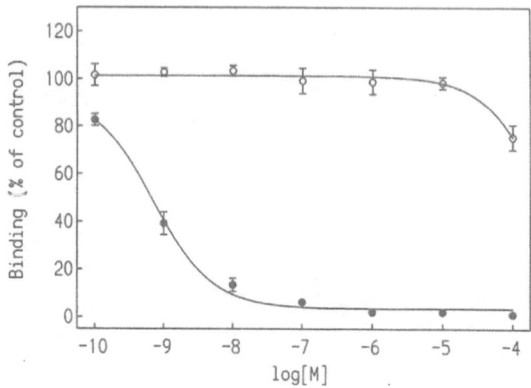

Figure 8. Displacement of [^{125}I]-Sar^1-ANG II binding from anterior cerebral arteries of 2-week-old rats. Displacement curves obtained from consecutive brain sections incubated in the presence of 5×10^{-10} M [^{125}I]-Sar^1-ANG II and increasing concentrations of losartan (O) and CGP 42112A (●). Values represent means ± SE for groups of six anterior cerebral arteries from three rats, measured individually. (From Tsutsumi and Saavedra.[73] Reprinted with permission.)

has been proposed to act via an endothelial-derived relaxing factor mechanism, perhaps nitric oxide.[72] However, ANG IV binding sites have not yet been reported to be present in arterioles, and the influence of ANG IV on nitric oxide formation has not been directly studied.

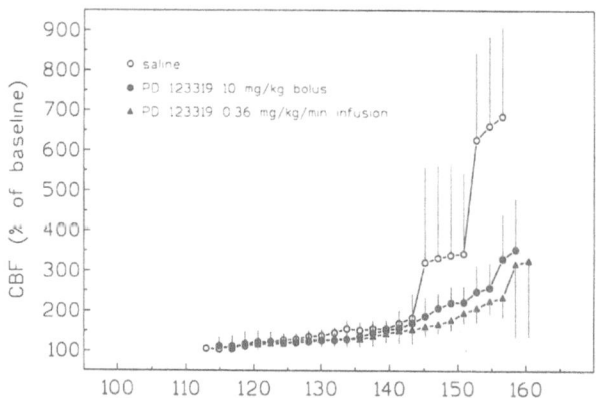

Figure 9. Effect of the AT$_2$ ligand PD123319 on cerebral blood flow during induced hypertension. Medial arterial blood pressure was increased by phenylephrine infusion, and the treatments were given 5 min before the infusion started. Each curve represents the means ± SD of five rats plotted at 2-mm Hg intervals. PD 123319 shifts the upper limit of autoregulation toward higher pressures. (From Strömberg et al.[74] Reprinted with permission.)

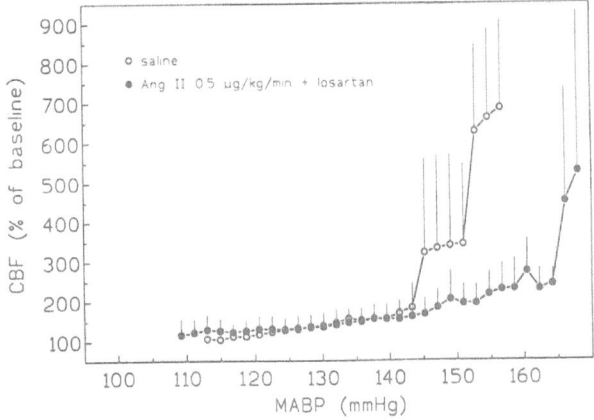

Figure 10. Effect of AT_2 receptor stimulation by ANG II infusion in the presence of losartan 4–10 mg/kg (mean 6.8 mg/kg). Medial arterial blood pressure was increased by phenylephrine infusion. Each curve represents the means of ± SD of five rats plotted at 2-mm Hg intervals. ANG II + losartan shifts the upper limit of autoregulation toward higher pressures. (From Strömberg *et al.*[74] Reprinted with permission.)

An additional complication stems from our results[74] of a similar shift in the autoregulatory curve by administration of the AT_1 blocker losartan. Since the effect of losartan on autoregulation could not be blocked by the nonselective blocker Sar[1],Ile[8]-ANG II, it is not possible to explain it by an indirect effect through AT_2 receptors since it releases endogenous ANG II.[78]

3. CONCLUSIONS

In this chapter we have analyzed our studies on the distribution, characterization, and development of the newly discovered ANG II AT_1 and AT_2 receptor subtypes in brain. These studies support the notion of a major role for AT_1 receptors in the classical actions of central ANG II. Our studies very strongly suggest, on the basis of their development in the brain, that AT_2 receptors may be linked to the development of specific brain functions, among those notably the sensory–motor systems. In addition, there is accumulating evidence to indicate that the AT_2 receptors could play a role in cerebrovascular control.[79]

Both our studies and those of others indicate that brain ANG II receptor subtypes are probably not limited to the AT_1 and AT_2 subtypes and suggest further heterogeneity.

Finally, the study of different models of tissue growth, both under normal and pathological conditions, points to a possible participation of both AT_1 and

AT_2 receptors during growth in peripheral vascular and connective tissues (see Chapter 11, this volume). Whether such a role is expressed during the process of brain growth is an open question worth investigating.

REFERENCES

1. Bickerton RK, Buckley JP: Evidence for a central mechanism of angiotensin induced hypertension. *Proc Soc Exp Biol Med* 106:834–836, 1961.
2. Volicer L, Loew CG: Penetration of angiotensin II into the brain. *Neuropharmacology* 10:631–636, 1971.
3. Harding JW, Sullivan MJ, Hanesworth JM, et al: Inability of [^{125}I]Sar^1,Ile^8 angiotensin II to move between the blood and cerebrospinal fluid compartments. *J Neurochem* 50:554–557, 1988.
4. Van Houten M, Schiffrin EL, Mann JFE, et al: Radioautographic localization of specific binding sites for blood-borne angiotensin II in the rat brain. *Brain Res* 186:480–485, 1980.
5. Bennett JP, Snyder SH: Angiotensin II binding to mammalian brain membranes. *J Biol Chem* 251:7423–7430, 1976.
6. Sirett NE, McLean AS, Bray JJ, et al: Distribution of angiotensin II receptors in rat brain. *Brain Res* 122:299–312, 1977.
7. Bennett JP, Snyder SH: Receptor binding interactions of the angiotensin II antagonist, ^{125}I-[sarcosine^1, isoleucine^8] angiotensin II, with mammalian brain and penpheral tissues. *Eur J Pharmacol* 67:11–25, 1980.
8. Sirett NE, Thornton SN, Hubbard JI: Angiotensin binding and pressor activity in the rat ventricular system and midbrain. *Brain Res* 166:139–148, 1979.
9. Mann JFE, Schiffrin EL, Schiller PW, et al: Brain receptor binding and central actions of angiotensin analogs in rats. *Am J Physiol* 241:R124–R129, 1981.
10. Harding JW, Stone LP, Wright JW: The distribution of angiotensin II binding sites in rodent brain. *Brain Res* 205:265–274, 1981.
11. Mendelsohn FAO, Quirion R, Saavedra JM, et al: Autoradiographic localization of angiotensin II receptors in rat brain. *Proc Natl Acad Sci USA* 81:1575–1579, 1984.
12. Saavedra JM: Brain and pituitary angiotensin. *Endocr Rev* 13:329–380, 1992.
13. Shigematsu K, Saavedra JM, Plunkett LM, et al: Angiotensin II binding sites in the anteroventral-third ventricle (AV3V) area and related structures of the rat brain. *Neurosci Lett* 67:37–41, 1986.
14. Castrén E, Saavedra JM: Repeated stress increases the density of angiotensin II binding sites in rat paraventricular nucleus and subfornical organ. *Endocrinology* 122:370–372, 1988.
15. Castrén E, Saavedra JM: Angiotensin II receptors in paraventricular nucleus, subfornical organ, and pituitary gland of hypophysectomized, adrenalectomized, and vasopressin-deficient rats. *Proc Natl Acad Sci USA* 86:725–729, 1989.
16. Castrén E, Saavedra JM: Effect of corticoids on brain angiotensin II receptors. In Velasco M, Israel A, Romero E, Silva H (eds): *Recent Advances in Pharmacology and Therapeutics.* Elsevier, Amsterdam, 1989, pp. 211–214.
17. Mendelsohn FAO, Allen AM, Chai SY, et al: The brain angiotensin system. Insights from mapping its components. *Trends Endocrinol Metab* 1:189–198, 1990.
18. Gehlert DR, Speth RC, Wamsley JK: Distribution of [^{125}I]angiotensin II binding sites in the rat brain: A quantitative autoradiographic study. *Neuroscience* 18:837–856, 1986.
19. Speth RC, Wamsley JK, Gehlert DR, et al: Angiotensin II receptor localization in the canine CNS. *Brain Res* 326:137–143, 1985.

20. Healy DP, Printz MP: Localization of angiotensin II binding sites in rat septum by autoradiography. *Neurosci Lett* 44:167–172, 1984.

21. Mendelsohn FAO, Allen AM, Clevers J, et al: Localization of angiotensin II receptor binding in rabbit brain by in vitro autoradiography. *J Comp Neurol* 270:372–384, 1988.

22. Saavedra JM, Israel A, Plunkett LM, et al: Quantitative distribution of angiotensin II binding sites in rat brain by autoradiography. *Peptides* 7:679–687, 1986.

23. Walters DE, Speth RC: Neuronal localization of specific angiotensin II binding sites in the rat inferior olivary nucleus. *J Neurochem* 50:812–817, 1988.

24. Diz DI, Barnes KL, Ferrario CM: Contribution of the vagus nerve to angiotensin II binding sites in the canine medulla. *Brain Res Bull* 17:497–505, 1986.

25. Lewis SJ, Allen AM, Verberne AJM, et al: Angiotensin II receptor binding in the rat nucleus tractus solitarii is reduced after unilateral nodose ganglionectomy or vagotomy. *Eur J Pharmacol* 125:305–307, 1986.

26. Rowe BP, Kalivas PW, Speth RC: Autoradiographic localization of angiotensin II receptor binding sites on noradrenergic neurons of the locus coeruleus of the rat. *J Neurochem* 55:533–540, 1990.

27. Israel A, Correa FMA, Niwa M, et al: Quantitative determination of angiotensin II binding sites in rat brain and pituitary gland by autoradiography. *Brain Res* 322:341–345, 1984.

28. Israel A, Correa FMA, Niwa M, et al: Quantitative measurement of angiotensin II (AII) receptors in discrete regions of rat brain, pituitary and adrenal gland by autoradiography. *Clin Exp Hypertens* A6:1761–1984, 1984.

29. Israel A, Plunkett LM, Saavedra JM: Quantitative autoradiographic characterization of receptors for angiotensin II and other neuropeptides in individual brain nuclei and peripheral tissues from single rats. *Cell Mol Neurobiol* 5:211–222, 1985.

30. Nazarali AJ, Gutkind JS, Saavedra JM: Calibration of ^{125}I-polymer standards with ^{125}I-brain paste standards for use in quantitative receptor autoradiography. *J Neurosci Methods* 30:247–253, 1989.

31. Peach MJ, Dostal DE: The angiotensin II receptor and the actions of angiotensin II. *J Cardiovasc Pharmacol* 16(Suppl 4):S25–S30, 1990.

32. Pobiner BF, Hewlett EL, Garrison JC: Role of Ni in coupling angiotensin receptors to inhibition of adenylate cyclase in hepatocytes. *J Biol Chem* 260:16200–16209, 1985.

33. Gunther S: Characterization of angiotensin II receptor subtypes in rat liver. *J Biol Chem* 259:7622–7629, 1984.

34. Chiu AT, Duncia JV, McCall DE, et al: Nonpeptide angiotensin II receptor antagonists. III. Structure–function studies. *J Pharmacol Exp Ther* 250:867–874, 1989.

35. Whitebread S, Mele M, Kamber B, et al: Preliminary biochemical characterization of two angiotensin II receptor subtypes. *Biochem Biophys Res Commun* 163:284–291, 1989.

36. Dudley DT, Panek RL, Major TC, et al: Subclasses of angiotensin II binding sites and their functional significance. *Mol Pharmacol* 38:370–377, 1990.

37. Timmermans PBMWM, Carini DJ, Chiu AT, et al: The discovery of a new class of highly specific nonpeptide angiotensin II receptor antagonists. *Am J Hypertens* 4:275S–281S, 1991.

38. Timmermans PBMWM, Carini DJ, Chiu AT, et al: Nonpeptide angiotensin II receptor antagonists: A novel class of antihypertensive agents. *Blood Vessels* 27:295–300, 1990.

39. Timmermans PBMWM, Wong PC, Chiu AT, et al: Nonpeptide angiotensin II receptor antagonists. *Trend Pharmacol Sci* 12:55–62, 1991.

40. Bumpus FM, Catt KJ, Chiu AT, et al: Nomenclature for angiotensin receptors. A report of the nomenclature committee of the Council for High Blood Pressure Research. *Hypertension* 17:720–721, 1991.

41. Gibson RE, Thorpe HH, Cartwright ME, et al: Angiotensin-II receptor subtypes in renal cortex of rats and Rhesus monkeys. *Am J Physiol* 261:F512–F518, 1991.

42. Balla T, Baukal AJ, Eng S, et al: Angiotensin-II receptor subtypes and biological responses in the adrenal cortex and medulla. *Mol Pharmacol* 40:401–406, 1991.

43. Chiu AT, McCall DE, Price WA, et al: Nonpeptide angiotensin II receptor antagonists. VII. Cellular and biochemical pharmacology of DuP 753, an orally active antihypertensive agent. *J Pharmacol Exp Ther* 252:711–718, 1990.

44. Wong PC, Hart SD, Zaspel AM, et al: Functional studies of nonpeptide angiotensin II receptor subtype-specific ligands: DuP 753 (AII-1) and PD 123177 (AII-2). *J Pharmacol Exp Ther* 255:584–592, 1990.

45. Wright GB, Alexander RW, Ekstein LS, et al: Sodium, divalent cations, and guanine nucleotides regulate the affinity of the rat mesenteric artery angiotensin II receptor. *Circ Res* 50:462–469, 1982.

46. Viswanathan M, Tsutsumi K, Correa FMA, et al: Changes in expression of angiotensin receptor subtypes in the rat aorta during development. *Biochem Biophys Res Commun* 179:1361–1367, 1991.

47. Pucell AG, Hodges JC, Sen I, et al: Biochemical properties of the avarian granulosa cell type-2 angiotensin II receptor. *Endocrinology* 128:1947–1959, 1991.

48. Chiu AT, Herblin WF, McCall DE, et al: Identification of angiotensin receptor subtypes. *Biochem Biophys Res Commun* 165:196–203, 1989.

49. Rogg H, Schmid A, de Gasparo M, et al: Identification and characterization of angiotensin II receptor subtypes in rabbit ventricular myocardium. *Biochem Biophys Res Commun* 173:416–422, 1990.

50. Tsutsumi K, Strömberg C, Viswanathan M, et al: Angiotensin-II receptor subtypes in fetal tissues of the rat—autoradiography, guanine nucleotide sensitivity, and association with phosphoinositide hydrolysis. *Endocrinology* 129:1075–1082, 1991.

51. Rowe BP, Grove KL, Saylor DL, et al: Discrimination of angiotensin II receptor subtype distribution in the rat brain using non-peptidic receptor antagonists. *Regul Pept* 33:45–53, 1991.

52. Song K, Allen AM, Paxinos G, et al: Angiotensin II receptor subtypes in rat brain. *Clin Exp Pharmacol Physiol* 18:93–96, 1991.

53. Leung KH, Smith RD, Timmermans PBMWM, et al: Regional distribution of the two subtypes of angiotensin II receptor in rat brain using selective nonpeptide antagonists. *Neurosci Lett* 123:95–98, 1991.

54. Gehlert DR, Gackenheimer SL, Reel JK, et al: Nonpeptide angiotensin II receptor antagonists discriminate subtypes of ^{125}I-angiotensin II binding sites in the rat brain. *Eur J Pharmacol* 187:123–126, 1990.

55. Wamsley JK, Herblin WF, Alburges ME, et al: Evidence for the presence of angiotensin II-type 1 receptors in brain. *Brain Res Bull* 25:397–400, 1990.

56. Tsutsumi K, Saavedra JM: Characterization and development of angiotensin II receptor subtypes (AT$_1$ and AT$_2$) in rat brain. *Am J Physiol* 261:R209–R216, 1991.

57. Tsutsumi K, Saavedra JM: Increased dithiothreitol-insensitive, type 2 angiotensin II receptors in selected brain areas of young rats. *Cell Mol Neurolbiol* 11:295–299, 1991.

58. Tsutsumi K, Saavedra JM: Quantitative autoradiography reveals different angiotensin II receptor subtypes in selected rat brain nuclei. *J Neurochem* 56:348–351, 1991.

59. Tsutsumi K, Saavedra JM: Differential development of angiotensin II receptor subtypes in the rat brain. *Endocrinology* 128:630–632, 1991.

60. Raizada MK, Yang JW, Fellows RE: Rat brain cells in primary culture characterization of angiotensin binding sites. *Brain Res* 207:343–355, 1981.

61. Millan MA, Carvallo P, Izumi S, et al: Novel sites of expression of functional angiotensin II receptors in the late gestation fetus. *Science* 244:1340–1342, 1989.

62. Tsutsumi K, Viswanathan M, Strömberg C, et al: Type-1 and type-2 angiotensin-II receptors in fetal rat brain. *Eur J Pharmacol* 198:89–92, 1991.

63. Baxter CR, Horvath JS, Duggin GG, et al: Effect of age on specific angiotensin II-binding sites in rat brain. *Endocrinology* 106:995–999, 1980.

64. Tsutsumi K, Saavedra JM: Angiotensin II receptor subtypes in median eminence and basal forebrain areas involved in regulation of pituitary function. *Endocrinology* 129:3001–3008, 1991.

65. Kakar SS, Riel KK, Neill JD: Differential expression of angiotensin II receptor subtype mRNAs (AT-1A and AT-1B) in the brain. *Biochem Biophys Res Commun* 185:688–692, 1992.

66. Iwai N, Inagami T, Ohmichi N, et al: Differential regulation of rat AT1a and AT1b receptor mRNA. *Biochem Biophys Res Commun* 188:298–303, 1992.

67. Seltzer A, Pinto JEB, Viglione PN, et al: Estrogens regulate angiotensin-converting enzyme and angiotensin receptors in female rat anterior pituitary. *Neuroendocrinology* 55:460–467, 1992.

68. Seltzer A, Tsutsumi K, Shigematsu, et al: Reproductive hormones modulate angiotensin II AT_1 receptors in the dorsomedial arcuate nucleus of the female rat. *Endocrinology* 133:939–941, 1993.

69. Tsutsumi K, Saavedra JM: Heterogeneity of angiotensin II AT_2 receptors in the rat brain. *Mol Pharmacol* 41:290–297, 1992.

70. Tsutsumi K, Zorad S, Saavedra JM: The AT_2 subtype of the angiotensin II receptors has differential sensitivity to dithiothreitol in specific brain nuclei of young rats. *Eur J Pharmacol Mol Pharmacol* 226:169–173, 1992.

71. Hall KL, Hanesworth JM, Ball AE, et al: Identification and characterization of a novel angiotensin binding site in cultured vascular smooth muscle cells that is specific for the hexapeptide (3-8) fragment of angiotensin II, angiotensin IV. *Regul Pept* 44:225–232, 1993.

72. Swanson GN, Hanesworth JM, Sardinia MF, et al: Discovery of a distinct binding site for angiotensin II (3-8), a putative angiotensin IV receptor. *Regul Pept* 40:409–419, 1992.

73. Tsutsumi K, Saavedra JM: Characterization of AT_2 angiotensin-II receptors in rat anterior cerebral arteries. *Am J Physiol* 261:H667–H670, 1991.

74. Strömberg C, Näveri L, Saavedra JM: Nonpeptide angiotensin AT_1 and AT_2 receptor ligands modulate the upper limit of cerebral blood flow autoregulation in rats. *J Cereb Blood Flow Metab* 13:298–303, 1993.

75. Näveri L, Strömberg C, Saavedra JM: Angiotensin II AT_2 receptor stimulation extends the upper limit of cerebral blood flow autoregulation: Agonist effects of CGP 42112A and PD 123319. *J Cereb Blood Flow Metab* 14:38–44, 1994.

76. Edvinsson L, MacKenzie ET, McCulloch J: *Cerebral Blood Flow and Metabolism.* Raven Press, New York, 1993.

77. Haberl RL, Decker PJ, Einhaupl KM: Angiotensin degradation products mediate endothelium-dependent dilation of rabbit brain arterioles. *Circ Res* 68:1621–1627, 1991.

78. Mizuno K, Tani M, Niimura S, et al: Losartan, a specific angiotensin II receptor antagonist, increases angiotensin I and angiotensin II release from isolated rat hind legs: Evidence for locally regulated renin–angiotensin system in vascular tissue. *Life Sci* 50:PL209–PL214, 1992.

79. Saavedra JM, Tsutsumi K, Strömberg C, et al: Localization, characterization, development, and function of brain angiotensin II receptor subtypes. In Raizada MK, Phillips MI, Sumners C (eds): *Cellular and Molecular Biology of the Renin–Angiotensin System.* CRC Press, Boca Raton, Florida, 1993, p. 357–378.

9

Second Messengers for Brain Angiotensin Receptor Subtypes

Frank M. J. Heemskerk and Juan M. Saavedra

1. INTRODUCTION

1.1. Angiotensin II in the Brain

1.1.1. Central Angiotensin II

Angiotensin II (ANG II) is an octapeptide produced in many tissues from its precursor molecules by subsequent cleavage by the enzymes renin and angiotensin converting enzyme.[1-3] In the brain all components of this renin–angiotensin system are present.[1,2,4] ANG II is involved in the central regulation of cardiovascular function, drinking and fluid metabolism, salt appetite, pituitary hormone release, and stress.[1,2,4-6]

Using immunocytochemical techniques, the localization of ANG II was described in specific brain structures,[7,8] mainly in neurons and not in glial cells.[9-11] ANG II was found in synaptic vesicles in nerve terminals[7] forming contacts with other neurons and also in other nerve endings without synaptic specializations terminating on fenestrated blood vessels of the circumventricular organs.[7]

A neurotransmitterlike function for ANG II is also supported by the finding that the peptide excites neurons at very low doses and with high specific-

Frank M. J. Heemskerk and Juan M. Saavedra • Section on Pharmacology, Laboratory of Clinical Science, National Institute of Mental Health, National Institutes of Health, Bethesda, Maryland 20892.

Angiotensin Receptors, edited by Juan M. Saavedra and Pieter B.M.W.M. Timmermans. Plenum Press, New York, 1994.

ity.[2,12–16] Furthermore, central administration of ANG II increases the release of catecholamines[17–23] and several neuropeptides.[20,24–26]

1.1.2. Central ANG II Receptors

Receptors for ANG II have been located and pharmacologically characterized in many areas in the brain of a number of species including human.[1,5,6] Although the existence of multiple receptor subtypes for ANG II was suggested in peripheral tissues earlier,[27,28] only recently, with the availability of new pharmacological ligands,[29,30] at least two subtypes, AT_1 and AT_2,[31] can be distinguished clearly. AT_1 receptors are characterized by high affinity for losartan (formerly DuP 753) and their sensitivity to reducing agents. AT_2 receptors selectively bind with high affinity to PD 123177, PD 123319, and the peptidic ligand CGP 42112. There is accumulating evidence that even more subtypes may be prevalent (e.g., ref. 32; see also Section 4.2 and refs. 5, 6, 33). The receptor subtypes are not only pharmacologically distinct, but have a heterogenous distribution as well. Many tissues, including brain, contain both subtypes, each with a unique localization. In addition there are interspecies differences in localization. Especially the expression of the AT_2 receptor subtype seems to be developmentally regulated.[1,34,35]

Since the circumventricular organs and the hypothalamic–pituitary axis are involved in fluid homeostasis and the regulation of (reproductive) hormone release, most of the ANG II research until now has focused on these regions in the brain. However, ANG II receptors were found also in many other brain areas (see reviews mentioned above) in which a possible function of ANG II receptors has not yet been defined.

Both receptor subtypes seem to have physiological roles in the brain. Although most of the central effects of ANG II that are currently known seem to be exerted through stimulation of the AT_1 receptor,[1,2] there are a few new studies suggesting a role of AT_2 receptors in cerebral flow regulation,[36,37] drinking behavior and salt intake,[17,38] hormone (luteinizing hormone and prolactin) release,[39] and vasopressin release.[17] In addition, there might be involvement of both subtypes of ANG II receptors in learning and memory processes.[40,41]

Considering its role in cardiovascular control, fluid homeostasis, and hormone release, it is clear that ANG II has important multiple roles in the central nervous system. However, it is not clear at all exactly how ANG II exerts its various roles. Intracellular signaling pathways are unknown for cells in most of the brain regions that were implicated in the effects of ANG II described above. With a few exceptions, which will be discussed below, very few studies have dealt with freshly dissociated neural cells, brain slices *in vitro*, or even with homogenates from central tissue.

1.2. Second Messenger Systems

1.2.1. Most Information Is from Peripheral Systems

In contrast to peripheral tissue/systems, where signal transduction mechanisms in response to ANG II have been described in considerable detail, still little is known about the mechanisms through which ANG II affects neural cells. Several reasons might be underlying this disparity in knowledge of cellular signaling of ANG II in peripheral and central systems.

First, it was not recognized in early studies that circulating ANG II, being a peptide, could have a central role. Only after the discovery of a renin–angiotensin system inside the blood–brain barrier, more research was directed at the central effects of ANG II.

Second, the level of expression of ANG II receptors is much lower in most brain areas than in many peripheral tissues. This made the analysis of the localization and pharmacology of the receptors difficult. Also, only at the time when the developmental regulation of expression of ANG II receptors[1,34,35,42] was described was it realized that central AT_2 receptors were generally much more abundant early in development.

Third, especially in brain tissue, ANG II is rapidly metabolized, causing high "nonspecific" binding of radioligands, and some of the metabolites are physiologically active themselves.[2]

Fourth, the heterogenous localization of the receptors in discrete areas of the brain, together with the recent discovery of the existence of at least two receptor subtypes, have to be taken into account in experiments on the cellular signaling of ANG II. This limits the techniques available to investigate, e.g., extensive biochemical studies are often impossible for reasons of heterogeneity and/or small size of the sample.

With this in mind it is no wonder that most investigators chose to work with established cell lines or primary cell cultures. Even these systems have their own drawbacks. The ratio of ANG II receptor subtypes can change during culture and the culture conditions will invariably influence the intracellular response of the cell to receptor stimulation.

1.2.2. Peripheral Second Messenger Systems

The most common signal transduction systems of ANG II receptors in nonneuronal cells are coupling to adenylate cyclase and phospholipase C through inhibitory and stimulatory G-proteins, respectively.[1,5,43] With a few notable exceptions (e.g., ref. 79), a second messenger response has not been firmly linked yet to AT_2 receptor stimulation.[44,45] In some tissues, as in the kidney and the liver, more than one physiological response can be found. Although in hepatocytes essentially only one receptor subtype has been described (AT_1), it is not clear whether both responses—the coupling to phospholipase C and the increase in glucagon metabolism—are coupled to the same receptor or are mediated by distinct receptors of a similar subtype (possibly on the same or on different cells).[46] ANG II was found to stimulate phospholipase C and to inhibit adenylate cyclase even within one cell line.[47] Similarly, in kidney/mesangial cells, subtypes of AT_1 receptor (AT_{1A} and AT_{1B}) were described apparently coupled to different signal transduction mechanisms.[48] To add to the confusion, these receptor subtypes are probably not the same subtypes as the AT_{1A} and AT_{1B} as defined at the mRNA level by Iwai and Inagami.[33]

Other intracellular signals affected by stimulation of AT_1 receptors generally involve stimulation of protein kinase C (in response to elevated levels of Ca^{2+}_i and diacylglycerol) and altered tyrosine protein phosphorylation,[49-55] possibly as a result of changes in Ca^{2+}_i levels. Changes in these second messengers and in protein phosphorylation could be a major pathway through which ANG II influences early gene expression and cellular growth (see Chapter 11, this volume).

1.3. Aim of This Chapter

This chapter will focus on the signal transduction mechanisms affected by ANG II in the central nervous system and in cells of neural origin. A broad definition of neural cells will be used. This includes studies in the brain (*in vivo*) as well as in brain slices *in vitro,* and in primary cultures as well as in astrocyte, pituitary cell, and neuroblastoma cell lines. Hopefully new research in this field will be stimulated and provoked through an overview of the lack of detailed knowledge about the intracellular signaling systems influenced by ANG II in neural cells. In order to improve our knowledge of the central effects of ANG II, we will have to understand the intracellular actions in much more detail than we currently do.

2. CALCIUM AND INOSITOL PHOSPHATE TURNOVER

2.1. Brain

The few reports available point to a coupling of central AT_1 receptors to phospholipase C much as they do in peripheral systems. A recent study in slices from cerebral cortex and hippocampus described the stimulation by ANG II of the formation of inositol tris- and tetraki-phosphate. The response to ANG II was more evident in slices from senescent (30 months old) rats than in younger rats.[56] In this study no attempt was made to see which receptor subtype was involved in this response. However, the formation of these higher inositol phosphates is most likely linked to the AT_1 subtype, since the predominant ANG II receptor subtype in the hippocampus and cerebral cortex of adult rats is the AT_1 receptor[34,57] and the expression of the AT_2 receptor in many other brain areas decreases during development.

In (unpublished) studies in our laboratory we also found a small (30% above control) stimulation of inositol phosphate formation and inositol phospholipid turnover in hippocampal slices. In microdissected slices from two other brain areas expressing mainly AT_2 receptors (the medial geniculate nucleus and the inferior olivary complex), ANG II induced no changes. Carbachol and quisqualate were used as internal positive controls in these experiments and stimulated inositol monophosphate accumulation severalfold. Hence, it seems that in the hippocampus AT_1 receptors are coupled to phospholipase C, while at least in the medial geniculate nucleus and the inferior olive AT_2 receptors do not use this pathway.

Studies in several laboratories, including ours, have shown the presence of AT_1 receptors in the pituitary, the median eminence, and isolated cervical ganglia.[1,58–61] In the median eminence stimulation of AT_1 receptors results in accumulation of inositol phosphates. This response to ANG II was sensitive to losartan and DTT and a pertussis-toxin-insensitive G-protein seems to be involved (A. Seltzer personal communication).

2.2. Neural Cells in Culture

Earlier studies with cultured astrocytes isolated from rat brain showed that ANG II stimulated inositol phosphate formation and this effect was blocked by (Sar^1Ile^8) ANG II.[62] At that time, however, receptor specific ligands were not yet available.

More detailed studies were performed in several laboratories with cells in culture. The hybrid cell line NG108-15, which is of neuroblastoma x glioma origin, has been used by several investigators to study the effects of ANG II *in vitro*.[63–66] Similarly, the N1E-115 neuroblastoma cell line has been used by others[67,68] for the same purpose. These two studies described losartan-sensitive stimulation of AT_1 receptors resulting in increased inositol phosphate formation. Monck *et al.*[67] also described increases in intracellular calcium levels, as measured with Fura-2, after stimulation of the AT_1 receptors. Apparently the response was coupled through a pertussis-toxin-insensitive G-protein. The increase in intracellular calcium was relatively independent of extracellular calcium levels, indicating involvement of stimulation of phospholipase C, formation of inositol triphosphate, and Ca^+ release from intracellular stores. Similar results were described in NG108-15 cells.[65]

Buisson *et al.*[66] described in undifferentiated NG108-15 cells the presence of AT_2 receptors that are not coupled to inositol phosphate responses. Instead this receptor subtype seems to mediate a decrease in a T-type Ca^+ current. Interestingly, the AT_2-selective ligand CGP 42112 (nicotinic acid-Tyr-(*N*-benzoyloxycarbonyl-Arg)-Lys-His-Pro-Ile-OH) acts as an agonist in this system. It will be worthwhile to investigate whether other AT_2-selective ligands, such as PD 123177 [1-(4-amino-3-methylphenyl) methyl-5-diphenylacetyl-4,5,6,7-tetrahydro-1H-imidazo[4,5-c]pyridine-6-carboxylic acid-2HCl] have agonistic or antagonistic effects in this model.

NG108-15 cells have the interesting ability to differentiate *in vitro* in response to the combination of fetal bovine serum and dimethylsulfoxide.[63,64] Differentiated cells apparently differ from undifferentiated ones in the extent and subtype of ANG II receptor expression.[29,64,69] In an early study, Carrithers *et al.*[63] reported ANG II stimulating inositol phosphate formation in differentiated NG108-15 cells; however, no attempt was taken to determine the receptor subtype involved. In another study[65] with the same cells it is not clear whether the cells were differentiated or not.

Ransom *et al.*[64] reported the presence of ANG II receptors on NG108-15 cells coupled to inositol phosphate turnover, regardless of the state of differentiation, although the responses were small. Subsequently, these investigators isolated a subclone with larger responses to ANG II, which were inhibited by nanomolar concentrations of the AT_1 receptor antagonist losartan but not by the AT_2-selective ligand CGP 42112. Apparently CGP 42112 alone did not alter inositol phosphate metabolism. It is noteworthy that in two studies[63,64] ANG III was more potent than ANG II in stimulating inositol phosphate formation. This is somewhat different from binding studies to AT_1 receptors where ANG III generally has a lower affinity for the receptor than ANG II.

The PC12W cell line, upon stimulation with nerve growth factor, has the capability to differentiate into cells with many characteristics of neurons. ANG II receptors on these cells are only of the AT_2 subtype and apparently are not coupled to inositol phosphate formation or changes in intracellular calcium.[5,70,71] Also, anterior pituitary cells in primary culture express ANG II receptors that induce an increase in inositol phosphate turnover.[59]

While neurons isolated from immature rat brain express AT_2 receptors mediating changes in cGMP levels (see Section 4.2), astrocytes in these primary cultures have ANG II receptors that stimulate inositol phosphate formation.[72] The receptor subtype involved is clearly AT_1 since the response was losartan sensitive and was inhibited by dithiothreitol. This is supported by another study[73] showing AT_1 receptors coupled to inositol phosphate formation in the C6 glioma cell line and in astrocytes derived from neoplasms in human brain.

3. cAMP AND ADENYLATE CYCLASE

3.1. Brain

There is only one study investigating effects of ANG II on cAMP metabolism in tissue isolated from the brain. In this study ANG II stimulated adenylate cyclase activity in rat brain homogenate enriched in microvessels.[74] Although these results were obtained in a broken cell preparation, apparently primary cultures of endothelial cells gave similar results.[74] In the anterior pituitary, in contrast to experiments using membrane preparations, it is likely that in intact cells ANG II receptors are not coupled to changes in cAMP.[1,60,61]

3.2. Neural Cells in Culture

No changes in adenylate cyclase after AT_2 receptor stimulation were found in PC12W cells by two groups,[70,71] and no response was found in primary neuron or astrocyte cultures either.[75]

4. cGMP, GUANYLATE CYCLASE, AND NITRIC OXIDE

4.1. Brain

To our knowledge there are no published reports yet directly investigating changes in cGMP levels or guanylate cyclase activity in response to ANG II receptor stimulation in brain tissue *in vivo* or *in vitro*.

4.2. Neural Cells in Culture

One earlier study using neuroblastoma N1E-115 cells reported a stimulation of cGMP release by ANG II and ANG III.[76] At that time it was not possible to determine the receptor subtype involved. Only recently more studies have appeared analyzing the effect of ANG II on cGMP levels. Zarahn et al.[68] (see also ref. 77) described a fast (30-sec) increase in cGMP levels in N1E-115 cells in response to ANG II. It was suggested that this response was partially due to stimulation of cGMP through AT_2 receptors since CGP 42112 counteracted some of the effect of ANG II. CGP 42112 alone had no effect. This would suggest that CGP 42112 in this system would act as an antagonist to the AT_2 receptor. In the same cells ANG II in the lower nanomolar range raised Ca^{2+}_i levels through stimulation of phospholipase C as well as an increase in cGMP levels through AT_1 receptors (as both effects were sensitive to losartan). The authors hypothesize that the elevated Ca^{2+}_i levels result in stimulation of calmodulin, enhanced nitric oxide synthase activity, enhanced soluble guanylate cyclase activity, and result in the increase of cGMP. Apparently ANG II did not stimulate particulate guanylate cyclase in membrane fractions of these cells. This paper also shows that above 1 μM CGP 42112 cross-reacts on the AT_1 receptor response.

In two other studies[72,78] using primary culture from 1-day-old rat brain (hypothalamus and brainstem), neurons were shown to express both ANG II receptor subtypes. In neurons ANG II appeared to inhibit basal cGMP through AT_2 receptors (IC_{50} 1–10 nM) since this effect was fully blocked by (Sar^{1}Ile^8)

ANG II, PD 123177, or CGP 42112, and not by losartan. This would imply that CGP 42112 acts as an antagonist in this system. The response was rather slow, at least 1–5 min. On the other hand, ANG II receptors on astrocytes (AT_1) increased cGMP within 30 sec. The effect of ANG II in neuronal cultures was inhibited by Cd^{2+}, nifedipine, and was IBMX sensitive, suggesting that Ca^{2+} entry and subsequent stimulation of phosphodiesterase were involved. ANG II had no effect on particulate or soluble guanylate cyclase or on cGMP release.

In contrast to the studies mentioned above, another group[79] described AT_2-mediated inhibition of basal and atrial natriuretic peptide-stimulated cGMP formation through inhibition of particulate guanylate cyclase in PC12W cells, which express only AT_2 receptors. In this system CGP 42112 behaved as an agonist: it decreased cGMP levels as well. The ANG II response was not sensitive to IBMX, suggesting phosphodiesterase activity was not involved. The authors suggested that AT_2 receptors might influence the particulate guanylate cyclase indirectly through dephosphorylation on tyrosine residues. This was confirmed in a recent study.[80] However, these results are still controversial since other groups[70,71] using the same cells have not found any response after AT_2 receptor stimulation.

Finally, recently a new ANG II receptor subtype has been described by Chaki and Inagami[81] in neuro-2A cells. In these cells ANG II induced a dose-dependent increase in cGMP within 30 sec. This response was sensitive to (Sar^1Ile^8) ANG II; however, neither losartan nor PD 123319 had any effect, indicating a non-AT_1, non-AT_2 receptor subtype. No more details about the mechanism were provided in that study.

5. OTHER SIGNAL TRANSDUCTION MECHANISMS

5.1. Arachidonic Acid and Related Substances

There is some evidence that prostaglandins might be involved in induction of endothelium-dependent vasodilation of rat cerebral arterioles by ANG II since it is blocked by the cyclo-oxygenase inhibitor indomethacin.[82]

In the anterior pituitary ANG II receptor stimulation is associated with arachidonate and stearate release from phospholipids through phospholipases A_1, A_2 and/or diacylglycerol lipase.[1,83–85]

Human astrocytes from brain neoplasms in culture express both AT_1 and AT_2 receptors.[73] AT_1 receptors seem to be coupled to changes in inositol phosphates and Ca^{2+}_i as well as increases in prostaglandin (PGE_2) release. Stimulation of AT_2 receptors, on the other hand, resulted in an increase of prostacyclin (PGI_2) release. The latter response was mimicked by CGP 42112, thus acting as an agonist in this system. Similar results were observed in the C6 glioma cell line. An increase in prostacyclin release has also been reported in NG108-15 cells.[69] However, in C6 glioma cells AT_1 receptors are present also that are coupled to a stimulation of PGE_2 release.[86] Moreover, both losartan and (Sar^1Ile^8) ANG II are able to act as agonists on prostacyclin release.[73,87] These observations warrant caution about interpretation of results in terms of specific ANG II receptor subtypes involved, especially since similar results have been obtained in smooth muscle cells that have only AT_1 receptors.[51] In at least two other systems, no coupling of AT_2 receptors to the arachidonic metabolic pathway could be found.[70,71]

5.2. (De)-Phosphorylation

Currently there are no reports investigating whether ANG II induces any changes in phosphorylation in the brain. Considering the literature about AT_1 coupling to inositol phosphate metabolism and changes in Ca^{2+}_i levels (see Section 2), one would expect that at least the AT_1 receptor will influence protein phosphorylation through protein kinase C and/or calmodulin-dependent kinase(s).

In the developing nervous system these processes are essential for normal development. In the mature brain phosphorylation processes are more likely to play important roles in regular cell function such as neurotransmitter release.

There are a number of reports describing modulation of neurotransmitter release by ANG II (e.g., refs. 18–22, 88, 89). It would be interesting to know whether ANG II stimulation results in modification of phosphorylation of any of the well-described phosphoproteins involved in regulation of neurotransmitter release.

Up to now only one group has reported changes in protein phosphorylation in neuronal or neuronlike cells in culture. Bottari et al.[5,79] showed that in the PC12W cells they use ANG II enhances dephosphorylation of a number of plasma membrane proteins in vitro. The authors argue that the main effect is on phosphotyrosine residues as sodium orthovanadate was able to block the response, which was not sensitive to okadaic acid, an inhibitor of serine/threonine phosphatases. The dephosphorylation was apparent on Western blots probed with anti-phosphotyrosine antibodies. No receptor specific ligands were used, however, so no information is available about how specific this response is for ANG II or which receptor subtypes are involved. Two other groups[70,71] using the same cell line were unable to find any differences in protein (de)-phosphorylation due to AT_2 receptor stimulation.

6. CONCLUSIONS

In several types of neural cells in culture, and possibly in brain slices too, the AT_1 receptor subtype seems to induce formation of inositol phosphates and changes in intracellular calcium, while there is no evidence that AT_2 receptor might use this pathway. In at least one study AT_2 receptors have been shown to affect Ca^{2+} currents,[66] but further details about the exact mechanism involved are not available yet.

In contrast to peripheral ANG II receptors there is little evidence yet to suspect that central receptors are coupled to adenylate cyclase. It is important to consider the differences in results obtained in membrane preparations and those in intact cells.

There seems to be accumulating evidence, although entirely based on cells in culture, for changes in cGMP metabolism after ANG II receptor stimulation. Multiple mechanisms may be involved. At this moment it is hard to judge in general which receptor subtype is coupled to which signal transduction mechanism. It seems that at least in some cells AT_1 receptors *elevate* cGMP indirectly through increased Ca^{2+}_i levels, similar to what has been described in peripheral systems. Whether a pathway through stimulation of nitric oxide synthase and a change in activity of soluble guanylate cyclase might be involved or some other mechanism could be dependent on the cell type and has to be investigated more specifically.

There seems to be some consensus that at least in some cells ANG II is capable of *decreasing* cGMP levels through AT_2 receptors. Exactly how AT_2 receptors might influence cGMP levels is still an open question. Conflicting evidence from different groups suggest either a change in Ca^{2+} entry resulting in more phosphodiesterase activity or an inhibition of a particulate guanylate cyclase. More direct experiments need to be done to affirm whether any of these responses is directly coupled to ANG II receptors, using receptor specific ligands, and whether any of these mechanisms occur in brain tissue. Especially in brain vasculature it is of great interest to know more about the intracellular mechanisms used by ANG II receptors in the light of a possible role in cerebral blood flow regulation.

Most of the evidence available now supports coupling of AT_1 receptors to an increase in prostaglandin formation in some cells. Whether AT_2 receptors might influence the formation of any of the arachidonate metabolites is not firmly established yet. It is not known whether any of these responses is the result of direct coupling of a receptor to one of the phospholipases or through a more indirect pathway.

It is clear from the studies discussed in previous paragraphs that CGP 42112 seems to act as an agonist in some systems and apparently as an antagonist in others. It is important to use the proper concentrations since micromolar doses of CGP 42112 might act through binding to AT_1 receptors instead.

We lack adequate information about interaction between ANG II responses and other neurotransmitter/hormone/peptide systems in the brain. Often the focus of many studies with ANG II has been whether ANG II by itself has an effect on the parameter studied. The question whether ANG II receptor stimulation might modulate the responses of other receptors has hardly been addressed yet.

REFERENCES

1. Saavedra JM: Brain and pituitary angiotensin. *Endocr Rev* 13:329–380, 1992.
2. Wright JW, Harding JW: Regulatory role of brain angiotensins in the control of physiological and behavioral responses. *Brain Res Rev* 17:227–262, 1992.
3. Phillips MI, Speakman EA, Kimura B: Levels of angiotensin and molecular biology of the tissue renin angiotensin systems. *Regul Pept* 43:1–20, 1993.
4. Phillips MI: Functions of angiotensin in the central nervous system. *Annu Rev Physiol* 49:413–435, 1987.
5. Bottari SP, de Gasparo M, Muscha-Steckelings U, et al: Angiotensin II receptor subtypes: Characterization, signalling mechanisms, and possible physiological implications. In Ganong WF, Martini L (eds): *Frontiers in Neuroendocrinology,* ed. 14. Raven Press, New York, 1993.
6. Timmermans PBMWM, Wong PC, Chiu AT, et al: Angiotensin II receptors and angiotensin II receptor antagonists. *Pharmacol Rev* 45:205–251, 1993.
7. Pickel VM, Chan L, Ganten D: Dual peroxidase and colloidal gold labelling study of angiotensin converting enzyme and angiotensin-like immunoreactivity in the rat subfornical organ. *J Neurosci* 6:2457–2469, 1986.
8. Imboden H, Harding JW, Ganten D, et al: Comparison of angiotensin II staining in rat brain using affinity purified and crude antisera. *Clin Exp Hypertens* [A]9:1133–1139, 1987.
9. Lind RW, Swanson LW, Ganten D: Organization of angiotensin II immunoreactive cells and fibers in the rat central nervous system. *Neuroendocrinology* 40:2–24, 1985.
10. Haas HL, Felix D, Celio MR, et al: Angiotensin II in the hippocampus. A histochemical and electrophysiological study. *Experientia* 36:1394–1395, 1980.
11. Lind RW, Swanson LW, Ganten D: Angiotensin II immunoreactivity in the neural afferents and efferents of the subfornical organ of the rat. *Brain Res* 321:209–215, 1984.
12. Knowles WD, Phillips MI: Angiotensin II responsive cells in the organum vasculosum lamina terminalis (OVLT) recorded in the hypothalamic brain slices. *Brain Res* 197:256–259, 1980
13. Akaishi T, Negoro H, Kobayasi S: Electrophysiological evidence for multiple sites of actions of angiotensin II for stimulating paraventricular neurosecretory cells in the rat. *Brain Res* 220:386–390, 1981.
14. Legendre P, Simmonet G, Vincent JD: Electrophysiological effects of angiotensin II on cultured mouse spinal cords neurons. *Brain Res* 297:287–296, 1984.
15. Ishibashi S, Oomura Y, Gueguen B, et al: Neuronal responses in subfornical organ and other regions to angiotensin II applied by various routs. *Brain Res Bull* 14:307–313, 1985.
16. Ambühl P, Felix D, Imboden H, et al: Effects of angiotensin II and its selective antagonists on inferior olivary neurons. *Regul Pept* 41:19–26, 1992.
17. Hogarty DC, Speakman EA, Puig V, et al: The role of angiotensin, AT_1 and AT_2 receptors in the pressor, drinking and vasopressin responses to central angiotensin. *Brain Res* 586:289–294, 1992.
18. Nahmod VE, Finkielman S, Benarroch EE, et al: Angiotensin regulates release and synthesis of serotonin in brain. *Science* 202:1091–1093, 1978.
19. Qadri F, Badoer E, Stadler T, et al: Angiotensin II-induced noradrenaline release from anterior hypothalamus in conscious rats: a brain microdialysis study. *Brain Res* 563:137–141, 1991.
20. Diz DI, Pirro N: Differential actions of angiotensin II and angiotensin-(1-7) on transmitter release. *Hypertension* 19(Suppl II):II41–II48, 1992.

21. Stadler T, Veltmar A, Qadri F, et al: Angiotensin II evokes noradrenaline release from the paraventricular nucleus in conscious rats. *Brain Res* 569:117–122, 1992.

22. Braszko JJ, Holy ZZ, Kupryszewski G, et al: Effect of angiotensin II, its 2-8, 3-8, 4-8 fragments and saralasin on the level and turnover of dopamine in the striatum and olfactory tubercle of the rat. *Asian Pacif J Pharmacol* 6:243–247, 1991.

23. Badoer E, Würth H, Türck D, et al: Selective local action of angiotensin II on dopaminergic neurons in the rat hypothalamus in vivo. *Naunyn Schmiedebergs Arch Pharmacol* 340:31–35, 1989.

24. Keil LC, Summy-Long J, Severs WB: Release of vasopressin by angiotensin II. *Endocrinology* 96:1063–1065, 1975.

25. Rettig R, Gertsberger R, Meyer J-U, et al: Central effect of angiotensin II in conscious hamsters: Drinking, pressor response, and release of vasopressin, *Comp Physiol B* 158:703–709, 1989.

26. Ferguson AV, Kasting NW: Angiotensin acts at the subfornical organ to increase plasma oxytocin concentrations in the rat. *Regul Pept* 23:343–352, 1988.

27. Gunther S: Characterizations of angiotensin receptor subtypes in rat liver. *J Biol Chem* 259:7622–7629, 1984.

28. Peach MJ, Dostal DE: The angiotensin II receptor and the actions of angiotensin II. *J Cardiovasc Pharmacol* 16(Suppl 4):S25–S30, 1990.

29. Whitebread SE, Mele M, Kamber B, et al: Preliminary biochemical characterization of two angiotensin II receptor subtypes. *Biochem Biophys Res Commun* 163:284–291, 1989.

30. Chiu AT, Herblin WF, McCall DE, et al: Identification of angiotensin II receptor subtypes. *Biochem Biophys Res Commun* 165:196–203, 1989.

31. Bumpus FM, Catt KJ, Chiu AT, et al: Nomenclature for angiotensin receptors. *Hypertension* 17:720–721, 1991.

32. Tsutsumi K, Saavedra JM: Heterogeneity of angiotensin II AT_2 receptors in the rat brain. *Mol Pharmacol* 41:290–297, 1992.

33. Iwai N, Inagami T: Identification of two subtypes in the rat type I angiotensin II receptor. *FEBS Lett* 298:257–260, 1992.

34. Tsutsumi K, Saavedra JM: Characterization and development of angiotensin II receptor subtypes (AT_1 and AT_2) in rat brain. *Am J Physiol* 261:R209–R216, 1991.

35. Tsutsumi K, Saavedra JM: Differential development of angiotensin II receptor subtypes in the rat brain. *Endocrinology* 128:630–634, 1991.

36. Strömberg C, Näveri L, Saavedra JM: Angiotensin AT_2 receptors regulate cerebral blood flow in rats. *Neuroreport* 3:703–704, 1992.

37. Strömberg C, Näveri L, Saavedra JM: Nonpeptide angiotensin AT_1 and AT_2 receptor ligands modulate the upper limit of cerebral blood flow autoregulation in rats. *J Cereb Blood Flow Metab* 13:298–303, 1993.

38. Rowland NE, Rozelle A, Riley PJ, et al: Effect of nonpeptide angiotensin receptor antagonists on water intake and salt appetite. *Brain Res Bull* 29:389–393, 1992.

39. Stephenson KN, Steele MK: Brain angiotensin II receptor subtypes and the control of luteinizing hormone and prolactin secretion in female rats. *J Neuroendocrinol* 4:441–447, 1992.

40. Barnes NM, Champagneria S, Costall B, et al: Cognitive enhancing actions of DuP 753 detected in a mouse habituation paradigm. *Neuroreport* 1:239–242, 1990.

41. Barnes NM, Costall B, Kelly ME, et al: Cognitive enhancing actions of PD 123177 detected in a mouse habituation paradigm. *Neuroreport* 2:351–353, 1991.

42. Millan MA, Carvallo P, Izumi S-I, et al: Novel sites of expression of functional angiotensin II receptors in the late gestational fetus. *Science* 244:1340–1342, 1989.

43. Smith RD, Chiu AT, Wong PC, et al: Pharmacology of non-peptide angiotensin II receptor antagonists. *Annu Rev Pharmacol Toxicol* 32:135–165, 1992.

44. Dudley DT, Hubbell SE, Summerfelt RM: Characterization of angiotensin II (AT$_2$) binding sites in R3T3 cells. *Mol Pharmacol* 40:360–367, 1991.

45. Tsutsumi K, Strömberg C, Viswanathan M, et al: Angiotensin-II receptor subtypes in fetal tissues of the rat: Autoradiography, guanine nucleotide sensitivity, and association with phosphoinostide hydrolysis. *Endocrinology* 129:1075–1082, 1991.

46. Garcia-Sainz JA, Macias-Silva M: Angiotensin II stimulates phosphoinositide turnover and phosphorylase through AII-1 receptors in isolated hepatocytes. *Biochem Biophys Res Commun* 172:780–785, 1990.

47. Ohnishi J, Ishido M, Shibata T, et al: The rat angiotensin II AT$_{1A}$ receptor couples with three different signal transduction pathways. *Biochem Biophys Res Commun* 186:1094–1101, 1992.

48. Zhou J, Ernsberger P, Douglas JG: Angiotensin II receptor subtypes in rat renal mesangial cells. *FASEB J* 5:A870, 1991.

49. Force T, Kyriakis JM, Avruch J, et al: Endothelin, vasopressin, and angiotensin II enhance tyrosine phosphorylation by protein kinase C-dependent and -independent pathways in glomerular mesangial cells. *J Biol Chem* 266:6650–6656, 1991.

50. Huckle WR, Prokop CA, Dy RC, et al: Angiotensin II stimulates protein–tyrosine phosphorylation in a calcium-dependent manner. *Mol Cell Biol* 10:6290–6298, 1990.

51. Lang U, Vallotton MB: Effects of angiotensin II and of phorbol ester on protein kinase C activity and on prostacyclin production in cultured rat aortic smooth-muscle cells. *Biochem J* 259:477–484, 1989.

52. Kawahara Y, Sunako M, Tsuda T, et al: Angiotensin II induces expression of the *cFOS* gene through protein kinase C activation and calcium ion mobilization in cultured vascular smooth muscle cells. *Biochem Biophys Res Commun* 150:52–59, 1988.

53. Chardonnens D, Lang U, Capponi AM, et al: Comparison of the effects of angiotensin II and vasopressin on cytosolic calcium concentrations, protein kinase C activity and prostacyclin production in cultured rat aorta and mesenteric smooth muscle cells. *J Cardiovasc Pharmacol* 14:S39–S44, 1989.

54. Salles J-P, Gayral-Taminh M, Fauvel J, et al: Sustained effect of angiotensin II on tyrosine phosphorylation of annexin I in glomerular mesangial cells. *J Biol Chem* 268:12805–12811, 1993.

55. Molloy CJ, Taylor DS, Weber H: Angiotensin II stimulation of rapid protein tyrosine phosphorylation and protein kinase activation in rat aortic smooth muscle cells. *J Biol Chem* 268:7338–7345, 1993.

56. Kurian P, Narang N, Crews FT: Decreased carbachol-stimulated inositol 1,3,4,5-tetrakisphosphate formation in senescent rat cerebral cortical slices. *Neurobiol Aging* 13:521–526, 1992.

57. Song K, Allen AM, Paxinos G, et al: Mapping of angiotensin II receptor subtype heterogeneity in rat brain. *J Comp Neurol* 316:467–484, 1992.

58. Strömberg C, Tsutsumi K, Viswanathan M, et al: Angiotensin II AT$_1$ receptors in rat superior cervical ganglia: Characterization and stimulation of phosphoinositide hydrolysis. *Eur J Pharmacol* 208:331–336, 1991.

59. Enjalbert A, Sladeczek F, Guillon G, et al: Angiotensin II and dopamine modulate both cAMP and inositol phosphate productions in anterior pituitary cells. *J Biol Chem* 261:4071–4075, 1986.

60. Enjalbert A: Multiple transduction mechanisms of dopamine, somatostatin and angiotensin receptors in anterior pituitary cells. *Horm Res* 31:6–12, 1989.

61. Bockaert J, Journot L, Enjalbert A: Second messengers associated with the action of AII and dopamine D2 receptors in anterior pituitary. Relationship with prolactin secretion. *J Recept Res* 8:225–243, 1988.

62. Raizada MK, Phillips MI, Crews FT, et al: Distinct angiotensin II receptors in primary cultures of glial cells from rat brain. *Proc Natl Acad Sci USA* 84:4655–4659, 1987.

63. Carrithers MD, Raman VK, Masuda S, et al: Effect of angiotensin II and III on inositol polyphosphate production in differentiated NG108-15 hybrid cells. *Biochem Biophys Res Commun* 167:1200–1206, 1990.

64. Ransom JT, Sharif NA, Dunne JF, et al: AT_1 angiotensin receptors mobilize intracellular calcium in a subclone of NG108-15 neuroblastoma cells. *J Neurochem* 58:1883–1888, 1992.

65. Wang JF, Sun XJ, Yang HF, et al: Mobilization of calcium from intracellular store as a possible mechanism underlying the anti-opioid effect of angiotensin II. *Neuropeptides* 22:219–222, 1992.

66. Buisson B, Bottari S, de Gasparo M, et al: The angiotensin AT_2 receptor modulates T-type calcium current in non-differentiated NG108-15 cells. *FEBS Lett* 309:161–164, 1992.

67. Monck JR, Williamson RE, Rogulja I, et al: Angiotensin II effects on the cytosolic free Ca^{2+} concentration in N1E-115 neuroblastoma cells: Kinetic properties of the Ca^{2+} transient measured in single Fura-2-loaded cells. *J Neurochem* 54:278–287, 1990.

68. Zarahn ED, Ye X, Ades AM, et al: Angiotensin-induced cyclic GMP production is mediated by multiple receptor subtypes and nitric oxide in N1E-115 neuroblastoma cells. *J Neurochem* 58:1960–1963, 1992.

69. Tallant EA, Diz DI, Khosla MC, et al: Identification and regulation of angiotensin II receptor subtypes on NG108-15 cells. *Hypertension* 17:1135–1143, 1991.

70. Webb ML, Liu EC-K, Cohen RB, et al: Molecular characterization of angiotensin II type II receptors in rat pheochromocytoma cells. *Peptides* 13:499–508, 1992.

71. Leung KH, Roscoe WA, Smith RD, et al: Characterization of biochemical responses of angiotensin II (AT_2) binding sites in the rat pheochromocytoma PC12W cells. *Eur J Pharmacol Mol Pharmacol* 227:63–70, 1992.

72. Sumners C, Tang W, Zelezna B, et al: Angiotensin II receptor subtypes are coupled with distinct signal-transduction mechanisms in neurons and astrocytes from rat brain. *Proc Natl Acad Sci USA* 88:7567–7571, 1991.

73. Jaiswal N, Tallant A, Diz D, et al: Subtype 2 angiotensin receptors mediate prostaglandin synthesis in human astrocytes. *Hypertension* 17:1115–1120, 1991.

74. Baranczyk-Kuzma A, Audus KL, Guillot FL, et al: Effects of selected vasoactive substances on adenylate cyclase activity in brain, isolated brain, isolated brain microvessels, and primary cultures of brain microvessel endothelial cells. *Neurochem Res* 17:209–214, 1992.

75. Sumners C, Myers LM, Kalberg CJ, et al: Physiological and pharmacological comparisons of angiotensin II receptors in neuronal and astrocyte glial cultures. *Prog Neurobiol* 34:355–385, 1990.

76. Gilbert JA, Pfenning MA, Richelson E: Angiotensin I, II and III stimulated formation of cyclic GMP in murine neuroblastoma clone N1E-115. *Biochem Pharmacol* 33:2527–2530, 1984.

77. Reagan LP, Ye X, Maretzksi CH, et al: Down-regulation of angiotensin II receptor subtypes and desensitization of cyclic GMP production in neuroblastoma N1E-115 cells. *J Neurochem* 60:24–31, 1993.

78. Sumners C, Myers L: Angiotensin II decreases cGMP levels in neuronal cultures from rat brain. *Am J Physiol* 260:C79–C87, 1991.

79. Bottari SP, King IN, Reichlin S, et al: The angiotensin AT_2 receptor stimulates protein tyrosine phosphatase activity and mediates inhibition of particulate guanylate cyclase. *Biochem Biophys Res Commun* 183:206–211, 1992.

80. Brechler V, Jones PW, Levens NR, et al: Agonistic and antagonistic properties of angiotensin analogs at the AT_2 receptor in PC12W cells. *Regul Pept* 44:207–213, 1993.

81. Chaki S, Inagami T: A newly found angiotensin receptor subtype mediates cyclic GMP

formation in differentiated Neuro-2A cells. *Eur J Pharmacol Mol Pharmacol* 225:355–356, 1992.

82. Haberl RL, Anneser F, Vilringer A, et al: Angiotensin II induces endothelium-dependent vasodilation of rat cerebral arterioles. *Am J Physiol* 258:H1840–H1846, 1990.

83. Lamberts SWJ, MacLeod RM: Regulation of prolactin secretion at the level of the lactotroph. *Physiol Rev* 70:279–318, 1990.

84. Judd AM, Ross PC, Spangelo BL, et al: Angiotensin II increases pituitary cell prolactin release and arachidonate liberation. *Mol Cell Endocrinol* 57:115–121, 1988.

85. Canonico PL: Angiotensin increases arachidonate metabolism in cultured anterior pituitary cells. *Gynecol Endocrinol* 3:165–177, 1989.

86. Leung KH, Chang RS, Lotti WJ, et al: AT$_1$ receptors mediate the release of prostaglandins in porcine smooth muscle cells and rat astrocytes. *Am J Hypertens* 5:648–656, 1992.

87. Jaiswal N, Diz DI, Tallant EA, et al: The nonpeptide angiotensin II antagonist DuP 753 is a potent stimulus for prostacyclin synthesis. *Am J Hypertens* 4:228–233, 1991.

88. Barnes JM, Barnes NM, Costall B, et al: Angiotensin II inhibits the release of [^{3}H]acetylcholine from rat entorhinal cortex in vitro. *Brain Res* 491:136–143, 1989.

89. Jumblatt JE, Hackmiller RT: Potentiation of norepinephrine secretion by angiotensin II in the isolated rabbit iris-ciliary body. *Cur Eye Res* 9:169–176, 1990.

10

Angiotensin Receptor Subtypes and Cerebral Blood Flow

Christer Strömberg, Liisa Näveri, and Juan M. Saavedra

1. INTRODUCTION

The brain receives approximately 15% of the cardiac output to maintain a blood flow rate of around 50 ml/100 g tissue per min. The cerebral circulation is secured by two separate vascular channels: the carotid arteries and the vertebrobasilar system. These systems meet at the base of the brain in a vascular ring known as the circle of Willis.[1] Due to this complex anatomy, cerebral blood flow (CBF) can be maintained even if one of the vascular territories fails. The cerebral arteries are innervated by sympathetic, parasympathetic, and sensory nerve fibers, each type containing several potential neurotransmitters.[1] Furthermore, CBF is controlled by local metabolic, or chemical, factors.[1]

Increases or decreases in cerebral perfusion pressure are counteracted by fast constriction or dilation of cerebral resistance vessels. This mechanism, CBF autoregulation, is a physiological adaptation to maintain constant blood flow despite alterations in systemic arterial pressure.[1,2] CBF autoregulation is usually seen in normotensive animals and humans in a mean arterial pressure range between 60 and 140 mm Hg.[1]

CBF and its autoregulation may be impaired or altered in several pathological conditions or they may be affected by drug treatment.[1–3] This chapter attempts to summarize the effects of angiotensin II (ANG II) and the role of its receptor subtypes in the regulation of CBF.

Christer Strömberg • National Agency for Medicines, Pharmacological Department, 00301 Helsinki, Finland.   *Liisa Näveri* • Department of Pharmacology and Toxicology, University of Helsinki, 00170 Helsinki, Finland.   *Juan M. Saavedra* • Section on Pharmacology, Laboratory of Clinical Science, National Institute of Mental Health, National Institutes of Health, Bethesda, Maryland 20892.

Angiotensin Receptors, edited by Juan M. Saavedra and Pieter B.M.W.M. Timmermans. Plenum Press, New York, 1994.

2. ANG II RECEPTORS AND ANGIOTENSIN-CONVERTING ENZYME IN CEREBRAL BLOOD VESSELS

2.1. Microvessels

ANG II binding sites have been found in cerebral microvessels from dog[4] and rat,[5–7] but their subtype has not yet been clarified. The number of microvascular ANG II receptors is higher in adult spontaneously hypertensive rats (SHR) than in normotensive rats,[5,6] whereas 4-week-old SHRs do not differ from the control strain.[6] Thus, the ANG II receptor density seems to be increased only in SHRs with sustained hypertension. Microvessels also contain angiotensin-converting enzyme (ACE), suggesting local ANG II synthesis.[8–14] In contrast to ANG II receptors, however, the activity of ACE is reduced in SHRs.[12]

ANG II stimulates phosphatidylinositol hydrolysis and diacylglycerol formation in cerebral endothelial cells constituting the blood–brain barrier.[7] These being known signal transduction pathways for AT$_1$ receptors, but not described for AT$_2$ receptors, may suggest that the brain capillary ANG II receptors are of the AT$_1$ subtype. ANG II also interacts with atrial natriuretic factor in the brain capillary endothelium by potentiating the atrial natriuretic factor-stimulating increase in cGMP.[7] In addition, atrial natriuretic factor decreases ANG II binding to microvessels.[7]

Since this chapter deals with the actions of ANG II on CBF, it should be stressed that in many of the above-cited studies it is not clear whether preparations of small arteries or pure capillaries were used. Cerebral capillaries do not contribute to cerebrovascular resistance, but they are essential for the blood–brain barrier function.

2.2. Resistance Arteries

The existence of ACE activity in large cerebral arteries from various species, including humans, has been known for 10 years.[15] Therefore, it is surprising that specific ANG II binding sites in these arteries were found only recently.[16] The interesting discovery that these receptors in rat belonged only to the AT$_2$ subtype[16] (Fig. 1) suggested that ANG II may act on cerebral arteries in a way not seen in other vascular beds.

3. ACTIONS OF ANG II ON CEREBRAL ARTERIES

3.1. In Vitro Preparations

ANG II and ANG I contract cerebral artery preparations from various species by a partially endothelium-dependent mechanism possibly involving

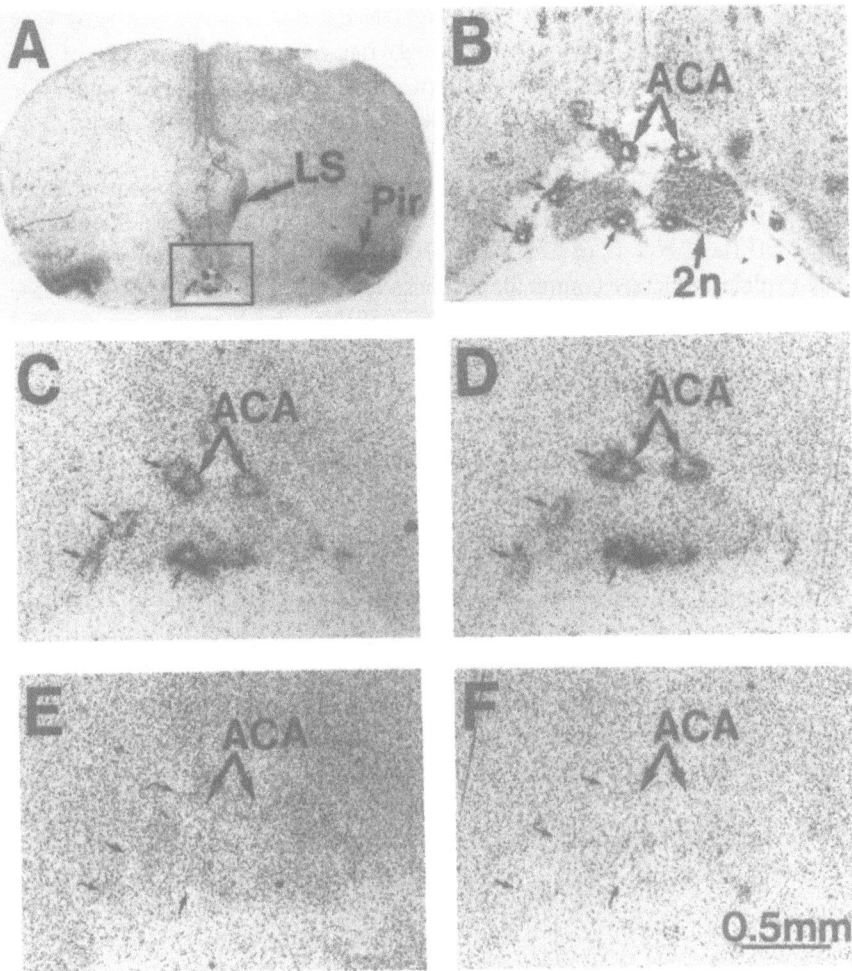

Figure 1. Autoradiography of ANG II binding in cerebral arteries of 2-week-old rats. (A and C) Total binding, incubated in presence of 0.5 nM [^{125}I]-Sar^1-ANG II. (B) View corresponding to C–F, stained with toluidine blue (framed area in A). (D) Displaced with 10 μM losartan (AT$_2$ receptors visualized). (E) Displaced with 100 nM CGP 42112 (AT$_1$ receptors visualized). (F) Displaced with 5 μM unlabeled ANG II (nonspecific binding). ACA, anterior cerebral arteries (orbital segment; small arrows show front-basal branches of anterior cerebral arteries); Pir, piriform cortex; LS, lateral septum; 2n, optic nerve. (From Tsutsumi and Saavedra.[16] Reprinted with permission.)

release of vasoconstrictor prostaglandins, but also stimulation of thromboxane A$_2$ formation has been implicated.[17–25] The contractions can be antagonized by Sar^1,Ile^8-ANG II or Sar^1,Ala^8-ANG II[19,24,25] and the ANG I-induced contraction is inhibited by captopril.[21,22,24] Human, cat, and rabbit cerebral arteries seem to be much more sensitive to ANG II contraction than those of dogs or cows.[19] Whether the ANG II contraction is developmentally regulated is at present controversial. Young (2-week-old) rats have higher numbers of AT_2 receptors in the anterior cerebral arteries than adult (8-week-old) rats.[16] On the other hand, ANG II induced similar contractions of basilar arteries from rabbits that were 15 to 360 days old.[26]

As an exception to the above, in dog cerebral artery strips precontracted with prostaglandin $F_{2\alpha}$, ANG II induced a relaxation that can be antagonized with Sar[1],Ile[8]-ANG II or Sar[1],Ala[8]-ANG II.[23,27] This relaxation component may explain the lesser contractile response to ANG II in this artery.[19] Release of prostacyclin is implicated as a mediator of this relaxation.[27]

3.2. In Vivo Models

ANG II has been shown to constrict pial arteries of all sizes.[28–31] Also in these models ANG I elicits contractions that are attenuated by captopril.[31] In agreement with these findings, ANG II has been shown to increase resistance of pial arteries and, consequently, to decrease cerebral microvascular pressure.[32,33]

In rabbit and rat cerebral arteries ANG II also appears to induce endothelium-dependent pial artery dilation,[34,35] and a recent report indicates that this dilation can be antagonized by PD 123319 but to some extent also by losartan.[36] This suggests that AT_2 receptors and possibly also AT_1 receptors may be involved in the endothelium-dependent dilation. It was also suggested that the dilatory effect is mediated by the natural degradation product of ANG II, ANG II(3–8).[35] However, a recently discovered endothelial binding site for ANG II(3–8) does not bind PD 123177, CGP 42112, or losartan.[37]

4. ANG II EFFECTS OF CBF

Some earlier studies have concluded that ANG II, when infused in the internal carotid artery in humans, has no direct effect on CBF.[38–40] Although this contention may not be entirely true, it now appears that systemic ANG II has only minimal effects on normal CBF because of the autoregulatory mechanisms.[33,41–46]

On the other hand, there is a vast amount of data suggesting that ANG II has a modulatory role on CBF. Inhibition of ACE with captopril shifts the limits of autoregulation toward lower blood pressures and narrows the autoregulatory range in normotensive rats, but does not affect resting CBF.[47–49] This effect can be attenuated by sympathetic stimulation.[50] Captopril appears to affect the CBF autoregulation in humans similarly.[51,52]

Our recent experiments with peptide and nonpeptide ANG II receptor ligands have shown that PD 123319 and CGP 42112 shift the limits of CBF autoregulation toward higher blood pressures in rats[45,46,53,54] (Fig. 2). This effect was antagonized by Sar[1],Ile[8]-ANG II, indicating that it was mediated through ANG II AT$_2$ receptors, and that PD 123319 and CGP 42112 act as agonists in this model. In these studies, losartan showed a similar but less pronounced response, implicating not only AT$_2$ but possibly also AT$_1$ receptors as mediators of the CBF modulation by ANG II. However, the effect of losartan on CBF autoregulation may be mediated indirectly through AT$_2$ receptors since losartan is known to increase plasma ANG II levels severalfold,[55] and the endogenous ANG II so released could stimulate AT$_2$ receptors. The observation that the effect of losartan on CBF autoregulation was not antagonized by Sar[1],Ile[8]-ANG II[54] may suggest that it exerted its effect through a mechanism unrelated to ANG II. It may also be possible that losartan affects the CBF regulation by an action within the brain, although this seems unlikely, since losartan does not penetrate the blood–brain barrier in physiologically relevant amounts.[56] The inability of intracerebroventricularly administered captopril to modulate the CBF autoregulation[48,57] further argues against a role for the brain renin–angiotensin system in the modulation of CBF.

It has been proposed that ANG II maintains a physiological tone in large

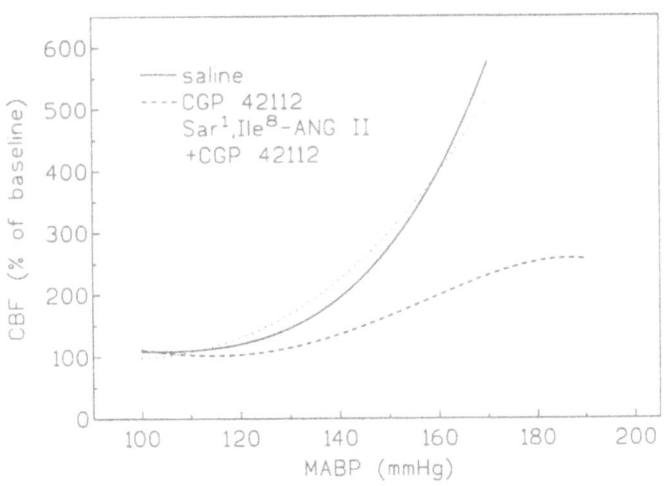

Figure 2. Schematic illustration of AT$_2$-receptor-mediated modulation of CBF autoregulation. Blood pressure (MABP) was increased by phenylephrine (PE) infusion. The solid line shows the CBF response when saline was given with PE. When the AT$_2$ receptor agonist CGP 42112 (1 mg/kg per min) was infused with PE (dashed line), the autoregulation was maintained at higher blood pressures, but basal CBF (left end of curve) was not altered. Infusion of the nonselective antagonist Sar[1], Ile[8]-ANG II (4 µg/kg per min) together with CGP 42112 (dotted line) blocked the effect of CGP 42112.

cerebral arteries and that ACE inhibition would abolish this tone.[47–49,58–60] We have used intravenous infusion of ANG II after administration of losartan to selectively stimulate AT_2 receptors. In this case the limits of autoregulation are again shifted toward higher pressures.[45,46] These findings suggest that AT_2 receptors are responsible for the vascular ANG II tone.

5. EFFECTS OF ANG II ON CBF IN DISEASE MODELS

5.1. Hypertension

CBF autoregulation is shifted toward higher blood pressures in untreated hypertensive rats and humans.[2,3,58,59] ACE inhibition decreases cerebrovascular resistance[61] and causes in SHRs and apparently also in hypertensive patients a similar shift toward lower blood pressures, as has been seen in normotensives.[47–52,59,62] This effect can be regarded as beneficial, because when the blood pressure is normalized, a concomitant downward shift in the autoregulatory range tends to preserve CBF. On the other hand, if the blood pressure suddenly increased during ACE inhibitor treatment, the cerebral circulation would be more vulnerable. Although sympathetic stimulation also modulates the CBF autoregulation,[2] it appears that the ANG II receptors responsible for the autoregulatory effect are not located in sympathetic nerve endings, since the effect of captopril was not abolished in SHRs after sympathetic denervation.[63]

That circulating ANG II may have clinically important effects on CBF is highlighted by two case reports, where normalization of blood pressure apparently may have caused lethal cerebral hypoperfusion in patients with hyperangiotensinemic hypertension.[64,65] This is in agreement with an ANG II-induced shift of CBF autoregulation toward higher pressures.[45,46]

5.2. Chronic Heart Failure

Patients with chronic heart failure often have high plasma renin activity[66] and lower than normal resting CBF.[67,68] Interestingly, low CBF was normalized after captopril treatment in patients with severe heart failure.[67] Thus, it is possible that the activation of the peripheral renin–angiotensin system contributes to the decreased CBF in these patients. In agreement with this idea, ACE inhibition was reported to increase resting CBF in sodium-depleted dogs with high plasma renin activity.[69]

5.3. Stroke

It has been hypothesized that ANG II may be beneficial in the protection against strokes.[70] On the other hand, hypertensive patients with low plasma renin activity appear to be protected from stroke.[71] Experimental stroke models have given equally conflicting results in this respect. ANG II, independently of its hypertensive effect, reduces mortality in gerbils that have undergone unilateral carotid artery ligation.[72] In a similar model, mortality was increased by treatment with enalaprilat or Sar[1], Ala[8]-ANG II.[73] The mechanism behind the increased survival after ANG II is not known, but it was suggested that ANG II may stimulate the development of collateral circulation and hence reduce the severity of brain ischemia.[72,73] Another explanation suggests that delayed platelet aggregation after ANG II treatment could play a role in its beneficial effect.[74]

In one study, captopril was reported to reduce brain injury after ischemia in rats.[75] In this study the ischemic model involved hemorrhagic hypotension. Therefore, it is possible that animals treated with captopril had higher levels of CBF during the experiment,[47–49] and thus in fact were protected from ischemia. In acute stroke patients, captopril did not increase regional CBF when given in a single dose 0.5–5 days after the incident.[76]

Chronic studies have shown more consistent results, since chronic treatment with either ACE inhibitors or losartan reduces stroke related mortality in stroke prone SHRs.[77–79] Whether this effect is mainly due to control of the hypertension is a matter of debate.

5.4. Subarachnoid Hemorrhage

One of the major complications after acute subarachnoid hemorrhage is vasospasm of the cerebral arteries.[80] Inhibition of ACE was shown to reverse acute and delayed cerebral vasospasm after experimental subarachnoid hemorrhage in dogs.[81,82] In a clinical study, an increase in plasma renin activity was related to poor prognosis after subarachnoid hemorrhage, whereas the absolute renin levels had no prognostic significance.[83] These findings suggest that ANG II may be a factor in the pathophysiology of vasospasm.

6. CONCLUSION AND CLINICAL IMPLICATIONS

The action of ANG II on cerebral vessels is well established in several species including humans, and there are clear indications that these actions contribute to the regulation of CBF. Although ANG II has been shown to play a part in certain models of cerebrovascular pathology, most notably stroke and subarachnoid hemorrhage, it has not been conclusively established whether the actions of ANG II are beneficial or detrimental for the outcome. Interestingly, ANG II has practically no effect on resting CBF.

It now appears that the AT_2 receptor subtype is involved in the cerebrovascular effects of ANG II. Further studies should focus on the impact of AT_2

receptor stimulation and antagonism in pathophysiological models of the cerebral circulation. The relative scarcity of AT_2-receptor-mediated physiological events suggests that treatment with AT_2 selective drugs could be implemented without overt adverse effects.

REFERENCES

1. Edvinsson L, MacKenzie ET, McCulloch J: *Cerebral Blood Flow and Metabolism*. Raven Press, New York, 1993.
2. Paulson OB, Strandgaard S, Edvinsson L: Cerebral autoregulation. *Cerebrovasc Brain Metab Rev* 2:161–192, 1990.
3. Herpin D: The effects of antihypertensive drugs on the cerebral blood flow and its regulation. *Prog Neurolbiol* 35:75–83, 1990.
4. Speth RC, Harik SI: Angiotensin II receptor binding sites in brain microvessels. *Proc Natl Acad Sci USA* 82:6340–6343, 1985.
5. Grammas P, Diglio C, Giacomelli F, et al: Cerebrovascular angiotensin II receptors in spontaneously hypertensive rats. *J Cardiovasc Pharmacol* 13:227–232, 1989.
6. Ibaragi M, Niwa M: Atrial natriuretic peptide and angiotensin II binding sites in cerebral capillaries of spontaneously hypertensive rats. *Cell Mol Neurobiol* 9:221–231, 1989.
7. Grammas P, Giacomelli F, Bessert D, et al: Angiotensin II and atrial natriuretic factor receptor interactions at the blood–brain barrier. *Brain Res* 562:93–97, 1991.
8. Brecher P, Tercyak A, Gavras H, et al: Peptidyl dipeptidase in rabbit brain microvessels. *Biochim Biophys Acta* 526:537–546, 1978.
9. Wigger HJ, Stalcup SA: Distribution and development of angiotensin converting enzyme in the fetal and newborn rabbit. *Lab Invest* 38:581–585, 1978.
10. Gimbrone MA Jr, Majeau GR, Atkinson WJ, et al: Angiotensin-converting enzyme activity in isolated brain microvessels. *Life Sci* 25:1075–1084, 1979.
11. Brecher P, Tercyak A, Chobanian AV: Properties of angiotensin-converting enzyme in intact cerebral microvessels. *Hypertension* 3:198–204, 1981.
12. Kobayashi H, Take K, Wada A, et al: Angiotensin-converting enzyme activity is reduced in brain microvessels of spontaneously hypertensive rats. *J Neurochem* 42:1655–1658, 1984.
13. Jandeleit K, Jackson B, Perich R, et al: Angiotensin-converting enzyme in macro- and microvessels of the rat. *Clin Exp Pharmacol Physiol* 18:353–356, 1991.
14. Perich R, Jackson B, Paxton D, et al: Characterization of angiotensin converting enzyme in isolated cerebral microvessels from spontaneously hypertensive and normotensive rats. *J Hypertens* 10:149–153, 1992.
15. Miyazaki M, Okunishi H, Nishimura K, et al: Vascular angiotensin-converting enzyme activity in man and other species. *Clin Sci* 66:39–45, 1984.
16. Tsutsumi K, Saavedra JM: Characterization of AT_2 angiotensin II receptors in rat anterior cerebral arteries. *Am J Physiol* 261:H667–H670, 1991.
17. Uchida E, Bohr DF, Hoobler SW: A method for studying isolated resistance vessels from rabbit mesentery and brain and their responses to drugs. *Circ Res* 21:525–536, 1967.
18. Acar U, Pickard JD: Effect of angiotensin II on pial arterioles. *J Physiol* 284:8–58P, 1978.
19. Toda N, Miyazaki M: Regional and species differences in the response of isolated arteries to angiotensin II. *Japan J Pharmacol* 28:495–497, 1978.
20. Edvinsson L, Hardebo JE, Owman C: Effects of angiotensin II on cerebral blood vessels. *Acta Physiol Scand* 105:381–383, 1979.

21. Whalley ET, Wahl M: The effects of kininase II inhibitors on the response of feline cerebral arteries to bradykinin and angiotensin. *Pflugers Arch* 398:175–177, 1983.

22. Whalley ET, Amure YO, Lye RH: Analysis of the mechanism of action of bradykinin on human basilar artery in vitro. *Naunyn Schmiedebergs Arch Pharmacol* 335:433–437, 1987.

23. Toda N: Hemolysate inhibits cerebral artery relaxation. *J Cereb Blood Flow Metab* 8:46–53, 1988.

24. Manabe K, Shirahase H, Usui H, et al: Endothelium-dependent contractions induced by angiotensin I and angiotensin II in canine cerebral artery. *J Pharmacol Exp Ther* 251:317–320, 1989.

25. Toda N, Ayaziki K, Okamura T: Modifications by endogenous prostaglandins of angiotensin II-induced contractions in dog and monkey cerebral and mesenteric arteries. *J Pharmacol Exp Ther* 252:374–379, 1990.

26. Toda N, Hayashi S: Age-dependent alteration in the response of isolated rabbit basilar arteries to vasoactive agents. *J Pharmacol Exp Ther* 211:716–721, 1979.

27. Toda N, Miyazaki M: Angiotensin-induced relaxation in isolated dog renal and cerebral arteries. *Am J Physiol* 240:H247–H254, 1981.

28. Wei EP, Kontos HA, Patterson JL: Vasoconstrictor effect of angiotensin on pial arteries. *Stroke* 9:487–489, 1978.

29. Joyner WL, Young R, Blank D, et al: In vivo microscopy of the cerebral microcirculation using neonatal allografts in hamsters. *Circ Res* 63:758–766, 1988.

30. Mayhan WG, Amundsen SM, Faraci FM, et al: Responses of cerebral arteries after ischemia and reperfusion in cats. *Am J Physiol* 255:H879–H884, 1988.

31. Whalley ET, Wahl M: Cerebrovascular reactivity to angiotensin and angiotensin-converting enzyme activity in cerebrospinal fluid. *Brain Res* 438:1–7, 1988.

32. Reynier-Rebuffel AM, Pinard E, Aubineau PF, et al: Generalized cerebral vasoconstriction induced by intracarotid infusion of angiotensin II in the rabbit. *Brain Res* 269:91–101, 1983.

33. Faraci FM, Mayhan WG, Schmid PG, et al: Effects of arginine vasopressin on cerebral microvascular pressure. *Am J Physiol* 255:H70–H76, 1988.

34. Haberl R, Anneser F, Villringer A, et al: Angiotensin II induces endothelium-dependent vasodilation of rat cerebral arterioles. *Am J Physiol* 258:H1840–H1846, 1990.

35. Haberl RL, Decker PJ, Einhäupl KM: Angiotensin degradation products mediate endothelium-dependent dilation of rabbit brain arterioles. *Circ Res* 68:1621–1627, 1991.

36. Brix J, Haberl RL: The AT_2-receptor mediates endothelium-dependent dilation of rat brain arterioles. *FASEB J* 6:A1264, 1992.

37. Swanson GN, Hanesworth JM, Sardinia MF, et al: Discovery of a distinct binding site for angiotensin II(3-8), a putative angiotensin IV receptor. *Regul Pept* 40:409–419, 1992.

38. Agnoli A, Battistini N, Bozzao L, et al: Drug action on regional cerebral blood flow in cases of acute cerebro-vascular involvement. *Acta Neurol Scand* 41(Suppl 14):142–144, 1965.

39. Greenfield JC Jr, Tindall GT: Effect of norepinephrine, epinephrine, and angiotensin on blood flow in the internal carotid artery of man. *J Clin Invest* 47:1672–1684, 1968.

40. Olesen J: The effect of intracarotid epinephrine, norepinephrine, and angiotensin on the regional cerebral blood flow in man. *Neurology* 22:978–987, 1972.

41. Hardebo JE, Nilsson B, Owman C: Circulating angiotensin decreases cerebral blood flow; relevance to studies on the upper limit of autoregulation and blood–brain barrier function. *Acta Neurol Scand* 60(Suppl 72):604–605, 1979.

42. Tuor UI, Kondysar MH, Harding RK: Effect of angiotensin II and peptide YY on cerebral and circumventricular blood flow. *Peptides* 9:141–149, 1988.

43. Maktabi MA, Heistad DD, Faraci FM: Effects of angiotensin II on blood flow to choroid plexus. *Am J Physiol* 258:H414–H418, 1990.

44. Tamaki K, Saku Y, Ogata J: Effects of angiotensin and atrial natriuretic peptide on the cerebral circulation. *J Cereb Blood Flow Metab* 12:318–325, 1992.

45. Strömberg C, Näveri L, Saavedra JM: Nonpeptide angiotensin AT_1 and AT_2 receptor ligands modulate the upper limit of cerebral blood flow autoregulation in rats. *J Cereb Blood Flow Metab* 13:298–303, 1993.

46. Näveri L, Strömberg C, Saavedra JM: Stimulation of angiotensin II AT_2 receptors modulates the lower limit of cerebral blood flow autoregulation in rats. *FASEB J* 7:A1952, 1993.

47. Barry DI, Jarden JO, Paulson OB, et al: Cerebrovascular aspects of converting-enzyme inhibition I: Effects of intravenous captopril in spontaneously hypertensive and normotensive rats. *J Hypertens* 2:589–597, 1984.

48. Barry DI, Paulson OB, Jarden JO, et al: Effects of captopril on cerebral blood flow in normotensive and hypertensive rats. *Am J Med* 76(Suppl. 5B):79–85, 1984.

49. Paulson OB, Waldemar G, Andersen AR, et al: Role of angiotensin in autoregulation of cerebral blood flow. *Circulation* 77(Suppl. I):I-55–I-58, 1988.

50. Waldemar G, Paulson OB, Barry DI, et al: Angiotensin converting enzyme inhibition and the upper limit of cerebral blood flow autoregulation: Effect of sympathetic stimulation. *Circ Res* 64:1197–1204, 1989.

51. Paulson OB, Vorstrup S, Andersen AR, et al: Converting enzyme inhibition resets cerebral autoregulation at lower blood pressure. *J Hypertens* 3(Suppl 3):S487–S488, 1985.

52. Waldemar G, Schmidt JF, Andersen AR, et al: Angiotensin converting enzyme inhibition and cerebral blood flow autoregulation in normotensive and hypertensive man. *J Hypertens* 7:229–235, 1989.

53. Strömberg C, Näveri L, Saavedra JM: Angiotensin AT_2 receptors regulate cerebral blood flow in rats. *Neuroreport* 3:703–704, 1992.

54. Näveri L, Strömberg C, Saavedra JM: Angiotensin II AT_2 receptor stimulation extends the upper limit of cerebral blood flow autoregulation: Agonist effects of CGP 42112 and PD 123319. *J Cereb Blood Flow Metab* 14:38–44, 1994.

55. Mizuno K, Tani M, Niimura S, et al: Losartan, a specific angiotensin II receptor antagonist increases angiotensin I and angiotensin II release from isolated rat hind legs: Evidence for locally regulated renin–angiotensin system in vascular tissue. *Life Sci* 50:PL209–PL214, 1992.

56. Bui JD, Kimura B, Phillips MI: Losartan potassium, a nonpeptide antagonist of angiotensin II, chronically administered p.o. does not readily cross the blood–brain barrier. *Eur J Pharmacol* 219:147–151, 1992.

57. Jarden JO, Barry DI, Juhler M, et al: Cerebrovascular aspects of converting-enzyme inhibition II: Blood–brain barrier permeability and effect of intracerebroventricular administration of captopril. *J Hypertens* 2:599–604, 1984.

58. Waldemar G, Paulson OB: Angiotensin converting enzyme inhibition and cerebral circulation—a review. *Br J Clin Pharmacol* 28:177S–182S, 1989.

59. Paulson OB, Waldemar G: Role of the local renin–angiotensin system in the autoregulation of the cerebral circulation. *Blood Vessels* 28:231–235, 1991.

60. Postiglione A, Bobkiewicz T, Vinholdt-Pedersen E, et al: Cerebrovascular effects of angiotensin converting enzyme inhibition involve large artery dilatation in rats. *Stroke* 22:1362–1368, 1991.

61. Bouthier JD, Safar ME, Benetos A, et al: Haemodynamic effects of vasodilating drugs on the common carotid and brachial circulations of patients with essential hypertension. *Br J Clin Pharmacol* 21:137–142, 1986.

62. Bray L, Lartaud I, Muller F, et al: Effects of the angiotensin I converting enzyme inhibitor perindopril on cerebral blood flow in awake hypertensive rats. *Am J Hypertens* 4:246S–252S, 1991.

63. Waldemar G: Acute sympathetic denervation does not eliminate the effect of angiotensin

converting enzyme inhibition on CBF autoregulation in spontaneously hypertensive rats. *J Cereb Blood Flow Metab* 10:43–47, 1990.

64. Haas DC, Streeten DHP, Kim RC, et al: Death from cerebral hypoperfusion during nitroprusside treatment of acute angiotensin-dependent hypertension. *Am J Med* 75:1071–1076, 1983.

65. Haas DC, Anderson GH, Streeten DHP: Role of angiotensin in lethal cerebral hypoperfusion during treatment of acute hypertension. *Arch Intern Med* 145:1922–1924.

66. Brown JJ, Davies DL, Johnson VW, et al: Renin relationships in congestive cardiac failure, treated and untreated. *Am Heart J* 80:329–342, 1970.

67. Rajagopalan B, Raine AEG, Cooper R, et al: Changes in cerebral blood flow in patients with severe congestive cardiac failure before and after captopril treatment. *J Hypertens* 2(Suppl. 3):555–558, 1984.

68. Paulson OB, Jarden JO, Vorstrup S, et al: Effect of captopril on the cerebral circulation in chronic heart failure. *Eur J Clin Invest* 16:124–132, 1986.

69. Gavras H, Liang CS, Brunner HR: Redistribution of regional blood flow after inhibition of the angiotensin-converting enzyme. *Circ Res* 43(Suppl. I):I-59–I-63, 1978.

70. Brown MJ, Brown J: Does angiotensin-II protect against strokes? *Lancet* 2:427–429, 1986.

71. Brunner HR, Laragh JH, Baer L, et al: Essential hypertension: Renin and aldosterone, heart attack and stroke. *N Engl J Med* 286:441–449, 1972.

72. Fernandez LA, Spencer DD, Kaczmar T Jr: Angiotensin II decreases mortality rate in gerbils with unilateral carotid ligation. *Stroke* 17:82–85, 1986.

73. Kaliszewski C, Fernandez LA, Wicke JD: Differences in mortality rate between abrupt and progressive carotid ligation in the gerbil: Role of endogenous angiotensin II. *J Cereb Blood Flow Metab* 8:149–154, 1988.

74. Rosenblum WI, El-Sabban F, Hirsh PD: Angiotensin delays platelet aggregation after injury of cerebral arterioles. *Stroke* 17:1203–1205, 1986.

75. Werner C, Hoffman WE, Kochs E, et al: Captopril improves neurologic outcome from incomplete cerebral ischemia in rats. *Stroke* 22:910–914, 1991.

76. Waldemar G, Vorstrup S, Andersen AR, et al: Angiotensin-converting enzyme inhibition and regional cerebral blood flow in acute stroke. *J Cardiovasc Pharmacol* 14:722–729, 1989.

77. Stier CT Jr, Benter IF, Ahmad S, et al: Enalapril prevents stroke and kidney dysfunction in salt-loaded stroke-prone spontaneously hypertensive rats. *Hypertension* 13:115–121, 1989.

78. von Lutterotti N, Camargo MJF, Campbell WG Jr, et al: Angiotensin II receptor antagonist delays renal damage and stroke in salt-loaded Dahl salt-sensitive rats. *J Hypertens* 10:949–957, 1992.

79. Ogiku N, Sumikawa H, Hashimoto Y, et al: Prophylactic effect of imidapril on stroke in stroke-prone spontaneously hypertensive rats. *Stroke* 24:245–252, 1993.

80. Sengupta RP, McAllister VL: *Subarachnoid Haemorrhage*. Springer-Verlag, Berlin, New York, 1985.

81. Gavras H, Andrews P, Papadakis N: Reversal of experimental delayed cerebral vasospasm by angiotensin-converting enzyme inhibition. *J Neurosurg* 55:884–888, 1981.

82. Andrews P, Papadakis N, Gavaras H: Reversal of experimental acute cerebral vasospasm by angiotensin converting enzyme inhibition. *Stroke* 13:480–483, 1982.

83. Hamann G, Stober T, Schimrigk K: Has plasma renin activity a prognostic significance in subarachnoid haemorrhage? *Acta Neurochir (Wien)* 100:25–30, 1989.

11

Angiotensin II Receptor Subtypes and Growth

Mohan Viswanathan and Juan M. Saavedra

1. ANGIOTENSIN II AND GROWTH

It is becoming increasingly clear that the renin–angiotensin system has a profound influence on growth besides the well-known physiological action on blood pressure regulation and control of fluid volume.[1-4] Angiotensin II (ANG II) induces the expression of growth-related proto-oncogenes c-fos[5-7] and c-myc.[8] Expression of growth factors like platelet-derived growth factor (PDGF)[8,9] and extracellular matrix components like thrombospondin[10] and tenascin[11,12] is also induced by ANG II. Recently, hepatocyte nuclei have been demonstrated to express binding sites for ANG II,[13,14] indicating that the peptide may also act through nuclear receptors. ANG II therefore may exert a direct action on growth or in some cases mediate its effect by interacting with growth factors.

Until recently, most of the studies on the action of ANG II on growth have been conducted on cardiovascular tissues. The recent observation that fetal tissues express enhanced levels of ANG II receptors[15] and the discovery of ANG II receptor subtypes[16] have prompted several groups of investigators to explore the role of ANG II receptor subtypes in growth and development. The aim of this chapter is to specifically examine the available evidence linking ANG II receptor subtypes and growth.

Mohan Viswanathan and Juan M. Saavedra • Section on Pharmacology, Laboratory of Clinical Science, National Institute of Mental Health, National Institutes of Health, Bethesda, Maryland 20892.

Angiotensin Receptors, edited by Juan M. Saavedra and Pieter B.M.W.M. Timmermans. Plenum Press, New York, 1994.

2. ANG II RECEPTOR SUBTYPE EXPRESSION
DURING DEVELOPMENT

2.1. Fetus

ANG II receptors are present in large numbers in several tissues of the fetus including skeletal muscle and connective tissue.[15] The number of receptors in the muscle and connective tissue decrease dramatically soon after birth.[15] Recent studies using quantitative autoradiography reveal that the majority of the ANG II receptors in these sites is of the AT_2 subtype[17,18] (Fig. 1). Tsutsumi *et al.*[17] were also able to demonstrate that the small number of AT_1 receptors present in the same tissue (< 5% of the total number) were coupled to phosphoinositide hydrolysis. Although the transient expression of AT_2 receptors during development strongly suggests a role for this receptor subtype in growth, we still do not understand the functional relevance of the enhanced expression of AT_2 receptors in the fetus.

2.2. Aorta

One of the well-recognized physiological functions of vascular ANG II receptors is to mediate vasoconstriction and regulate blood pressure. Although vascular ANG II receptors were believed to be exclusively AT_1, it was demonstrated recently that the adult rat aorta expresses a small but significant amount of AT_2 receptors as well[19,20] (Fig. 2). We also found that the proportion of AT_2 receptors was markedly higher in the aorta of fetal and young rats, and this predominance of AT_2 receptors is reversed during development.[19] It is now known that AT_2 receptors in the aorta of the neonate do not influence ANG II-induced vascular tone.[21] However, there appears to be an association between organogenesis and AT_2 receptors.

2.3. Kidney

In the kidney of the adult rat and the human, AT_1 receptors predominate.[22–24] Our recent studies reveal that in rat fetus and the neonate, immature glomeruli in the form of comma- and S-shaped bodies located in the nephrogenic zone of the renal cortex express only AT_2 receptors.[25] On the other hand, juxtamedullary glomeruli that were at a more advanced stage of development in these rats and fully developed glomeruli in older rats express only AT_1 receptors (Fig. 3). Therefore, during renal development, AT_2 receptor expression seems to be limited to immature structures.

2.4. Heart

Both AT_1 and AT_2 receptors are expressed in the myocardium of the rabbit[26] and the rat.[27] Ventricular myocardium in the adult rat contains twofold

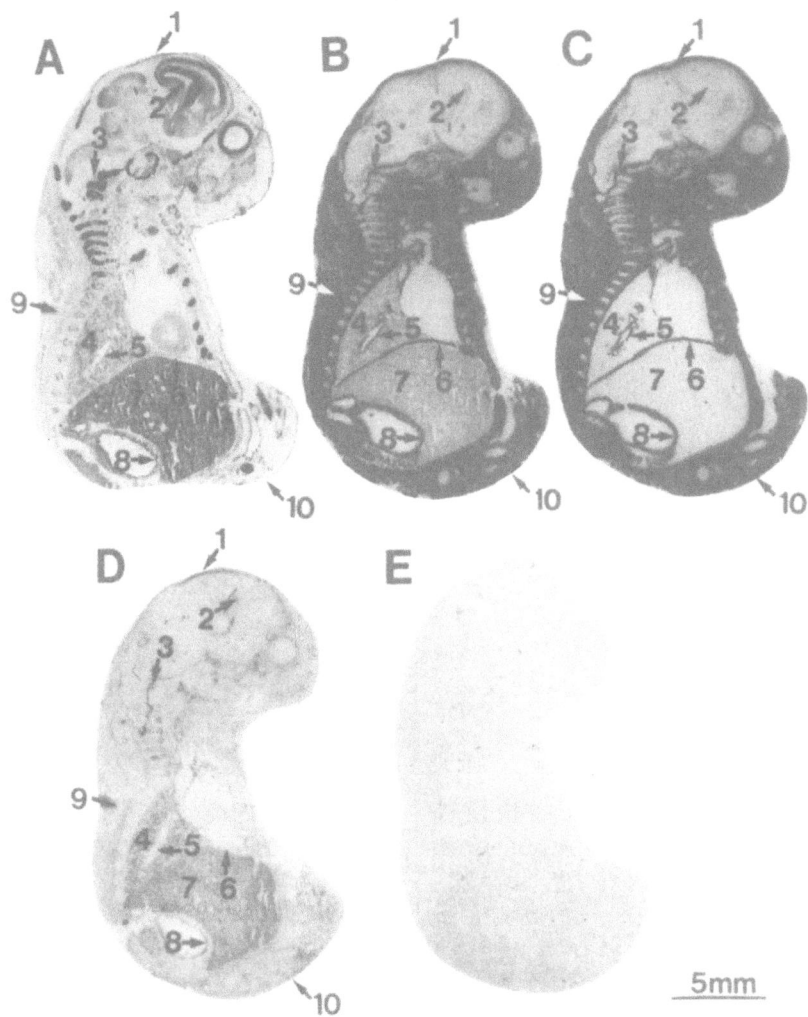

Figure 1. Autoradiography of ANG II receptors in rat fetus. (A) Sagittal section stained with tol-uidine blue. (B) Consecutive section showing total ANG II binding after incubation with 5×10^{-10} M [^{125}I]-Sar^1-ANG II. (C) Consecutive section incubated as in B in the presence of 10^{-5} M losartan. (D) Consecutive section incubated as in B in the presence of 10^{-7} M CGP 42112A. (E) Consecutive section incubated as in B in the presence of 5×10^{-6} ANG II. 1, Skin; 2, choroid plexus; 3, meninges; 4, lung; 5, bronchi; 6, diaphragm; 7, liver; 8, stomach; 9, skeletal muscle; 10, hindlimb. Magnification $\times 3.7$ (bar = 5 mm). (From Tsutsumi *et al.*[17] Reprinted with permission.)

higher number of AT$_1$ receptors than AT$_2$ receptors.[27] Unlike in the case of fetal aorta[19] and kidney glomeruli,[25] fetal rat myocardium has been reported to express equal densities of AT$_1$ and AT$_2$ receptors.[27] The cell type(s) express-

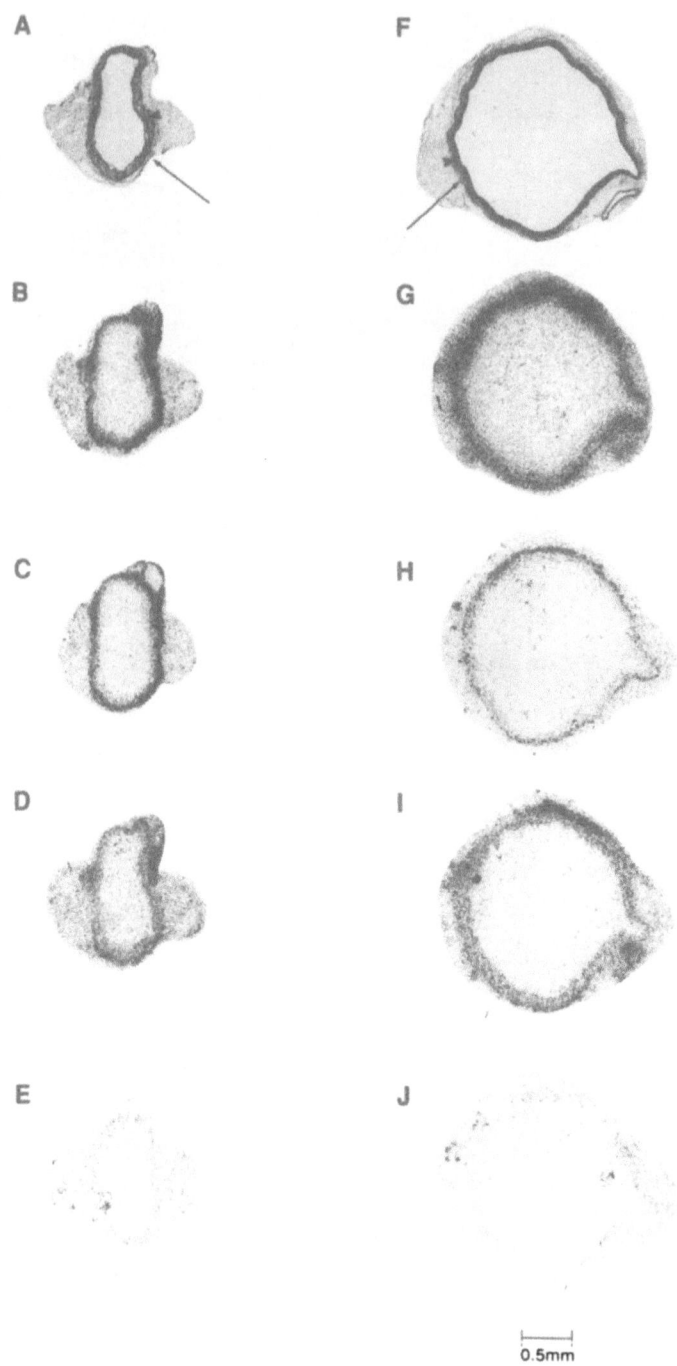

0.5mm

ing each of the receptor subtypes in this tissue has not been identified. This information is needed to assess the physiological functions mediated by the receptor subtypes. ANG II-mediated growth in the myocardium is inhibited by AT_1 receptor antagonist, losartan.[28] This suggests that although both receptor subtypes are present in the myocardium, AT_1 receptors mediate the growth-promoting action of ANG II.

2.5. Fibroblasts

Fetal skin fibroblasts maintained in culture express mainly AT_2 receptors.[29] One of the unique features of the ANG II receptor expression in the fetal fibroblasts is that while in freshly isolated cells a major proportion of ANG II receptors is of the AT_2 subtype, the proportion of AT_1 receptors increases during culture.[29] A fibroblast cell line (Swiss 3T3, R3T3) has also been reported to express exclusively AT_2 receptors.[30] The expression of AT_2 receptors is very low in growing R3T3 cells, while in confluent, quiescent cells, the AT_2 receptor expression is high.[31] These *in vitro* models may become useful for further studies to understand the physiological role of AT_2 receptors.

2.6. Brain

Fetal rat brain contains both AT_1 and AT_2 receptors.[32] While AT_1 receptors are expressed in the nucleus of the solitary tract and the choroid plexus, AT_2 receptors are expressed in high numbers in the inferior olive, paratrigeminal, and hypoglossal nuclei and in the meninges.[32] Comparison of ANG II receptor subtype expression in the brain nuclei of 2-week-old and adult rats reveals that while AT_1 receptor expression is similar in both ages, AT_2 receptor expression is markedly higher in young animals.[33] The subject of brain ANG II receptor subtypes is dealt with in more detail elsewhere in this volume (Chapter 8).

3. ANG II RECEPTOR SUBTYPES AND RESPONSE TO INJURY

The predominance in the expression of AT_2 receptors in tissues undergoing rapid maturation has suggested an association between this receptor

---

Figure 2. Autoradiography of ANG II receptors in young and adult rat aorta. (A–E) Consecutive sections of aorta from a 2-week-old rat. (F–J) Consecutive sections from an 8-week-old rat. (A and F) Sections stained with hematoxylin and eosin. (B and G) Sections showing binding after incubation with 5×10^{-10} M [^{125}I]-Sar^1-ANG II (total binding). (C and H) Sections incubated as in B and G, in the presence of 10^{-5} M losartan. (D and I) sections incubated as in B and G, in the presence of 10^{-5} M PD 123177. (E and J) Sections incubated as in B and G, in the presence of 5×10^{-6} M ANG II (nonspecific binding). Thin arrow points to tunica adventitia, and arrowhead to tunica media. (From Viswanathan *et al.*[19] Reprinted with permission.)

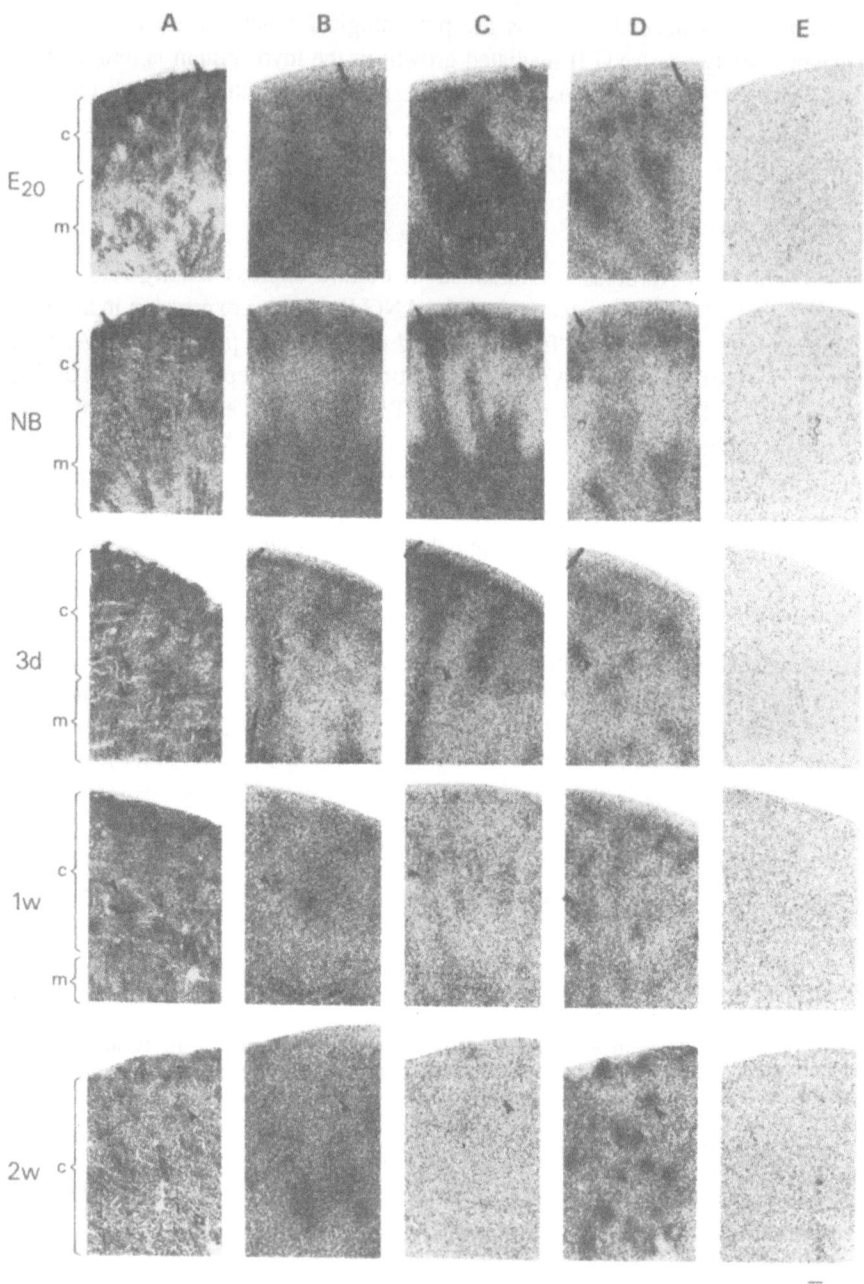

0.1mm

subtype and growth. Several models of injury or inflammation and resulting repair and growth has been studied to assess the role of ANG II receptor subtypes in these processes.

3.1. Wound Healing

Using quantitative autoradiography, both AT_1 and AT_2 receptors could be localized in the skin of young rats.[34] During experimental wound healing, the expression of ANG II receptors increase significantly in a localized band of tissue within the superficial dermis of the skin surrounding the wound (Fig. 4). The major proportion of this increase is due to AT_2 receptors[34] (Fig. 5). Wound-healing studies performed in adult rats have found that AT_1 receptors predominate in the skin and that there is decrease in AT_1 receptors during the initial 12–24 hr after wounding.[35] Both AT_1 and AT_2 receptors seem to be involved in the process of wound healing. However, the precise role of ANG II in skin or in wound healing is not known.

3.2. Vascular Injury

In response to vascular injury using a balloon catheter, medial smooth muscle cells proliferate and migrate to the intima and accumulate there to form neointima.[36,37] Such a phenomenon is believed to be responsible for the initiation of atherosclerosis, and therefore factors that sustain the replication of vascular smooth muscle leading to the formation of neointima are of importance. ANG II is thought to be one such factor.[38,39]

The findings that AT_2 receptors are present in adult aorta and that they predominate in the aorta of fetal and young rats[19] along with the reports that neointimal smooth muscle cells may share some of the phenotypic characteristics of vascular smooth muscle cells isolated from newborn rats[40] raise the possibility that AT_2 receptors may mediate the action of ANG II during phases of rapid smooth muscle replication. However, the neointimal smooth muscle cells formed in the rat aorta or in the carotid artery after balloon injury express

---

Figure 3. Autoradiography of $[^{125}I]$-Sar^1-ANG II binding sites in the kidney and subtype characterization. (A) Sections stained with hematoxylin and eosin. (B–E) Sections adjacent to those in column A. (B) Binding after incubation with 5×10^{-10} M $[^{125}I]$-Sar^1-ANG II (total binding). (C) Binding in the presence of 10^{-5} M losartan (AT_2 receptors). (D) Binding in the presence of 10^{-7} M CGP 42112A (AT_1 receptors). (E) Binding in the presence of 5×10^{-6} M ANG II (nonspecific binding). Regions are indicated as cortex (c) and medulla (m); arrows indicate undeveloped (comma, S-shaped bodies); and arrowheads indicate developing (3-day-old) or mature glomeruli. E_{20}, embryonic day 20; NB, newborn; 3d, 3-day-old; 1w, 1-week-old; 2w, 2-week-old. (From Ciuffo *et al.*[25] Reprinted with permission.)

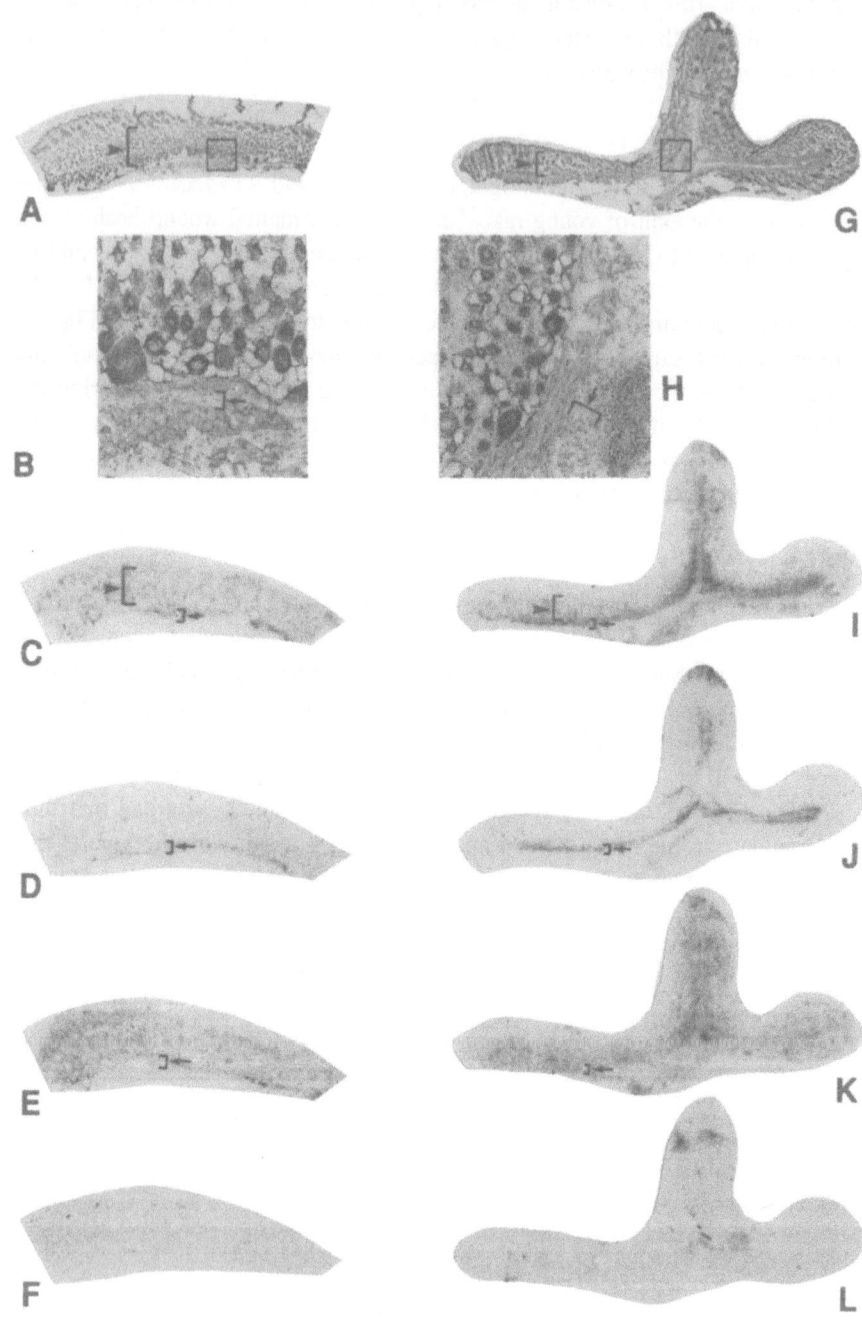

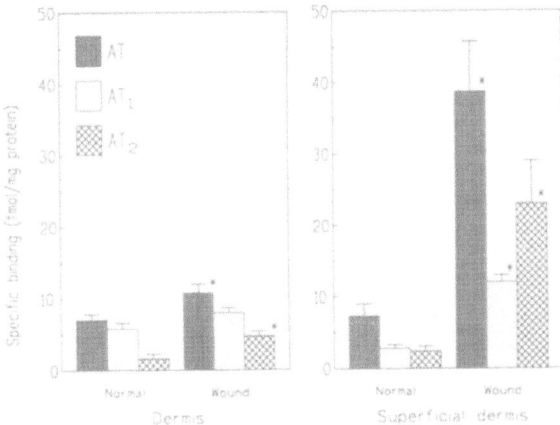

Figure 5. Apparent amount of ANG II receptor subtypes in the dermis (left) and fibroblast-rich band in the superficial dermis (right) in normal skin and in skin containing the healing wound. AT, specific [^{125}I]-Sar^1-ANG II binding; AT$_1$ and AT$_2$, specific binding not displaced by 10^{-5} M PD 123177 or losartan, respectively. Values are means ± SEM obtained from five rats, measured individually. * = $p < 0.05$ vs. normal. (From Viswanathan *et al.*[34] Reprinted with permission.)

exclusively AT$_1$ receptors[41,42] (Figs. 6 and 7). We have also noticed that the expression of AT$_1$ receptors is enhanced in the neointima and that the higher expression is limited to the proliferating cells in the neointima.[42] These data suggest that AT$_1$ receptors may have a significant role in injury-induced growth of vascular smooth muscle. Therapeutic strategies using AT$_1$ receptor blockade seem to have the potential to be effective in the intervention of restenosis after angioplasty.

Figure 4. Autoradiography of ANG II receptors in 2-week-old rat skin. (A and C–F) Consecutive sections of normal skin. (G and I–L) Consecutive sections of skin containing a healing wound. (A and G) Sections stained with hematoxylin and eosin (magnification × 9). (B and H) Enlarged (magnification × 200) views of areas marked in A and G, respectively; arrow points to fibroblast-rich area in superficial dermis. (C and I) Sections showing binding after incubation with 5×10^{-10} M [^{125}I]-Sar^1-ANG II (total binding). (D and J) Sections incubated as in C and I, in the presence of 10^{-5} M losartan. (E and K) Sections incubated as in C and I, in the presence of 10^{-5} M PD 123177. (F and L) Sections incubated as in C and I, in the presence of 5×10^{-6} M ANG II (nonspecific binding). Arrow points to fibroblast-rich band below the dermis named superficial dermis, and arrowhead to dermis. (From Viswanathan *et al.*[34] Reprinted with permission.)

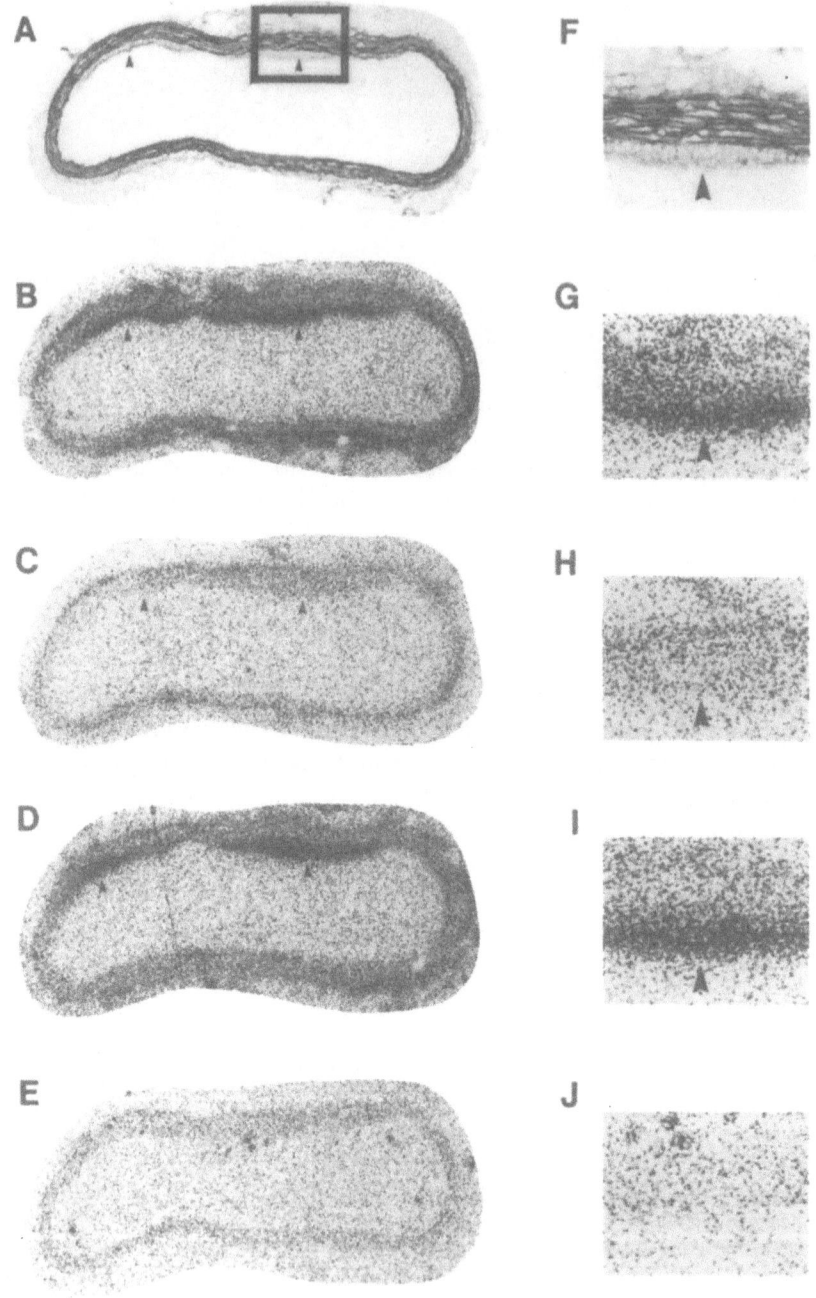

Figure 7. Apparent amount of ANG II receptor subtypes in the media of aorta from sham-operated and balloon-catheterized rats, and in the neointima. AT, specific $[^{125}I]$-Sar^1-ANG II binding; AT$_1$ and AT$_2$, specific binding not displaced by 10^{-5} M PD 123177 or losartan, respectively. Values are mean ± SEM obtained from five rats, measured individually. ANG II receptors (AT) in the neointima were significantly higher (*$p < 0.001$; one-way ANOVA followed by Bonferroni test) than in the aortic media of either sham-operated or balloon-catheterized rats, and this was due to increase in AT$_1$ receptors. (From Viswanathan *et al.*[41] Reprinted with permission.)

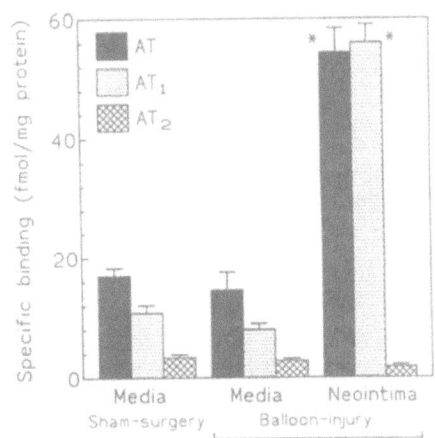

3.3. Tumor

ANG II has been proposed to play an active role in the regulation of proliferation of malignant cells.[43,44] A recent study from our laboratory examined the distribution of ANG II receptor subtypes in medroxyprogesterone-induced adenocarcinomas of the mammary gland in BALB/c mice.[45] Normal mouse mammary gland contains no detectable ANG II receptors. However, enhanced expression of AT$_1$ receptors was found in ductal adenocarcinomas, while AT$_2$ receptor expression was seen in the peritumoral connective tissue around lobular adenocarcinomas (Fig. 8). Further studies are needed to address the function of ANG II receptor subtypes in carcinogenesis.

←——————————————————————————————————

Figure 6. Autoradiography of ANG II receptors in rat aorta injured by balloon catheter. (A–E) Consecutive sections of aorta with arrowheads pointing to neointima (× 37). (F–J) Enlarged (× 100) view of area marked in A, from corresponding sections on their left. Arrowhead points to neointima. (A and F) Sections stained with hematoxylin and eosin. (B and G) Sections showing binding after incubation with 5×10^{-10} M $[^{125}I]$-Sar^1-ANG II (total binding). (C and H) Section incubated as in B, in the presence of 10^{-5} M losartan. (D and I) Section incubated as in B, in the presence of 10^{-5} M PD 123177. (E and J) Sections incubated as in B, in the presence of 5×10^{-6} M ANG II (nonspecific binding). (From Viswanathan *et al.*[41] Reprinted with permission.)

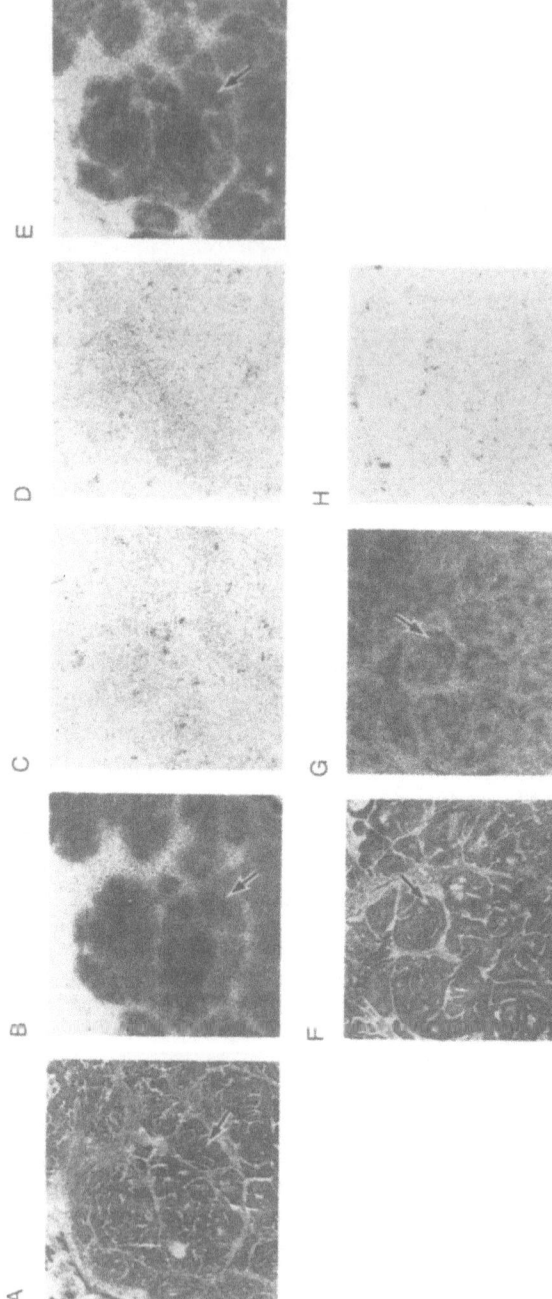

Figure 8. Comparative autoradiography of [^{125}I]-Sar^1-ANG II (upper row) and angiotensin-converting enzyme (ACE) (lower row) distribution to a ductal tumor (tumor 3). (A, F) Sections stained with hematoxylin and eosin. (B–H) Adjacent sections to those in A. (B) Binding after incubation with 5×10^{-10} M [^{125}I]-Sar^1-ANG II (total binding). (C) Binding in the presence of 5×10^{-5} M of ANG II (nonspecific). (D) Binding in the presence of 5×10^{-5} M of losartan (AT$_2$ receptors). (E) Binding in the presence of 1×10^{-5} M of PD 123177 (AT$_1$ receptors). (G) Binding in the presence of 2×10^{-9} M [^{125}I]-351A (total binding). (H) Binding in the presence of 1×10^{-6} M of MK 521 (nonspecific). Arrows point to the epithelial tumoral (neoplastic tissue) (magnification $\times 13$). (From Guerra *et al.*[45] Reprinted with permission.)

4. ANG II AND ANGIOGENESIS

There is compelling evidence that ANG II stimulates angiogenesis in several animal models. ANG II induces neovascularization in the rabbit cornea[46] and the chick embryo chorioallantoic membrane.[47] Formation of collateral vessels to the kidney after aortic stenosis is also stimulated by ANG II.[48] The chick embryo chorioallantoic membrane is the only model in which an attempt was made to characterize the ANG II receptor subtype mediating the angiogenic effect.[49] In this avian model, functional and radioligand binding studies demonstrate that the ANG II receptor has different characteristics than the mammalian receptor subtypes, AT_1 and AT_2.

5. CONCLUSIONS

Is one particular ANG II receptor subtype involved in growth and development? It is obvious that no final conclusions can be drawn from these initial studies. All of the available evidence suggest that both AT_1 and AT_2 receptors have an active role in growth, and that the receptor subtype that is selectively expressed varies with the model. However, several important questions have emerged and remain unanswered. Primarily, what are the factors regulating the differential expression of ANG II receptor subtypes? We still do not understand the functional role and signal transduction mechanisms of the AT_2 receptors. The recent purification of AT_2 receptors from the newborn rat kidney[50] is an exciting advance toward understanding the nature of this receptor subtype.

REFERENCES

1. Schelling P, Fischer H, Ganten D: Angiotensin and cell growth: A link to cardiovascular hypertrophy? *J Hypertens* 9:3–15, 1991.
2. Re R: Angiotensin and the regulation of cellular growth. *Am J Hypertens* 4:217S–219S, 1991.
3. Katz AM: Angiotensin II: Hemodynamic regulator or growth factor? *J Mol Cell Cardiol* 22:739–747, 1990.
4. Heagerty AM: Angiotensin II: Vasoconstrictor or growth factor? *J Cardiovasc Pharmacol* 18:S14–S19, 1991.
5. Kawahara Y, Sunako M, Tsuda T, et al: Angiotensin II induces expression of the *c-fos* gene through protein kinase C activation and calcium ion mobilization in cultured vascular smooth muscle cells. *Biochem Biophys Res Commun* 150:52–59, 1988.
6. Taubman MB, Berk BC, Izumo S, et al: Angiotensin II induces *c-fos* mRNA in aortic smooth muscle: Role of Ca^{2+} mobilization and protein kinase C activation. *J Biol Chem* 264:526–530, 1989.
7. Naftilan AJ, Pratt RE, Eldridge CS, et al: Angiotensin II induces *c-fos* expression in smooth muscle via transcriptional control. *Hypertension* 13:706–711, 1989.
8. Naftilan AJ, Pratt RE, Dzau VJ: Induction of platelet-derived growth factor A-chain and

c-myc gene expressions by angiotensin II in cultured rat vascular smooth muscle cells. *J Clin Invest* 83:1419–1424, 1989.

9. Araki S, Kawahara Y, Kariya K, et al: Stimulation of platelet derived growth factor-induced DNA synthesis by angiotensin II in rabbit vascular smooth muscle cells. *Biochem Biophys Res Commun* 168:350–357, 1990.

10. Scott-Burden T, Bühler FR: Regulation of smooth muscle proliferative phenotype by heparainoid-matrix interactions. *Trends Pharmacol Sci* 9:94–98, 1988.

11. Sharifi BG, LaFleur DW, Pirola CJ, et al: Angiotensin II regulates tenascin gene expression in vascular smooth muscle cells. *J Biol Chem* 267:23910–23915, 1992.

12. Mackie EJ, Scott-Burden T, Hahn AW, et al: Expression of tenascin by vascular smooth muscle cells: Alterations in hypertensive rats and stimulation by angiotensin II. *Am J Pathol* 141:377–388, 1992.

13. Booz GW, Conrad KM, Hess AL, et al: Angiotensin II-binding sites on hepatic nuclei. *Endocrinology* 130:3641–3649, 1992.

14. Tang SS, Rogg H, Schumacher R, et al: Characterization of nuclear angiotensin II-binding sites in rat liver and comparison with plasma membrane receptors. *Endocrinology* 131:374–380, 1992.

15. Millan MA, Caravallo P, Izumi S-I, et al: Novel sites of expression of functional angiotensin II receptors in the late gestation fetus. *Science* 244:1340–1342, 1989.

16. Bumpus FM, Catt KJ, Chiu AT, et al: Nomenclature for angiotensin receptors: A report of the nomenclature committee of the council for high blood pressure research. *Hypertension* 17:720–721, 1991.

17. Tsutsumi K, Strömberg C, Viswanathan M, et al: Angiotensin II receptor subtypes in fetal tissues of the rat: Autoradiography, guanine nucleotide sensitivity, and association with phosphoinositide hydrolysis. *Endocrinology* 129:1075–1082, 1991.

18. Grady EF, Sechi LA, Griffin CA, et al: Expression of AT_2 receptors in the developing rat fetus. *J Clin Invest* 88:921–933, 1991.

19. Viswanathan M, Tsutsumi K, Correa FMA, et al: Changes in expression of angiotensin receptor subtypes in the rat aorta during development. *Biochem Biophys Res Commun* 179:1361–1367, 1991.

20. Chang RS, Lotti VJ: Angiotensin receptor subtypes in rat, rabbit and monkey tissues. Relative distribution and species dependency. *Life Sci* 49:1485–1490, 1991.

21. Keiser JA, Major TC, Lu GH, et al: Is there a functional cardiovascular role for AT_2 receptors? *Drug Dev Res* 29:94–99, 1993.

22. Gröne H-J, Simon M, Fuchs E: Autoradiographic characterization of angiotensin receptor subtypes in fetal and adult human kidney. *Am J Physiol* 262:F326–F331, 1992.

23. Sechi LA, Grady EF, Griffin CA, et al: Characterization of angiotensin II receptor subtypes in the rat kidney and heart using the non-peptide antagonists DuP 753 and PD 123177. *J Hypertens* 9:S224–S225, 1991.

24. Sechi LA, Grady EF, Griffin CA, et al: Distribution of angiotensin II receptor subtypes in rat and human kidney. *Am J Physiol* 262:F236–F240, 1992.

25. Ciuffo GM, Viswanathan M, Seltzer AM, et al: Glomerular angiotensin II receptor subtypes during development of rat kidney. *Am J Physiol* 265:F254–F271, 1993.

26. Rogg H, Schmid A, de Gasparo, M: Identification and characterization of angiotensin II receptor subtypes in rabbit ventricular myocardium. *Biochem Biophys Res Commun* 173:416–422, 1990.

27. Sechi LA, Griffin CA, Grady EF, et al: Characterization of angiotensin II receptor subtypes in rat heart. *Circ Res* 71:1482–1489, 1992.

28. Dostal DE, Baker KM: Angiotensin II stimulation of left ventricular hypertrophy in adult rat heart: Mediation by the AT_1 receptor. *Am J Hypertens* 5:276–280, 1992.

29. Johnson MC, Aguilera G: Angiotensin II receptor subtypes and coupling to signaling systems in cultured fetal fibroblasts. *Endocrinology* 129:1266–1274, 1991.

30. Dudley DT, Hubell SE, Summerfelt RM: Characterization of angiotensin II (AT₂) binding sites in R3T3 cells. *Mol Pharmacol* 40:360–367, 1991.

31. Dudley DT, Summerfelt RM: Regulated expression of angiotensin II (AT₂) binding sites in R3T3 cells. *Regul Pept* 44:199–206, 1993.

32. Tsutsumi K, Viswanathan M, Strömberg C, et al: Type-1 and type-2 angiotensin II receptors in fetal rat brain. *Eur J Pharmacol* 198:89–92, 1991.

33. Tsutsumi K, Saavedra JM: Characterization and development of angiotensin II receptor subtypes (AT₁ and AT₂) in rat brain. *Am J Physiol* 261, R209–R215, 1991.

34. Viswanathan M, Saavedra JM: Expression of angiotensin II AT₂ receptors in the rat skin during experimental wound healing. *Peptides* 13:783–786, 1992.

35. Kimura B, Sumners C, Phillips MI: Changes in skin angiotensin II receptors in rats during wound healing. *Biochem Biophys Res Commun* 187:1083–1090, 1992.

36. Ross R: The pathogenesis of atherosclerosis: An update. *N Engl J Med* 314:488–500, 1986.

37. Schwartz SM, Campbell GR, Campbell JH: Replication of smooth muscle cells in vascular disease. *Circ Res* 58:427–444, 1986.

38. Powell JS, Clozel JP, Müller RKM, et al: Inhibitors of angiotensin-converting enzyme prevent myointimal proliferation after vascular injury. *Science* 245:186–188, 1989.

39. Daemen MJAP, Lombardi DM, Bosman FT, et al: Angiotensin II induces smooth muscle cell proliferation in the normal and injured rat arterial wall. *Circ Res* 68:450–456, 1991.

40. Majesky MW, Schwartz SM: Smooth muscle diversity in arterial wound repair. *Toxicol Pathol* 18:554–559, 1990.

41. Viswanathan M, Strömberg C, Seltzer A, et al: Balloon angioplasty enhances the expression of angiotensin II AT₁ receptors in neointima of rat aorta. *J Clin Invest* 90:1707–1712, 1992.

42. Viswanathan M, Saavedra JM: Expression of angiotensin II AT₁ receptors during development of neointima in the carotid artery of rats. *FASEB J* 7:A331, 1993.

43. Chen L, Re RN, Prakash O, et al: Angiotensin-converting enzyme inhibition reduces neuroblastoma cell growth rate. *Proc Soc Exp Biol Med* 196:280–283, 1991.

44. Ariza A, Fernandez LA, Inagami T, et al: Renin in glioblastoma multiforme and its role in neovascularization. *Am J Clin Pathol* 90:437–441, 1988.

45. Guerra FK, Ciuffo GM, Elizalde PV, et al: Enhanced expression of angiotensin II receptor subtypes and angiotensin converting enzyme in medroxyprogesterone-induced mouse mammary adenocarcinomas. *Biochem Biophys Res Commun* 193:93–99, 1993.

46. Fernandez LA, Twickerl J, Mead A: Neovascularization produced by angiotensin II. *J Lab Clin Pathol* 105:141–145, 1985.

47. Le Noble FAC, Hekking JWM, Van Straaten HWM, et al: Angiotensin II stimulates angiogenesis in the chorioallantoic membrane of the chick embryo. *Eur J Pharmacol* 195:305–306, 1991

48. Fernandez LA, Casride VJ, Twickler J, et al: Renin–angiotensin and development of collateral circulation after renal ischemia. *Am J Physiol* 243:H869–H875, 1982.

49. Le Noble FAC, Schreurs NHJS, Van Straaten HWM, et al: Evidence for a novel angiotensin II receptor involved in angiogenesis in chick embryo chorioallantoic membrane. *Am J Physiol* 264:R460–R465, 1993.

50. Ciuffo GM, Heemskerk FMJ, Saavedra JM: Purification and characterization of angiotensin AT₂ receptors from newborn rat kidney. *Proc Natl Acad Sci USA* 90:11009–11013, 1993.

12

Inhibiting the Effects
of Angiotensin II on
Cardiovascular Hypertrophy
in Experimental Hypertension

James J. Morton

1. INTRODUCTION

Much of the information concerning angiotensin II (ANG II) and its effects on cardiovascular hypertrophy derives from therapy studies in experimental models of hypertension, particularly the spontaneously hypertensive rat (SHR). Angiotensin-converting enzyme (ACE) inhibitors seem to be particularly effective in that they have persistent effects on blood pressure after treatment is stopped. More recently, similar findings have been found using the angiotensin AT_1 receptor antagonist losartan. It has been suggested that the persistent effects of these agents might be related to their ability to modify vascular structure by blocking a direct tropic effect of ANG II on blood vessels. Many questions remain, however, concerning the mechanism involved. In particular, the precise role played by ANG II remains to be clarified.

2. THE DEVELOPMENT AND FUNCTION OF THE VASCULATURE

The primary function of the vasculature is to deliver blood from the heart to the capillaries and organs of the body as required for optimum tissue

James J. Morton • Medical Research Council Blood Pressure Unit, Western Infirmary, Glasgow G11 6NT, Scotland.

Angiotensin Receptors, edited by Juan M. Saavedra and Pieter B.M.W.M. Timmermans. Plenum Press, New York, 1994.

metabolism. For this to happen, blood vessels must develop structurally in a manner that enables them to both withstand the pressure of blood within the vessel and regulate flow as required by the metabolic needs of the tissue. The former is dependent on the characteristics and architecture of the vessel wall[1] and the latter is related principally to the size of the lumen[2] and its autoregulatory control by local myogenic and vasoconstrictor agents. As the diameter of the vessel falls to less than about 200 μm (arterioles and capillaries), resistance increases. This serves as a mechanism to control flow within the vascular beds. Resistance in conjunction with cardiac output determines the level of systemic blood pressure, which in turn determines the structural requirements of the vessel wall. The vascular beds of most organs have a large capacity for autoregulation to keep blood flow constant. Consequently, any persistent increase in peripheral resistance, as in hypertension, probably reflects adaptive changes to the structure of the vessel in order to regulate flow in the face of increased pressure. Hence, the development of vascular hypertrophy in hypertension is a secondary and compensatory response to the increase in wall stress brought about by the raised pressure. The implication of this is that reduction of blood pressure to normal levels is in itself insufficient and should be accompanied by a normalization of vascular structure if blood flow is to be maintained at an optimum level.[3] This is not generally the case with antihypertensive treatment where reduction of blood pressure is not accompanied by a reversal of vascular hypertrophy unless prolonged treatment is used.[4]

3. VASCULAR HYPERTROPHY

As mentioned, all forms of established hypertension in both humans and animals is associated with altered vascular structure[5,6] as characterized by a thickening of the vessel wall and narrowing of the lumen with a resultant increase in the media/lumen ratio. According to Poiseuille's Law, this altered structure will result in an increased peripheral resistance at maximum dilatation with a reduction in blood flow.[1] The increase in wall thickness can result from an increase in smooth muscle cell size (hypertrophy), an increase in cell number (hyperplasia), or a rearrangement of existing cells (same size and number) such that there are more layers of cells within the vessel wall (remodeling) (Fig. 1). Remodeling is distinguished from hypertrophy and hyperplasia in that it necessarily results in a reduction of the vessel diameter. In addition, it is likely that there are alterations in the nature and amount of extracellular matrix proteins. Which of these alternatives or possible combinations is actually present in hypertension is not known for sure. For the purposes of this chapter, all will be referred to under the general heading of vascular hypertrophy.

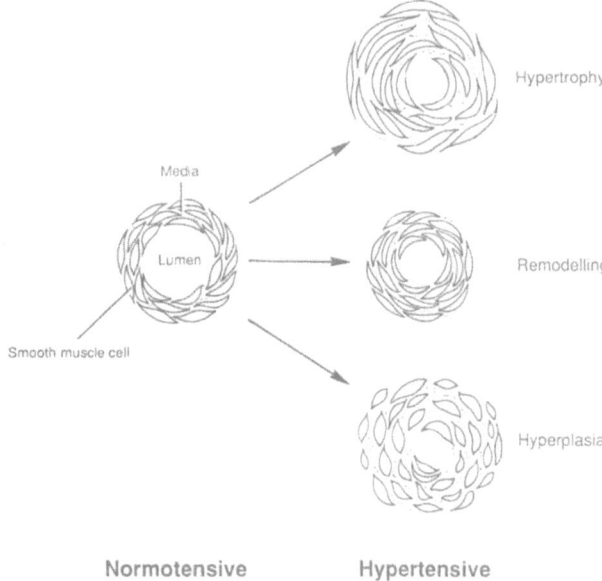

Figure 1. Diagrammatic representation of a normotensive blood vessel that has undergone hypertensive structural change resulting in increased media thickness and narrowed lumen. The possible forms this might take are shown as hypertrophy (increased cell size), hyperplasia (increased cell number), and remodeling (rearrangement of existing cells).

4. THE VASCULAR AMPLIFIER

Folkow[7] and Korner and his colleagues[8,9] proposed that this altered structural geometry leads to a fixed hyperresponsiveness of the arterial vascular system and acts as an amplifier further aggravating the hypertension. Hypertrophy as a secondary response might therefore contribute to the further development of the hypertension through a structure–function interaction. Folkow[5] proposed that essential hypertension was initiated in early life following complex interactions between environmental, genetic, and neurohumoral factors, resulting in intermittent pressure rises that then trigger the cardiovascular hypertrophic response. Take-up by the amplifier turns this normal response into a pathological mechanism, leading to a slow progressive rise in arterial pressure. Thus hypertrophy can be both a consequence and a cause of hypertension.[6] If this is the case, then prevention of this early rise in pressure by the use of effective antihypertensive therapy would block the onset of hypertrophy, disarm the amplifier, and prevent the development of hypertension. This is an attractive hypothesis, but the evidence of this existence in hypertension remains contradictory.

5. LOAD-INDEPENDENT VASCULAR HYPERTROPHY

The classical view that cardiovascular hypertrophy develops as a secondary response to the load of increased pressure has been extended to include the concept of load-independent hypertrophy. This was originally proposed by Lee and Smeda[10] who suggested that there may be a primary abnormality of the blood vessel structure in hypertension that develops independent of pressure. This could trigger the vascular amplifier and be a mechanism by which hypertension is initiated. In support of this, they reported[11,12] the presence of cardiovascular hypertrophy in very young SHR at a time before pressure had started to rise, i.e., at a prehypertensive stage. Adams *et al.* have reported a similar finding. While other researchers[14] have also found vascular hypertrophy at this very early age in the SHR, however, they were unable to confirm the existence of a prehypertensive stage. These very young animals had an intra-arterial pressure 35 mm Hg higher than equivalent normotensive control rats.[14] The existence of a prehypertensive stage in the SHR is therefore open to question, and increased levels of pressure have been reported in rats as young as 1 day old.[15-17] Certainly as far as the SHR is concerned, the suggestion of hypertrophy being present in early life independent of raised pressure remains questionable.

6. ANG II AND VASCULAR GROWTH

Lever[18] has also suggested that hypertrophy could develop independent of pressure as a consequence of an abnormality of growth and thus be a primary cause of hypertension. He proposed that this could involve a tropic property of ANG II acting directly on smooth muscle cells within the media. Such a mechanism has been promoted as the explanation of ANG II's well-documented slow pressor effect.[18]

There is accumulating evidence that ANG II under the right circumstances can act to promote cell growth.[19] Evidence from *in vitro* cell culture experiments is strongest. There is, however, some question as to whether it produces an increase in cell size[20] or number.[21,22] Angiotensin has also been shown to stimulate the production of cellular nuclear proto-oncogenes thought to be important in the regulation of cell growth.[23,24] These effects are inhibited by both saralasin[22] and more specifically losartan,[25] indicating that the intracellular mechanisms stimulating growth are mediated via the AT_1 receptor. Vascular smooth muscle cells in culture dedifferentiate to a proliferating phenotype and are consequently more susceptible to growth stimuli. *In vivo* evidence for a direct growth effect of ANG II on vascular smooth muscle cells comes from studies in which ANG II was infused into rats at subpressor doses, producing a slow progressive rise in pressure over 5–6 days accompanied by the development of cardiac and vascular hypertrophy.[26] Suppressing the rise in pressure with hydralazine during ANG II infusion did not prevent the development of vascular hypertrophy, implying a non–load-dependent mechanism. Also, administration of ANG II in a slow-release form into the cornea of the eye produced new vessel formation,[27] indicating an angiogenic property. Indirect evidence of ANG IIs importance in cell growth regulation *in vivo* comes from studies showing that ACE inhibition can prevent myointimal proliferation after vascular injury.[28] There is therefore increasing evidence of an ANG II-mediated effect on cell growth. However, the involvement of these phenomena and their contribution to the etiology of essential hypertension remain obscure.

7. EFFECT OF ANG II INHIBITION ON HYPERTENSION AND VASCULAR HYPERTROPHY

Studies examining the effects of various drug treatments during early development in the SHR have provided some of the best evidence of ANG II's involvement in hypertensive development and its associated cardiovascular structural changes. These have suggested a therapy-specific effect in that while a wide variety of antihypertensive agents are able to prevent the rise in blood pressure in the SHR, the effect on pressure after stopping treatment appears to depend on the particular class of agents used. Although beta blockers, vasodilators, and calcium antagonists are effective in preventing the rise in pressure during treatment, pressure returns to that seen in untreated hypertensive animals whenever treatment is stopped.[29–35] These drugs were also relatively ineffective in preventing the associated cardiovascular structural changes that develop. In contrast, therapy with ACE inhibitors have proved particularly effective in preventing structural change and have effects on blood pressure that persist long after treatment stopped. It has been suggested that these persistent effects of ACE inhibitors on blood pressure could be related to their ability to prevent the onset of the vascular amplifier[36] by specifically blocking a trophic effect of ANG II on blood vessels,[26,37] which because of the close relation between structure and pressure[1] would have long-term consequences for hypertensive levels.[6] In other words, long-term blood pressure levels are related to and determined by ANG II-mediated vascular hypertrophy in early life. It is the specific ability of ACE inhibitors to prevent the latter that determines the level of the former. This theory, though interesting, remains unproven and there is indeed some evidence to the contrary.

The original work that led to the idea of ACE-specific therapy came from Freslon and Giudicelli.[29] They showed that captopril and hydralazine were equally effective as antihypertensive agents but that only captopril[38] had a long-lasting effect in preventing the functional and morphological vascular alterations seen during hypertensive development. Following this, similar studies using other ACE inhibitors including enalapril[36] and perindopril[37,39,40] showed comparable persistent effects on vascular structure and blood pressure. Long-term converting enzyme inhibition in the SHR attenuated or prevented the onset of medial hypertrophy in both resistance vessels and the thoracic aorta and reduced the number of polyploid muscle cells compared with control animals.[33,41] Also, captopril treatment in renal hypertensive rats reduced aortic and microvascular growth.[38] In this respect, hydralazine was less potent, and β-adrenoreceptor blockade had little influence on the morphological changes of the vasculature.[33] Similarly, long-term treatment with calcium antagonists pinacidil[30] and felodipine[31] had little or no effect on the vascular structural changes that occur during hypertensive development in the SHR in spite of achieving a significant lowering of blood pressure. While the effects of increased wall stress was not disputed, it was suggested that ANG II had an action over and above this to bring about hypertrophy either directly or via enhanced sympathetic activity.[33] The myocardial hypertrophy in the SHR could also be prevented or reversed by chronic converting enzyme inhibition[42–44] but not by treatment with a calcium antagonist or vasodilator.[30] The accumulating evidence of these studies and the fact that all ACE inhibitors were effective provides mounting evidence in favor of an ANG II-mediated involvement in the early onset of hypertension in this genetic model of hypertension, possibly via a direct myo- and cardiogenic property.

More recent evidence implicating ANG II comes from studies using the ANG II receptor antagonist losartan.[45] This is a nonpeptide antagonist specific for the ANG II AT_1 receptor.[46,47] Chronic treatment with losartan and captopril were equally effective in preventing the onset of hypertension and cardiovascular hypertrophy in the young SHR.[45] When given for 4 and 10 weeks, losartan prevented not only the development of hypertension during treatment but also attenuated the redevelopment of hypertension after treatment was stopped (Fig. 2). While losartan given for 10 weeks also prevented the onset of vascular hypertrophy as judged by a reduced media/lumen, there was no effect on media/lumen ratio compared with control SHR following 4 weeks treatment. Captopril given for 4 weeks also had no effect on vascular structure.

Circulating levels of renin and ANG II are not abnormal in the SHR, being either lower or the same as that for the control normotensive WKY.[14] It is unlikely, therefore, that ACE inhibitors work by inhibiting the effects of circulating ANG II. It is possible that they bring about their structural and functional effects by inhibiting a growth or tonic property of locally generated ANG II. There is now evidence that both the heart and blood vessels contain the genetic

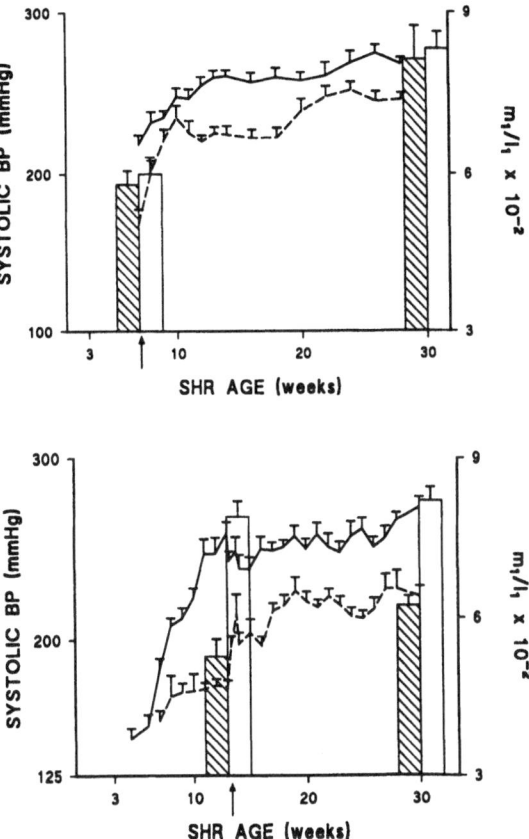

Figure 2. Mean systolic blood pressure in spontaneously hypertensive rats given water (—) or losartan (– –) 15 mg/kg per day in the drinking water for 4 weeks from age 3 to 7 weeks (top) and for 10 weeks from age 3 to 13 (bottom) weeks. Arrows indicate end of treatment periods. After stopping treatment, pressures for both the 4- and 10-week losartan-treated groups remained significantly below those of SHR controls (ANOVA, $p < 0.001$). The columns show the effect of losartan (shaded) on mesenteric-resistance artery structure (media/lumen ratio) at the end of treatment and at age 30 weeks after stopping treatment. Treatment with losartan for 4 weeks did not prevent the onset of hypertrophy despite having a persistent effect on blood pressure. In contrast, treatment for 10 weeks had a highly significant inhibitory effect ($p < 0.001$) on the development of resistance artery hypertrophy which persisted throughout the posttreatment period up to age 30 weeks. The lack of effect after 4 weeks may be dose-related.

material and all the components of the renin–angiotensin system necessary for the generation of ANG II.[48,49] It is likely that blood vessels both synthesize and release ANG II, but that *de novo* synthesis makes relatively little contribution, the majority being generated from renin taken up by vessels from the circulation.[50] A recent ACE inhibitor study has indicated the importance of this system to the maintenance of high blood pressure.[51]

8. DOES VASCULAR STRUCTURE DETERMINE BLOOD PRESSURE?

The idea of early structural changes to the arterial system (possibly involving ANG II) influencing long-term hypertensive levels has engaged the interest of researchers in the field of hypertension for a number of years. While an attractive idea and supported by much circumstantial evidence, it remains essentially unproven. Indeed, not all work is supportive. A recent report from Mulvany *et al.*[52] has indicated that there may be important differences between genetic models of hypertension with respect to the effects of ACE inhibitors. In contrast to the SHR, ACE inhibition with perindopril did not alter long-term pressure levels in the Milan hypertensive rat despite having an effect on both vascular structure and blood pressure during treatment. Christensen *et al.*,[53] in a study comparing the effect of different doses of perindopril in the SHR, showed that on treatment perindopril had a dose-dependent effect on both pressure and structure, but that the persistent effect on pressure off treatment was no longer related to either the dose or to resistance vessel media thickness at the end of the treatment period. In another study comparing the effects of captopril, hydralazine, isradipine, and metoprolol[34] the same workers demonstrated the importance of pressure in determining prevailing vascular and cardiac structure. Similarly in the study of Morton *et al.*[45] with losartan it was possible to produce long-term persistent effects on blood pressure after stopping treatment without affecting vascular structure. At age 30 weeks after the withdrawal of treatment, blood pressure and vascular structure were correlated (Fig. 3). These data demonstrate a dissociation between structure and function, suggesting that the former is not necessarily a good predictor of the latter, and emphasize the primacy of pressure in determining both vascular and cardiac structure.

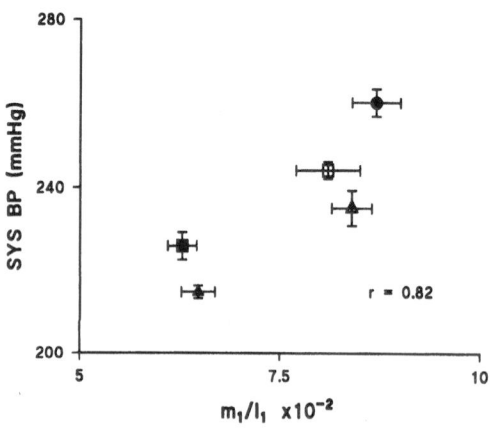

Figure 3. Relation between media/lumen ratio and systolic blood pressure at age 30 weeks posttreatment in control SHR (●), SHR given captopril (△) or losartan (□) from age 3 to 7 weeks, and SHR given captopril (▲) or losartan (■) from age 3 to 13 weeks ($r = 0.82$, $p = 0.095$).

9. PULSE PRESSURE AND VASCULAR HYPERTROPHY

A recent and interesting study by Christensen[54] has indicated that pulse pressure and heart rate are of particular importance in defining arterial structure. He found in a study involving five different antihypertensive therapies, which included two ACE inhibitors, a calcium antagonist, a β-blocker, and a vasodilator, that overall media/lumen ratio correlated best with pulse pressure (Fig. 4) followed by systolic, mean, and diastolic pressures in that order. The results showed that the two ACE inhibitors were most effective in reducing both pulse pressure and structure, the β-blocker was least effective, while the others fell between these two extremes. These data indicate that a reduction in the amplitude and frequency (heart rate) of pulse pressure may be particularly important in preventing the development of abnormal resistance artery structure in hypertension. It also suggests that variation in these intralumenal mechanical factors can account for most of the variation in vessel media/lumen ratio, even in studies using different classes of hypertensive drugs. In other words, it argues against drug-specific effects on vessel structure independent of intravascular pressure.

The relevance of these considerations to regression of resistance artery hypertrophy in established hypertension comes from the accepted finding that reduction of blood pressure does not normalize vascular structure unless very long periods of antihypertensive therapy are used.[4] It would appear that under these conditions the normal relation between pressure and structure is altered,[34] and to achieve regression of hypertrophy requires that pressure be lowered to

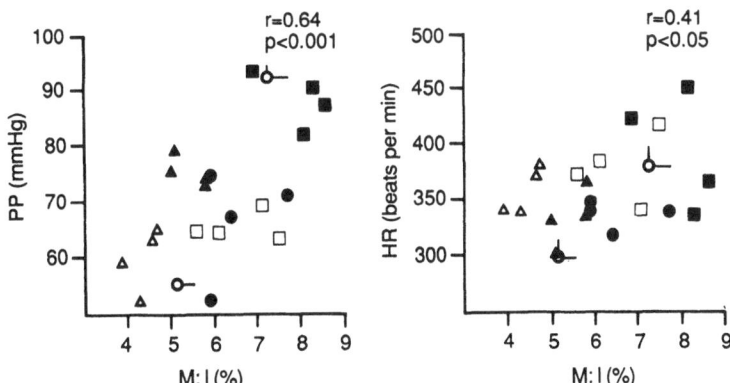

Figure 4. Scatterplots showing relation between resistance vessel media/lumen ratio (m : l) and 24-hr measurement of pulse pressure (pp) and heart rate (HR) in SHR at the end of 20 weeks treatment with hydralazine (●), metoprolol (■), isradipine (▲), perindopril (△), and captopril (□). ○, Group mean ± SEM for untreated WKY and SHR control rats. (From Mulvany and Aalkjaer.[55] Reprinted with permission of the author and the American Heart Association, Inc.)

hypotensive levels. It will be of interest to see if future therapy targeted not just to the reduction of pressure per se but more specifically to the reduction of pulse pressure and heart rate is more successful in this respect.

10. CONCLUSION

As discussed in the early part of this chapter, regression of vascular hypertrophy to maintain tissue blood flow is an important goal in hypertensive therapy. There is now strong experimental evidence that therapeutic regimens that interfere with the renin–angiotensin system (ACE inhibitors and ANG II receptor antagonists) are particularly effective, though not necessarily unique, in preventing the development of both hypertension and cardiovascular hypertrophy. Although recent interest has centered on the possibility that this might involve inhibition of a direct trophic effect of angiotensin on blood vessels, conclusive evidence of this and the role of vascular structure in determining pressure levels is still awaited. In addition, there is now evidence[55,56] that altered vascular structure in hypertension might in some instances involve a remodeling process; that is, an increase in media/lumen ratio resulting from a redistribution of existing vascular material without an increase in mass (Fig. 1). This would not require growth. In view of this and the strong relation between pressure and structure, particularly pulse pressure, pathophysiologically more acceptable alternative mechanisms cannot be ruled out. This might include the ability of these agents to alter these intraluminal mechanical forces by interfering with a tonic effect of locally generated ANG II or by correcting an early abnormality of ANG II-mediated sodium absorption within the kidney. Many questions remain to be answered.

REFERENCES

1. Folkow B: Physiological aspects of primary hypertension. *Physiol Reviews* 62:347–504, 1982.
2. Langille BL, O'Donnell F: Reductions in arterial diameter produced by chronic decreases in blood flow are endothelium-dependent. *Science* 231:405–407, 1986.
3. Mulvany MJ, Aalkjaer C: Calcium metabolism and structure in the peripheral vasculature: Implications for hypertension. *J Cardiovasc Pharmacol* 12(Suppl. 5):s134–s138, 1988.
4. Hartford M, Wendelhag I, Berglund G, et al: Cardiovascular and renal effects of long-term antihypertensive treatment. *JAMA* 259:2553–2557, 1988.
5. Folkow B: Vascular changes in hypertension—a review and recent animal studies. In Berglund G, Hansson L (eds): *Pathophysiology and Management of Arterial Hypertension.* L Lidgren and Soner, Molndall, Sweden, 1975, pp. 95–113.
6. Mulvany MJ: Role of vascular structure in blood pressure development of the spontaneously hypertensive rat. *J Hypertens* 4(Suppl. 3):S61–S63, 1986.
7. Folkow B: Structural adaptation: Its role in the initiation and maintenance of primary hypertension. *Clin Sci* 55:3s–22s, 1978.

8. Korner PI: Causal and homeostatic factors in hypertension. *Clin Sci* 63:5s–26s, 1982.

9. Wright CE, Angus JA, Korner PI: Vascular amplifier properties in renovascular hypertension in conscious rabbits: Hindquarter responses to constrictor and dilator stimulii. *Hypertension* 9:122–131, 1987.

10. Lee RMKW, Smeda JS: Primary versus secondary structural changes of the blood vessels in hypertension. *Can J Physiol Pharmacol* 63:392–401, 1985.

11. Lee RMKE: Vascular changes at the prehypertensive phase in the mesenteric arteries from spontaneously hypertensive rats. *Blood Vessels* 22:105–126, 1985.

12. Smeda JS, Lee RMKW, Forrest JB: Structural and reactivity alterations of the renal vasculature of spontaneously hypertensive rats prior to and during established hypertension. *Circ Res* 63:518–533, 1988.

13. Adams MA, Bobik A, Korner PI: Differential development of vascular and cardiac hypertrophy in genetic hypertension: Relation to sympathetic function. *Hypertension* 14:191–202, 1989.

14. Morton JJ, Beattie EC, Griffin SA, et al: Vascular hypertrophy, renin and blood pressure in the young spontaneously hypertensive rat. *Clin Sci* 79:523–530, 1990.

15. Gray SD: Spontaneous hypertension in the neonatal rat. *Clin Exp Hypertens* 6:755–781, 1984.

16. Bruno L, Agar S, Weller D: Absence of a prehypertensive stage in postnatal kyoto hypertensive rats. *Jpn Heart J* 20(Supp. 1):90–92, 1985.

17. Gray SD: Pressure profiles in neonatal spontaneously hypertensive rats. *Biol Neonate* 45:25–32, 1984.

18. Lever AF: Slow pressor mechanisms in hypertension: A role for hypertrophy of resistance vessels. *J Hypertens* 4:515–524, 1986.

19. Schelling P, Fischer H, Ganten D: Angiotensin and cell growth: A link to cardiovascular hypertrophy? *J Hypertens* 9:3–15, 1991.

20. Geisterfer AAT, Peach MJ, Owens GK: Angiotensin II induces hypertrophy, not hyperplasia, of cultured rat aortic smooth muscle cells. *Circ Res* 62:749–756, 1988.

21. Campbell-Boswell M, Robertson AL: Effects of angiotensin II and vasopressin on human smooth muscle cells in vitro. *Exp Mol Pathol* 35:265–276, 1981.

22. Lyall F, Morton JJ, Lever AF, et al: Angiotensin II activates Na^+-H^+ exchange and stimulates growth in cultured vascular smooth muscle cells. *J Hypertens* 6(Suppl. 4):S438–S441, 1988.

23. Naftilan AJ, Pratt RE, Eldridge CS, et al: Angiotensin II induces *c-fos* expression in smooth muscle via transcriptional control. *Hypertension* 13:706–711, 1989.

24. Lyall F, Gillespie D, Morton JJ: Angiotensin II stimulates *c-jun* expression in cultured vascular smooth muscle cells: Superinduction by emetine. *Eur J Intern Med* 2:271–273, 1992.

25. Lyall F, Dornan ES, McQueen J, et al: Angiotensin II increases proto-oncogene expression and phosphoinositide turnover in vascular smooth muscle cells via the angiotensin II AT_1 receptor. *J Hypertens* 10:1463–1469, 1992.

26. Griffin SA, Brown WCB, MacPherson F, et al: Angiotensin II causes vascular hypertrophy in part by a non-pressor mechanism. *Hypertension* 17:626–635, 1991.

27. Fernandez LA, Caride VJ, Twickler J, et al: Renin–angiotensin and development of collateral circulation after renal ischemia. *Am J Physiol* 243:H869–H875, 1982.

28. Powell JS, Clozel JP, Muller RKM, et al: Inhibitors of angiotensin-converting enzyme prevent myointimal proliferation after vascular injury. *Science* 245:186–188, 1989.

29. Freslon JL, Giudicelli JF: Compared myocardial and vascular effects of captopril and dihydralazine during hypertension development in spontaneously hypertensive rats. *Br J Pharmacol* 80:533–543, 1983.

30. Jespersen LT, Nyborg NCB, Pedersen OL, et al: Cardiac mass and peripheral vascular structure in hydralazine-treated spontaneously hypertensive rats. *Hypertension* 7:734–741, 1985.

31. Nyborg NCB, Mulvany MJ: Lack of effect of anti-hypertensive treatment with felodipine on

cardiovascular structure of young spontaneously hypertensive rats. *Cardiovasc Res* 19:528–536, 1985.

32. Jespersen LT, Baandrup U, Nyborg NCB, et al: Aggressive long-term antihypertensive therapy with pinacidil does not cause regression of cardiovascular hypertrophy in the spontaneously hypertensive rat. *J Hypertens* 4:223–227, 1986.

33. Owens GK: Influence of blood pressure on development of aortic medial smooth muscle hypertrophy in spontaneously hypertensive rats. *Hypertension* 9:178–187, 1987.

34. Christensen KL, Jespersen LT, Mulvany MJ: Development of blood pressure in spontaneously hypertensive rats after withdrawal of long-term treatment related to vascular structure. *J Hypertens* 7:83–90, 1989.

35. Smeda JS, Lee RMKW: Effect of hydralazine on the mesenteric vasculature of hypertensive rats. *Hypertension* 17:526–533, 1991.

36. Adams MA, Bobik A, Korner PI: Enalapril can prevent vascular amplifier development in spontaneously hypertensive rats. *Hypertension* 16:252–260, 1990.

37. Harrap SB, Van der Merwe WM, Griffin SA, et al: Brief angiotensin converting enzyme inhibitor treatment in young spontaneously hypertensive rats reduces blood pressure long-term. *Hypertension* 16:603–614, 1990.

38. Wang KH, Prewitt RL: Captopril reduces aortic and microvascular growth in hypertensive and normotensive rats. *Hypertension* 15:68–77, 1990.

39. Harrap SB, Nicolaci JA, Doyle AE: Persistent effects on blood pressure and renal haemodynamics following chronic angiotensin converting enzyme inhibition with perindopril. *Clin Exp Pharmacol Physiol* 13:753–765, 1986.

40. Cadihac M, Guidicelli JR: Myocardial and vascular effects of perindopril, a new converting enzyme inhibitor, during hypertension development in spontaneously hypertensive rats. *Arch Int Pharmacodyn Ther* 286:114–125, 1986.

41. Black MJ, Adams MA, Bobik A: Effect of enalapril on aortic smooth muscle cell polyploidy in the spontaneously hypertensive rat. *J Hypertens* 7:997–1003, 1989.

42. Sen S, Tarazi RC, Bumpus FM: Effect of converting enzyme inhibitor (SQ 24,225) on myocardial hypertrophy in spontaneously hypertensive rats. *Hypertension* 2:169–176, 1980.

43. Clozel JP, Kuhn H, Hefti F: Effects of chronic ACE inhibition on cardiac hypertrophy and coronary vascular reserve in spontaneously hypertensive rats with developed hypertension. *J Hypertens* 7:267–275, 1989.

44. Linz W, Scholkens BA, Ganten D: Converting enzyme inhibition specifically prevents the development and induces regression of cardiac hypertrophy in rats. *Clin Exp Hypertens* 7:1325–1350, 1989.

45. Morton JJ, Beattie EC, MacPherson F: Angiotensin II receptor antagonist losartan has persistent effects on blood pressure in the young spontaneously hypertensive rat: Lack of relation to vascular structure. *J Vasc Res* 29:264–269, 1992.

46. Chiu AT, Herblin WF, McCall DE, et al: Identification of angiotensin II receptor subtypes. *Biochem Biophys Res Commun* 165:196–203, 1989.

47. Chiu AT, Duncia JV, McCall DE, et al: Nonpeptide angiotensin II receptor antagonists. III Structure–function studies. *J Pharmacol Exp Ther* 250:867–874, 1989.

48. Dzau VJ: Vascular angiotensin pathways: A new therapeutic target. *J Cardiovasc Pharmacol* 10:S9–S16, 1987.

49. Gohlke P, Bunning P, Unger T: Distribution and metabolism of angiotensin I and II in the blood vessel wall. *Hypertension* 20:151–157, 1992.

50. Swales JD, Heagerty AM: Vascular renin–angiotensin system: Some unanswered questions. *J Hypertens* 5:S1–S5, 1987.

51. Mizuno K, Tani M, Niimura S, et al: Effect of delapril on the vascular angiotensin II release in isolated hind legs of the spontaneously hypertensive rat: Evidence for potential relevance

of vascular angiotensin II to the maintenance of hypertension. *Clin Exp Pharmacol Physiol* 18:619–625, 1991.

52. Mulvany MJ, Persson AEG, Andresen J: No persistent effect of angiotensin converting enzyme inhibitor treatment in Milan hypertensive rats despite regression of vascular structure. *J Hypertens* 9:589–593, 1991.

53. Christensen HRL, Nielsen H, Christensen KL: Long-term hypotensive effects of an angiotensin converting enzyme inhibitor in spontaneously hypertensive rats: Is there a role for vascular structure? *J Hypertens* 6(Suppl. 3):S27–S31, 1988.

54. Christensen KL: Reducing pulse pressure in hypertension may normalize small artery structure. *Hypertension* 18:722–727, 1991.

55. Mulvany MJ, Aalkjaer C: Structure and function of small arteries. *Physiol Rev* 70:921–961, 1990.

56. Heagerty AM, Aalkjaer C, Bund SJ, et al: Small artery structure in hypertension. Dual processes of remodeling and growth. *Hypertension* 21:391–397, 1993.

13

Inhibiting the Effects of Angiotensin on Cardiovascular Hypertrophy

Kaj P. Metsärinne, Monika Stoll, Mechthild Falkenhahn, Peter Gohlke, and Thomas Unger

The rapid increase in blood pressure brought about by angiotensin II (ANG II) is due to vasoconstriction. In addition to this rapid effect, low doses of ANG II increase blood pressure by a slow process, which may be due to the development of vascular hypertrophy.[1,2] Abnormal vascular smooth muscle cell growth, which consists of hypertrophy and/or hyperplasia, is a characteristic feature in arteries from hypertensive patients and animals.[3-5] This condition may even lead to chronic hypertension by means of a vascular amplifier mechanism,[6] implying enhanced inherent susceptibility of hypertrophied vasculature to endogenous constricting agents. The cardiac response to increased blood pressure is a combination of myocardial/perivascular fibrosis and hypertrophy of the cardiomyocytes.[7] What is the role of ANG II in the development and maintenance of cardiovascular (CV) hypertrophy? In this chapter, we will seek for the answer by reviewing the current data on the prevention and regression of CV hypertrophy in association with pharmacologically induced inhibition of ANG II effects.

Kaj P. Metsärinne • Minerva Foundation Institute for Medical Research, and IVth Department of Medicine, Helsinki University Hospital, SF-00250 Helsinki, Finland. *Monika Stoll, Mechthild Falkenhahn, Peter Gohlke, and Thomas Unger* • German Institute for High Blood Pressure Research and Pharmacological Institute, University of Heidelberg, D-6900 Heidelberg, Germany.

Angiotensin Receptors, edited by Juan M. Saavedra and Pieter B.M.W.M. Timmermans. Plenum Press, New York, 1994.

1. DEFINITION OF CARDIOVASCULAR HYPERTROPHY

Vascular hypertrophy encompasses an increase in vascular smooth muscle cell (VSMC) mass. This may involve: (1) an increase in VSMC size (hypertrophy), which may or may not be associated with polyploidy, i.e., an increase in the cellular DNA content, and/or (2) an increase in VSMC number (hyperplasia). VSMC growth is accompanied by extracellular matrix changes consisting of an increase in arterial collagen, elastin, and proteoglycans.[8] Hypertensive vasculopathy may involve both hypertrophy and hyperplasia,[9–12] while the atherosclerotic lesion formation is associated with intimal migration and proliferation of medial VSMC (hyperplasia), with little or no hypertrophy at all.[13] Neointima formation, or myointimal hyperplasia, the characteristic lesion responsible for restenosis after experimental and clinical balloon angioplasty, or graft atherosclerosis in transplanted hearts, also involves intimal migration and proliferation of medial VSMC.[13,14]

Cardiac hypertrophy, when induced by high systemic blood pressure, consists of hypertrophy of the myocytes and accumulation of collagen (fibrosis) in the left ventricle.[7] A growing body of evidence has accumulated to indicate a role for the renin–angiotensin system (RAS) in the development of hypertension and CV hypertrophy. As the effector molecule of circulating and local RAS, ANG II may be involved in the development of the above-mentioned forms of CV hypertrophy (for review, see ref. 15).

1.1. Effects of ANG II on Growth of Vascular Smooth Muscle Cells

1.1.1. In Vitro Studies

In vitro studies have shown that ANG II stimulates both protein synthesis (hypertrophy) and cellular proliferation (hyperplasia) in VSMCs from man[16] and rats, the mode of reaction in the rats (hypertrophy or hyperplasia) being dependent on (1) strain, (2) tissue, and (3) the presence or absence of cofactors, i.e., fetal calf serum or growth factors.[9–12,17–21] Differences in ANG II-induced growth responses in VSMCs from different rat strains may also be attributed to differential phenotypic modulation *in vitro*. This implies the reversion of VSMCs from the contractile to the proliferative/secretory phenotype.[22] Thus, ANG II treatment promotes hyperplasia in cultured aortic VSMCs derived from spontaneously hypertensive rats (SHR),[11,12,19] but induces hypertrophy in aortic VSMCs derived from normal Sprague-Dawley (S-D) rats.[9,10,18,21] However, in mesenteric VSMCs derived from normal rats, ANG II treatment (in the presence of 10% fetal calf serum) was associated with hyperplasia.[17] In addition to promoting the growth of VSMCs, ANG II stimulates extracellular matrix synthesis by cultured VSMCs.[11] ANG II-induced hypertrophy and hyperplasia of VSMC, as well as extracellular matrix synthesis, were blocked by the ANG II receptor antagonist Sar[1],Ile[8] ANG II (saralasin).[9–12,17]

The early signal transduction events of ANG II-induced trophic effects (both hypertrophy and hyperplasia) in VSCMs include stimulation of phospholipase C-mediated hydrolysis of membrane phosphoinositides, leading to the formation of inositol phosphates and diacylglycerol (DAG), followed by an increase in intracellular calcium and activation of protein kinase C (for review, see ref. 15). Protein kinase C is the main cellular mediator of increased expression of immediate early genes (IEG) like *c-fos, c-myc,* and *c-jun.* Treatment with ANG II results in rapid activation of *c-fos* and *c-myc* genes,[12,23–25] as well as increased autocrine expression of growth factors like platelet-derived growth factor (PDGF) and transforming growth factor beta (TGFβ) in cultured VSMCs.[19,24,26,27] This sequential activation of IEG growth factor genes may be important for ANG II-induced vascular hypertrophy and hyperplasia.[28] ANG II-induced effects on inositol phosphate production and proto-oncogene expression were inhibited by losartan (DuP 753), an AT_1 receptor antagonist, but not by an AT_2 receptor binding compound.[25,29]

1.1.2. In Vivo Studies

Morphometric studies have shown that the increase in VSMC mass in the SHR consists of hyperplasia in mesenteric resistance vessels[30] and hypertrophy and hyperploidy in the aorta.[3,31,32] A similar VSMC hypertrophy and hyperploidy is seen in two-kidney, one-clip (2K-1C) hypertension,[33] a condition associated with high and normal circulating ANG II levels in the acute and chronic phases, respectively.[34] A postulated local RAS in the vascular wall,[35,36] which may give rise to locally increased ANG II levels as shown by Morishita *et al.,*[34] could point to a role for local ANG II in the development of vascular hypertrophy in spite of normal (or even low) circulating ANG II levels.[37] Infusion of low doses of ANG II, below the threshold of its rapid vasoconstrictor effect, leads to a slow and progressive increase in blood pressure, associated with development of vascular hypertrophy.[2,38] Simultaneously administered hydralazine inhibited the pressor and cardiac effects of ANG II,[2] while prazosin, an α_1-adrenoceptor inhibitor, reduced ANG II-induced VSMC DNA synthesis.[38] Infusing ANG II into rats subjected to balloon catheterization injury results in greater neointima formation and more marked DNA synthesis in the neointima smooth muscle cells than normal smooth muscle cells.[39,40]

1.2. Effects of ANG II on Growth of Cardiomyocytes

Twenty years ago it was shown by Khairallah *et al.*[41] that infusion with ANG II resulted in significant left ventricular hypertrophy in the rat although the increased blood pressure response to ANG II was blocked. Direct hypertrophic effect of ANG II on the heart was shown by Aceto and Baker,[42] who found increased rate of protein synthesis in cultured embryonic chick myocytes in response to treatment with sarcosine–ANG II ([Sar¹]-ANG II). This increased protein synthesis was inhibited in the presence of [Sar¹,Ile⁸]-ANG II (saralasin), an ANG II receptor antagonist. The findings of these studies suggest that ANG II-induced effects on cardiac cell growth may, in part, be direct and independent of increased afterload.

1.3. Effects of ANG II on Development of Cardiac Fibrosis

The structural remodeling of the hypertrophied myocardium in various forms of hypertension consists of an alteration in collagen architecture (fibrosis) based on growth of cardiac fibroblasts.[7] The presence of interstitial and perivascular fibrosis in the myocardium seems to be the hallmark of pathological hypertrophy.[43] ANG II stimulates growth of 3T3 mouse fibroblast cell line in a dose-dependent manner.[44] This mitogenic action was dependent on the presence of fetal calf serum and could be prevented by simultaneously applied ANG II receptor antagonist, [Sar¹,Ile⁸]-ANG II. High levels of ANG II (10^{-7} M) reportedly increase cell number as well as collagen synthesis in isolated adult rat cardiac fibroblasts.[45]

2. INHIBITING THE EFFECTS OF THE RAS ON VASCULAR HYPERTROPHY

Prevention and regression of hypertensive vascular hypertrophy and neointimal hyperplasia by drug treatment have been subject to ardent research. In theory, ANG II-induced effects can be blocked by treatment with angiotensin-converting enzyme (ACE) inhibitors and angiotensin-receptor antagonists. Concerning ACE inhibitors, it should be kept in mind that besides blocking the formation of ANG II, these drugs concomitantly potentiate the effects of bradykinin.[46] [Sar¹,Ile⁸]-ANG II (saralasin), a nonselective antagonist of the angiotensin receptors, and losartan (DuP 753), the nonpeptide antagonist of angiotensin type 1 receptor (AT_1 receptor), are the most frequently used angiotensin receptor antagonists in experimental research so far.

2.1. ACE Inhibitors

2.1.1. Experimental Studies

A significant prevention or regression of vascular hypertrophy, whether measured as decreased DNA synthesis in VSMCs[5] or morphometrically as decreased smooth muscle mass,[5,37,47–54] is brought about by ACE inhibitors in different models of genetic and experimental hypertension, including

SHR,[47–49,51–54] 2K-1C,[50] and one-kidney, one-clip (1K-1C)[37] rats. Henrichs *et al.*[47] treated spontaneously hypertensive rats (SHRs) with captopril for two generations (the parent generation during complete gestational and lactational periods and the offspring up to 14 weeks of age). The captopril-treated rats had been normotensive during their complete life span. A significant increase in the cross-sectional area of arterial media in the thoracic aorta, abdominal aorta, renal artery, and intrarenal arteries down to the fourth division, as well as in peripheral renal arteries, was observed in the untreated SHRs compared with captopril-treated SHR or normotensive Wistar-Kyoto rats. Captopril treatment completely prevented the vascular hypertrophy in all the above-mentioned arteries. The findings of the study of Henrichs *et al.*[47] suggest that the structural change of the vasculature in SHR is an adaptive change associated with high blood pressure rather than a genetically determined causative factor in hypertension. A 4-week period of perindopril treatment from 6 to 10 weeks of age in the SHR resulted in a significant reduction of vascular hypertrophy at 32 weeks of age.[55] Simultaneously administered ANG II abolished the effect of perindopril on vascular hypertrophy and even aggravated hypertrophy compared with untreated SHR. These studies clearly show that treatment with ACE inhibitors reduce and prevent the hypertensive vascular hypertrophy in the rat. In addition, ACE inhibitor treatment already *in utero* and in young SHR seems to be associated with long-term beneficial effects on vascular structure, even after discontinuation of treatment.[47,55]

Arterial injury by balloon catheterization in experimental animals leads to migration and proliferation of medial VSMCs, forming a neointima, the characteristic lesion seen clinically as restenosis after coronary angioplasty. Powell *et al.*[56] were the first to show a prevention of neointima formation by treatment with the ACE inhibitors cilazapril and captopril in rats after carotid balloon injury. ACE inhibitor treatment was started 6 days before balloon catheterization and lasted for 14 days, when there was an 80% reduction in neointima formation compared with placebo. As shown in Fig. 1, the beneficial effects of cilazapril are dose-dependent and maintained for at least 8 weeks after stopping the treatment.[26] Cilazapril and heparin given together may be more effective than cilazapril alone.[57] Farhy *et al.*[58] also reported on an 80% reduction in neointimal area with ramipril treatment after balloon injury in rats. However, if ramipril was given together with a specific bradykinin B_2 receptor antagonist, HOE 140, the reduction in neointima formation was only 43% (Fig. 2). This finding indicates that kinins are also involved in the inhibitory effect of ACE inhibitors on neointimal proliferation. ACE inhibitor-induced suppression of neointima formation after balloon injury has also been shown in rabbit[59,60] and guinea pig.[60] In baboons and pigs, however, cilazapril did not prevent neointima formation.[61,62]

The exact mechanism for the ACE inhibitor-induced reduction in vascular hypertrophy and neointima formation in rats is unknown. It has been attributed

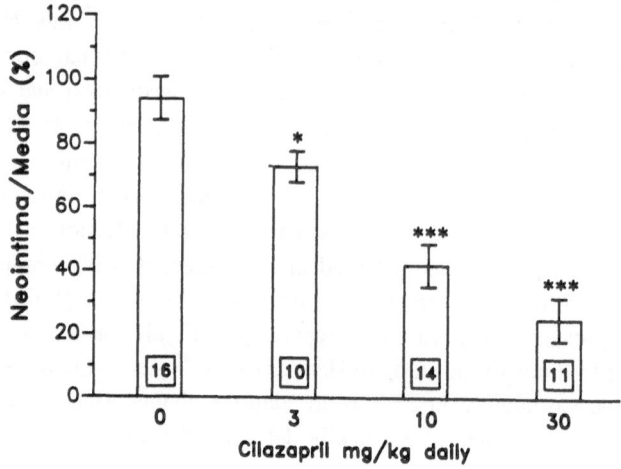

Figure 1. Inhibition of neointima formation by different doses of the ACE inhibitor, cilazapril, in the rat model of vascular injury by balloon catheterization. In this experiment, cilazapril was administered, at the doses shown mixed with food, from 6 days before to 14 days after balloon catheterization. All animals were analyzed at 14 days. The number in the bars is the number of animals used in each group. Results are expressed as the mean ± SEM; *$p < 0.05$, ***$p < 0.001$ compared by Student's two-tailed t test with the parallel placebo group. (From Powell *et al.*[26] Reprinted with permission.)

to decreased ANG II and, possibly, as shown recently,[58] to potentiation of the bradykinin effect. The role of ANG II in view of these studies remains elusive.

2.1.2. Clinical Studies

Results of MERCATOR, the first placebo-controlled clinical study on the effect of cilazapril in angiographic restenosis prevention after percutaneous transluminal coronary angioplasty (PTCA), were reported on recently.[63] Patients were treated with cilazapril 5 mg bid for 6 months, but no prevention of restenosis was seen, nor was there any favorable effect on clinical outcome after PTCA. The lack of efficiency by cilazapril to reduce restenosis may be due to differences in dose, time relation, and species, as pointed out by the authors of the MERCATOR study. Smooth muscle cells in an already-formed neointimal lesion respond to ANG II treatment by more marked DNA synthesis than smooth muscle cells in the underlying media or normal vessel wall.[39,40] Dilation of an atherosclerotic vessel may, therefore, lead to a stronger proliferative response of VSMCs compared with normal vessels. The dose of 5 mg bid cilazapril used in the MERCATOR study may have been too small. The ongoing American/Canadian sister trial to MERCATOR, MARCATOR, which randomizes between 1, 5, and 10 mg cilazapril bid, will provide us with further

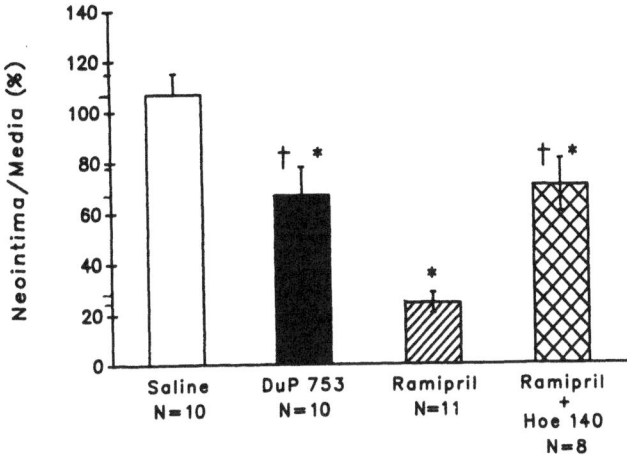

Figure 2. Changes in neointima formation 2 weeks after carotid artery ballon deendothelization in rats chronically infused with vehicle (saline) ($n = 10$), losartan 10 mg/kg per day ($n = 10$), ramipril 5 mg/kg per day ($n = 11$), and ramipril plus HOE 140 70 µg/kg per day ($n = 8$). $*p < 0.05$ compared with vehicle, $\dagger p < 0.05$ compared with ramipril. (From Farhy et al.[58] Reprinted with permission.)

information regarding the effect of ACE inhibitor dose on neointima formation in man.

2.2. Angiotensin Receptor Antagonists

The ACE inhibitor-induced prevention and regression of vascular hypertrophy has been amply verified. The development of specific angiotensin receptor antagonists and bradykinin receptor antagonists has made it possible to assess the relative impact of decreased ANG II and increased bradykinin on ACE inhibitor effects on development of vascular changes.

2.2.1. Experimental Studies

Ten weeks' treatment of 3- to 4-week-old SHR with losartan, an AT_1 receptor antagonist, resulted in a significant 31% reduction in mesenteric vascular hypertrophy measured as media/lumen ratio[54] and prevention of functional vascular structure/amplifier mechanisms.[64] The preventive effect of losartan on vascular hypertrophy was no less than that of captopril, and persisted at least up to 30 weeks[54] (Table I).

Losartan brings about a 40–60% reduction in the development of neointima formation after balloon injury in the rabbit[65] and rat.[39,58] As shown in Table II and Fig. 2, the effect of losartan was comparable to [39] or less than [58] that of cilazapril.

Table I. Preventive Effect of Losartan
and Captopril on Vascular Hypertrophy[a]

Treatment period	Age (weeks)	BP (mm Hg)	$m_1/l_1 \times 10^{-2}$	$H/B \times 10^{-3}$
Controls	7	200.5 ± 26.7 (24)[b]	6.0 ± 1.2 (16)	3.66 ± 0.4 (16)
	13	258.9 ± 18.2^c	7.7 ± 0.8^c (8)	3.74 ± 0.2 (8)
	30	274.5 ± 18.6^c	8.6 ± 1.2^c (19)	3.72 ± 0.3 (19)
Losartan, 15 mg/kg per day				
Age 3–7 weeks	7	168.0 ± 23.4^d (11)	5.7 ± 1.1 (18)	3.58 ± 0.3 (20)
After treatment	30	246.7 ± 23.0^d	8.1 ± 0.1 (10)	3.39 ± 0.2 (10)
Age 3–13 weeks	13	175.0 ± 8.5^d (6)	5.3 ± 0.8^d (6)	3.21 ± 0.3^d (6)
After treatment	30	224.4 ± 13.6^d	6.2 ± 0.4^d (6)	3.16 ± 0.1^d (6)
Captopril, 100 mg/kg per day				
Age 3–7 weeks	7	142.1 ± 12.6^d (6)	5.1 ± 1.2 (7)	3.37 ± 0.2 (7)
After treatment	30	239.0 ± 26.0^d	8.4 ± 0.5 (5)	3.33 ± 0.3 (5)
Age 3–13 weeks	13	157.5 ± 12.6^d (6)	5.6 ± 0.8^d (6)	3.15 ± 0.2^d (6)
After treatment	30	217.0 ± 17.5^d	6.5 ± 0.5^d (6)	3.18 ± 0.2^d (6)

[a]Systolic blood pressure, mesenteric resistance vessel m_1/l_1 (media/lumen), ratio and heart/body weight ratio (H/B) in SHRs treated from the age of 3 to 7 weeks and 3 to 13 weeks with losartan and captopril and at the age of 30 weeks after stopping treatment (mean ± SD).
[b]Figures in parenthesis are numbers of rats studied.
[c]$p < 0.05$ vs. 7-week control group.
[d]$p < 0.05$ vs. equivalent control group.
From Morton et al.,[54] with permission.

Studies with losartan have shown that ANG II may contribute to the formation of neointimal hyperplasia and vascular hypertrophy in the rat. The reduction in VSMC mass with losartan treatment is comparable to that of ACE inhibitors.[39,54] However, as stated above, studies applying bradykinin B_2 receptor antagonist HOE 140 have shown that kinins also take part in the antiproliferative and antihypertrophic effects of ACE inhibitors.[58,66]

3. INHIBITING THE EFFECTS OF THE RAS ON CARDIAC HYPERTROPHY

Early studies showed a pharmacologically induced reversal of cardiac hypertrophy by drugs that inhibited the adrenergic system or the renal pressor system (for review, see ref. 67). All classes of antihypertensive agents have been studied as to the effect of reversing cardiac hypertrophy. Long-term clinical and experimental studies have shown that when treating patients with hypertension adequately for a long enough period of time, cardiac mass will decrease regardless of what drug is used.[68,69] The outcome is different in short-term studies. Drugs such as ACE inhibitors, calcium antagonists, β-adrenergic receptor antagonists, and centrally acting adrenergic agents are able to reduce cardiac mass in patients in as early as 4–8 weeks of treatment, while

Table II. Effects of Cilazapril and the ANG II Receptor Antagonist Losartan on Balloon Catheter-Induced Vascular Injury[a]

	Number of rats	Area of neointima (10^3 μm^2)	Neointima/media (%)	Coverage of IEL by neointima (%)
Control	12	99 ± 7	109 ± 7	100
Cilazapril	14	45 ± 7[b]	61 ± 9[b]	93 ± 3[c]
Losartan	8	49 ± 10[b]	59 ± 11[b]	84 ± 10[c]

[a]Values are mean ± SEM. Left carotid artery of rats was injured by balloon catheterization. Arteries were perfusion fixed and evaluated morphologically 14 days later. Cilazapril (10 mg/kg per day as food admix) and losartan (10 mg/kg PO twice daily) were given 1 hr before balloon injury up to 14 days later. IEL, internal elastic lamina.
[b]$p > 0.001$.
[c]$p < 0.05$ vs. control (Student's two-tailed t test).
From Osterrieder *et al.*,[39] with permission.

drugs like α_1-adrenergic receptor blockers, diuretics, and direct-acting smooth muscle cell relaxing vasodilators are not.[67,70] Therefore, the decrease in cardiac mass may not be due to correction of hypertension alone, since discrepancies between the control of blood pressure and regression of hypertrophy have been reported.[67,70–72]

3.1. ACE Inhibitors

3.1.1. Experimental Studies

ACE inhibitors reduce cardiac mass with short-term and prolonged therapy. This has been shown in the SHR treated with different ACE inhibitors.[5,48,49,52,53,73–79] ACE inhibitor-induced reduction in cardiac mass was shown also in the Dahl rats, 2K-1C hypertensive rats, as well as in rats subjected to experimental aortic stenosis.[80–82] Interestingly, in rats subjected to aortic banding, in which coarctation is applied to the abdominal aorta between the renal arteries, even a subantihypertensive dose of ramipril 10 μg/kg per day was able to prevent and to induce a regression of cardiac hypertrophy.[66,72] In the SHR, however, treatment with subantihypertensive doses of ramipril or zabicipril *in utero* and up to 20 weeks did not influence the development of cardiac hypertrophy, while the antihypertensive doses were effective in both prevention[78] (Fig. 3) and regression[79] of cardiac hypertrophy. A concomitant increase in left ventricular capillary length density by both antihypertensive and subantihypertensive doses of ramipril in the SHR[78] was observed (Fig. 4). This was accompanied by an increase in myocardial ATP and glycogen content and by a decrease in citric acid, reflecting improved myocardial metabolism associated with treatment by ACE inhibitors (Fig. 5). These beneficial effects of ACE inhibitors on cardiac metabolism and blood supply in the SHR were possibly due to potentiation of the kinin effect, leading to augmented myocar-

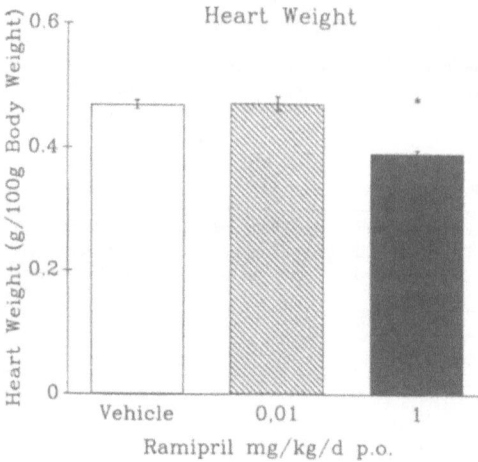

Figure 3. Bar graph shows effect of early onset treatment with two doses of ramipril on cardiac left ventricular weight (LVW) in SHR. Data are mean ± SEM; $n = 12$ per group; *$p < 0.05$. (From Unger et al.[78] Reprinted with permission.)

dial blood flow.[78] However, decreased levels of ANG II may also contribute to the ACE inhibitor-induced increased angiogenesis *in vivo*. This view could be supported by our novel finding of ANG II acting as a strong antiproliferative agent for SHR cardiac capillary endothelial cells *in vitro*.[83] Ramipril treatment in rats with banding of the aortic arch did not result in prevention of cardiac hypertrophy,[84] nor did treatment of stenosis of the ascending aorta in rats by quinapril show any preventive effect.[82] Recently, Linz and Schölkens[66] re-

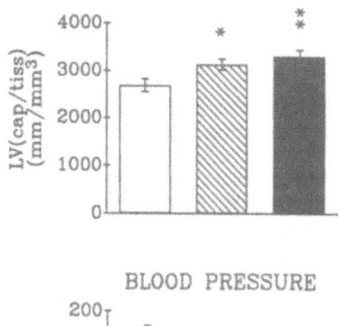

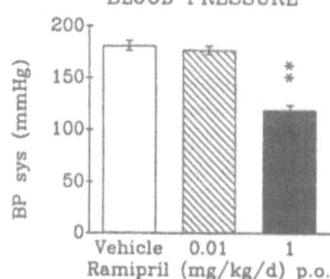

Figure 4. Bar graph shows effect of early onset long-term treatment with two oral doses of ramipril in SHR on myocardial capillary length density (LV cap/tiss) as determined with the orientator method, and blood pressure. Data are mean ± SEM; $n = 12$ per group; *$p < 0.05$; **$p < 0.001$. (From Unger et al.[78] Reprinted with permission.)

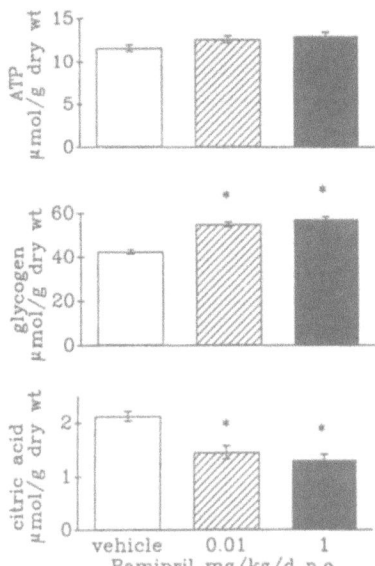

Figure 5. Bar graphs show effect of early-onset long-term treatment with two oral doses of ramipiril on parameters of myocardial metabolism in SHR. Data are mean ± SEM; $n = 12$ per group; *$p < 0.05$. (From Unger *et al.*[78] Reprinted with permission.)

ported on disappearance of the beneficial effects of ramipril on development of cardiac hypertrophy in rats with aortic banding by simultaneous treatment with the specific bradykinin B_2 receptor antagonist, HOE 140 (Fig. 6). Thus, in rats with aortic banding and cardiac hypertrophy, the ACE inhibitor-induced potentiation of bradykinin contributes to the salutary effects of ACE inhibitors.

These studies show that regression of cardiac hypertrophy brought about by ACE inhibitors is not necessarily due to reduced blood pressure per se, but may involve additional mechanisms like inhibition of the cardiac RAS, leading to reduced local formation of ANG II, potentiation of bradykinin effects, or "switching off" of specific IEGs and growth factors.

The ACE inhibitor-induced decrease in cardiac mass is accompanied by a decline in myocardial collagen synthesis, implying attenuation of fibrous tissue accumulation and medial thickening in intramyocardial coronary arteries.[73,77] These cardioreparative properties of ACE inhibitors may prove valuable in reversing left ventricular diastolic and systolic dysfunction in hypertensive cardiac hypertrophy.[77]

3.1.2. Clinical Studies

Studies in hypertensive patients have consistently shown that treatment with ACE inhibitors reduces left ventricular mass, measured by echocardiography or nuclear magnetic resonance imaging.[70,85–89] This effect is also seen in elderly patients treated for 3 months with captopril or enalapril.[90] The reduction

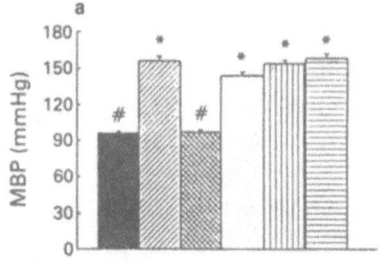

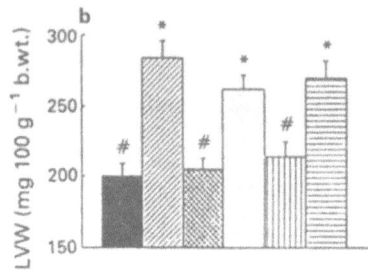

Figure 6. Effects of orally administered ramipril (1 mg kg^{-1} day^{-1} and 10 µg kg^{-1} day^{-1}) and coadministration of HOE 140 (500 µg kg^{-1} day^{-1} SC) on (a) mean blood pressure (MBP) and (b) left ventricular weight (LVW) in rats with aortic banding. (■) Sham; (▨) control vehicle; (▧) ramipril 1 mg kg^{-1}; (□) ramipril 1 mg + HOE 140; (▥) ramipril 10 µg kg^{-1}; (▤) ramipril 10 µg+ HOE 140. *$p < 0.05$ vs. sham, *$p < 0.05$ vs. control.

in left ventricular mass seems mainly to be due to decreased left ventricular afterload.[70,86] The impact of decreases in ANG II-dependent myocardial protein synthesis and ANG II-mediated adrenergic outflow is probably less.

3.2. Angiotensin Receptor Antagonists

Considering the indisputable effects of ACE inhibitors on regression of both experimental and human cardiac hypertrophy, the question of the relative contribution of decreased ANG II in this regard may be answered by studies using specific angiotensin receptor antagonists. To date, only a few experimental data exist.

3.2.1. Experimental Studies

Short-term treatment of SHR with the AT$_1$-receptor antagonist losartan resulted in a significant 11% regression of left ventricular hypertrophy.[91] This is a stronger regression than the one caused by enalapril in a comparable study.[76] Morton *et al.*[54] treated SHRs with losartan and captopril, starting the treatment at 3 weeks of age before established cardiac hypertrophy and continuing for 4 or 10 weeks. A significant equal prevention in the onset of cardiac hypertrophy was observed after 10 weeks treatment with both drugs[54] (Table I). In addition, the heart/body weight ratio did not increase after discontinuation of the treatments. The reductions in blood pressure and heart/body weight ratio after treatment with losartan or captopril at 30 weeks were also equal (Table I). Similar results regarding prevention of cardiac hypertrophy with losartan treatment have been obtained by Oddie *et al.*[64] in the SHR and Camargo *et al.*[92] in the stroke-prone SHR. The findings of Morton *et al.*[54] and Oddie *et al.*[64] with losartan are in line with data reported by others[48,55,93]: that a brief period of ACE inhibitor treatment in young SHRs prevents the full expression of cardiovascular hypertrophy. Thus, it has been shown that interrupting the activity of the RAS by two different treatment modalities may result in similar beneficial effects on cardiac hypertrophy. This seems to indicate that ANG II is important in the development of hypertension and cardiac hypertrophy in the SHR. Interestingly, in the study of Mizuno *et al.*,[91] the left ventricular ANG II content was reported to be decreased by losartan treatment, while circulating ANG II and plasma renin activity were increased. The reason for this apparent decrement in cardiac ANG II is unclear, but this effect of losartan would suggest that the local cardiac RAS rather than circulating RAS is important for development and maintenance of cardiac hypertrophy.

The efficacy of losartan in reducing pressure-induced left ventricular hypertrophy in experimental aortic stenosis in the rat has been reported by Kromer *et al.*[94] The reduction in left ventricular hypertrophy associated with losartan treatment was significant, but clearly weaker than the corresponding effect of the ACE-inhibitor quinapril[94] (33% vs. 80%). Linz *et al.*[95] compared the effects of losartan 3 mg/kg per day and a subantihypertensive dose of ramipril 10 μg/kg per day on prevention and regression of cardiac hypertrophy in rats subjected to aortic banding. Losartan treatment was associated with normalization of blood pressure and significant prevention and regression in cardiac hypertrophy. The effect of a subantihypertensive dose of ramipril on cardiac hypertrophy was more marked than that of losartan. The reductions in cardiac hypertrophy after treatment with ACE inhibitors or losartan in the SHR seem to be comparable, although it is not always clear whether or not stringent conditions were chosen for the comparison (e.g., maximal effective dose). ACE inhibitors are more effective in reducing cardiac hypertrophy than losartan in rats subjected to aortic stenosis or banding, suggesting that in these experimental models mechanisms other than decreased ANG II may contribute to the antihypertrophic effect of ACE inhibitors. Potentiation of kinin effect seems to be important in this context.[95]

4. CONCLUSIONS

The pathophysiological mechanisms leading to hypertensive CV hypertrophy are only partly identified. The direct ANG II-induced effects on growth of

VSMCs and cardiomyocytes in experimental studies are AT_1 receptor-mediated and seem to indicate a role for ANG II as a trophic factor in the cardiovascular system. A role of local RAS, in particular, for development of CV hypertrophy has been suggested.[91,96,97] Although treatment of experimental animals with ANG II results in CV hypertrophy, it does not follow that reductions in CV hypertrophy obtained with drugs like ACE inhibitors and angiotensin receptor antagonists would solely be due to inhibition of the ANG II effect. Thus, whereas treatment with ACE inhibitors is associated with reduction in CV hypertrophy, it does not allow us to make any far-reaching conclusions about the role of ANG II, let alone tissue ANG II, especially in view of reports that blocking the bradykinin effect by a bradykinin B_2 receptor antagonist, abolishes the favorable effect of ACE inhibitors on prevention of cardiac hypertrophy and neovascular hyperplasia.[58,66] The effects of ACE inhibitor-induced potentiation of other agents, like prostaglandins, nitric oxide, and substance P, remain to be more thoroughly studied. In addition, converting enzyme is not the only catalyst leading to formation of ANG II.[98,99] Further studies employing accurate and reliable methods for ANG II measurement are needed to assess the relative impact of decreased ANG II for ACE inhibitor effects.

More specificity is provided by the angiotensin receptor antagonists. The first studies have shown comparable effects by losartan and ACE inhibitors in reducing CV hypertrophy in the SHR.[39,54] This effect of losartan has been attributed to inhibition of the AT_1 receptor. Cardiac tissue seems to contain two receptor types, while VSMCs express AT_1 receptor only.[100,101] Further studies with specific antagonists of both AT_1 and AT_2 receptor subtypes will add to our knowledge of the mechanisms leading to hypertensive CV hypertrophy.

REFERENCES

1. Dickinson CJ, Lawrence JR: A slowly developing pressor response to small concentrations of angiotensin. *Lancet* 1:1354–1356, 1963.
2. Griffin SA, Brown WCB, MacPherson F, et al: Angiotensin II causes vascular hypertrophy in part by a non-pressor mechanism. *Hypertension* 17:626–635, 1991.
3. Aalkjaer C, Heagerty AM, Peterson KK, et al: Evidence for increased media thickness, increased neuronal amine uptake, and depressed excitation–contraction coupling in isolated resistance vessels from essential hypertensives. *Circ Res* 61:181–186, 1987.
4. Owens GK, Schwartz SM: Alterations in vascular smooth muscle cell mass in the spontaneously hypertensive rat. Role of cellular hypertrophy, hyperploidy, and hyperplasia. *Circ Res* 51:280–289, 1982.
5. Owens GK: Influence of blood pressure on development of aortic medial smooth muscle hypertrophy in spontaneously hypertensive rats. *Hypertension* 9:178–187, 1987.
6. Lever AF: Slow pressor mechanisms in hypertension: A role for hypertrophy of resistance vessels. *J Hypertens* 4:515–524, 1986.
7. Weber KT, Brilla CG: Pathological hypertrophy and cardiac interstitium. Fibrosis and renin–angiotensin–aldosterone system. *Circulation* 83:1849–1865, 1991.

8. Chobanian AV: 1989 Corcoran lecture: Adaptive and maladaptive responses of the arterial wall to hypertension. *Hypertension* 15:666–674, 1990.

9. Geisterfer AAT, Peach MJ, Owens GK: Angiotensin II induces hypertrophy, not hyperplasia, of cultured rat aortic smooth muscle cells. *Circ Res* 62:749–765, 1988.

10. Berk BC, Vekshtein V, Gordon HM, et al: Angiotensin II-stimulated protein synthesis in cultured vascular smooth muscle cells. *Hypertension* 13:305–314, 1989.

11. Scott-Burden T, Hahn AWA, Resink TJ, et al: Modulation of extracellular matrix by angiotensin II: Stimulated glycoconjugate synthesis and growth in vascular smooth muscle cells. *J Cardiovasc Pharmacol* 16(Suppl 4):S36–S41, 1990.

12. Paquet J-L, Baudouin-Legros M, Brunelle G, et al: Angiotensin II-induced proliferation of aortic myocytes in spontaneously hypertensive rats. *J Hypertens* 8:565–572, 1990.

13. Ross R: The pathogenesis of atherosclerosis: An update (review). *N Engl J Med* 314:488–500, 1986.

14. Clowes AW, Reidy MA, Clowes MM: Kinetics of cellular proliferation after arterial injury. *Lab Invest* 49:327–334, 1983.

15. Schelling P, Fischer H. Ganten D: Angiotensin and cell growth: A link to cardiovascular hypertrophy? *J Hypertens* 9:3–15, 1991.

16. Campbell-Boswell M, Robertson AL: Effects of angiotensin II and vasopressin on human smooth muscle cells in vitro. *Exp Mol Pathol* 35:265–276, 1981.

17. Lyall F, Morton JJ, Lever AF, et al: Angiotensin II activates Na^+-H^+ exchange and stimulates growth in cultured vascular smooth muscle cells. *J Hypertens* 6(Suppl 4):S438–S441, 1988.

18. Turla MB, Thompson MM, Corjay MH, et al: Mechanisms of angiotensin II- and arginine vasopressin-induced increases in protein synthesis and content in cultured rat aortic smooth muscle cells. Evidence for selective increases in smooth muscle isoactin expression. *Circ Res* 68:288–299, 1991.

19. Stouffer GA, Owens GK: Angiotensin II-induced mitogenesis of spontaneously hypertensive rat-derived cultured smooth muscle cells is dependent on autocrine production of transforming growth factor-β. *Circ Res* 70:820–828, 1992.

20. Krug LM, Berk BC: Na^+,K^+-adenosine triphosphatase regulation in hypertrophied vascular smooth muscle cells. *Hypertension* 20:144–150, 1992.

21. Berk BC, Elder E, Mitsuka M: Hypertrophy and hyperplasia cause differing effects on vascular smooth muscle cell Na^+-H^+ exchange and intracellular pH. *J Biol Chem* 265:19632–19637, 1990.

22. Chamley-Campbell JH, Campbell GR, Ross R: The smooth muscle cell in culture. *Physiol Rev* 59:1–61, 1979.

23. Kawahara Y, Sunako M, Tsuda T, et al: Angiotensin II induces expression of the c-fos gene through protein kinase C activation and calcium ion mobilisation in cultured vascular smooth muscle cells. *Biochem Biophys Res Commun* 150:52–59, 1988.

24. Naftilan AJ, Pratt RE, Dzau VJ: Induction of platelet-derived growth factor A-chain and c-myc gene expressions by angiotensin II in cultured rat vascular smooth muscle cells. *J Clin Invest* 83:1419–1424, 1989.

25. Lyall F, Boswell F, Dornan ES, et al: Angiotensin II increases proto-oncogene expression and inositol phosphate levels through the AT1 receptor (abstract). *J Hypertens* 10(Suppl 4):S152, 1992.

26. Powell JS, Müller RKM, Rouge M, et al: The proliferative response to vascular injury is suppressed by angiotensin-converting enzyme inhibition. *J Cardiovasc Pharmacol* 16(Suppl 4):S42–S49, 1990.

27. Nakahara K, Nishimura N, Kuro-o M, et al: Identification of three types of PDGF-A chain gene transcripts in rabbit vascular smooth muscle and their regulated expression during development and by angiotensin II. *Biochem Biophys Res Commun* 184:811–818, 1992.

28. Dzau VJ, Gibbons GH, Pratt RE: Molecular mechanisms of vascular renin–angiotensin system in myointimal hyperplasia. *Hypertension* 18(Suppl 2):S100–S105, 1991.

29. Sachinidis A, Ko Y, Görg A, et al: Losartan inhibits the angiotensin II (AII)-induced stimulation of the phosphoinositide (PI)-turnover and hypertrophy in vascular smooth muscle cells (abstract). *J Hypertens* 10(Suppl 4):S2, 1992.

30. Mulvany MJ, Baandrup U, Gundersen HJG: Evidence for hyperplasia in mesenteric resistance vessels of spontaneously hypertensive rats using a three-dimensional disector. *Circ Res* 57:794–800, 1985.

31. Owens GK, Rabinovitch PS, Schwartz SM: Smooth muscle cell hypertrophy versus hyperplasia in hypertension. *Proc Natl Acad Sci USA* 78:7759–7763, 1981.

32. Olivetti G, Melissari M, Marchetti G, et al: Quantitative structural changes in the rat thoracic aorta in early spontaneous hypertension. Tissue composition, and hypertrophy and hyperplasia of smooth muscle cells. *Circ Res* 51:19–26, 1982.

33. Owens GK, Schwartz SM: Vascular smooth muscle cell hypertrophy and hyperploidy in the Goldblatt hypertensive rat. *Circ Res* 53:491–501, 1983.

34. Morishita R, Higaki J, Miyazaki M, et al: Possible role of the vascular renin–angiotensin system in hypertension and vascular hypertrophy. *Hypertension* 19(Suppl 2):S62–S67, 1992.

35. Campbell DJ: Circulating and tissue angiotensin systems. *J Clin Invest* 79:1–6, 1987.

36. Samani NJ, Swales JD: Molecular biology of the vascular renin–angiotensin system. *Blood Vessels* 28:210–216, 1991.

37. Wang D, Prewitt RL: Captopril reduces aortic and microvascular growth in hypertensive and normotensive rats. *Hypertension* 15:68–77, 1990.

38. van Kleef EM, Smits JFM, De Mey JGR, et al: α_1-Adrenoceptor blockade reduces the angiotensin II-induced vascular smooth muscle cell DNA synthesis in the rat thoracic aorta and carotid artery. *Circ Res* 70:1122–1127, 1992.

39. Osterrieder W, Müller RKM, Powell JS, et al: Role of angiotensin II in injury-induced neointima formation in rats. *Hypertension* 18(Suppl 2):S60–S64, 1991.

40. Daemen MJAP, Lombardi DM, Bosman FT, et al: Angiotensin II induces smooth muscle cell proliferation in the normal and injured rat arterial wall. *Circ Res* 68:450–456, 1991.

41. Khairallah PA, Robertson AL, Davila D: Effects of angiotensin II on DNA, RNA and protein synthesis. In Genest J, Koiw E (eds): *Hypertension '72*. Springer-Verlag, Berlin, 1972, pp. 212–220.

42. Aceto JF, Baker KM: [Sar¹]angiotensin II receptor mediated stimulation of protein synthesis in chick heart cells. *Am J Physiol* 258:H806–H813, 1990.

43. Weber KT, Clark WA, Janicki JS, et al: Physiologic versus pathologic hypertrophy and the pressure-overlaod myocardium. *J Cardiovasc Pharmacol* 10:537–549, 1987.

44. Schelling P, Ganten D, Speck G, et al: Effects of angiotensin II and angiotensin II antagonist saralasin on cell growth and renin in 3T3 and SV3T3 cells. *J Cell Physiol* 98:503–514, 1979.

45. Brilla CG, Zhou G, Weber K: Angiotensin II and collagen synthesis in cultured adult rat cardiac fibroblasts (abstract). *J Hypertens* 10(Suppl 4):S125, 1992.

46. Wiemer G, Schölkens BA, Becker RHA, et al: Ramiprilat enhances endothelial autacoid formation by inhibiting breakdown of endothelium-derived bradykinin. *Hypertension* 18:558–563, 1991.

47. Henrichs KJ, Unger T, Berecek KH, et al: Is arterial media hypertrophy in spontaneously hypertensive rats a consequence of or a cause for hypertension. *Clin Sci* 59:331–333, 1980.

48. Freslon JL, Giudicelli JF: Compared myocardial and vascular effects of captopril and dihydralazine during hypertension development in spontaneously hypertensive rats. *Br J Pharmacol* 80:533–543, 1983.

49. Limas C, Westrum B, Limas CJ: Comparative effects of hydralazine and captopril on the cardiovascular changes in spontaneously hypertensive rats. *Am J Pathol* 117:360–371, 1984.

50. Michel J-B, Dussaule J-C, Choudat L, et al: Effects of antihypertensive treatment in one-clip, two kidney hypertension in rats. *Kidney Int* 29:1011–1020, 1986.

51. Clozel J-P, Kuhn H, Hefti F: Decreases of vascular hypertrophy in four different types of arteries in spontaneously hypertensive rats. *Am J Med* 87(Suppl 6B):92–95, 1989.

52. Clozel J-P, Véniant M, Hess P, et al: Effects of two angiotensin converting enzyme inhibitors and hydralazine on coronary circulation in hypertensive rats. *Hypertension* 18(Suppl 2):S8–S14, 1991.

53. Lee RMKW, Berecek KH, Tsoporis J, et al: Prevention of hypertension and vascular changes by captopril treatment. *Hypertension* 17:141–150, 1991.

54. Morton JJ, Beattie EC, MacPherson F: Angiotensin II receptor antagonist losartan has persistent effects on blood pressure in the young spontaneously hypertensive rat: Lack of relation to vascular structure. *J Vasc Res* 29:264–269, 1992.

55. Harrap SB, Van der Merwe WM, Griffin SA, et al: Brief angiotensin converting enzyme inhibitor treatment in young spontaneously hypertensive rats reduces blood pressure long-term. *Hypertension* 16:603–614, 1990.

56. Powell JS, Clozel J-P, Müller RKM, et al: Inhibitors of angiotensin-converting enzyme prevent myointimal proliferation after vascular injury. *Science* 245:186–188, 1989.

57. Clowes AW, Clowes MM, Vergel SC, et al: Heparin and cilazapril together inhibit injury-induced intimal hyperplasia. *Hypertension* 18(Suppl 2):S65–S69, 1991.

58. Farhy RD, Ho K-L, Carretero OA, et al: Kinins mediate the antiproliferative effect of ramipril in rat carotid artery. *Biochem Biophys Res Commun* 182:283–288, 1992.

59. Bilazarian SD, Currier JW, Haudenschild C, et al: Angiotensin converting enzyme inhibition reduces restenosis in experimental angioplasty (abstract). *J Am Coll Cardiol* 17:268A, 1991.

60. Clozel J-P, Hess P, Michael C, et al: Inhibition of converting enzyme and neointima formation after vascular injury in rabbits and guinea pigs. *Hypertension* 18(Suppl 2):S55–S59, 1991.

61. Hanson SR, Powell JS, Dodson T, et al: Effects of angiotensin converting enzyme inhibition with cilazapril on intimal hyperplasia in injured arteries and vascular grafts in the baboon. *Hypertension* 18(Suppl 2):S70–S76, 1991.

62. Lam JYT, Lacoste L, Bourassa MG: Cilazapril and early atherosclerotic changes after balloon injury of porcine carotid arteries. *Circulation* 85:1542–1547, 1992.

63. The Multicenter European Research Trial With Cilazapril after Angioplasty to Prevent Transluminal Coronary Obstruction and Restenosis (MERCATOR) Study Group: Does the new angiotensin converting enzyme inhibitor cilazapril prevent restenosis after percutaneous transluminal coronary angioplasty? Results of the MERCATOR study: A multicenter, randomized, double-blind placebo-controlled trial. *Circulation* 86:100–110, 1992.

64. Oddie CJ, Dilley RJ, Bobik A: Long-term angiotensin II antagonism in spontaneously hypertensive rats: Effects on blood pressure and cardiovascular amplifiers. *Clin Exp Pharmacol Physiol* 19:392–395, 1992.

65. Azuma H, Niimi Y, Hamasaki H: Prevention of intimal thickening after endothelial removal by a nonpeptide angiotensin II receptor antagonist, losartan. *Br J Pharmacol* 106:665–671, 1992.

66. Linz W, Schölkens BA: A specific B2-bradykinin receptor antagonist HOE 140 abolishes the antihypertrophic effect of ramipril. *Br J Pharmacol* 105:771–772, 1992.

67. Frohlich ED: Regression of cardiac hypertrophy and left ventricular pumping ability post-regression. *J Cardiovasc Pharmacol* 17(Suppl 2):S81–S86, 1991.

68. Pfeffer JM, Pfeffer MA, Mirsky I, et al: Regression of left ventricular hypertrophy and prevention of left ventricular dysfunction by captopril in the spontaneously hypertensive rat. *Proc Natl Acad Sci USA* 79:3310–3314, 1982.

69. Kannel WB, D'Agostino RB, Levy D, et al: Prognostic significance of regression of left ventricular hypertrophy. *Circulation* 78(Suppl 2):S89, 1988.

70. Dzau VJ: Angiotensin converting enzyme inhibitors and the cardiovascular system. *J Hypertens* 10(Suppl 3):S3–S10, 1992.

71. Tsoporis J, Leenen FHH: Effects of arterial vasodilators on cardiac hypertrophy and sympathetic activity in rats. *Hypertension* 11:376–386, 1988.

72. Linz W, Schölkens B, Ganten D: Converting enzyme inhibition specifically prevents the development and induces regression of cardiac hypertrophy in rats. *Clin Exp Hypertens* A11(7):1325–1350, 1989.

73. Sen S, Tarazi RC, Bumpus MF: Effect of converting enzyme inhibitor (SQ 14,225) on myocardial hypertrophy in spontaneously hypertensive rats. *Hypertension* 2:169–176, 1980.

74. Pfeffer JM, Pfeffer MA, Fletcher PJ, et al: Favorable effects of therapy on cardiac performance in spontaneously hypertensive rats. *Am J Physiol* 242:766–784, 1982.

75. Clozel J-P, Kuhn H, Hefti F: Effects of chronic ACE inhibition on cardiac hypertrophy and coronary vascular reserve in spontaneously hypertensive rats with developed hypertension. *J Hypertens* 7:267–275, 1989.

76. Childs TJ, Adams MA, Mak AS: Regression of cardial hypertrophy in spontaneously hypertensive rats by enalapril and the expression of contractile proteins. *Hypertension* 16:662–668, 1990.

77. Brilla CG, Janicki JS, Weber KT: Cardioreparative effects of lisinopril in rats with genetic hypertension and left ventricular hypertrophy. *Circulation* 83:1771–1779, 1991.

78. Unger T, Mattfeldt T, Lamberty V, et al: Effect of early onset angiotensin converting enzyme inhibition on myocardial capillaries. *Hypertension* 20:478–482, 1992.

79. Gohlke P, Stoll M, Lamberty V, et al: Cardiac and vascular effects of chronic angiotensin converting enzyme inhibition at subantihypertensive doses. *J Hypertens* 10(Suppl 6):141–144, 1992.

80. Sharma JN, Fernandez PG, Kim BK, et al: Cardiac regression and blood pressure control in the Dahl rat treated with either enalapril maleate (MK 421, an angiotensin converting enzyme inhibitor) or hydrochlorothiazide. *J Hypertens* 1:251–256, 1983.

81. Sen S, Tarazi RC, Bumpus FM: Reversal of cardiac hypertrophy in renal hypertensive rats: Medical vs. surgical therapy. *Am J Physiol* 240:H408–H412, 1981.

82. Kromer EP, Riegger GAJ: Effects of long-term angiotensin converting enzyme inhibition on myocardial hypertrophy in experimental aortic stenosis in the rat. *Am J Cardiol* 62:161–163, 1988.

83. Metsärinne KP, Stoll M, Gohlke P, et al: Angiotensin II is antiproliferative for coronary endothelial cells in vitro. *Pharm Pharmacol Lett* 2:150–152, 1992.

84. Zierhut W, Zimmer H-G, Gerdes AM: Effect of angiotensin converting enzyme inhibition on pressure-induced left ventricular hypertrophy in rats. *Circ Res* 69:609–617, 1991.

85. Fouad FM, Tarazi RC, Bravo EL: Cardiac and haemodynamic effects of enalapril. *J Hypertens* 1(Suppl 1):S133–S142, 1983.

86. Sheiban I, Arcaro G, Covi G, et al: Regression of cardiac hypertrophy after antihypertensive therapy with nifedipine and captopril. *J Cardiovasc Pharmacol* 10(Suppl 10):S187–S191, 1987.

87. Asmar RG, Pannier B, Santoni JP, et al: Reversion of cardiac hypertrophy and reduced arterial compliance after converting enzyme inhibition in essential hypertension. *Circulation* 78:941–950, 1988.

88. Eichstaedt H, Danne O, Langer M, et al: Regression of left ventricular hypertrophy under ramipril treatment investigated by nuclear magnetic resonance imaging. *J Cardiovasc Pharmacol* 13(Suppl 3):S75–S80, 1989.

89. Schulte K-L, Meyer-Sabellec W, Liederwald K, et al: Relation of regression of left ventricular hypertrophy to changes in ambulatory blood pressure after long-term therapy with perindopril versus nifedipine. *Am J Cardiol* 70:468–473, 1992.

90. Nagano N, Iwatsubo H, Hata T, et al: Effects of antihypertensive treatment on cardiac hypertrophy and cardiac function in elderly hypertensive patients. *J Cardiovasc Pharmacol* 17(Suppl 2):S163–S165, 1991.

91. Mizuno K, Tani M, Hashimoto S, et al: Effects of losartan, a nonpeptide angiotensin II receptor antagonist, on cardiac hypertrophy and the tissue angiotensin II content in spontaneously hypertensive rats. *Life Sci* 51:367–374, 1992.

92. Camargo MJF, Campbell WG, Lutterotti N, et al: Chronic oral angiotensin II-blockade with losartan (DuP 753 or LOS) inhibits the development of hypertension and end-organ damage in stroke-prone spontaneously hypertensive rats (abstract). *J Hypertens* 10(Suppl 4):S190, 1992.

93. Adams MA, Bobik A, Korner PI: Enalapril can prevent vascular amplifier development in spontaneously hypertensive rats. *Hypertension* 16:252–260, 1990.

94. Kromer EP, Schunkert H, Ackermann B, et al: Effects of angiotensin II_1-receptor antagonism on pressure induced left ventricular hypertrophy (abstract). *J Hypertens* 10(Suppl 4):S252, 1992.

95. Linz W, Henning R, Schölkens BA: Role of angiotensin II receptor antagonism and converting enzyme inhibition in the progression and regression of cardiac hypertrophy in rats. *J Hypertens* 9(Suppl 6):S400–S401, 1991.

96. Re R, Rovigatti U: New approaches to the study of the cellular biology of the cardiovascular system. *Circulation* 77(Suppl 1):1–14, 1988.

97. Baker KM, Chernin MI, Wixson SK, et al: Renin–angiotensin system involvement in pressure-overload cardiac hypertrophy in rats. *Am J Physiol* 259:H324–H332, 1990.

98. Okunishi H, Myazaki M, Toda N: Evidence for a putatively new angiotensin II-generating enzyme in the vascular wall. *J Hypertens* 2:277–284, 1984.

99. Urata H, Kinoshita A, Misono KS, et al: Identification of a highly specific chymase as the major angiotensin II-forming enzyme in the human heart. *J Biol Chem* 265:22348–22357, 1990.

100. Scott AL, Chang RSL, Lotti VJ, et al: Cardiac angiotensin receptors: Effects of selective angiotensin II receptor antagonists, DuP 753 and PD 121981, in rabbit heart. *J Pharmacol Exp Ther* 261:931–935, 1992.

101. Whitbread S, Mele M, Kamber, et al: Preliminary biochemical characterization of two angiotensin II receptor subtypes. *Biochem Biophys Res Commun* 163:284–291, 1989.

14

Angiotensin Antagonists in Models of Heart Failure

Joseph M. Capasso

1. INTRODUCTION

Occlusion of the left main coronary artery in rats results in a segmental loss of viable myocardium from the anterior lateral aspect of the left ventricular free wall.[1-6] The extent and location of the tissue loss is the major determinant in the subsequent depression of cardiac performance following the acute event.[1-6] If myocardial necrosis is extensive, i.e., encompassing greater than 40% of the ventricular free wall, left ventricular failure rapidly ensues, often with substantial involvement of right side performance and cardiac congestive failure.[1-6] Anatomically, large infarctions precipitate considerable rearrangements in ventricular architecture as evidenced by wall thinning as a result of side-to-side slippage of myocytes and chamber dilatation in both the transverse as well as the longitudinal dimension.[4,7,8] In an attempt to maintain normal pump function, activation of the renin–angiotensin system during conditions of infarction-induced cardiac decompensation engenders increases in circulating levels of angiotensin II (ANG II), a major pressor agent.[9-12] To reduce the work load on this already compromised ventricle and possibly prevent the associated detrimental changes in chamber geometry, prevention of metabolic conversion of ANG I to ANG II by carboxypeptidase inhibitors (i.e., angiotensin-converting enzyme, or ACE, inhibitors) have been employed both clinically[12] and experimentally.[12-16] Although ACE inhibitors like captopril[12-16] have been

Joseph M. Capasso • Department of Anatomy, University of South Dakota School of Medicine, Vermillion, South Dakota 57069.

Angiotensin Receptors, edited by Juan M. Saavedra and Pieter B.M.W.M. Timmermans. Plenum Press, New York, 1994.

quite effective in ameliorating the damaging effects of myocardial infarction, undesirable side effects occur from the therapeutic use of these nonspecific kinases.[17–19] With this in mind, compounds have been sought that block specifically the ANG II receptor on cardiac cells. Blockade of this AT_1 receptor has been effectively accomplished by a 2-*n*-butyl-4-chloro-5-hydroxymethyl-1-[2′-(1H-tetrazol-5yl)biphenyl-4yl) methyl]imidazole potassium salt,[34] manufactured by DuPont (losartan or DuP 753).

Current investigations question whether the hypertrophic response of the unaffected cells represents a compensatory reaction or contributes to the chronic depression in ventricular dynamics.[7,8] The repair process following acute myocardial infarction creates problems in evaluating the function of the residual viable myocytes. The association of the remaining tissue with the damaged region vitiates evaluating its function *in vivo,* since the interaction of the infarcted area with the healthy myocardium places obligatory constraints on the later, complicating the estimation of the contribution of each component to ventricular pump function. Similarly, measurements of the contractile behavior of the hypertrophied myocardium *in vitro*[3,20,21] do not distinguish whether alterations occur in the myocyte population or in the extracellular compartment of the tissue, or both. These limitations become particularly relevant when the influence of drug interventions are analyzed in terms of their therapeutic efficacy. Importantly, ACE inhibitors improve cardiac performance after infarction by reducing the magnitude of ventricular dilation that takes place chronically following ischemic myocardial injury.[16,22] Moreover, captopril reduces collagen deposition in the overloaded heart,[23] adding to the complexity of discriminating the target structure(s) of this drug with infarction. Finally, it remains to be determined whether the amelioration of chamber and wall remodeling of the infarcted ventricle by ACE inhibitors is accompanied by a reduction in cellular hypertrophy in the unaffected myocytes.

It is generally believed that myocyte cellular hyperplasia cannot be induced in the mature heart even under extreme conditions of ventricular loading. However, experimental evidence suggests that adult atrial[24] and ventricular[25–27] myocytes may undergo DNA synthesis and nuclear hyperplasia *in vivo.* In addition, *in vitro* studies demonstrate that DNA synthesis can be evoked in these cells by various growth factors and tissue type plasminogen activator,[28] strengthening the concept that cellular proliferation may be induced *in vivo.* In this regard, adult ventricular cells express mRNA for proliferating cell nuclear antigen, which is associated with actively dividing cells.[29] Recent work has documented that a prolonged and sustained increase in work load on the myocardium produced by mechanical[25,30] or renal[26] hypertension results in myocyte nuclear and cellular hyperplasia. This response appears to characterize the phase of initial ventricular decompensation[30] and overt failure.[26,31] Moreover, myocyte cellular hyperplasia has been demonstrated in the senescent heart in combination with markedly depressed ventricular pump performance[27] and elevated transmural myocardial stress.[32] Therefore, extent of overload may play a primary role in the initiation of DNA synthesis and in triggering progression through the cell cycle in myocytes.

Therefore, the present review presents material that deals with the detrimental hemodynamic consequences of myocardial infarction and whether inhibition of the metabolic production of ANG II or blockade of its specific surface receptor is more effective in maintaining and/or improving normal cardiac pump performance. In addition, the mechanical and structural characteristics of ventricular myocytes were examined 1 week after myocardial infarction-induced failure to determine whether abnormalities in myocyte function developed in association with myocyte hypertrophy. In addition, in an attempt to establish whether ACE affects not only myocyte size and shape but also the contractile behavior of the surviving cells, the effect of captopril on myocyte morphometry and mechanical performance was determined. Finally, information is presented that establishes that the impairment in cardiac pump function produced by myocardial infarction is accompanied by stimulation of the replicatory apparatus of the remaining viable cells after coronary artery occlusion and the affect of ACE inhibitors on DNA synthesis.

2. METHODS AND MATERIALS

2.1. Animal Model of Myocardial Infarction

Ligation of the coronary artery was performed in male Sprague-Dawley rats at 2 months of age in order to produce infarcts associated with left ventricular failure.[4,8] Under ether anesthesia the thorax was opened, the heart exteriorized, and the left main coronary artery was ligated. The chest was closed, pneumothorax reduced by negative pressure, and the animals allowed to recover. Sham-operated control animals were treated similarly except that the ligature was not tied. Three subsets of infarcted animals were employed: (1) Infarcted animals treated with oral captopril (2 g/liter); (2) infarcted animals treated oral losartan (2 g/liter) to inhibit angiotensin binding to the AT$_1$ receptor; and (3) infarcted animals treated with placebo. Animals were evaluated with respect to global cardiac geometry, hemodynamic performance, myocyte mechanical contractile behavior, and flow cytometric analysis (see Section 2.2).

2.2. Global Cardiac Geometry: Ventricular Remodeling after Infarction

2.2.1. Cardiac Fixation

Two different anatomical states were generated following myocardial infarction: end diastole and peak systole. The first was achieved by injecting cadmium chloride,[33] which irreversibly arrests the heart in diastole. The second was obtained by injection of KCl, which induces a reversible diastolic arrest, and subsequently exposing the myocardium to $BaCl_2$ to induce contraction at peak systolic pressure.[4]

2.2.2 Chamber Geometry: Size and Shape

Following tissue fixation, hearts were excised and weights of the left ventricle including the septum and right ventricle were recorded. The major intracavitary axis of the left ventricle was then measured by inserting a calibrated probe in the ventricular chamber. Subsequently, the left ventricle inclusive of the septum was serial-sectioned into eight 1–2 mm thick rings, perpendicular to the longitudinal axis of the heart from base to apex. The mean thickness of each serial slice was determined by averaging ten separate measurements made at a magnification of $\times 28$ with a dissecting microscope having an ocular micrometer accurate to 0.03 mm. The individually numbered slices were then photographed at $\times 0.75$ and prints at $\times 12$ magnification obtained. These photographs representing eight uniformly spaced parallel section planes through the whole left ventricle were used for measurement of the areas occupied by necrotic and nonnecrotic myocardium. Moreover, the area represented by the cavitary lumen was determined. The septal and free wall areas were also examined. These values were obtained by analyzing each photograph with a computer-assisted image analysis system. Finally, the tracing of each wall area was divided by the mean length of the image profile to provide a direct estimate of mean wall thickness. The area measurements in each of the eight photographs of each ventricle were multiplied by the previously determined thickness of each tissue slice to yield the total volume of each parameter in each slice. By summation of each set of data derived from each serial section, absolute component volumes in the whole ventricle were obtained. In this way the volume of the ventricular chamber, wall thickness at eight different levels along the major intracavitary axis, and the relative amounts of infarcted and noninfarcted myocardium in the ventricles were determined in diastolic and systolic arrested hearts.[4]

2.3. Hemodynamic Evaluation

Hemodynamic performance profiles were obtained following anesthetization with chloral hydrate (300 mg/kg body weight, IP). Heart rate was obtained from electrocardiographic tracings of leads I, II, and III by placement of four individual subcutaneous limb electrodes. The external right carotid artery was exposed and cannulated with a microtip pressure transducer connected to an electrostatic chart recorder and systemic arterial blood pressures monitored and recorded. The microtip pressure transducer was then advanced into the left ventricle for the evaluation of left ventricular diastolic and systolic pressure dynamics. Moreover, the rates of rise and fall of left ventricular systolic pressure were derived by active analog differentiation of the pressure signal.[4,31,32]

2.3.1 Wall Stress

Circumferential stress at the mid-wall, σ_{cm}, is dependent on measurements of ventricular pressure (P), chamber radius (r), and wall thickness (h). If these values are obtained at a given time during the cardiac cycle, a two-dimensional curve relating wall stress to location along the longitudinal axis can be generated with wall stress highest at the equatorial region.[34] If pressure, chamber radius, and wall thickness are known throughout the cardiac cycle, however, a three-dimensional curve can be generated that relate mid-wall circumferential stress at each point during the cardiac cycle and for each region of the heart from base to apex. Instantaneous values for P, r, and h, were obtained as follows: P = left ventricular pressure was digitized and a value for pressure obtained every millisecond of the cardiac cycle and placed into a computer-based spreadsheet; r and h = end diastolic and peak systolic values for chamber radius and wall thickness were directly obtained from computer-assisted image analysis. Cardiac cycle time intervals, peak $\pm dP/dt$, and the intercept of the arterial and ventricular pressure waves were utilized to generate a best-fit trajectory between end diastole and peak systole and the subsequent curves digitized for analysis.[4]

Circumferential stress at the mid-wall for each of the eight layers of the ventricle from the base to the apex was derived as done previously.[4,31,32,34] In order to evaluate the stress envelope derived for the control and experimental groups, the total volume of stress for the two groups was determined. This volume measurement was defined by the upper and lower surfaces of the three-dimensional surface reconstruction of the two-dimensional plots of transmural circumferential stress for increasing distances from the uppermost section of the basal region of the ventricle and was calculated in cubic units by the trapezoidal rule.[31,32]

2.3.2. Myocyte Contractility

2.3.2a. Isolation of Single Myocytes. Calcium-tolerant myocytes were isolated according to the procedure previously described.[35–37] In brief, Sprague-Dawley rats were heparinized (500 U, IP) and killed by decapitation. Hearts were excised and placed on a stainless steel cannula for retrograde perfusion through the aorta.

The solutions were supplements of modified commercial MEM Eagle Joklik (K. C. Biological, DMC317). HEPES-MEM contained in millimoles: NaCl, 117; KCl, 5.7; NaHCO$_3$, 4.4; KH2CO4, 1.5; MgCl$_2$ 1.7; HEPES, 21.1;

glucose, 11.7; amino acids, 3; vitamins, 4; L-glutamine, 2; and insulin, 21 µ/ml ($69 \times 10\text{-}9$). The pH was adjusted to 7.2 with KOH. This solution is 292 mOsmoles, isomolar with rat serum, and in the standard procedure contained no added calcium although measured calcium activity was 5 µM. Resuspension medium was HEPES-MEM supplemented with 0.5% bovine serum albumin, 0.3 mM calcium chloride, and 10 mM taurine adjusted to 292 mOsmoles.

The cell isolation procedure consisted of three main steps:

1. Low calcium perfusion: Blood washout and collagenase (selected Worthington type II) perfusion of the heart was carried out at 32°C with HEPES-MEM gassed with 85% O_2, 15% N_2.
2. Mechanical tissue dissociation: The left ventricle inclusive of the interventricular septum was separated and minced. Collagenase-perfused tissue was subsequently shaken in resuspension medium containing creatine, collagenase, and 1.2 mM calcium chloride. Supernatant cell suspensions were washed and placed in resuspension medium.
3. Separation of intact cells: Intact cardiac cells were enriched by centrifugation through Percoll. Approximately 10^6 cells were suspended in 10 ml of isotonic Percoll (final concentrations 41% in resuspension medium) and centrifuged for 15 min at 34 g. Intact cells were recovered from the pellet, washed, and employed in the morphometric and mechanical evaluation of ventricular myocytes. Cell viability was assessed by Trypan blue exclusion.

2.3.2b. Myocyte Structure. General measurements of myocyte geometric dimensions were accomplished through the aid of a computerized image analysis system. The distribution of myocytes isolated from the left ventricle for each of the tree animal groups was divided according to number of cells in an established range of lengths. Frequency distribution histograms were constructed by plotting number of cells on the ordinate and cell length range on the abscissa. Following construction of the cell length histograms, a Gaussian curve was fit to the data by calculation of the probability density function of each ordinate value ($f(y)$) from the mean and standard deviation and plotting $f(y)$ against the range of cell lengths.[36,37]

2.3.2c. Myocyte Mechanics. The isolated muscle cells were placed in an open perfusion microincubator cell bath located on the stage of a Zeiss inverted microscope. The composition of the buffer solution was in millimoles: NaCl, 117; KCI, 5.7, $CaCl_2$, 1.2; $NaHCO_3$, 4.4; $KH2CO4$, 1.5; $MgCl_2$, 1.7; HEPES, 21.1; glucose, 11.7; amino acids, 3; vitamins, 4; L-glutamine, 2; and insulin, 21 µ/ml ($69 \times 10\text{-}9$). The pH was adjusted to 7.2. Temperature was maintained at 30 ± 0.2°C by a bipolar temperature controller. External bath calcium was maintained at 1.2 mM and myocytes were stimulated at 1.0 Hz by rectangular depolarizing pulses 5 msec in duration and twice diastolic threshold in intensity by platinum electrodes on either side of the cell and parallel with its long axis. Twenty myocytes from the left ventricle of each animal of the three groups of rats were evaluated mechanically as follows: A high speed camera was attached to the eyepiece of the inverted microscope and isolated myocytes were photographed at a final magnification of $\times 850$ during contraction at a rate of 500 frames/sec. In this way a real image of the cell was obtained ever 2 msec. This camera was synchronized with the stimulus pulse and unloaded (zero external load) peak shortening, velocity of cell shortening and relengthening, and time to peak shortening and time to one-half relengthening were obtained during still-frame playback by use of a computerized image analysis system. The entire apparatus necessary for measuring mechanics of isolated myocytes rests on an optical table that is supported by four pneumastable isolators to eliminate extraneous horizontal and vertical vibration.

2.4. DNA Synthesis in Viable Myocytes

2.4.1. Isolation of Myocyte Nuclei

Isolated myocytes were treated with hypotonic buffer (0.01 M HEPES, 1.5 mM $MgCl_2$, pH 7.4) for 5 min and then lysis buffer (3% glacial acetic acid, 5% ethylhexadecyldimethyl ammonium bromide in water) was added and tubes were shaken every 2 min for 10 min.[38] By this procedure, myocyte nuclei were fully dissociated from the myocyte cytoplasm.

2.4.2. Flow Cytometry

The suspension of isolated nuclei from each ventricle was filtered through a 53 µM nylon filter and nuclei were stained with the metachromatic dye acridine orange. In the first step, a 0.2 ml aliquot of nuclei suspension with approximately $1-2 \times 10^5$ nuclei was mixed with 0.4 ml of a solution containing 0.08 N HCl, 0.15 M NaCl, and 0.1% Triton X-100. The nuclei .were subsequently stained by adding 6 µg/ml acridine orange dissolved in 1.2 ml 0.2 M Na_2HPO_4–0.1 M citric acid buffer (pH 6.0), 1 mM EDTA, 0.15 M NaCl. All staining steps were performed at 0°C. The nuclei stained with acridine orange were excited with an argon laser at 488 nm and the green fluorescent emission at 530 nm was recorded by using the FACScan flow cytometer. The data were acquired and processed by the Consort 32 system. Data were transferred to a Zeos 386SX computer and analyzed using the Hicycle software.[39,40] The percent of nuclei in $G_0 + G_1$, S, and G_2M phases of the cell cycle were assessed in each preparation.[39,40]

3. DATA ANALYSIS

Results are present as mean ± SD computed from the average measurements obtained from each rat. Statistical significance for comparison between two measurements within the wall of each ventricle was determined using the paired Student's t test. Statistical significance for comparisons between two measurements in control and experimental animals was computed using the unpaired two-tailed Student's t test. Statistical significance in multiple comparisons among independent groups of data, in which analysis of variance and the F test indicated the presence of significant differences, was determined by the Scheffe method.[41,42] Values of $p < 0.05$ were considered to be significant.

4. RESULTS

Myocardial infarction for 7 days was associated with a decreased body weight in each experimental group compared to corresponding sham-operated control rats. Overall heart rate was seen to be reduced by captopril but not losartan in control animals, whereas neither drug affected heart rate following myocardial infarction. On the other hand, comparisons between similarly treated control and infarcted animals revealed no difference between untreated or treated controls and their infarcted counterparts.

Cardiac weights in control and experimental animals revealed a significant decrease in heart weight in control animals as a result of captopril treatment with no change following losartan therapy (Fig. 1A). On the other hand, losartan administration was associated with maintenance of heart weight in infarcted animals compared to untreated and captopril-treated infarcted animals (Fig. 1A). When comparisons were made between treated control and infarcted animals, it was seen that heart weight remained elevated in both drug-treated infarcted groups, whereas no difference was observed between untreated control and experimental animal groups (Fig. 1B). The effects of drug therapy on left ventricular weight was similar to that seen for overall heart weight in that captopril reduced left ventricular weight in the control group while losartan maintained an elevated value in infarcted animals (Fig. 1C). Losartan was also associated with an elevation of left ventricular weight compared to the sham-operated losartan-treated control group (Fig. 1D). No difference was noted in left ventricular weight between either untreated or captopril-treated controls and their respective experimental counterparts (Fig. 1D). Drug therapy revealed no effect on sham-operated or infarcted right ventricular weight (Fig. 1E), but was associated with an elevation of right ventricular weight in untreated,

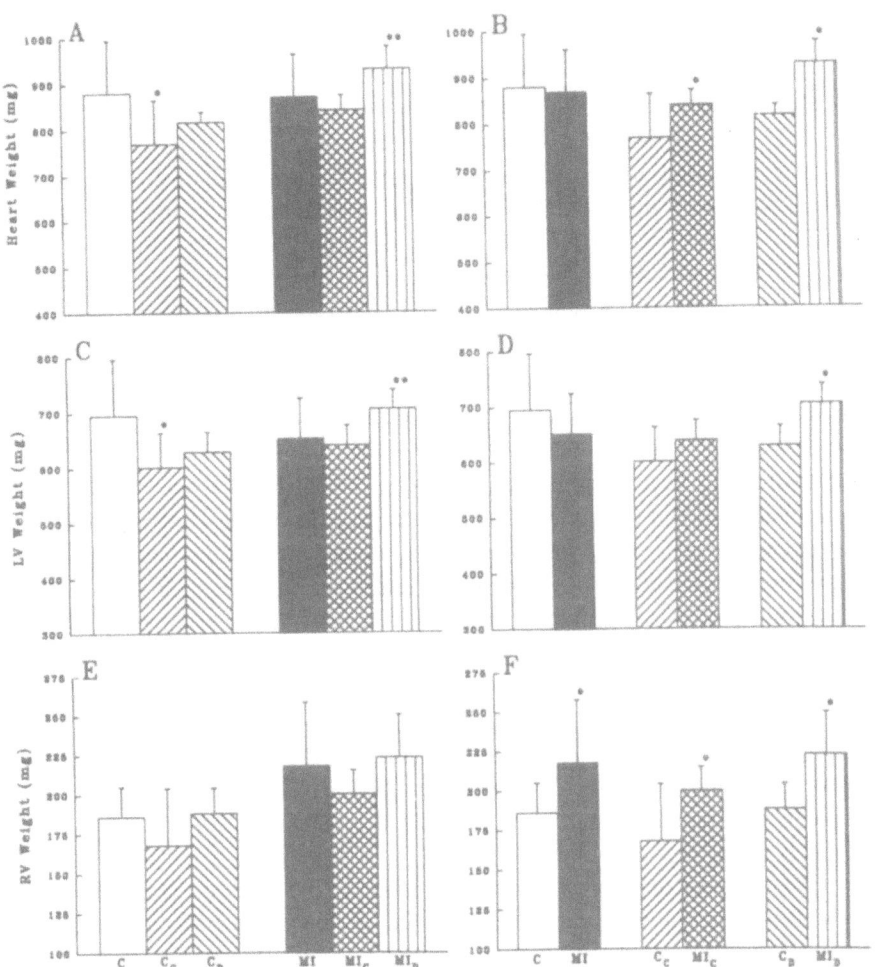

Figure 1. (A, B) Heart, (C, D) left ventricular, and (E, F) right ventricular weight 7 days after coronary artery ligation. C, Untreated sham-operated control group; C_C, sham-operated control group treated with oral captopril for 7 days prior to hemodynamic evaluation; C_D, sham-operated control group treated with oral losartan for 7 days prior to hemadynamic evaluation; MI, untreated myocardial infarcted groups; MI_C, myocardial infarcted group treated with captopril for 7 days prior to hemodynamic evaluation. MI_D, myocardial-infarcted group treated with losartan for 7 days prior to hemodynamic evaluation. Comparisons among control and infarcted groups are depicted in the left hand panels (A, C, E), while comparisons between control and infarcted groups are depicted in the right hand panels (B, D, F). In group comparisons: * indicates a difference that is statistically significant ($p < 0.05$) from the respective C or MI group. In comparisons between control and experimental groups: * indicates a difference which is statistically significant ($p < 0.05$) from the value in corresponding sham-operated animals. (From Capasso *et al.*[4] Reprinted with permission.)

captopril- and losartan-treated infarcted groups compared to the appropriate control groups (Fig. 1F).

Left ventricular peak systolic pressure was reduced in control animals as a result of both drugs, but no such change was seen in infarcted animals (Fig. 2A). This parameter was found to be consistently lower in infarcted animals compared to their corresponding controls regardless of drug treatment (Fig. 2B). However, losartan administration was associated with the smallest decrease in pressure-generating ability by the failing ventricles (Fig. 2B). End diastolic pressure on the other hand, revealed no difference as a result of either drug therapy in control animals (Fig. 2C) but did demonstrate a significant decrease in captopril-treated infarcted rats (Fig. 2C). However, each infarcted group revealed an elevation of this parameter with respect to the appropriately treated control rats (Fig. 2D). The difference between peak systolic pressure (Fig. 2A, B) and end diastolic pressure (Fig. 2C, D) is a measure of the ability of the heart to develop pressure and is depicted in Fig. 2E and F. Although control animals showed an identical decrease in their response to either unloading agent, no such effect was observed in similarly treated coronary occluded rats (Fig. 2E), resulting in a significant difference in developed pressure between all control and infarcted rats that was most pronounced between untreated control and infarcted rats (Fig. 2F).

In general, rats of left ventricular pressure rise and fall were reduced by captopril and losartan in control groups. Oral treatment of infarcted rats with captopril caused no change in these parameters. On the other hand, losartan treatment resulted in an increase in both positive and negative dP/dt in infarcted rats. In this regard, losartan reduced the depression in $\pm dP/dt$ in infarcted animals to the greatest extent.

Right ventricular hemodynamic performance revealed a decrease in central venous pressure in control animals following captopril administration, whereas losartan administration resulted in a central venous pressure that was not different from untreated controls. Although central venous pressure was elevated in untreated and losartan-treated infarcted rats, captopril therapy resulted in a significant decrease in this parameter in infarcted rats. However, central venous pressure remained elevated in untreated infarcted rats as well as in captopril- and losartan-treated animals compared to either treated or untreated control groups.

Right ventricular peak systolic pressure was reduced in and among control and experimental animals as a result of both drugs. This parameter showed no difference in untreated or captopril-treated control and infarcted rats, but losartan maintained peak systolic pressure in infarcted rats. Right ventricular end diastolic pressure decreased in control animals as a result of both drugs, while captopril alone resulted in a significant drop in this parameter in infarcted

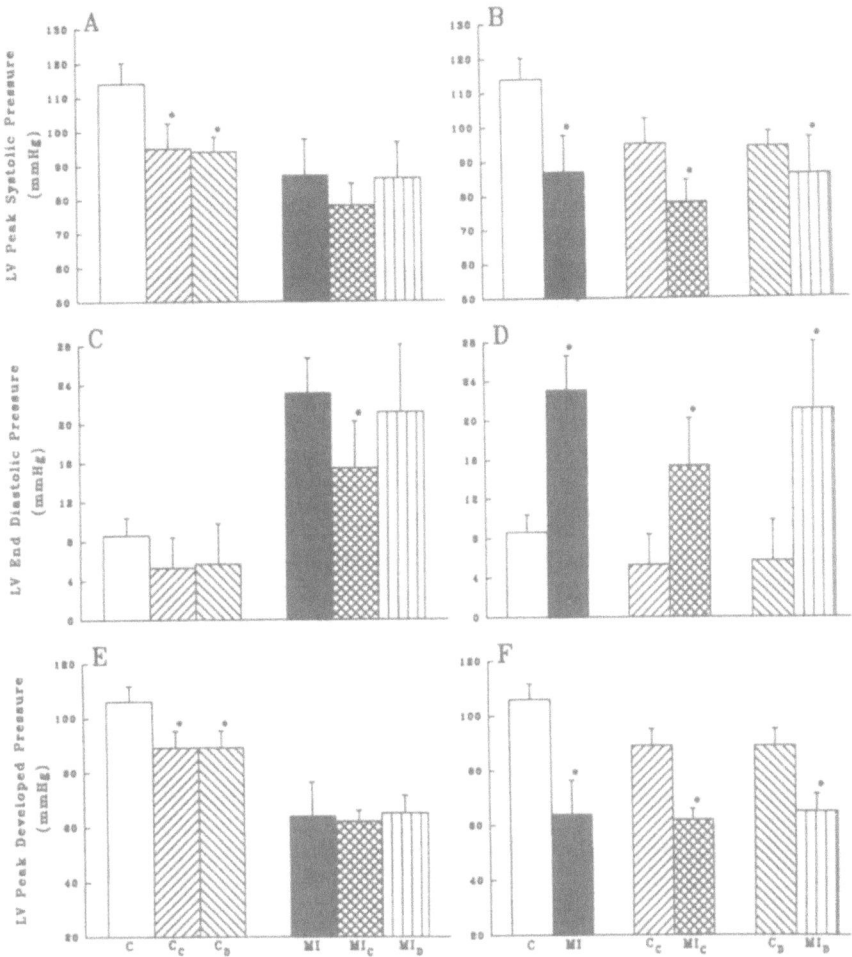

Figure 2. (A, B) Left ventricular peak systolic, (C) left ventricular end diastolic, and (D, F) left ventricular peak developed pressure 7 days after coronary artery ligation. C, Untreated sham-operated control group; C_C, sham-operated control group treated with oral captopril for 7 days prior to hemodynamic evaluation; C_D, sham-operated control group treated with oral losartan for 7 days prior to hemodynamic evaluation; MI, untreated myocardial infarcted groups; MI_C, myocardial-infarcted group treated with captopril for 7 days prior to hemodynamic evaluation; MI_D, myocardial-infarcted group treated with losartan for 7 days prior to hemodynamic evaluation. Comparisons among control and infarcted groups are depicted in the left hand panels (A, C, E), while comparisons between control and infarcted groups are depicted in the right hand panels (B, D, F). In group comparisons: * indicates a difference that is statistically significant ($p < 0.005$) from the respective C or MI group. In comparisons between control and experimental groups: * indicates a difference that is statistically significant ($p < 0.05$) from the value in corresponding sham-operated animals. (From Capasso *et al.*[4] Reprinted with permission.)

rats. However, each infarcted group revealed a significant elevation of end diastolic pressure compared to its appropriate control groups.

Peak rate of right ventricular pressure rise revealed no difference among control or infarcted rats as a function of drug therapy, but differences between control and infarcted rats showed a significant depression of this parameter in untreated and captopril-treated control and experimental animals. Importantly, no difference was observed between control and infarcted animals following losartan administration for 7 days. Peak rate of pressure fall in the right ventricle decreased in control and infarct groups as a result of captopril therapy, whereas losartan produced no change in controls and an increase in rate of relaxation in infarcted tissue.

4.1. Regional Variation in Ventricular Anatomy

The recognition of differences in the values of wall thickness and chamber diameter along the major axis of the heart is relevant to the understanding of the variability in stress distribution on the ventricular myocardium normally and after infarction. In this regard, Fig. 3 illustrates wall thickness measurements obtained at eight levels from the base to the apex in diastolic and systolic

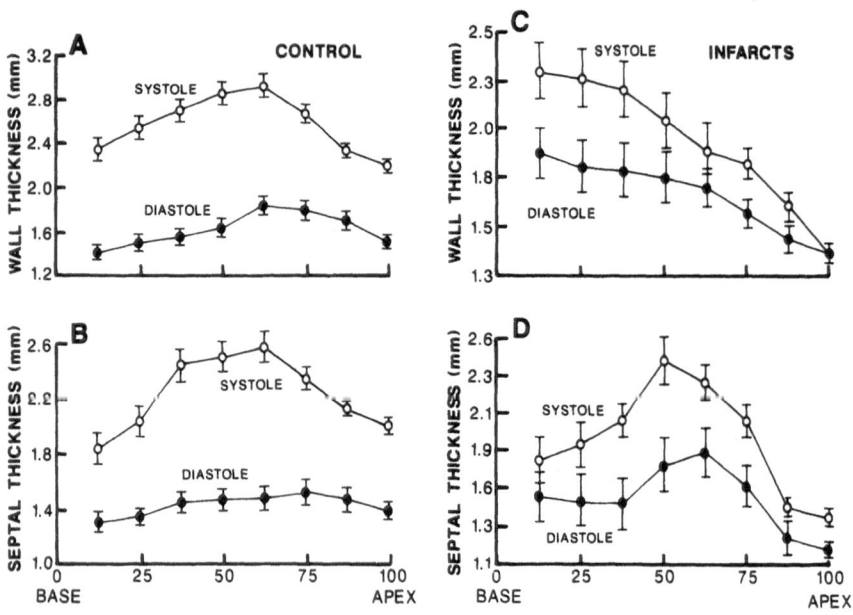

Figure 3. (A, C) Left ventricular wall and (C, D) septal thickness as a function of anatomical location (base = 0, apex = 100) and contractile state (systolic contraction = open circles; diastole = filled circles) 7 days after the creation of myocardial infarction. (A, B) Sham-operated control and (C, D) infarcted animals. Results are presented as mean ± SD.

arrested hearts of control and experimental animals. Normally, the left ventricle in diastole was found to possess a wall that was at the equator 16% ($p < 0.01$) and 8% (0.05) thicker than that at the base and apex, respectively. In systole, these differences increased, becoming 22% ($p < 0.01$) and 30% ($p < 0.001$). In the diastolic interventricular septum, septal thickness at the equator was 13% ($p < 0.05$) and 6% (NS) greater than that at the base and apical regions. Following systolic contracture, these differences were seen to be 37% ($p < 0.001$) and 25% ($p < 0.001$). Moreover, the left ventricular free wall was consistently thicker than that of the septum in all eight rings examined either following diastolic or systolic arrest. When the heterogeneity in systolic dimensions was examined separately in the wall and septum, it was observed that the free wall thickened by 65% at the base, 72–74% half way between the base and the apex, and 45% at the apex. Corresponding values for the interventricular septum were 41, 70–74, and 44%. All these changes were statistically significant ($p < 0.001$).

Myocardial infarction increased the variability in thickness of the surviving tissue of the free wall and interventricular septum from the base to the apex (Fig. 3C, D). In diastolic arrested hearts, mid-wall thickness was 7% ($p < 0.05$) thinner than that at the base and 28% ($p < 0.01$) thicker than the wall at the apical region. As a result of systolic contracture, these variations became 11% ($p < 0.05$) and 49% ($p < 0.001$). In the diastolic septum, the mid-portion was 14% ($p < 0.05$) and 43% ($p < 0.01$) thicker than the basal and apical regions, respectively. After systolic arrest, these differences were 38% ($p < 0.005$) and 74% ($p < 0.0001$). Moreover, these changes led to comparable thickness values in the wall and septum at several levels along the major axis of the heart, both in systole and diastole (Fig. 3C, D). In terms of the dissimilarity of shortening within the wall, the myocardium was found to thicken by 22% at the base, 17–24% at the equator, and 0% at the apex. In the septum, these changes were 16, 37–40, and 18%.

An analysis of cavitary diameters similar to that performed for wall thickness changes was conducted in control and experimental animals. Because of the shape configuration of the heart, the major and minor chamber diameters in each transverse ring were measured and their geometric mean calculated. In controls, reductions in average luminal diameter of 23, 50–51, and 44% were measured during systolic contracture at the base, equator, and apex, respectively. Corresponding percent changes after infarction were 4, 3–9, and 5% (data not shown).

Figure 4 shows in a composite form the relationship of myocardial tissue and chamber size of control and infarcted ventricles in diastole and systole. For simplicity, the necrotic region of the infarcted myocardium was not illustrated since barium contracture produced no effects on scarring dimension. This graphical representation documents that the infarcted ventricle lost

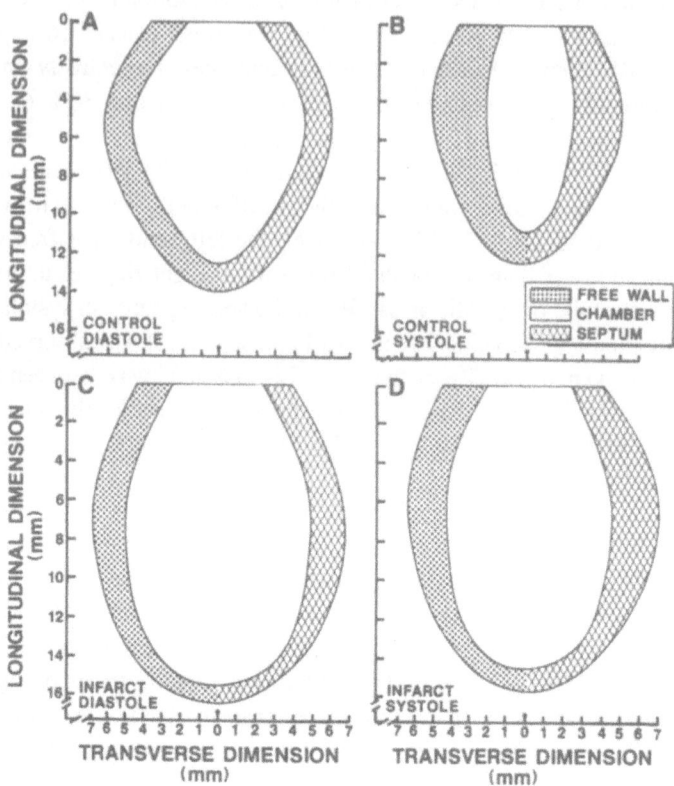

Figure 4. Schematic representation from average planimetric measurements of a cross-section through the ventricular chamber during (A, C) diastole and (B, D) following barium-induced systolic contraction in sham-operated control (A, B) and infarcted (C, D) animals 7 days after coronary artery ligation. The scarred area in the infarcted heart is not depicted since its dimensions did not change during the cardiac cycle. (From Capasso *et al.*[4] Reprinted with permission.)

its original elliptical configuration and assumed a more cylindrical shape. Because of minimal shortening in systole, this cylindrical profile was maintained after systolic arrest. Such a configurational change was the consequence of a greater increase in chamber dimension at the lower two thirds of the ventricle toward the apical region of the heart. Moreover, the reduced thickening of the surviving myocardium of the free wall in systole is shown in Fig. 4 from the base to apex. This phenomenon is contrasted with the response of the interventricular septum. Septal thickening, however, did not result in a major change in chamber transverse diameter at any level along the longitudinal axis of the heart.

4.2. Left Ventricular Wall Stress

Figure 5 illustrates mid-wall circumferential stress in the left ventricle as a function of anatomical location, from the base to the apex, and during the cardiac cycle. In control hearts, wall stress was found to be maximal at the equator, intermediate at the base, and minimal at the apex (Fig. 5A). Moreover, the trough in the center of Fig. 5A demonstrates that during systole, when chamber diameter was at its smallest dimension and the wall its thickest, wall stress was markedly reduced. Following myocardial infarction (Fig. 5B), the magnitude of overall wall stress increased [control, $3.5 \pm 0.13 \times 10^6$; infarct, $6.7 \pm 0.82 \times 10^6$ (dynes/mm^2)^3; $p < 0.001$]. However, diastolic stress augmented significantly more than systolic stress [control, $0.48 \pm 0.023 \times 10^5$; infarct, $3.27 \pm 0.37 \times 10^5$ (dynes/mm^2)^3; $p < 0.0001$]. It should also be apparent that there was a redistribution of wall stress in infarcted hearts, since similar values were found from the equatorial region to the apex of the ventricle. This was the consequence of the anatomical changes discussed in Fig. 4. The increase in the radius of curvature at the apical region of the heart resulted in a marked elevation of stress at this site.

Figure 6 illustrated, in a composite manner, the impact of myocardial

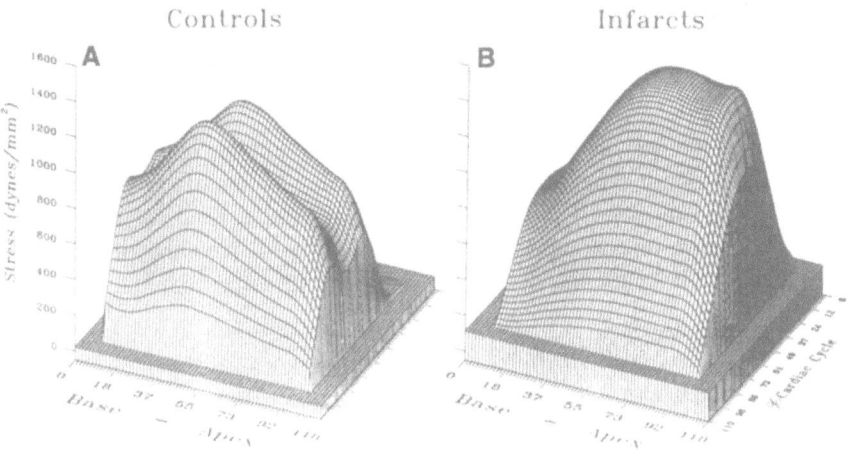

Figure 5. Mid-wall circumferential stress is depicted for the left ventricle as a function of anatomical location on the longitudinal axis (base to apex) and during the cardiac cycle (0 to 100 = minimal diastolic pressure through contraction and relaxation phases of the cardiac cycle and back to the starting point at minimal diastolic pressure) in (A) sham-operated control and (B) infarcted animals 7 days after the creation of myocardial infarction. Peak stress in infarcted animals is elevated at all phases of the cardiac cycle resulting in a significantly elevated envelope of total stress on the infarcted ventricle. (From Capasso *et al.*[4] Reprinted with permission.)

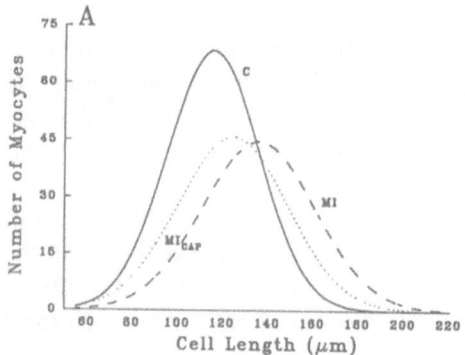

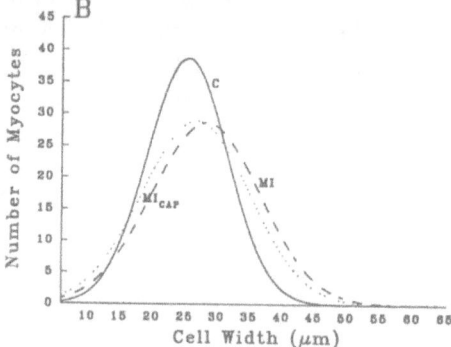

Figure 6. Superimposed Gaussian distribution of (A) cell lengths and (B) cell widths of enzymatically dissociated myocytes from the left ventricle of control animals (C), myocardial infarcted animals (MI), and myocardial infarcted animals treated with captopril (MI_{cap}). Gaussian distribution curves move to the right with myocardial infarction, whereas captopril treatment partially reversed this shift in distribution. (From Capasso *et al.*[37] Reprinted with permission.)

infarction alone and in combination with captopril administration on myocyte length and width. It should be apparent that the distribution of myocyte lengths (Fig. 6A) and myocyte widths (Fig. 6B) changed as a result of myocardial infarction. However, captopril therapy was capable of reducing, in part, the increases in cell length and diameter associated with large infarctions.

4.3. Myocyte Contractility

Differences were observed in overall cell length among the three groups of animals that resulted in statistically significant variations in the absolute amounts of myocyte cell shortening between the sham-operated and infarcted groups (C, 11.15 ± 0.96 μm; MI, 7.74 ± 0.88 μm; MI_{cap}, 10.15 ± 0.93 μm). Therefore, due to the observed differences in the linear dimension of myocyte longitudinal axis, myocyte shortening was expressed as a fraction of overall myocyte length (Fig. 7). In this regard, the peak percent cell shortening decreased by 40.1% in enzymatically isolated left ventricular myocytes from myocardial infarcted hearts compared to sham-operated controls (Fig. 7A). Captopril treatment prevented the decrease in shortening, but peak shortening was still

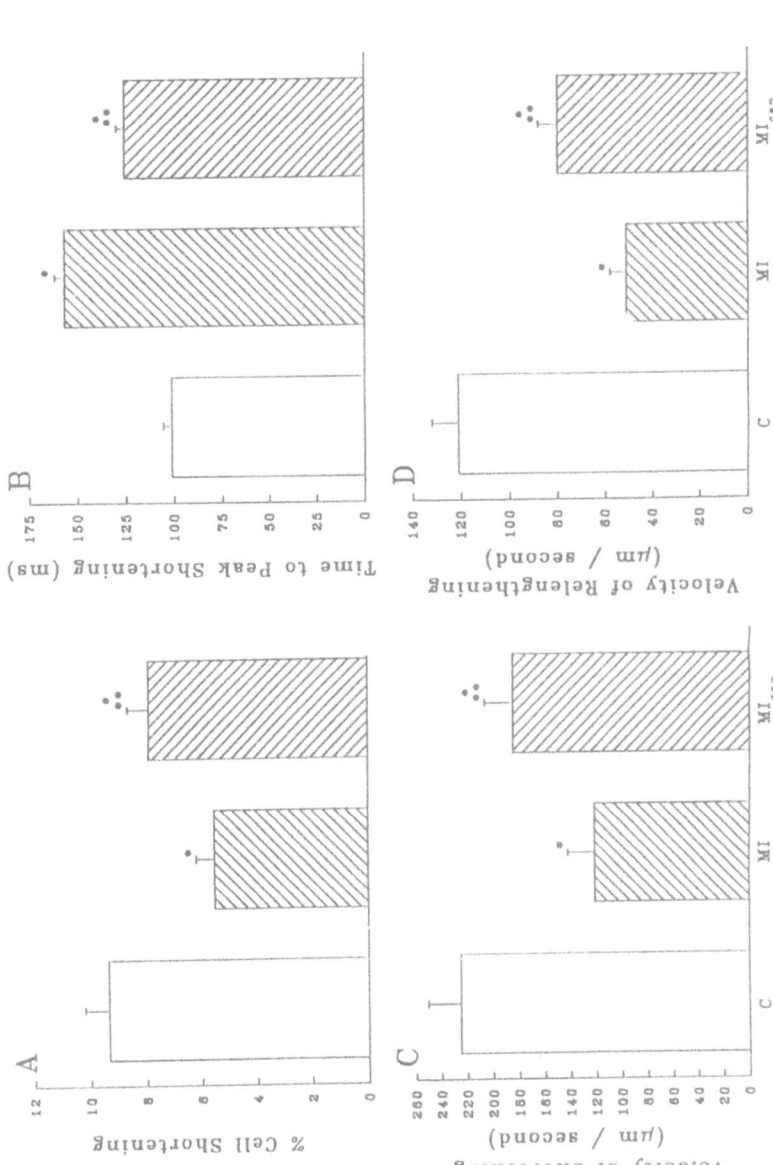

Figure 7. Changes in (A) cell shortening, (B) time to peak shortening, (C) velocity of cell shortening, and (D) velocity of cell relengthening of isolated myocytes from control (C), myocardial-infarcted (MI), and myocardial-infarcted animals treated with captopril (ML$_{cap}$). *, ** Indicates values that are significantly different ($p < 0.05$) from the corresponding result C and MI animals, respectively. (From Capasso *et al.*[37] Reprinted with permission.)

depressed by 16% compared to sham-operated controls. Moreover, MI rats displayed a 30% reduction in shortening compared to MI_{cap} animals.

The time allocated by the contractile machinery from the initiation of cell shortening to a point that represents maximum shortening is seen in the measurement of time to peak shortening (Fig. 7B). Left ventricular myocytes from MI required a 54 and 25% longer time to reach peak shortening than C and MI_{cap} rats, respectively. However, captopril therapy prevented only, in part, the prolongation of contraction duration since the difference in time to peak shortening between C and MI_{cap} animals, 24%, was statistically significantly.

The speed at which the isolated myocyte shortens is a critical factor in determining peak shortening and consequently ejection of blood in the intact heart. This parameter decreased by 46% in MI and 18% in MI_{cap} rats compared to C animals (Fig. 7C). It follows that a significant difference in shortening velocity was observed between MI and MI_{cap} rats. Captopril was seen to maintain velocity of shortening closer to control levels.

The velocity of muscle cell relengthening can be viewed to be dependent on cross-bridge interactions and contractile protein regulation as well as elastic recoil due to external or internal restoring elements. In a manner similar to that seen for velocity of shortening, peak unloaded velocity of isotonic relengthening was seen to decrease in MI (58%) and MI_{cap} (34%) animal groups compared to controls. However, a 56% increase in this parameter was seen in myocytes exposed to captopril compared to untreated myocytes isolated from infarcted hearts.

Figure 8 shows the effects of myocardial infarction and left ventricular failure on the amount of DNA present in nuclei isolated from the remaining viable myocytes 7 and 30 days after coronary artery occlusion. One week after infarction, the percentage of myocyte nuclei in the $G_0 + G_1$ phase of the cell cycle decreased by 7% (Fig. 8A), whereas the fraction of nuclei in G_2M increased 163% (Fig. 8C). An augmentation in nuclei in the S phase was also seen, but this change was not statistically significant (Fig. 8B). When the fractions of nuclei in the S and G_2M phases were combined to yield the total percentage of nuclei displaying DNA synthesis, myocardial infarction was found to be associated with a 144% increase in this parameter (Fig. 8D). In contrast, at 30 days after coronary artery occlusion, the fraction of nuclei in $G_0 + G_1$, S, and G_2M returned to control values. Thus, with respect to the 7-day time period, statistically significant decreases in these quantities occurred at 30 days (Fig. 8).

An identical analysis performed on myocyte nuclei isolated from the right ventricle demonstrated that the fraction of nuclei in $G_0 + G_1$ decreased 5% in this side of the heart at 7 days after infarction (Fig. 9A). On the other hand, the percentage of nuclei in G_2M augmented 139% at this time (Fig. 9C) with little variation in the portion of myocyte nuclei in the S phase (Fig. 9B).

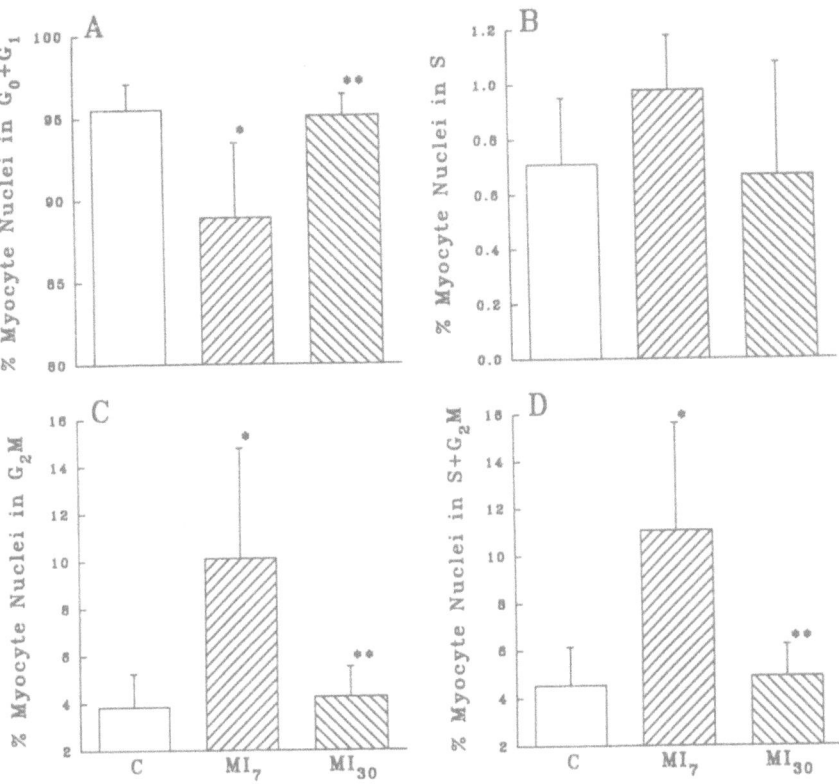

Figure 8. Fraction of left ventricular myocyte nuclei in the (A) $G_0 + G_1$, (B) S, (C) G_2M, and (D) $S + G_2M$ phases of the cell cycle at 7 (MI_7) and 30 (MI_{30}) days after infarction. *, ** Indicate values that are significantly different ($p < 0.05$) from the corresponding result in control and MI_7 animals, respectively. (From Capasso *et al.*[71] Reprinted with permission.)

Moreover, myocyte nuclei in $S + G_2M$ was found to be elevated by 121% in experimental animals (Fig. 9D). At 30 days following infarction, the fraction of nuclei in the S phase was found to be greater than that seen in control (152%) and 1 week after coronary artery occlusion (101%) (Fig. 9B). With respect to the early time period, coronary occlusion at 30 days was characterized by a decrease in the percent of nuclei in G_2M by 37% (Fig. 9C). However, this value was still 50% higher than that in sham-operated control rats.

In summary, the fraction of myocyte nuclei progressing through S and G_2M increased in the left and right ventricles 7 days after infarction but decreased at 30 days. However, in comparison with control values, DNA synthesis in myocyte nuclei remained elevated in the right ventricle but returned to normal in the left ventricle.

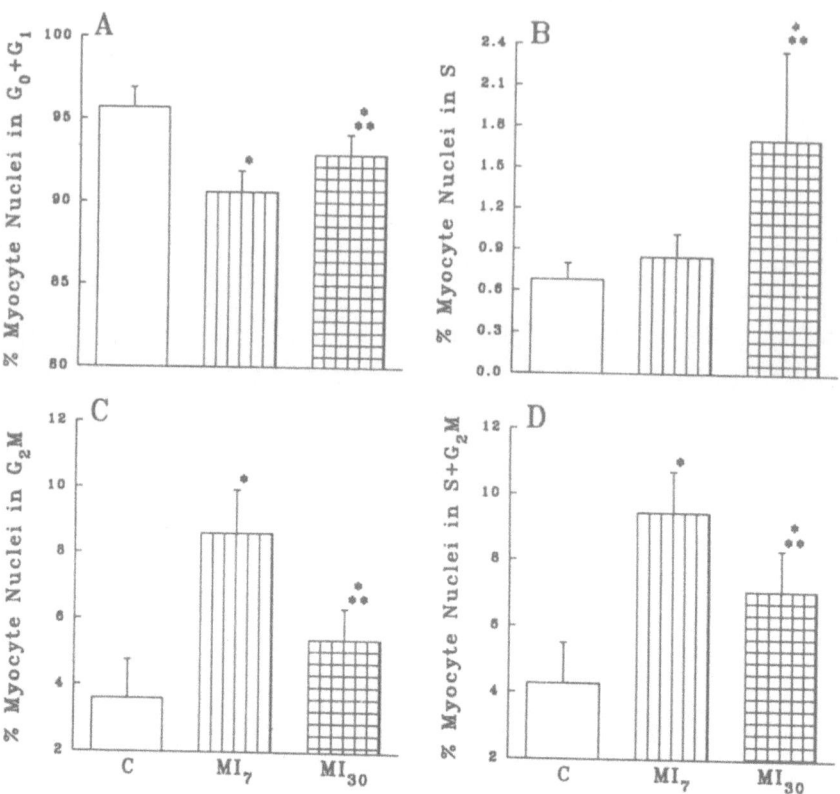

Figure 9. Fraction of right ventricular myocyte nuclei in the (A) $G_0 + G_1$, (B) S, (C) G_2M, and (D) $S + G_2M$ phases of the cell cycle at 7 (MI_7) and 30 (MI_{30}) days after infarction. *, ** Indicate values that are significantly different ($p < 0.05$) from the corresponding result in control and MI_7 animals, respectively. (From Capasso et al.[71] Reprinted with permission.)

4.4. DNA Synthesis

Figures 10 and 11 illustrate the effects of captopril on the distribution of myocyte nuclei in the phases of the cell cycle, under control conditions, and 7 and 30 days after infarction and left ventricular failure. In the presence of the ACE inhibitor, increases in the fraction of nuclei in the S, G_2M, and $S + G_2M$ phase were seen in the left ventricle at 7 days, but these changes were not statistically significant (Fig. 10). A similar pattern was observed at 30 days, although the percentage of nuclei in the S phase was 224% and 153% higher than that measured in controls and in rats at 7 days after infarction, respectively. Captopril administration had an impact on the right ventricle comparable to that described in the left ventricle (Fig. 11). The fraction of myocyte nuclei in G_2M became elevated 7 days after coronary artery occlusion, but

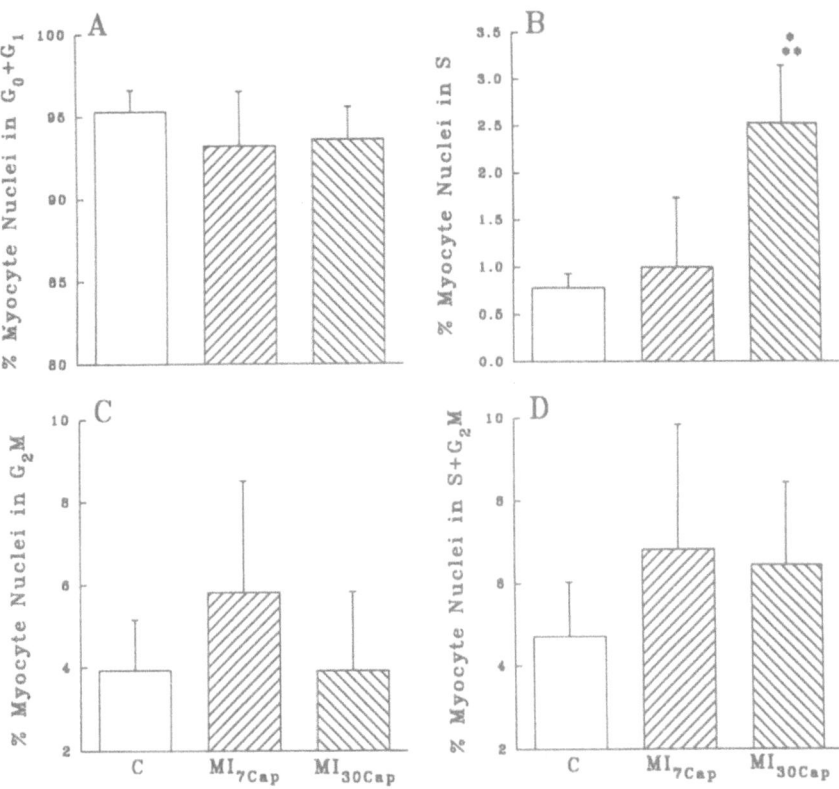

Figure 10. Fraction of left ventricular myocyte nuclei in the (A) $G_0 + G_1$, (B) S, (C) G_2M, and (D) $S + G_2M$ phases of the cell cycle after infarction and administration of captopril immediately after surgery and maintained for 7 (MI_{7Cap}) and 30 (MI_{30Cap}) days. *, ** Indicate values that are significantly different ($p < 0.05$) from the corresponding result in control and MI_{7Cap} animals, respectively. (From Capasso *et al.*[71] Reprinted with permission.)

decreased at 30 days, concomitant with an augmentation in the percentage of nuclei in the S phase.

In summary, captopril attenuated the magnitude of DNA synthesis in myocytes following infarction acutely and chronically.

5. DISCUSSION

5.1. ACE Inhibition versus AT₁ Blockade

Recent studies in the literature have demonstrated that unloading of the infarcted myocardium by ACE inhibitors can serve to (1) improve car-

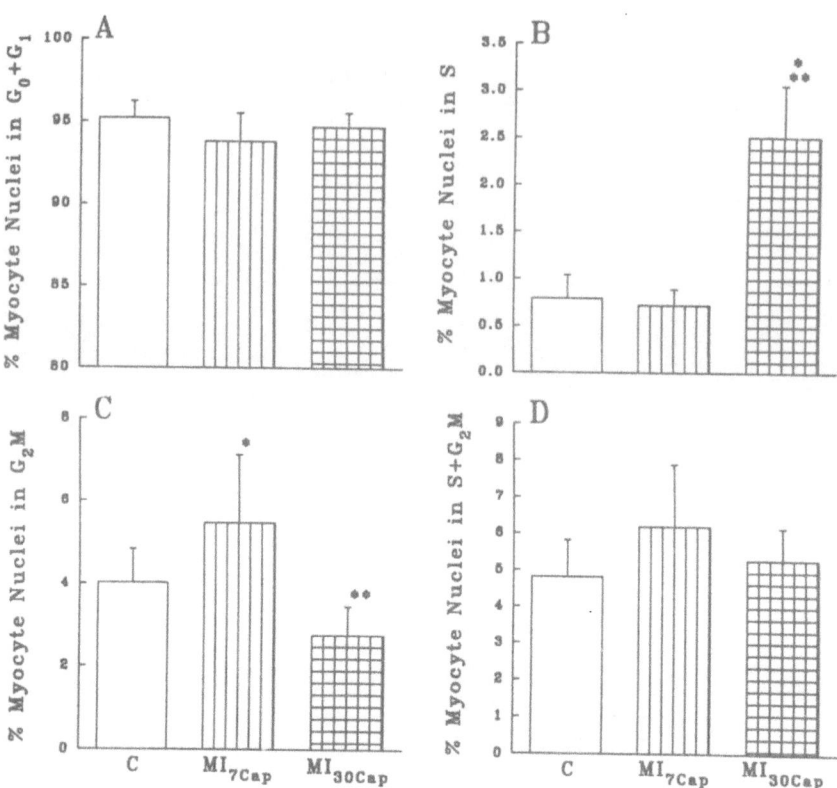

Figure 11. Fraction of right ventricular myocyte nuclei in the (A) $G_0 + G_1$, (B) S, (C) G_2M, and (D) $S + G_2M$ phases of the cell cycle after infarction and administration of captopril immediately after surgery and maintained for 7 (MI_{7Cap}) and 30 (MI_{30Cap}) days. *, ** Indicate values that are significantly different ($p < 0.05$) from the corresponding result in control and MI_{7Cap} animals, respectively. (From Capasso *et al.*[71] Reprinted with permission.)

diac pump performance; (2) reduce the detrimental effects of chamber and wall remodeling; and (3) decrease long-term mortality after myocardial infarction.[6,14–16] It has also been demonstrated that ANG I and ANG II conversion in human heart tissue is controlled by membrane-bound serine proteinases[43–47] that are not inhibited by ACE (i.e., carboxypeptidases) inhibition.[43–47] In fact, an enhanced inotropic effect to ANG I has been shown in hamster myocardium in the presence of captopril.[43] In this regard, it would be expected that captopril would reduce afterload and preload while ANG II would still be produced by the myocardial tissue per se.[43–47] This myocardial ANG II could benefit cardiac pump performance by exerting a positive inotropic effect on the myocardium itself, as has been shown in several species.[12] Increased circulating levels of ANG I following captopril administration would serve to potentiate the positive inotropic effect of myocardial ANG II by supplying increased levels of ANG I for conversion to the heart.[43] This could serve as the basis for a potential advantage of ACE inhibition compared to ANG II receptor blockade. However, the inotropic state of the infarcted rat heart following captopril was slightly depressed compared to hearts of losartan-treated animals. Although AT_1 blockade may be expected to be associated with a negative inotropic action if the above scheme is correct, no such effect was observed here in control or infarcted rats. Recent work from our laboratory has shown that ANG II exerts a negative inotropic effect on the myocardium of control animals and that this response is exacerbated in the remaining viable myocardium from animals with large infarctions of the anterior lateral wall of the left ventricle.[48] It should be pointed out that although captopril was more effective at unloading the infarcted heart, it was clear that losartan potentiated the inotropic state of the heart more than ACE inhibition, possibly by inhibiting the impact of ANG II on the myocardium itself.[43–47]

5.2. Chamber Remodeling and Myocardial Infarction

Coronary artery occlusion-induced heart failure has been shown to be associated with remodeling of cellular and global aspects of the left ventricle.[4,7,8,16] At the cellular level, loss of a large segment of the myocardium increases the diastolic and systolic LaPlace loads on the remaining viable myocardium, engendering side-to-side slippage of myocytes in the ventricular wall.[7,8] In addition, side-to-side slippage of myocytes will cause a reduction in ventricular mural cell number, wall thinning, tranverse chamber dilation, and an increase in cell number from base to apex, resulting in longitudinal chamber dilation.[4] Although preload elevations may be reduced by ventricular dilatation while end diastolic volume is augmented, the resultant increased diastolic wall stress[4] may compromise ejection of blood as the heart moves downward and to the right on the load-velocity and load-shortening curves.[10] Altered loading conditions in this setting will result in reactive cellular concentric and eccentric hypertrophy as myocytes increase in size longitudinally to accommodate a new chamber size and radially to increase force development and maintain ventricular and systemic pressures.[1,2,4,7,8,16] Myocyte enlargement, however, has detrimental effects because this change in cell shape has been shown to be associated with decreased speed and extent of shortening, despite a prolongation of shortening duration.[37,49,50] Additionally, myocardial hypertrophy will result in a shift from the V_3 or slower form of myosin ATPase, accounting, in part, for the slower and more prolonged contraction duration.[5,49,50] Since the present study has documented an increase in the inotropic state of the heart following losartan, it may be proposed that losartan administration has reduced the extent of chamber and wall remodeling and/or prevented shifts in the isoenzymic composition of the contractile proteins.

5.3. Hemodynamic Performance of Captopril and Losartan

In the present investigation, cardiac hemodynamic performance in untreated infarcted animals indicated the presence of overt heart failure and thus the loss of a large transmural region of the left ventricle.[1,2,4,6,16] Both drugs improved cardiac function in the failing infarcted heart. However, the mechanism of cardiac functional augmentation occurred differently depending on the particular therapeutic agent employed. Captopril, as with other ACE inhibitors,[12,18] improved pump function by significant reductions in preload and afterload, thus theoretically allowing cardiac output to increase. On the other hand, losartan decreased preload and afterload by only a modest amount but ameliorated the contractile state of the heart, possibly by preventing the negative inotropic effect of ANG II on the rat heart.[48] Although the doses of both drugs utilized were established in line with previous studies in which total ANG II inhibition or blockade was achieved,[4,14-16] it cannot be excluded that alterations in serum ANG II levels and/or AT_1 receptor number and affinity occurred with infarction resulting in a drug dosage that was insufficient and only partially effective.

5.4. Cellular Morphometry and Myocyte Contractility

Globally, compensated eccentric hypertrophy is associated with normal or improved myocardial mechanical performance, whereas compensated concentric hypertrophy is evidenced in a depressed contractile state of the myocardium.[52,53] These observations raise the possibility that at the cellular level these events may be reflected by enhanced contractility in long cells and depressed contractile ability in wide cells. These phenomena have been documented here since the decrease in unloaded velocity of shortening and peak myocyte shortening was found to be linearly related to cell diameter. In this regard, a general relationship has been found in the current study between cell length and diameter, on the one hand, and isotonic contractile properties on the other. Longer cells with no change in diameter possess greater velocity of shortening under unloaded conditions. In contrast, wider cells of identical length exhibit depressed shortening velocity. These characteristics apply to control and experimental animals. However, following myocardial infarction, the remaining viable myocytes increased both in diameter and length but the augmentation in the lateral dimension provoked a much greater impact on cell mechanical behavior than the increase in the longitudinal axis of the cell.

The mechanism underlying the length-dependent enhanced shortening of myocytes cannot be established at present. However, a shift in myosin isoenzyme from the slower V_3 to the faster V_1 isoform may account for the improved speed of contraction.[49] Moreover, changes at the regulatory protein level affecting calcium responsiveness at the myofilaments may modulate isotonic properties of myocytes.[54] Importantly, it has been documented that pressure overload hypertrophy results in a nonuniform distribution of myosin isoenzymes across the ventricular wall.[55] Such a nonhomogeneous localization of contractile protein enzyme activity may have significant implications in the abnormalities in cardiac pump function *in vivo*. Myocytes from discrete regions of the wall may display distinct mechanical behavior sustained by their dimensional and biochemical properties. Following myocardial infarction there is a wider distribution of myocyte lengths and widths that may reflect variations in diastolic loads on the ventricle and the proximity of myocytes to the necrotic area. Although these factors may shed light on the alterations in the distribution of cell length and width following infarction, they may also serve to produce greater heterogeneity of cell mechanical performance within the infarcted heart and therefore enhance the possibility of regions in the ventricular wall in which cells may function in a noncoordinated fashion.

It should be pointed out that myocytes from infarcted hearts may function at a different loading state than those in sham-operated animals and a load velocity profile should have been established. However, two factors mitigate against this type of approach:

1. Wall stress during the cardiac cycle is elevated in infarcted rats,[4] resulting in considerably higher loads on the individual myocytes *in vivo*.
2. The depression of unloaded velocity of shortening has been shown to be maintained at increasing viscous loads on the individual cells.[56]

Thus, it may be concluded that the depression in myocyte contractile performance seen in the present study *in vitro* would be worsened under the increased loading conditions of the infarcted heart *in vivo* throughout the range of physiologic loads.

5.5. Captopril and Myocyte Properties after Infarction

In the last several years the beneficial effects of ACE inhibition on the recovery and survival after myocardial infarction have been demonstrated in human and animals models.[22,57] Such a positive influence of captopril treatment, however, has been confined to its ability to reduce the magnitude of ventricular dilatation following ischemic myocardial injury,[22,57] although its mechanism at the cellular level remained unknown. The observations made here provide direct documentation of the potential sites of action of captopril therapy. Drug treatment reduced the magnitude of loading on the ventricle and attenuated the lateral and longitudinal expansion of myocytes, preserving, in part, the isotonic contractile behavior of the surviving cells. Thus, the decrease in load-dependent cellular hypertrophic response may constitute the basis of the lesser increase in cavitary volume demonstrated in the infarcted heart following captopril treatment.[22,57] Moreover, the relatively maintained shortening velocity of myocytes may explain the improvement in global cardiac performance with captopril after infarction.[22,57]

5.6. Ventricular Failure and DNA Synthesis in Myocytes

The current study suggests that adult, fully differentiated ventricular myocytes retain their ability to synthesize DNA when challenged by an abnormal elevation in wall stress produced by myocardial infarction. This contention is supported by the observation that the percentage of nuclei in the $S + G_2M$ phases of the cell cycle markedly increased at 7 days and subsequently decreased at 30 days. A reduction of this nature may be accounted for by myocyte mitotic division from 7 to 30 days, with a consequent decrease in the absolute number of cells in the $S + G_2M$ phases at the later time interval. However, myocytes entering the cell cycle, early after the imposition of the overload, may have died with the progression of healing producing a similar effect. This phenomenon is unlikely since quantitative morphological studies have indicated that cell loss does not appear to occur in the surviving left and right ventricular myocardium after infarction.[7,8,58]

Observations in this investigation are consistent with recent results that document in the presence of infarction and left ventricular failure, atrial myocytes exhibit DNA synthesis and nuclear mitotic division.[24] In addition, thymidine labeling has been detected in ventricular cells adjacent to the healing myocardium.[59] Similarly, conditions of overload in the adult heart following banding of the pulmonary artery,[25,30] renal hypertension,[26] myocardial aging,[27,35] and nutritional anemia[60] have been found to be characterized by myocyte nuclear and cellular hyperplasia through mitotic division of preexisting contractile cells.[35] Identical findings have been claimed in the hypertrophied human heart.[61–63] Importantly, proliferating cell nuclear antigen mRNA has been seen to be expressed not only in fetal and young myocytes, but also in adult differentiated cells.[29] Thus, myocytes express a gene essential for DNA replication and nuclear mitotic division at all stages of postnatal life. On this basis, the possibility may be advanced that cardiac myocytes may retain their capacity to proliferate throughout life and this growth reserve mechanism may become operative in response to severe myocardial dysfunction and overt failure.

It should be recognized that the fraction of myocyte nuclei found in G_2M cannot be equated with the magnitude of myocyte nuclei completing the cell cycle 7 days after infarction. The time required for a myocyte to divide is unknown and the percentage of cells actually completing mitotic division remains to be established. In both ventricles, early after infarction, the fraction of nuclei in the S phase was not significantly increased, implying a short S phase and/or that only a few cells were experiencing DNA synthesis. Moreover, the possibility of a prolonged G_2M phase cannot be disregarded. Were these phenomena to occur, the absolute number of myocyte nuclei completing the cell cycle at a given time could be a very small portion of that found in G_2M. This conclusion is supported by quantitative results that have indicated that myocyte cellular hypertrophy seems to be the primary mechanism of myocyte growth after infarction.[7,8] On the other hand, the simultaneous presence of myocyte loss with coronary occlusion complicates estimations of the actual amount of newly formed cells in the myocardium by any methodological procedure currently available. Myocyte loss leads to an underestimation of myocyte cellular hyperplasia in the tissue, whereas myocyte hyperplasia leads to an underestimation of the extent of myocyte death in the myocardium.[27] Such an effect has been shown to be operative in various pathologic states of the heart in animal models[64] and humans.[61,65] However, regional sampling in morphometric studies, assessing cellular responses at the equatorial plane of the infarcted ventricle, may have provided information limited to this region.[2] DNA synthesis in myocytes might have been present in the inferior two thirds of the heart, toward the apex, where the greater increase in wall stress has recently been documented to develop following large infarcts and ventricular failure at 7 days.[4]

In summary, these multiple results tend to suggest that DNA synthesis in myocyte nuclei and myocyte cellular hypertrophy may contribute significantly to the recovery of tissue mass and function acutely after infarction. Later in the evolution of the process, the expansion in cardiac mass may reduce the loading condition of the myocardium, attenuating the DNA synthetic reaction of the surviving cells. Our flow cytometric data at 30 days, in combination with previous measurements of myocyte size in this animal model,[7,8,65] support this conclusion. On the other hand, the long-term effects of myocardial infarction on ventricular remodeling and cardiac hemodynamics[16,66] may stimulate additional DNA synthesis and further cellular hypertrophy before end stage failure supervenes.

5.7. Captopril Treatment, Myocardial Infarction, and DNA Synthesis in Myocyte Nuclei

This investigation indicates that oral administration of an ACE inhibitor reduced biventricularly the number of myocyte nuclei found in the G_2M phase of the cell cycle 7 days after infarction. At 1 month, this difference was no longer present. On the other hand, in captopril-treated infarcted rats at 30 days, the percentage of nuclei in the S phase was greater than that in the correspond-ing untreated experimental rats and control animals. These observations point to the possibility that drug therapy was capable of reducing the DNA synthesis in surviving myocytes, by affecting the magnitude of the overload on these cells. Alternatively, captopril may have exerted a direct action on this nuclear response in spite of the changes in the hemodynamic state of the heart produced by this compound.[14] This latter effect appears unlikely, since a close relation-ship has been found between the unloading impact of captopril and the mag-nitude of cardiac hypertrophy accompanying arterial hypertension[66,67] and myocardial infarction.[14,15] However, subpressor doses of ACE inhibitors have been shown to prevent the development of hypertrophy following renal[13] and mechanical hypertension,[9] suggesting that induced myocardial growth may not be solely load-dependent. Opposite results have recently been described fol-lowing banding of the aortic arch in rats.[68] Whether this difference is related to the nature of the hypertrophic stimulus or whether the magnitude of the load varies according to the location of the constriction remains to be determined. Importantly, the incomplete normalization of ventricular loading at 30 days after myocardial infarction in the present study was associated with an increase in the percentage of nuclei in the S phase biventricularly, supporting the contention of load modulated DNA synthesis in stressed myocytes.

5.8. Myocardial Infarction, Ventricular Function, and Cardiac Hypertrophy

The findings in the present study demonstrate that the magnitude of myocardial hypertrophy increased from 7 to 30 days after infarction. This expansion in ventricular mass was accompanied by an amelioration of left and right ventricular pump function, although various hemodynamics indices of myocardial dysfunction and failure were still present. Captopril reduced the extent of hypertrophy and improved cardiac dynamics. These observations are consistent with previous results in which pump performance has been seen not to return to control values following large infarcts[6,15,30] in spite of an almost complete reconstitution of functioning tissue through cellular growth mecha-nisms in the remaining viable myocytes.[2,8] The protective impact of captopril has been documented to occur via a reduction in cavitary volume, coupled with an improvement in cardiac output and ejection fraction.[14] Nevertheless, the question remained whether a large stress, evoking a marked tissue reaction, led to hypertrophy that from the onset, rather than being a compensatory tissue reaction, represented an additional step toward depressed myocardial function.[8] The current results tend to support the contention that reactive hypertrophy after infarction has, at least in part, a beneficial influence on cardiac dynamics, strengthening observations made in humans.[69] Cardiac hypertrophy 1 month after infarction has been shown to be characterized by increases in cell diameter and length, consistent with a combination of pressure and volume overload stress on the myocardium.[70] On the other hand, the loading state at this time point was seen to shift to a prevailing elevation in diastolic stress with practically no change in systolic wall stress.[8] Thus, with time, this form of pathological hypertrophy may lead to a progressive increase in myocyte length, ventricular dilation, and the appearance of end-stage failure. Whether the long-term evolution of ischemic cardiomyopathy may reverse the attenuation in the DNA synthetic activity of myocytes seen at 1 month, initiating additional DNA synthesis, remains to be determined. Recent data on the failing senescent myocardium[27,32,35] suggest that this phenomenon may be operative late after infarction.

REFERENCES

1. Anversa P, Loud AV, Levicky V, et al: Left ventricular failure induced by myocardial infarction. I. Myocyte hypertrophy. *Am J Physiol* 248:H876–H882, 1985.
2. Anversa P, Beghi C, Kikkawa Y, et al: Myocardial infarction in rats: Infarct size, myocyte hypertrophy and capillary growth. *Cir Res* 58:26–37, 1986.
3. Bing OHL, Brooks WW, Conrad CH, et al: Myocardial mechanics of infarcted and hypertrophied non-infarcted myocardium following experimental coronary artery occlusion. In Jacob R, Gulch RW, Kissling G, (eds): *Cardiac Adaptation to Hemodynamic Overload, Training and Stress.* Dr. Dietrich Steinkopff, Darmstadt, 1983, pp. 265–276.
4. Capasso JM, Li P, Zhang X, et al: Heterogeneity of ventricular remodeling after acute myocardial infarction in rats. *Am J Physiol* 262:H486–H495, 1992.
5. Fletcher PJ, Pfeffer JM, Pfeffer MA, et al: Left ventricular diastolic pressure-volume relations in rats with healed myocardial infarction. *Circ Res* 49:618–626, 1981.
6. Pfeffer MA, Pfeffer JM, Fishbein MC, et al: Myocardial infarct size and ventricular function in rats. *Circ Res* 44:503–512, 1979.
7. Olivetti G, Capasso JM, Sonnenblick EH, et al: Side-to-side slippage of myocytes participates in ventricular wall remodeling acutely after myocardial infarction in rats. *Circ Res* 67:23–34, 1990.
8. Olivetti G, Capasso JM, Meggs LG, et al: Cellular basis of ventricular remodeling after myocardial infarction in rats. *Circ Res* 68:856–869, 1991.
9. Baker KM, Chernin MI, Wixon SK, et al: Renin–angiotensin system involvement in pressure-overload cardiac hypertrophy in rats. *Am J Physiol* 259:H324–H332, 1990.
10. Honig CR: Cardiac Mechanics. In Honig CR (ed): *Modern Cardiovascular Physiology.* Little, Brown, Boston, 1988, pp. 3–49.
11. Hirsch AT, Tolsness CE, Schunkert H, et al: Tissue specific activation of cardiac angiotensin converting enzyme in experimental heart failure. *Circ Res* 69:475–482, 1991.
12. Lindpaintner K, Ganten D: The cardiac renin-angiotensin system. An appraisal of present experimental and clinical evidence. *Circ Res* 68:905–921, 1991.
13. Linz W, Schoelkens BA, Ganten D: Converting enzyme inhibition specifically prevents the development and induces the regression of cardiac hypertrophy in rats. *Clin Exp Hypertens* 11:1325–1350, 1989.

14. Pfeffer JM, Pfeffer MA, Braunwald E: Influence of chronic captopril therapy on the infarcted left ventricle of the rat. *Circ Res* 57:84–95, 1985.

15. Pfeffer MA, Pfeffer JM, Steinberg C, et al: Survival after an experimental myocardial infarction: Beneficial effects of long-term therapy with captopril. *Circulation* 72:406–412, 1985.

16. Pfeffer JM, Pfeffer MA, Fletcher PJ, et al: Progressive ventricular remodeling in rat with myocardial infarction. *Am J Physiol* 260:H1406–H1414, 1991.

17. Johnston CI, Clappison BH, Anderson WP, et al: Effect of angiotensin converting enzyme inhibition on circulating and local kinin levels. *Am J Cardiol* 49:1401–1404, 1982.

18. Mimran A, Targhetta R, Laroche B: The antihypertensive effect of captopril. Evidence for an influence of kinins. *Hypertension* 2:732–737, 1980.

19. Zusman RM: Renin and nonrenin mediated antihypertensive action of converting enzyme inhibition. *Kidney Int* 25:969–983, 1984.

20. Buttrick P, Perla C, Malhotra A, et al: Effects of chronic dobutamine on cardiac mechanics and biochemistry after myocardial infarction in rats. *Am J Physiol* 260:H473–H484, 1991.

21. Geenen DL, Malhotra A, Scheuer J: Regional variation in rat cardiac myosin isoenzymes and ATPase activity following infarction. *Am J Physiol* 256:H745–H750, 1989.

22. Pfeffer MA, Braunwald E: Ventricular remodeling after myocardial infarction: Experimental observations and clinical implication. *Circulation* 81:1161–1172, 1990.

23. Jalil JE, Janicki JS, Pick R, et al: Coronary vascular remodeling and myocardial fibrosis in the rat with renovascular hypertension. Response to captopril *Am J Hypertens* 4:51–55, 1991.

24. Oberpriller JW, Ferrans VJ, Carroll RJ: Changes in DNA content, number of nuclei and cellular dimension of young rat atrial myocytes in response to left coronary artery ligation. *J Mol Cell Cardiol* 15:31–42, 1983.

25. Olivetti G, Ricci R, Anversa P: Hyperplasia of myocyte nuclei in long-term cardiac hypertrophy in rats. *J Clin Invest* 80:1818–1822, 1987.

26. Anversa P, Palackal T, Sonnenblick EH, et al: Hypertensive cardiomyopathy: Myocyte nuclei hyperplasia in the mammalian heart. *J Clin Invest* 85:994–997, 1990.

27. Anversa P, Palackal T, Sonnenblick EH, et al: Myocyte cell loss and myocyte cellular hyperplasia in the hypertrophied aging rat heart. *Circ Res* 67:871–885, 1990.

28. Claycomb WC, Moses RL: Growth factor and TPA stimulate DNA synthesis and alter the morphology of cultured terminally differentiated adult rat cardiac muscle cells. *Dev Biol* 127:257–265, 1988.

29. Marino TA, Haldar S, Williamson EC, et al: Proliferating cell nuclear antigen in developing and adult rat cardiac muscle cells. *Circ Res* 69:1353, 1360, 1991.

30. Olivetti G, Ricci R, Lagrasta C, et al: The cellular basis of wall remodeling in long-term pressure overload induced right ventricular hypertrophy in rats. *Circ Res* 63:648–657, 1988.

31. Capasso JM, Palackal T, Olivetti G, et al: Left ventricular failure induced by long-term hypertension in rats. *Circ Res* 66:1400–1412, 1990.

32. Capasso JM, Palackal T, Olivetti G, et al: Severe myocardial dysfunction induced by ventricular remodeling in the aging rat heart. *Am J Physiol* 259:H1086–H1096, 1990.

33. Grimm AF, Lin HL, Grimm BR: Left ventricular free wall and intraventricular pressure sarcomere length distribution. *Am J Physiol* 239:H101–H107, 1980.

34. Mirsky I: Ventricular and arterial wall stresses based on large deformation analyses. *Biophys J* 13:1141–1159, 1973.

35. Anversa P, Fitzpatrick D, Argani S, et al: Myocyte mitotic division in the aging mammalian rat heart. *Circ Res* 69:1159–1164, 1991.

36. Capasso JM, Fitzpatrick D, Anversa P: Cellular mechanisms of ventricular failure: Myocyte kinetics and geometry with age. *Am J Physiol* 262:H1770–H1781, 1992.

37. Capasso JM, Anversa P: Mechanical performance and morphometric characteristics of spared

myocytes after acute myocardial infarction in rats: Effects of captopril treatment. *Am J Physiol* 263:H841–H849, 1992.

38. Butler WB: Preparing nuclei from cells in monolayer cultures suitable for counting and for following synchronized cells through the cell cycle. *J Cell Physiol* 141:70–73, 1984.

39. Darzynkiewicz Z: Differential staining of DNA and RNA in intact cells and isolated cell nuclei with acridine orange. In Darzynkiewicz Z (ed): *Methods in Cell Biology.* Academic Press, New York, 1991, pp. 285–298.

40. Bruno S, Darzynkiewicz Z: Cell cycle dependent expression and stability of the nuclear protein detected by Ki-67 antibody in HL-60 cells. *Cell Prolif* 25:457–467, 1992.

41. Snedecor S, Cochran WG: *Statistical Methods,* 7th ed. Iowa University Press, Ames, 1980.

42. Wallenstein S, Zucker CL, Fleiss FL: Some statistical methods useful in circulation research. *Circ Res* 47:1–9, 1980.

43. Hirakata H, Fouad-Tarazi FM, Bumpus FM, et al: Angiotensins and the failing heart. Enhanced positive inotropic response to angiotensin I in cardiomyopathic hamster heart in the presence of captopril. *Circ Res* 66:891–899, 1990.

44. Urata U, Healy B, Stewart RW, et al: Angiotensin II forming pathways in normal and failing human hearts. *Circ Res* 66:883–890, 1990.

45. Urata U, Healy B, Stewart RW, et al: Angiotensin II receptors in normal and failing human hearts. *J Clin Endocrinol Metab* 69:54–66, 1989.

46. Urata U, Kinoshita A, Misono KS, et al: Identification of a highly specific chymase as the major angiotensin II forming enzyme in the human heart. *J Biol Chem* 265:22348–22357, 1990.

47. Urata U, Kinoshita A, Perez DM, et al: Cloning of the gene and cDNA for human heart chymase. *J Biol Chem* 266:17173–17179, 1991.

48. Capasso JM, Li P, Meggs LG, et al: Alterations in angiotensin II responsiveness in left and right myocardium after infarction induced heart failure in rats. *Am J Physiol* 264:H2056–H2067, 1993.

49. Capasso JM, Malhotra A, Scheuer J, et al: Myocardial biochemical, contractile and electrical performance after imposition of hypertension in young and old rats. *Circ Res* 58:445–460, 1986.

50. Capasso JM, Strobeck JE, Malhotra A, et al: Contractile behavior of rat myocardium after reversal of hypertensive hypertrophy. *Am J Physiol* 242:H882–H889, 1982.

51. Barany M: ATPase activity of myosin correlated with speed of muscle shortening. *J Gen Physiol* 50:197–216, 1967.

52. Grossman W, Jones D, McLaurin LP: Wall stress patterns of hypertrophy in the human left ventricle. *J Clin Invest* 56:56–64, 1975.

53. Sonnenblick EH, Strobeck JE, Capasso JM, et al: Ventricular hypertrophy: Models and methods. *Perspect Cardiovasc Res* 8:13–20, 1983.

54. Malhotra A: Regulatory proteins in hamster cardiomyopathy. *Circ Res* 66:1302–1309, 1990.

55. Bugaisky LB, Anderson PG, Hall RS, et al: Differences in myosin isoform expression in the subepicardial and subendocardial myocardium during cardiac hypertrophy in the rat. *Circ Res* 66:1127–1132, 1990.

56. Kent RL, Mann DL, Urabe Y, et al: Contractile function of isolated feline cardiocytes in response to viscous loading. *Am J Physiol* 257:H1717–H1727, 1989.

57. Pfeffer MA, Lamas GA, Vaughan DE, et al: Effect of captopril on progressive ventricular dilatation after anterior myocardial infarction. *N Engl J Med* 319:80–86, 1988.

58. Anversa P, Beghi C, McDonald SL, et al: Morphometry of right ventricular hypertrophy induced by myocardial infarction in the rat. *Am J Pathol* 116:504–513, 1984.

59. Rumyantsev PP, Kassem AM: Cumulative indices of DNA synthesizing myocytes in different compartments of the working myocardium and conductive system of the rat heart mus-

cle following extensive left ventricular infarction. *Virchows Arch B Cell Pathol* 20:329–342, 1976.

60. Olivetti G, Quaini F, Lagrasta C, et al: Myocyte cellular hypertrophy and hyperplasia contribute to ventricular wall remodeling in anemia-induced myocardial dysfunction in rats. *Am J Pathol* 141:227–240, 1992.

61. Linzbach AJ: Heart failure from the point of view of quantitative anatomy. *Am J Cardiol* 5:370–382, 1960.

62. Astorri E, Chizzola A, Visioli O, et al: Right ventricular hypertrophy: A cytometric study on 55 human hearts. *J Mol Cell Cardiol* 2:99–110, 1971.

63. Astorri E, Bolognesi R, Colla B, et al: Left ventricular hypertrophy: A cytometric study on 42 human hearts. *J Mol Cell Cardiol* 9:763–775, 1977.

64. Anversa P, Zhang Z, Li P, et al: Chronic coronary artery constriction leads to moderate myocyte loss and left ventricular dysfunction and failure in rats. *J Clin Invest* 89:618–629, 1992.

65. Olivetti G, Melissari M, Capasso JM, et al: Cardiomyopathy of the aging human heart: Myocyte loss and reactive cellular hypertrophy. *Circ Res* 68:1560–1568, 1991.

66. DeFelice A, Frering R, Horan P: Time course of hemodynamic changes in rats with healed severe myocardial infarction. *Am J Physiol* 257:H289–H296, 1989.

67. Pfeffer JM, Pfeffer MA, Mirsky I, et al: Regression of left ventricular hypertrophy and prevention of left ventricular dysfunction by captopril in the spontaneously hypertensive rat. *Proc Natl Acad Sci USA* 79:3310–3314, 1982.

68. Zierhut W, Zimmer H-G, Gerdes AM: Effect of angiotensin converting enzyme inhibition on pressure-induced left ventricular hypertrophy in rats. *Circ Res* 69:609–617, 1991.

69. Ginzton LE, Conant R, Rodriques DM, et al: Functional significance of hypertrophy of the noninfarcted myocardium after myocardial infarction in humans. *Circulation* 80:816–822, 1989.

70. Anversa P, Sonnenblick EH: Ischemic cardiomyopathy: Pathophysiologic mechanisms. *Prog Cardiovasc Dis* 33:49–70, 1990.

71. Capasso MJ, et al: Ventricular loading is coupled with DNA synthesis in adult cardiac myocytes after acute and chronic myocardial infarction in rats. *Circ Res* 71:1379–1389, 1992.

15

Angiotensin II Receptor Antagonism in an Ovine Model of Heart Failure
Comparison with ACE and Renin Inhibition

Michael Andrew Fitzpatrick, Miriam Rademaker, and Eric A. Espiner

1. INTRODUCTION

Over the last three decades, great strides have been made in our understanding of the pathophysiology of congestive heart failure, particularly with regard to the central role of the renin–angiotensin system (RAS), which is activated early in the course of heart failure. These findings are highlighted by recent studies that demonstrate improved survival in patients with heart failure and left ventricular dysfunction treated with angiotensin-converting enzyme (ACE) inhibitors.[1–3] However, questions remain about the specificity of these agents for blocking the RAS. Angiotensin II (ANG II), which is central to the activity of this important system, exerts numerous effects on a variety of vascular and nonvascular tissues due to the activation of specific ANG II receptors on the cell surface.[4–6] The recent development of specific AT_1 receptor antagonists (AT_1A) and the ability to analyze receptor numbers allow investigation of the central role of AT_1 receptors in modulating the effects of activation of the RAS in heart failure.

This chapter reviews our current understanding of the pathophysiology of heart failure with emphasis on the role of RAS as determined by observations using an ovine model of heart failure induced by rapid ventricular pacing.

Michael Andrew Fitzpatrick, Miriam Rademaker, and Eric A. Espiner • Departments of Cardiology and Endocrinology, The Princess Margaret Hospital, Christchurch, New Zealand.

Angiotensin Receptors, edited by Juan M. Saavedra and Pieter B.M.W.M. Timmermans. Plenum Press, New York, 1994.

Recent experiments investigating the acute hemodynamic, hormonal, and renal effects of AT_1 receptor antagonism in this model will then be described and compared with the effects of ACE and renin inhibition.

2. PATHOPHYSIOLOGY OF HEART FAILURE— AN OVERVIEW WITH OBSERVATIONS FROM THE RAPID VENTRICULAR PACING MODEL

Chronic heart failure may be viewed as a vicious cycle[7] initiated by any one of a number of pathological processes that impair cardiac performance. The resultant fall in cardiac output may initially be compensated by activation of neurohumoral systems, a response that eventually elevates systemic vascular resistance, in turn increasing the resistance to ventricular outflow and further depressing cardiac output. While the rate of progression of this cycle is highly variable, a ventricular arrhythmia may occur at any stage causing sudden death before the "vicious cycle" of heart failure has run its full course.

Renal retention of sodium and water is enhanced by activation of the renal sympathetic nerves, the renin–angiotensin–aldosterone axis, and by increased release of vasopressin. While the increase in preload produced by these compensatory changes is useful initially in sustaining cardiac output, the Starling forces are altered such that retained fluid cannot be contained within the intravascular compartment. Consequently, fluid collects unchecked in extravascular compartments, often producing life-threatening pulmonary edema, or troublesome ankle edema.

The rapid ventricular pacing model has been useful for observing the sequential pathophysiological changes during the onset and offset of heart failure[8] and has facilitated investigation of the effects of manipulation of neurohumoral systems in heart failure.[9–13] Rapid left ventricular (LV) pacing induces an early fall in arterial pressure and a progressive decline in cardiac output and rise in left atrial pressure. These hemodynamic changes are associated with activation of the renin–angiotensin–aldosterone and sympathetic systems, as well as with avid sodium retention and a decline in glomerular filtration rate[8] despite a significant rise in plasma atrial natriuretic peptide (ANP). Similar findings have been observed in dogs.[14,15] Cardiac changes include LV chamber dilation,[16–18] impaired systolic,[17,18] and diastolic LV function[18] as well as ultrastructural changes observed in other forms of heart failure.[19] Of interest, termination of pacing results in rapid reversal of the hemodynamic, hormonal, and renal changes,[8,17] and return of LV systolic function.[17] However, cardiac dilation, hypertrophy, and impaired diastolic function persist for at least 1 month after cessation of pacing.[17,18]

3. PHARMACOLOGICAL BLOCKADE OF THE RAS

As indicated in earlier chapters, pharmacological blockade of the RAS may occur by:

- Inhibition of renin activity by monoclonal anti-renin antibodies or angiotensinogen analogues (renin inhibitors);
- ACE inhibitors; or
- Antagonists of ANG II receptors.

ACE inhibitors are least specific in their action on the RAS. Enzymes, other than ACE, allow generation of ANG II[20,21] and may be physiologically important. Furthermore, ACE inhibitors not only block the formation of ANG II from ANG I, but also inhibit the degradation of bradykinin and prostaglandins. These less specific actions may contribute to the pharmacological efficacy of these drugs[22,23] and also to their side effects.[24] Renin inhibitors appear to be more specific in their action on the RAS. However, enzymes other than renin may enhance the formation of ANG I from angiotensinogen.[25] Furthermore, poor oral bioavailability of all renin inhibitors studied to date limits the clinical utility of these agents. Early peptide angiotensin analogues such as saralasin acted as competitive antagonists for ANG II. However, partial agonist activity and the need for intravenous administration limited their use. The advent of orally active, nonpeptide, specific AT_1 receptor antagonists provides a tool for exploring the specific blocking effect of ANG II regardless of the pathway of its formation.

Losartan is one of a new class of highly specific, nonpeptide ANG II-type (AT_1) receptor antagonists. Recent studies have demonstrated the existence of two distinct populations of ANG II receptors, AT_1 and ANG II-type 2 (AT_2).[5,26] It is now recognized that AT_1 receptors mediate virtually all of the known effects of ANG II (vasoconstriction, smooth muscle contraction, induction of hypertrophy, aldosterone secretion, catecholamine release, and thirst[4,6,27]). The function of the AT_2 receptors has not yet been identified because they do not appear to be involved in any of the known ANG II-induced responses and are not likely to be storage–clearance sites for ANG II.[6] Losartan blocks all of the ANG II-induced responses detailed above,[6] while displaying no affinity for receptors of other nonrelated polypeptides such as adrenocorticotropic hormone (ACTH), enkephalin, bradykinin, and vasopressin or norepinephrine.[28,29] Raya *et al.*[30] have recently documented the chronic hemodynamic effects of losartan in the rat infarct model of heart failure and compared these effects with the ACE inhibitor captopril. However, without the documentation of hormonal and renal effects, a comparison of the two agents is incomplete. Accordingly, we have investigated the acute hemodynamic, hormonal, and renal effects of ANG II antagonism[12] and compared them with the effects of ACE inhibition in the ovine model of heart failure. Similar experiments comparing the effects of renin and ACE inhibition are also described.

4. ACUTE EFFECTS OF ANG II RECEPTOR ANTAGONISM IN OVINE HEART FAILURE

4.1. Experimental Design

As previously described,[12] six Coopworth ewes (body weight 35–57 kg) were prepared surgically, placed in metabolic cages, and allowed to recover for ≥ 14 days before commencing the study protocol. During all experiments, the animals had free access to water and ate a normal laboratory diet (Na ~ 40 and K ~ 200 mmole/day) supplemented with an additional 40 mmole sodium administered with an applicator orally each morning as NaCl tablets. This allowed the maintenance of a minimum intake of sodium during the pacing period when dietary intake could not be guaranteed.

After the 2-week postoperative recovery period, rapid LV pacing was begun and continued for 6 days before the acute studies were performed. Heart failure was induced by rapid LV pacing at 225 beats/min. At this rate, sheep develop the hemodynamic, hormonal, and metabolic hallmarks of low cardiac output heart failure over a period of 7 days. The protocol called for the administration of losartan or vehicle in random order on two study days separated by a rest day. On the active treatment day, two bolus doses of losartan (1 mg/kg dissolved in 2 ml saline, period A; and 10 mg/kg dissolved in 20 ml saline, period B) were administered via a left atrial line at 1000 and 1200 hr, respectively, whereas vehicle (saline) was administered on the control day. At 1400 hr on both study days, a 12.5-mg bolus of captopril was administered (diluted in 5 ml saline, period C). Hemodynamic measurements[12] performed during the study included systolic, diastolic, and mean arterial pressure (SAP, DAP, and MAP), respectively; left atrial pressure (LAP); cardiac output (CO); and heart rate (HR). Hemodynamic recordings were performed at 15-min intervals for 1 hr before the bolus at 1000 hr (baseline) and during each 6-hr study period. All measurements were made with the sheep standing quietly in the metabolic cage.

Blood samples for the assay of plasma hormones were drawn via a left atrial catheter at selected times immediately after the hemodynamic recordings. Baseline samples were collected 30 min before and immediately before administration of the bolus at 1000 hr. After each bolus, blood samples were collected at 15, 30, 60, and 120 min. The blood was drawn into tubes on ice, centrifuged at 4°C immediately, then stored at 80°C until analyzed. Hormones[12] assayed included plasma renin, ANG II, ANP, and aldosterone. Blood samples for the assay of plasma cortisol levels were collected at 60 and 120 min after each bolus. All samples from each animal were measured in the same assay to avoid interassay variability. Blood samples for biochemical analysis of plasma sodium, potassium, and creatinine were taken immediately before the first bolus on each study day. Two-hour urine collections were performed during periods A, B, and C of each study day. The urine samples were measured for concentrations of sodium, potassium, and creatinine. The study protocol was approved by the Animal Ethics Committee of our institution.

Repeated-measures analysis of variance (ANOVA) was used to compare the effects of losartan and vehicle. The mean of the five baseline hemodynamic and two baseline hormonal recordings on each study day was used as a covariate to adjust the treatment means for the effect of any baseline difference. Uncorrected values are given in Table I and in the text; adjusted means are shown in Figs. 1 and 2. Time-matched data for the control and treatment periods A, B, and C were compared using two-day ANOVA for repeated measures. If significant treatment effects or time–treatment interactions were found, statistical comparisons of measurements at specific times were performed using t tests with the appropriate mean square error term from ANOVA. ANOVA demonstrated that the order of control and active treatment days did not affect any variable; consequently, this factor was not considered further. Statistical significance was assumed to be present when $p < 0.05$. All results are expressed as means ±SE.

4.2. Baseline Measurements

After 6 days of rapid ventricular pacing, the sheep in this study exhibited the hemodynamic and hormonal hallmarks of established heart failure (Table I). When compared with the normal values for our laboratory, MAP and CO were significantly reduced, whereas LAP was elevated. Plasma ANP was elevated, while activation of the RAS was evidenced by the elevation of plasma renin activity (PRA) and ANG II levels. There were no significant differences in any of the hormonal and hemodynamic parameters for the baseline period measured on the control and active treatment day.

4.3. Acute Hormonal Effects of Losartan

The effect of the ANG II receptor antagonist losartan on the hormonal indices in six sheep is shown in Fig. 1. Administration of losartan produced an increase in PRA and ANG II levels with both hormones peaking 15–60 min after each bolus. Compared with time-matched control recordings, a rise in PRA was observed after each bolus of losartan. PRA rose from a baseline of 1.0 ± 0.2 to a peak of 2.0 ± 0.4 nmole/liter per hour at 15 min after the 1 mg/kg bolus and to 3.3 ± 0.8 nmole/liter per hour ($p < 0.001$) 60 min after the 10

Table I. Hemodynamic, Hormonal, and Metabolic Data[a]

	Prepacing laboratory normal (n = 14)	Period A		Period B		Period C	
		Control Vehicle (n = 6)	Active Losartan (1 mg/kg)	Control Vehicle	Active Losartan (10 mg/kg)	Control Captopril only	Active Captopril after losartan
Hemodynamic data							
MAP (mm Hg)	86 ± 3[b]	60 ± 3	53 ± 2[d]	60 ± 4	46 ± 2[d]	46 ± 3[f]	42 ± 3[g]
LAP (mm Hg)	4 ± 1[b]	24 ± 1	22 ± 1[c]	24 ± 1	20 ± 1[d]	19 ± 1[f]	20 ± 1[g]
CO (liter/min)	5.9 ± 0.4[b]	1.7 ± 0.1	2.1 ± 0.2[c]	1.7 ± 0.1	2.0 ± 0.1	2.1 ± 0.2[e]	1.7 ± 0.1
Hormone data							
Ang II (pmole/liter)	12 ± 1[b]	50 ± 4	141 ± 37[d]	50 ± 4.3	227 ± 48[d]	35 ± 7	77 ± 26
PRA (nmole/liter per hr)	0.4 ± 0.1[b]	0.8 ± 0.1	1.8 ± 0.4	0.7 ± 0.1	3.1 ± 0.8[d]	3.0 ± 0.8[f]	6.5 ± 1.9[f,h]
ALDO (pmole/liter)	362 ± 41[a]	470 ± 119	477 ± 73	323 ± 76	446 ± 87	278 ± 69	347 ± 111
ALDO/PRA	—	767 ± 226	318 ± 69[d]	574 ± 159	330 ± 153[c]	110 ± 26[f]	114 ± 51
ANP (pmole/liter)	12 ± 1[b]	202 ± 14	176 ± 12	258 ± 28	174 ± 9[d]	214 ± 24[e]	204 ± 23[e]
CORT (nmole/liter)	220 ± 34	231 ± 32	216 ± 71	97 ± 12	160 ± 38	127 ± 32	140 ± 21

[a]Values are means ± SE of all hemodynamic and hormonal data recorded in six sheep for 15–45 min after administration of vehicle (control), losartan [1 mg/kg bolus (period A) and 10 mg/kg bolus (period B)], and the captopril bolus (period C). For comparison, normal values for our laboratory in 14 sheep before rapid LV pacing also are given. Abbreviations of variables are listed in the text. Significant differences between laboratory normal and vehicle control data: $^a p < 0.01$, $^b p < 0.001$ (unpaired t tests). Significant differences between the corresponding control period (vehicle) and active treatment (losartan): $^c p < 0.05$, $^d p < 0.001$ [determined from the repeated measures analysis of variance (ANOVA)]. Captopril effects can be observed by comparing period B vehicle values with period C captopril only values: $^e p < 0.05$, $^f p < 0.001$ determined from repeated measures ANOVA). Significant differences between period C captopril only and period C captopril after losartan: $^g p < 0.05$, $^h p < 0.001$ determined from repeated measures ANOVA).

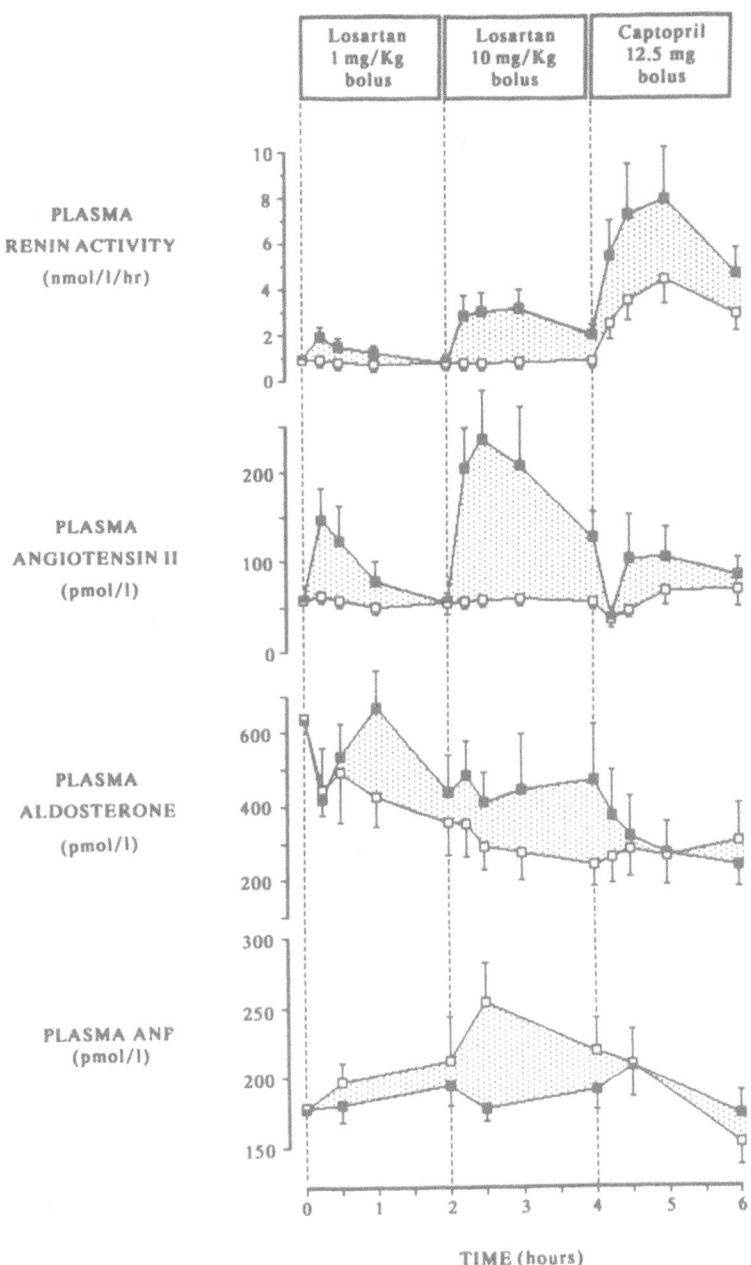

Figure 1. Hormonal response after intravenous bolus administration of vehicle and captopril (open symbols, control date) and losartan (1 mg/kg), losartan (10 mg/kg), and captopril (solid symbols, active treatment day). Data plots are corrected values from analysis of variance (ANOVA) using mean of preatreatment baseline levels as covariate (uncorrected data are given in Table I and in text). Baseline data point represents mean of control and active baseline values. Results are means ± SE. Significance of changes is discussed in text. (From Fitzpatrick *et al.*[12] Reprinted with permission.)

mg/kg bolus. Similarly, plasma ANG II levels rose after the 1 kg/mg bolus from 60 ± 10 to 153 ± 35 pmole/liter at 15 min ($p < 0.001$) and from 60 ± 12 to peak at 242 ± 54 pmole/liter ($p < 0.001$) 30 min after the 10 kg/mg bolus. The effects of the 10 mg/kg dose were sustained for ≥ 120 min. Two hours after the 1 kg/mg bolus, both PRA and ANG II levels had returned to the prebolus baseline levels. However, a more sustained response was observed after the 10 mg/kg bolus (Fig. 1), with PRA and ANG II still significantly higher than control levels 2 hr after the 10 mg/kg bolus ($p < 0.05$ and $p < 0.001$, respectively). The plasma aldosterone : PRA ratio fell from a baseline of 809 ± 303 to 258 ± 61 15 min after the 1 mg/kg bolus ($p < 0.001$). At 2 hr, the aldosterone : PRA ratio had not returned to baseline levels (574 ± 193, $p < 0.05$). After the 10 mg/kg bolus, the ratio declined further ($p < 0.01$) to reach a nadir of 224 ± 116 at 60 min. Overall, a close correlation existed between the mean PRA and ANG II ($r = 0.98$) in response to the losartan. On an individual basis, five of the six sheep had a significant correlation between these variables (range of r values in these 5 sheep: 0.86–0.99; $p < 0.05$–0.001).

As with ACE inhibitors, an increase in PRA occurred, presumably due to reduced feedback inhibition together with increased baroreceptor stimulation as arterial pressure falls. Similarly, plasma ANG II levels rose in a dose-dependent fashion with losartan administration in concert with the increase in PRA. Despite clear inhibition of RAS activity by losartan, no significant effect on plasma aldosterone levels was observed in this study. Similar findings have been observed with ACE inhibitors in heart failure[10] and also with losartan in normal volunteers.[31] Indeed, plasma aldosterone concentrations tended to rise after losartan administration. Despite this finding, the aldosterone : PRA ratio declined significantly during losartan treatment compared with the data from the control day, confirming blockade of ANG II-mediated aldosterone secretion. It should be noted that there was a trend for plasma cortisol to rise with losartan treatment. Clearly, the hormonal interrelationships are complex because hypotension (induced by losartan) may stimulate ACTH secretion[32] as well as impair steroid metabolic clearance,[33] offsetting any possible inhibiting effects from ANG II blockade. Reduced plasma ANP (due to the fall in LAP) also may have influenced plasma aldosterone[34] in this study.

4.4. Acute Hemodynamic Effects of Losartan

Hemodynamic effects of losartan are shown in Fig. 2. Compared with time-matched control recordings, a rapid drop in MAP was observed with each bolus of losartan. MAP dropped from the baseline of 60 ± 4 to 50 ± 4 mm Hg 15 min after the 1 mg/kg bolus ($p < 0.001$) and to a nadir of 45 ± 5 mm Hg 30 min after the 10 mg/kg bolus ($p < 0.001$) (Fig. 2). Similarly, LAP dropped after the 1 mg/kg losartan bolus from a baseline of 25 ± 2 to 22 ± 2 mm Hg at 15 min ($p < 0.001$) and from $23 \pm$ to 19 ± 1 mm Hg 75 mm after the 10 mg/kg

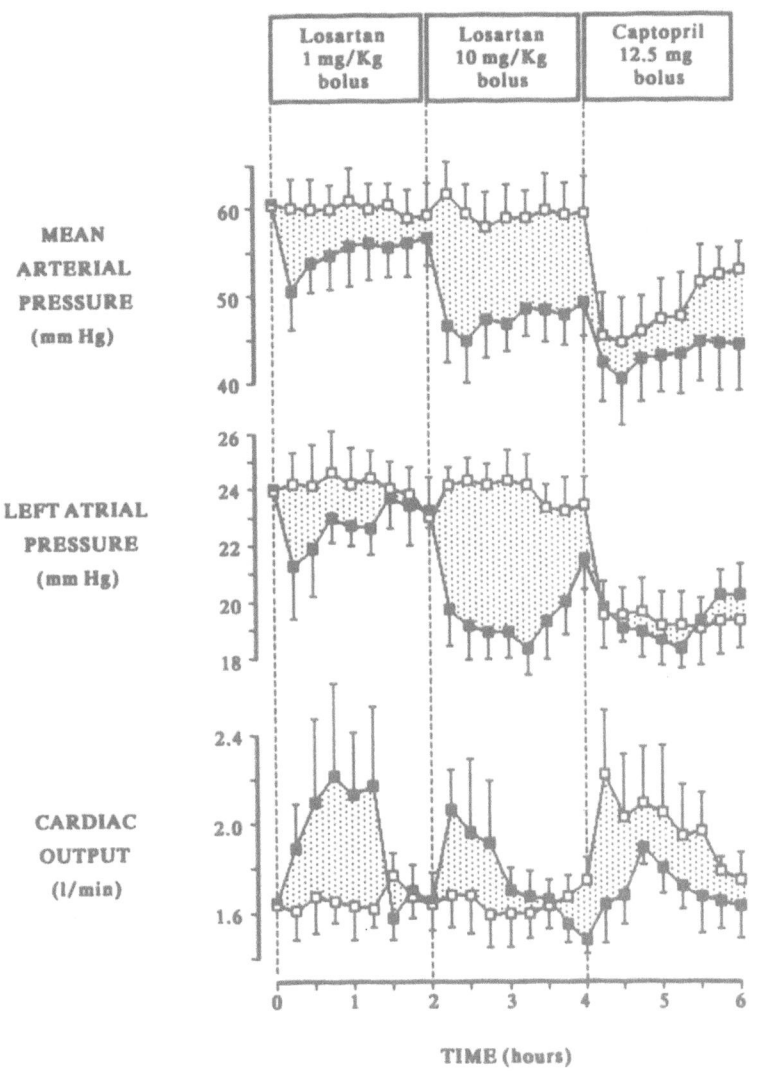

Figure 2. Hemodynamic response after intravenous bolus administration of vehicle and losartan (1 mg/kg), losartan (10 mg/kg), and captopril (12.5 mg) as per Fig. 1. Data plots are corrected values from ANOVA using mean of pretreatment baseline levels as covariate (uncorrected data are given in Table I). Baseline data point represents mean of control and active baseline values. Results are means ± SE. Significance of changes is discussed in text. (From Fitzpatrick *et al.*[12] Reprinted with permission.)

bolus ($p < 0.001$). The duration of action of the 10 mg/kg bolus of losartan exceeded that of the 1 mg/kg bolus. Both MAP and LAP were significantly lower than baseline and lower than time-matched control recordings 2 hr after the 10 mg/kg bolus ($p < 0.001$ and $p < 0.05$, respectively). Compared with time-matched control recordings, a rise in CO was observed after the 1 mg/kg losartan bolus, increasing from a baseline of 1.6 ± 0.1 to 2.2 ± 0.4 liter/min 45 min postbolus ($p < 0.01$) and from 1.6 ± 0.1 to 2.0 ± 0.2 liter/min 15 min after the 10 mg/kg bolus ($p < 0.05$). However, the effects on CO were not sustained and did not follow the dose–response pattern observed in MAP and LAP.

These data clearly show the potent vasodilator properties of losartan in heart failure. Raya *et al.*[30] demonstrated similar effects with chronic treatment but did not document the dose response or duration of action of losartan in heart failure. In the present study, bolus administration of losartan (1 and 10 mg/kg) clearly blocked the vasoconstrictive effect of ANG II, inducing a dose-related decrease in MAP and LAP in association with an improvement in CO. The peak hemodynamic effects were observed within 15–60 min of bolus administration, with the effect of the higher dose being more sustained over the subsequent 120-min observation period. These results indicate that intravenous losartan has a rapid onset and half-life similar to that of captopril. In a high-renin hypertensive rat model, similar reductions in blood pressure have been observed when losartan was given both intravenously and orally.[6,27,28,35] Wong *et al.*[35] demonstrated in the renal hypertensive rat that intravenous losartan (3 mg/kg) reduced blood pressure significantly at 15–90 min, with maximum effects seen at 60 min. On the other hand, antihypertensive effects were not detected in studies with normotensive rats or low-renin hypertensive models,[26,28,29,35] suggesting that the specificity of losartan's hypertensive action was mediated by inhibition of the RAS. With oral administration of the same dose, the duration of action was much longer (peak effect at 8 hr), and significant effects were still observed 24 hr after administration of the drug.[26,28,29,35] The comparable oral and intravenous antihypertensive potencies of losartan, as well as its long duration of action, may be due to good oral bioavailability and/or the generation of active metabolites.[29]

Compared with sustained changes in MAP and LAP, CO increases were short and did not rise in a dose-related fashion. It is possible that the tachycardia model of heart failure is more dependent on preload for maintaining CO. Thus a reduction in LAP (induced by inhibition of the RAS) may counteract the beneficial action of vasodilation on CO. Further studies, using higher doses of losartan, are required to confirm this finding.

4.5. Renal Effects of Losartan

No significant effects on renal function were observed with losartan. In the 2-hr collection period after the 10 mg/kg bolus, urine sodium excretion was similar to the time-matched period on the control day (8 ± 4 vs. 10 ± 4 mmole/hr, respectively), while endogenous creatinine clearance was also similar in both periods (83 ± 13 vs. 76 ± 9 ml/min, respectively). In microperfusion studies in normotensive rats, Xie *et al.*[36] found that losartan (10 mg/kg) increased renal blood flow and glomerular filtration rate (GFR) and that it also had a potent inhibitory effect on ANG II-induced sodium bicarbonate and sodium chloride reabsorption in the proximal convoluted tubule. This produced a substantial diuresis and natriuresis, similar to that achieved with captopril administration, although chloruresis was greater. Wong *et al.*[28] found no acute diuretic effect from losartan in spontaneously hypertensive rat. Similarly, captopril administration in chronic heart failure has variable effects on renal function.[10,37] In chronic heart failure where renal blood flow and perfusion pressure is reduced, the residual glomerular filtration fraction may be principally or exclusively maintained by the constrictive action of ANG II on the efferent arteriole. Blockade of the sustaining effects of ANG II on GFR, along with reduced renal perfusion pressure below the threshold for renal autoregulation,[38] may counterbalance the tubular (natriuretic) effects observed in normals. It is relevant to note, however, that despite a major fall in MAP (and presumably renal perfusion pressure) with losartan, GFR and urinary sodium excretion remained unchanged in this study. Further studies of chronic administration will be required to clarify the effects of ANG II receptor inhibition on renal function in heart failure.

4.6. Dose Response

To assess the maximum response to RAS blockade, captopril was administered 120 min after the 10 mg/kg losartan bolus. Although the effects of the higher losartan bolus had diminished by the time the captopril bolus was given, the response to the captopril bolus on the active day was greater than that on the control day. PRA rose higher, whereas MAP declined further ($p < 0.05$). The 10 mg/kg bolus of losartan produced a peak fall in MAP that was $79 \pm 6\%$ of the peak response in the presence of captopril (period C of the active day). This indicates that the 10 mg/kg bolus of losartan is likely to be close to the maximum on the dose–response curve.

5. COMPARISON OF THE ACUTE EFFECTS OF AT₁ RECEPTOR ANTAGONISM AND ACE INHIBITION

5.1. Acute Effects of Captopril in Ovine Heart Failure

The hormonal and hemodynamic effects of captopril administered on the control day are shown in Table I (compare periods B and C on the control day) and Figs. 1 and 2. On the control day when losartan was not given, captopril induced a rise in PRA from 0.8 ± 0.2 (at the end of period B) to a peak of 4.4

± 1.0 nmole/liter per hour 60 min postbolus ($p < 0.001$), whereas plasma ANG II dropped from 50 ± 7 to reach a nadir of 30 ± 6 pmole/liter 15 min postbolus. Plasma ANP levels also fell after captopril administration from 224 ± 24 to 158 ± 16 pmole/liter 2 hr later ($p < 0.01$). Plasma aldosterone levels did not change significantly after the captopril bolus. However, the aldosterone : PRA ratio fell from 364 ± 114 to a nadir of 63 ± 12 at 60 min ($p < 0.001$). Plasma cortisol levels tended to rise after the captopril bolus, but these changes were not significant.

On the control day when losartan was not administered, captopril induced a fall in MAP from 60 ± 4 mm Hg at the end of period B to a nadir of 45 ± 5 mm Hg 30 min postbolus ($p < 0.001$). Similarly, LAP declined significantly from 23 ± 1 to 19 ± 1 mm Hg 75 min postbolus ($p < 0.001$). CO rose from 1.8 ± 0.1 liter/min to a peak of 2.2 ± 0.3 liter/min 15 min postbolus ($p < 0.05$). As occurred with losartan, the effects on MAP and LAP were sustained for 2 hr, whereas CO had returned to control levels. In the 2-hr postcaptopril urine collection period, urine sodium excretion and endogenous creatinine clearance were similar to those observed in the preceding control periods (A and B).

5.2. Comparison of Captopril and Losartan

Comparison of the effects of one dose of each agent was afforded by analysis of the effects of the 12.5-mg captopril bolus on the control day (period C) and the 10 mg/kg losartan bolus on the active day (period B). There was no significant difference between the effects of losartan (10 mg/kg) and captopril (12.5 mg) for any hormone except plasma ANG II. Plasma ANG II levels responded quite differently for the two drugs, rising from 60 ± 12 to a peak of 242 ± 54 pmole/liter 30 min after the losartan bolus ($p < 0.001$), whereas plasma ANG II fell from 50 ± 7 to 30 ± 6 pmole/liter 15 min after the captopril bolus. The relationship between the response in PRA and MAP for both agents was similar (Fig. 2). Regression lines for the mean value at each time point during the 0–120 min observation period after the 10 mg/kg bolus of losartan and captopril were similar (Fig. 3). Furthermore, the slopes and intercepts of the regression lines in individual sheep for each agent were not significantly different.

Despite the arbitrary dose of captopril chosen for the present study, the hemodynamic responses to losartan (10 mg/kg) and to the ACE inhibitor were strikingly similar. The fact that losartan specifically inhibits the RAS through AT_1 receptor blockade, whereas ACE inhibitors also have effects on bradykinin and prostaglandins,[22,23] suggests that ACE inhibitors produce their acute hemodynamic effects in heart failure predominantly through inhibition of the RAS. This contention is supported by the finding that chronic treatment with losartan and captopril in the rat infarct model of heart failure have similar hemodynamic effects.[30]

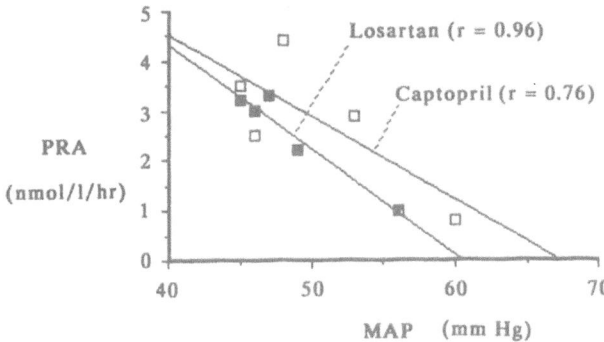

Figure 3. Regression lines representing the linked response of PRA and MAP to the 10/mg per kg bolus of losartan (closed symbols) and captopril (open symbols). Data plots represent the mean value for six sheep measured at each time point during the 0–120 min observation period (period B of the active day for losartan and period C of the control day for captopril). (From Fitzpatrick *et al.*[12] Reprinted with permission.)

Our study showed no significant acute hemodynamic differences between the two drugs, each having a similar efficacy and duration of action in the rapid-pacing model of heart failure. Apart from the expected differences in plasma ANG II responses, the hormonal responses to losartan (10 mg/kg) and captopril also were very similar. As expected, plasma ANG II concentrations responded quite differently to the two drugs. Losartan induced a clear-cut rise in plasma ANG II, whereas the failure to observe a more clear-cut fall in plasma ANG II with captopril was probably due to *in vitro* generation of angiotensin immunoreactive peptides after sample collection.[13,39] However, captopril clearly reduced plasma ANG II levels elevated by prior administration of losartan.

6. COMPARISON OF THE ACUTE EFFECTS OF RENIN AND ACE INHIBITION

In earlier experiments, we investigated the acute hemodynamic and hormonal effects of incremental doses of an ovine renin inhibitor (EMD 52 297 4 to 80 µg/kg per min) and captopril (0.1 to 1.6 mg/kg per min) in the rapid ventricular-pacing ovine model of heart failure.[13] Nine sheep were used in this study. Hemodynamic recordings and hormonal measurements were identical to those already described. After 5 days of rapid ventricular pacing, the sheep exhibited the hemodynamic and hormonal hallmarks of established heart failure.

Inhibition of the RAS by EMD 52 297 was evidenced by a significant decrease in PRA and ANG II levels (Fig. 4). PRA and ANG II levels

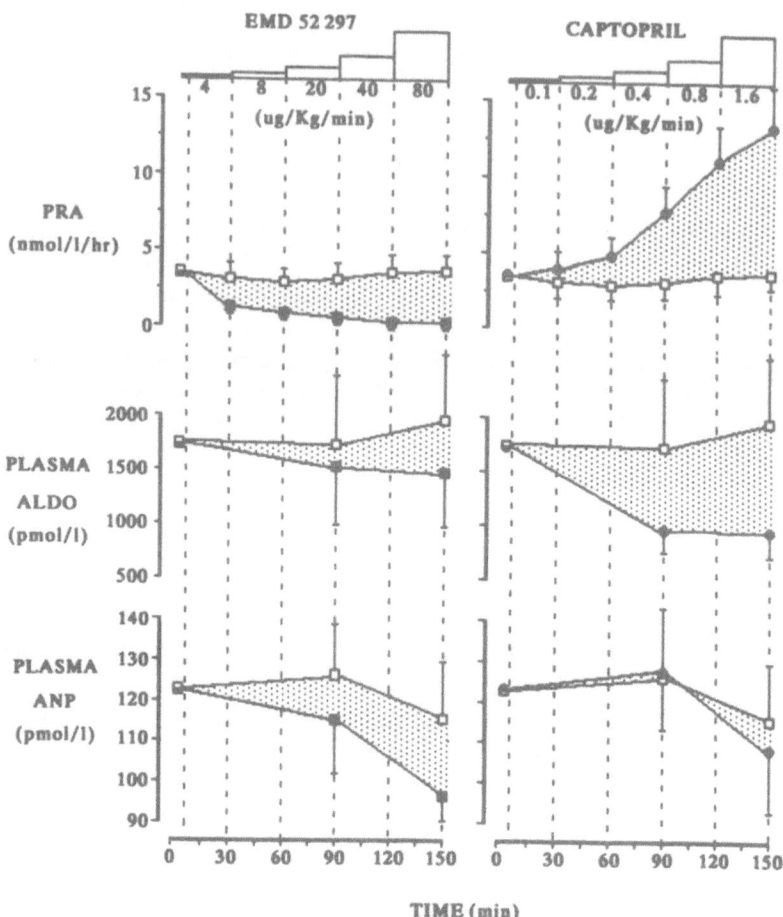

Figure 4. Hormonal response before and during incremental infusion of vehicle (open squares), EMD 52 297 (solid squares), and captopril (solid diamonds). Data plotted are the corrected values from the analysis of variance (ANOVA) using the mean of the preinfusion baseline levels as the covariate. Baseline data point represents the mean of the two baseline values before infusions were started. Results are the mean ± SEM. From the ANOVA, a significant treatment effect was observed for plasma aldosterone (ALDO), whereas there was a significant time–treatment interaction for PRA. EMD 52 297 produced a significant decrease in PRA from baseline to the recording at the highest dose ($p < 0.01$), ANP, atrial natriuretic peptide. (From Fitzpatrick *et al.*[13] Reprinted with permission.)

decreased from a baseline of 3.7± nmole/liter per hour and 229 ± 81 pmole/liter to 0.4 ± 0.1 nmole/liter per hour and 73 ± 20 pmole/liter ($p < 0.001$), respectively. Plasma aldosterone and cortisol did not change significantly. However, there was a slight but significant fall in plasma ANP levels (112 ± 12 to

90±pmole/liter, $p < 0.01$). As compared with the time-matched control recordings, a rapid decrease in MAP and LVSP was observed within 15 min after the EMD 52 297 infusion was started at the lowest dose (4 μg/kg per min) (Fig. 5). With incremental doses, both MAP and LVSP continued to decrease. MAP

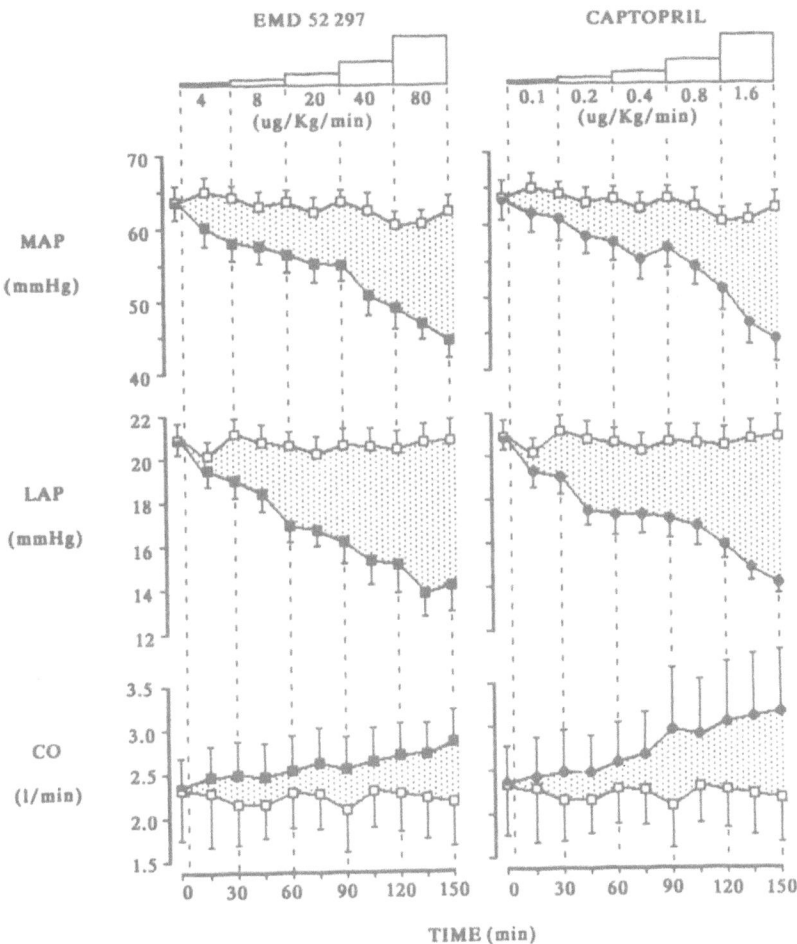

Figure 5. Hemodynamic response before and during incremental infusion of vehicle (open squares), EMD 52 297 (solid squares), and captopril (solid diamonds). Data plotted as described in legend to Fig. 1. Baseline data point represents the mean of the two baseline values before infusions were started. Uncorrected mean values are shown in Table I. Results are the mean ± SEM. Analysis of variance showed a significant treatment effect for mean arterial pressure (MAP, $p < 0.001$), left atrial pressure (LAP, $p < 0.01$), but not for cardiac output (CO, $0.05 < p < 0.10$). We observed no difference between EMD 53 397 and captopril. (From Fitzpatrick *et al.*[13] Reprinted with permission.)

decreased from 65 ± 2 during baseline to 45 ± 2 mm Hg at the end of the highest infusion rate ($p < 0.001$), and left ventricular systolic pressure (LVSP) decreased from a baseline value of 77 ± 3 to 59 ± 2 mm Hg ($p < 0.005$). A corresponding decrease in LAP was observed (20.8 ± 0.7 to 14.0 ± 1.2 mm Hg; ($p < 0.005$), whereas CO tended to increase (2 ± 0.4 to 2.5 ± 0.4 liter/mm; NS).

With captopril administration, a significant increase in PRA was observed (from a baseline of 3 ± 1 to 13 ± 3 nmole.1^{-1}.hr^{-1} at the highest infusion level, $p < 0.001$). Plasma aldosterone levels also decreased from a baseline of 1432 ± 411 to a nadir of 723 ± 225 pmole/liter ($p < 0.05$). Not shown in Fig. 4, an unexpected increase in plasma ANG II levels was observed at higher captopril infusion rates, which was attributed to *in vitro* generation of immunoreactive ANG II after sample collection.[13]

As compared with the time-matched control recordings, a decrease in MAP was observed with incremental captopril infusion: from a baseline value of 67.1 ± 3.0 to 47.4 ± 3.3 mm Hg ($p < 0.001$) at the end of the highest infusion rate. Similarly, LVSP decreased from 80.4 ± 3.5 to 65.6 ± 2.2 mm Hg ($p < 0.05$) and LAP decreased from 20.8 ± 0.66 to 13.9 ± 0.45 mm Hg ($p < 0.005$), while CO increased slightly from 2.17 ± 0.42 to 3.16 ± 0.71 liter/min ($0.05 < p < 0.1$).

As shown in Fig. 5, the hemodynamic effects of EMD 52 297 and captopril on preload and afterload were similar. This is further shown in Fig. 6 in which regression lines, relating change (decrease) in MAP to change (decrease) in LAP, are plotted for treatment with EMD 52 297 and captopril. Comparison of the mean intercept and mean slope of the regression line for each sheep demonstrated no significant difference (y intercept -3.27 ± 5.3 vs.

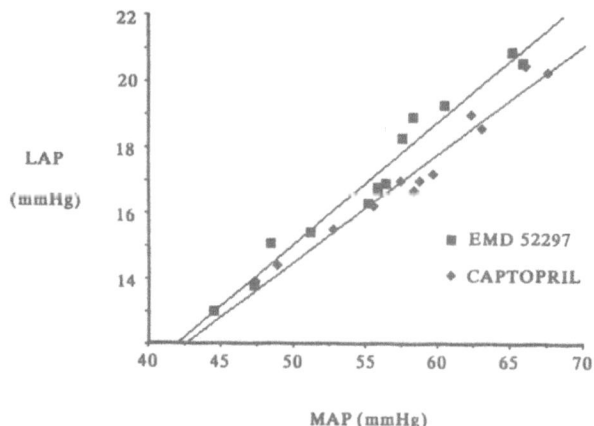

Figure 6. Relationship between the changes in mean arterial pressure (MAP) and left arterial pressure (LAP) during incremental infusion of EMD 52 297 (solid squares) and captopril (solid diamonds). (From Fitzpatrick *et al.*[13] Reprinted with permission.)

–2.59 ± 2.6 mm Hg; slope 0.36 ± 0.08 vs. 0.29 ± 0.05) for renin inhibition and captopril, respectively.

The major finding of this study was the striking similarity of the hemo-dynamic response to incremental doses of ACE inhibitors and renin inhibition. Because renin inhibitors specifically inhibit the RAS while ACE inhibitors also have effects on bradykinin and prostaglandins,[22,23] the results of this study are consistent with the view that ACE inhibitors produce their acute hemodynamic effects in heart failure largely through inhibition of the RAS. However, to confirm this, comparative levels of authentic plasma ANG II are required.

7. FUTURE RESEARCH DIRECTIONS

7.1. Distribution and Regulation of AT Receptors in Heart Failure

The recent development of techniques to analyze ANG II receptor num-bers has allowed documentation of the type, density, distribution, and activation of ANG II receptors in a variety of tissues and organs in the normal state. However, there is still a paucity of information for heart failure. Investigations to date have involved human heart failure[40] rather than animal models. The degree of response mediated by the ANG II receptors was thought to be controlled largely by the level of circulating ANG II. However, it is now thought that the ANG II receptors themselves may have important regulatory functions in their own right. The change in receptor density (mediated by up- or downregulation), level of response, or affinity of binding sites for Ang II may alter target organ responsiveness in diseased states.[41]

Urata et al.[40] have recently investigated the distribution of ANG II re-ceptor binding sites in human cardiac ventricles removed for cardiac transplan-tation. This study found ANG II receptors consistently in the myocardium, cardiac adrenergic nerves, and coronary vessels. The ventricular ANG II recep-tor density and affinity was found to be unaltered in hearts from patients with ischemic or idiopathic cardiomyopathy compared to normal hearts. In the normal human heart, the binding site density in the left ventricle was similar to that in the right ventricle, was relatively homogeneous within different regions of both ventricles, and showed a trend toward greater receptor density at the base than at the apex of the heart. ANG II receptor density in all regions of the right atrium was two- to fourfold higher than in the ventricular regions, and was higher still in the sinus node compared to the atrial myocardium. This pattern of distribution was not altered in the failing hearts removed for trans-plantation. This lack of alteration of AT receptor binding contrasts with the ventricular beta-adrenergic receptor system, which is compromised in the fail-ing heart[42] through receptor downregulation and development of tolerance to beta-adrenergic agonists. These findings suggest that the direct and neural angiotensinergic inputs to the myocardium may play a role in the regulation of cardiac function in man and that these inputs are preserved in the failing heart. Furthermore, they lend support to the notion that cardiac ANG II receptors are not regulated in heart failure. However, findings should be interpreted with caution since drug therapy administered prior to cardiac transplantation may have an important effect on receptor levels. It is of note that ACE inhibitors, used in these patients, have been shown to alter ANG II receptor binding in SHR,[41,43] with variation in response in different tissues. To date, AT receptor binding and regulation has not been investigated in animal models of heart failure where the confounding influence of drug therapy can be avoided. These studies are required to determine the importance of regulation of AT receptors in the overall control of RAS activity.

As discussed elsewhere in this book, tissue ANG II binding has been known to be regulated in a distinct and complex fashion in animal models of hypertension.[44] Furthermore, the pattern of regulation appears to be tissue specific with no consistent relationship between circulating levels of ANG II and receptor binding in various tissues. Factors involved in the regulation of receptor density, hormone affinity, and responsiveness in these hypertensive models require further investigation. Given these findings in hypertensive models, studies documenting ANG II binding or regulation are required in animal models of heart failure.

7.2. Effects of AT_1 Receptor Antagonists on Local RAS in Heart Failure

All the components of the RAS have been identified in many different tissues.[45] These locally active systems, particularly in renal, vascular, and cardiac tissues, are thought to have an important paracrine function and contribute to the pathophysiology of heart failure. In the rat infarct model,[46] renal angiotensinogen mRNA is elevated twofold, whereas renin and ACE levels remain similar to sham animals. ACE inhibition normalizes mRNA levels. This effect on renal RAS in heart failure may explain renal vasodilatation and natriuresis that occurs with ACE inhibition when activity of plasma renin levels are normal.[47] In cardiac tissue, ACE activity is increased, while serum ACE and renin levels are not affected.[48,49] Enhanced local activity of the RAS may have direct and indirect effects of cardiac function; direct effects of ANG II on inotropic state of the failing myocardium[50,51]; facilitation of norepinephrine release from sympathetic nerve terminals[52,53]; and induction of ventricular hypertrophy and remodeling with the growth promoting effects of ANG II.[54,55] Control of local RAS and the interaction with systemic renin activity is poorly understood in heart failure.[45] Furthermore, the effects of ANG II receptor antagonists on these systems is unknown and should provide an area for active research, particularly if antagonists are developed with a different predilection for certain tissues.

7.3. Long-Term Effects of AT_1 Receptor Antagonism and Clinical Studies in Heart Failure

Recently, Raya *et al.*[30] showed that 2 weeks of treatment with losartan had beneficial hemodynamic effects in the rat model of heart failure and that these effects were similar to those of captopril. Similar reduction in left ventricular end-diastolic pressure and volume index were observed with both agents, indicating that the beneficial action of ACE inhibitors is related, at least in part, to inhibition of the RAS. Further animal and clinical studies are required to confirm these findings. An alternative to ACE inhibition for interrupting the RAS is required. Evidence indicating the benefit of early treatment in states of isolated LV dysfunction means that less symptomatic patients are receiving therapy. However, since troublesome cough (a likely side effect of ACE inhibitors potentiating bradykinin) significantly limits the use of these drugs, alternative inhibitors of ANG II activity are required for clinical use.

8. CONCLUSION

In conclusion, the above study documents for the first time in heart failure the beneficial vasodilator response, preservation of renal function, and the hormonal changes induced by the specific AT_1 receptor antagonist losartan. The hemodynamic actions are remarkably similar to the effects of ACE and renin inhibitors, suggesting that all three classes of drug act primarily via the inhibition of ANG II formation or action. Losartan, a specific long-acting AT_1 antagonist with good oral bioavailability, may provide a useful alternative therapy to ACE inhibition in patients with heart failure. However, clinical studies will be required to demonstrate efficacy.

REFERENCES

1. CONSENSUS Trial Study Group: Effects of enalapril on mortality in severe congestive heart failure. *N Engl J Med* 316:1429–1435, 1987.
2. SOLVD Investigators: Effects of enalapril on survival in patients with reduced left ventricular ejection fraction and congestive heart failure. *N Engl J Med* 325:293–302, 1991.
3. Pfeffer MA, Braunwald E, Moye LA, et al: Effect of captopril on mortality and morbidity in patients with left ventricular dysfunction after myocardial infarction: Results of the Survival and Ventricular Enlargement Trial. *N Engl J Med* 327:669–677, 1992.

4. Rogg H, Schmid A, de Gaspara M: Identification and characterization of angiotensin II receptor subtypes in rabbit ventricular myocardium. *Biochem Biophys Res Commun* 173:416–422, 1990.

5. Tsutsumi K, Saavedra JM: Quantitative autoradiography reveals different angiotensin II receptor subtypes in selected rat brain nuclei. *J Neurochem* 56:348–351, 1991.

6. Wong PC, Hart SD, Zaspel AM, et al: Functional studies of nonpeptide angiotensin II receptor subtype-specific ligands: DuP 753 (AII-1) and PD 123177 (AII-2). *J Pharmacol Exp Ther* 255:584–592, 1990.

7. Franciosa JA. *Cardiology Update: Reviews for Physicians.* Churchill-Livingstone, Edinburgh, 1981.

8. Fitzpatrick MA, Nicholls MG, Espiner EA, et al: Neurohumoral changes during onset and offset of ovine heart failure: Role of ANP. *Am J Physiol* 256:H1052–1059, 1989.

9. Fitzpatrick MA, Rademaker MT, Frampton CM, et al: Hemodynamic and hormonal effects of renin inhibition in ovine heart failure. *Am J Physiol* 258:H1625–1631, 1990.

10. Fitzpatrick MA, Rademaker MT, Frampton CM, et al: Renal effects of ACE inhibition in ovine heart failure: A comparison of intermittent and continuous ACE inhibition. *J Cardiovasc Pharmacol* 16:629–635, 1990.

11. Fitzpatrick MA, Yandle TG, Espiner EA, et al: ANP infusion in the treatment of heart failure and comparison with ACE inhibition. *J Cardiovasc Pharmacol* 15:536–543, 1990.

12. Fitzpatrick MA, Rademaker MT, Charles CJ, et al: Angiotensin II receptor antagonism in ovine heart failure: Acute hemodynamic, hormonal and renal effects. *Am J Physiol* 263:H250–256, 1992.

13. Fitzpatrick MA, Rademaker MT, Yandle TG, et al: Comparison of the effect on renin inhibition and angiotensin-converting enzyme inhibition in heart failure. *J Cardiovasc Pharmacol* 19:169–175, 1992.

14. Riegger AJG, Liebau G: The renin–angiotensin–aldosterone system, antidiuretic hormone and sympathetic nerve activity in an experimental model of congestive heart failure in the dog. *Clin Sci* 62:465–469, 1982.

15. Riegger AJG, Liebau G, Holzschuh M, et al: Role of the renin–angiotensin system in the development of congestive heart failure in the dog as assessed by chronic converting enzyme blockade. *Am J Cardiol* 53:614–618, 1984.

16. Wilson JR, Douglas P, Hickey WF, et al: Experimental congestive heart failure produced by ventricular pacing in the dog: Cardiac effects. *Circulation* 75:857–867, 1987.

17. Howard RJ, Stopps TP, Moe GW, et al: Recovery from heart failure: Structural and functional analysis in a canine model. *Can J Physiol Pharmacol* 66:1505–1512, 1988.

18. Tomita M, Spinale FG, Crawford FA, et al: Changes in left ventricular volume, mass, and function during the development and regression of supraventricular-induced cardiomyopathy. *Circulation* 83:635–644, 1991.

19. Weber KT, Pick R, Silver MA, et al: Fibrillar collagen and remodelling of dilated canine left ventricle. *Circulation* 82:1387–1401, 1990.

20. Reilly CF, Tewksbury DA, Schechter NM, et al: Rapid conversion of angiotensin I to angiotensin II by neutrophil and mast cell proteinases. *J Biol Chem* 257:8619–8622, 1982.

21. Okunishi H, Miyazaki M, Toda N: Evidence for a putatively new angiotensin II generating enzyme in the vascular wall. *J Hypertens* 2:227–284, 1984.

22. Ondetti MA, Rubin B, Cushman DW: Design of specific inhibitors of angiotensin-converting enzyme: New class of orally active antihypertensive agents. *Science* 196:441–447, 1977.

23. Swartz SL, Williams GH: Angiotensin-converting enzyme inhibition and prostaglandins. *Am J Cardiol* 49:1405–1409, 1982.

24. Morice AA, Brown MJ, Lowry R, et al: Angiotensin-converting enzyme and the cough reflex. *Lancet* 2:1116–1118, 1987.

25. Ikeda M, Sasaguri M, Marutu H, et al: Formation of angiotensin II by tonin-inhibitor complex. *Hypertension* 11:63–70, 1988.

26. Wong PC, Price WA, Chiu AT, et al: Nonpeptide angiotensin II receptor antagonists. VIII characterization of functional antagonism displayed by DuP 753, an orally active antihypertensive agent. *J Pharmacol Exp Ther* 252:719–725, 1990.

27. Wong PC, Price Jr WA, Wexler RR, et al: Nonpeptide angiotensin II receptor antagonists. Studies with EXP 9270 and DuP 753. *Hypertension* 15:823–834, 1990.

28. Wong PC, Price WA Jr, Chiu AT, et al: Hypotensive action of DuP 753, an angiotensin II antagonist, in spontaneously hypertensive rats. Nonpeptide angiotensin II receptor antagonists: X. *Hypertension* 15:459–468, 1990.

29. Wong PC, Price WA, Chiu AT, et al: Nonpeptide angiotensin II receptor antagonists IX. Antihypertensive activity in rats of DuP 753, an orally active antihypertensive agent. *J Pharmacol Exp Ther* 252:726–732, 1990.

30. Raya TE, Fonken SJ, Lee RW, et al: Hemodynamic effects of direct angiotensin-II blockade compared to converting enzyme-inhibition in a rat model of heart failure. *Am J Hypertens* 4:S334–340, 1991.

31. Christen Y, Waeber B, Nussberger J, et al: Oral administration of DuP 753, a specific angiotensin II receptor antagonist, to normal male volunteers: Inhibition of pressor response to exogenous angiotensin I and II. *Circulation* 83:1333–1343, 1991.

32. Wood CE: ACTH, cortisol, and renin responses to arterial hypotension in sheep. *Am J Physiol* 251:R18–22, 1986.

33. Camargo CA, Dowdy AJ, Hancock EW, et al: Decreased plasma clearance and hepatic extraction of aldosterone in patients with heart failure. *J Clin Invest* 44:356–365, 1965.

34. Cuneo RC, Espiner EA, Nicholls MG, et al: Effect of physiological levels of atrial natriuretic peptide on hormone secretion: Inhibition of angiotensin-induced aldosterone secretion and renin release in normal man. *J Clin Endocrinol Metab* 66:1–8, 1987.

35. Wong PC, Price WA Jr, Chiu AT, et al: Nonpeptide angiotensin II receptor antagonists XI. Pharmacology of EXP3174: An active metabolite of DuP 753, on orally active antihypertensive agent. *J Pharmacol Exp Ther* 255:211–217, 1990.

36. Xie MH, Liu FY, Wong PC, et al: Proximal nephron and renal effects of DuP 753, a nonpeptide angiotensin II receptor antagonist. *Kidney Int* 38:473–479, 1990.

37. Packer M, Lee WH, Medina N, et al: Functional renal insufficiency during long-term therapy with captopril and enalapril in severe chronic heart failure. *Ann Intern Med* 106:346–354, 1987.

38. Kirchheim H, Ehmke H, Persson P: Physiology of the renal baroreceptor mechanism of renin release and its role in congestive heart failure. *Am J Cardiol* 62:68E–71E, 1988.

39. Nussberger J, Brunner DB, Waeber B, et al: In vitro renin inhibition to prevent generation of angiotensin during determination of angiotensin I and II. *Life Sci* 42:1683–1688, 1988.

40. Urata H, Healy B, Stewart RW, et al: Angiotensin II receptors in normal and failing human hearts. *J Clin Endocrinol Metab* 69(1):54–66, 1989.

41. Wilson KM, Magargal W, Berecek KH: Long-term captopril treatment. Angiotensin II receptors and responses. *Hypertension* 11(2 Pt 2):I148–152, 1988.

42. Bristow MR, Ginsburg R, Minobe W, et al: Decreased catecholamine sensitivity and β-adrenergic receptor density in failing human heart. *N Engl J Med* 307:205–211, 1982.

43. Nazarali AJ, Gutkind JS, Correa FM, et al: Decreased angiotensin II receptors in subfornical organ of spontaneously hypertensive rats after chronic antihypertensive treatment with enalapril. *Am J Hypertens* 3:59–61, 1990.

44. Wilson SK, Lynch DR, Ladenson PW: Angiotensin II and atrial natriuretic factor-binding sites in various tissues in hypertension: Comparative receptor localization and changes in different hypertension models in the rat. *Endocrinology* 124(6):2799–2808, 1989.

45. Dzau VJ, Hirsch AT: Emerging role of the tissue renin–angiotensin systems in congestive heart failure. *Eur Heart J* 11:65–71, 1990.
46. Schunkert H, Hirsch AT, Mankadi S, et al: Renal angiotensinogen gene expression in experimental heart failure: Effect of angiotensin converting-enzyme inhibition. *Clin Res* 37:584A, 1989.
47. Hostetter TH, Pfeffer JM, Pfeffer MA, et al: Cardiorenal hemodynamics and sodium excretion in rats with myocardial infarction. *Am J Physiol* 245:H98–103, 1983.
48. Hirsch AT, Talsness C, Lage A, et al: The effect of experimental myocardial infarction and chronic captopril treatment on plasma and tissue angiotensin converting-enzyme activity (abstract). *Clin Res* 37:266A, 1989.
49. Lorell BH, Schunkert H, Grice WN, et al: Alteration in cardiac angiotensin converting enzyme activity in pressure overload hypertrophy (abstract). *Circulation* 80(Suppl II):II297 1989.
50. Dempsey PJ, McCallum ZT, Kent KM, et al: Direct myocardial effects of angiotensin II. *Am J Physiol* 220:477–481, 1971.
51. Kobayashi M, Furukawa Y, Chiba S: Positive chronotopic and inotropic effects of angiotensin II in the dog heart. *Eur J Pharmacol* 50:17–25, 1978.
52. Van Zwieten PA, de Jonge A: Interaction between the adrenergic and renin–angiotensin–aldosterone systems. *Postgrad Med J* 62(Suppl 1):23–27, 1986.
53. Persson S: Endocrinology of cardiac failure. Pathophysiologic aspects: Haeodynamics. *Acta Med Scand Suppl* 707:7–14, 1986.
54. Naftilan AJ, Pratt RJ, Eldridge CS, et al: Angiotensin II induces *c-fos* expression in smooth muscle via transcriptional control. *Hypertension* 13:706–711, 1989.
55. Tarazi RC, Fouad FM: Reversal of cardiac hypertrophy. *Hypertension* 6:III140–145, 1984.

16

Angiotensin Antagonists in Models of Heart Failure

Thomas E. Raya, Eugene Morkin, and Steven Goldman

1. INTRODUCTION

The development of agents that inhibit the renin–angiotensin system by specific blockade of the angiotensin II (ANG II) receptor has provided investigators with a new pharmacological tool to examine cardiovascular pathophysiology. Before the development of these compounds, saralasin was the only agent that could block the angiotensin receptor. The disadvantages of saralasin are that it is a partial agonist as well as an antagonist and that it is active only if given intravenously, which precludes investigation of the chronic effects of ANG II blockade. By comparing the effects of ANG II blockers with converting enzyme inhibition, it also is now possible to examine the effects of the renin–angiotensin system more specifically. One interesting issue is whether there are different effects of renin–angiotensin blockade produced by inhibition of the converting enzyme versus blockade of the angiotensin receptor. This differentiation is particularly germane in congestive heart failure because this is a clinical syndrome in which the renin–angiotensin system is activated and converting enzyme inhibition is now conventional therapy for this disease.

Thomas E. Raya, Eugene Morkin, and Steven Goldman • University of Arizona Heart Center and Tucson VA Medical Center, Tucson, Arizona 85723.

Angiotensin Receptors, edited by Juan M. Saavedra and Pieter B.M.W.M. Timmermans. Plenum Press, New York, 1994.

2. ANGIOTENSIN RECEPTOR BLOCKADE IN ISCHEMIC CARDIOMYOPATHY

The beneficial effects of inhibition of the renin–angiotensin system in patients and in animal models of heart failure have been well documented. To date most studies have been performed using angiotensin-converting enzyme inhibitors. For example, with enzyme inhibitors improvements in hemodynamics, exercise tolerance, symptoms, and mortality have been reported.[1-5] The hemodynamic improvements that result from angiotensin-converting enzyme inhibition include decreases in left ventricular end-diastolic pressure, operating end-diastolic volume, total blood volume, mean circulatory filling pressure, systemic vascular resistance, and increases in venous compliance. As a consequence of these changes there is less left ventricular dilatation. While these improvements in cardiovascular function are thought to result from blockade of the renin–angiotensin system, with resultant decreases in circulating ANG II levels, recently investigators have suggested that the tissue renin–angiotensin system in the heart may also play a role in the hemodynamic responses to converting enzyme inhibition.[6]

Direct blockade of the renin–angiotensin system with inhibition of the ANG II receptor has been proposed as a more specific treatment for heart failure. To examine this approach experimentally, our laboratory used the rat coronary artery ligation model. This model of ischemic cardiomyopathy is useful because it has many characteristics in common with clinical heart failure. Ligation of the left coronary artery in rats creates an anterior wall infarction in which the changes in hemodynamics and ventricular remodeling are related to the size of the infarction.[4,7,8] In addition, pharmacological studies in this model have been useful in predicting the outcomes of similar trials in patients with ischemic cardiomyopathy.[1,9]

We compared the effects of losartan and captopril in rats with large myocardial infarction and heart failure after coronary artery ligation.[10] Without treatment, heart failure rats had left ventricular end-diastolic pressures greater than 26 mm Hg with an average infarct size of 45%. The changes in body weight, chamber weights, and hemodynamics are seen in Fig. 1–3. When treatment with captopril and losartan were compared to untreated heart failure control rats, there were no changes in ventricular weights or body weights, with the exception that losartan decreased body weight. Heart rate was unchanged with both treatments and there was a trend, although not significant, toward decreased right atrial and mean aortic pressures. Left ventricular end-diastolic pressure and left ventricular end-diastolic volume decreased significantly with captopril and losartan. Mean circulatory filling pressure decreased and venous compliance increased with both agents. There were no significant changes in blood volume. Our conclusions were that direct ANG II blockade with losartan

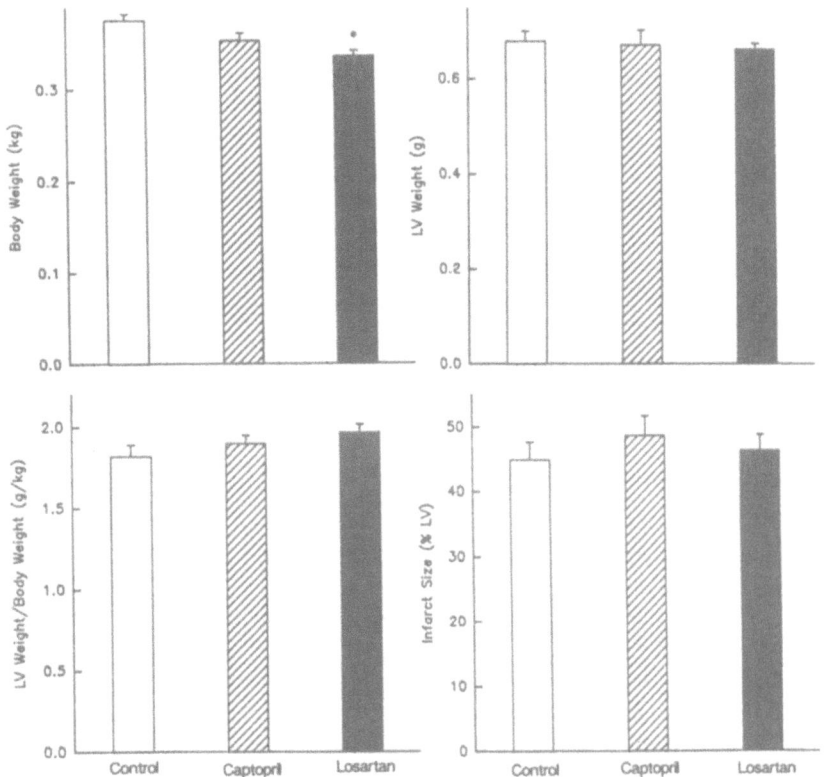

Figure 1. Changes in heart weight in rats with heart failure treated with captopril and losartan. $N = 9$ for control and captopril, $n = 10$ for losartan; $*p < 0.05$ compared to control.

or converting enzyme inhibition produced similar hemodynamic changes in rats with heart failure after myocardial infarction.

When losartan is given during and after completion of coronary ligation in rats, resting and volume loaded cardiac output does not change.[11] In this setting, however, losartan reduces hypertrophy as defined by decrease in heart weight and inhibition of collagen deposition.

The improvement in hemodynamics in heart failure rats treated with losartan would argue that the benefit that occurs in patients treated with converting enzyme inhibition is due to direct blockade of the renin–angiotensin system and not to some other action of these enzyme inhibitors. For example, converting enzyme inhibitors are known to potentiate the bradykinin–depressor response and may increase bradykinin levels. These agents also result in activation of prostaglandin synthesis.

The importance of the renin–angiotensin system in heart failure has been

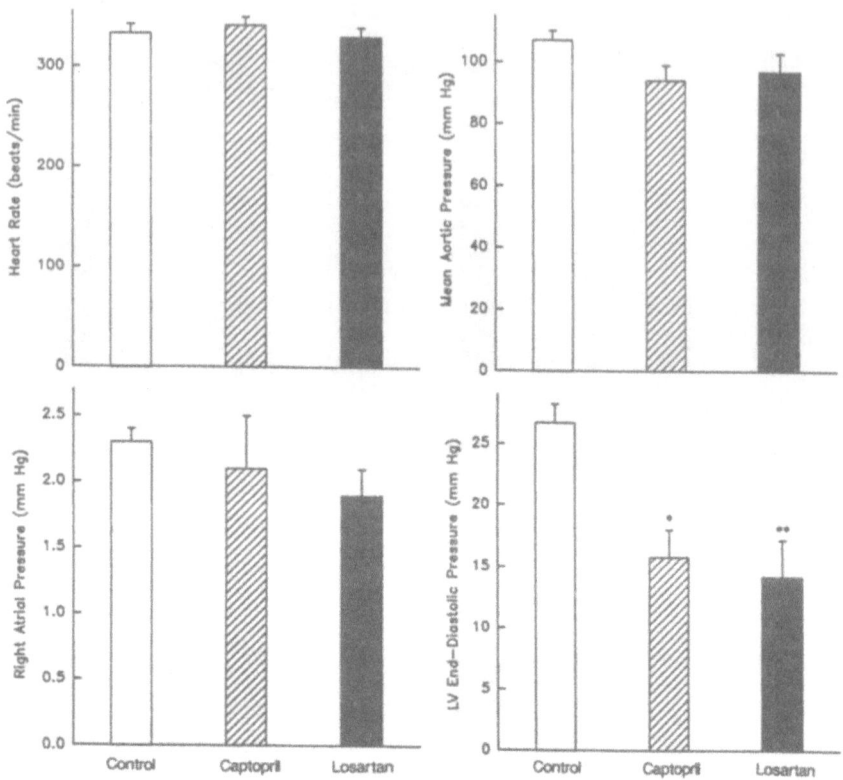

Figure 2. Changes in heart rate, right atrial pressure, mean aortic pressure, and left ventricular end-diastolic pressure in rats with heart failure treated with captopril and losartan. $N = 9$ for control and captopril, $n = 10$ for losartan; $*p < 0.05$ compared to control.

attributed to its systemic vasoconstricting properties. Even with normal ventricular function, increases in afterload with ANG II can create a preload–afterload mismatch such that cardiac output will fall.[12] As a result of vasoconstriction accompanied by aldosterone-stimulated sodium retention, blood volume may increase. By contrast, blockade of the renin–angiotensin system blocks vasoconstriction and prevents the increase in circulating or effective blood volume. Another important result of ANG II blockade is that venoconstriction, which is part of the pathophysiology of heart failure, is reversed.[9] The venodilatory effects of blocking either the enzyme or the receptor results in blood pooling in the periphery and a decrease in left ventricular preload, which in turn results in a decrease in left ventricular volume.

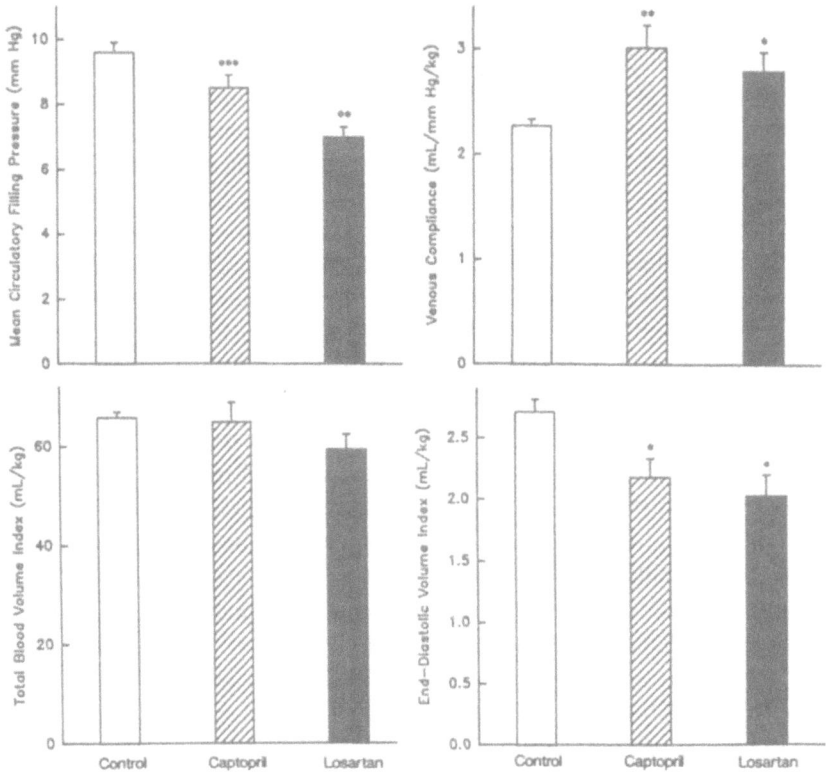

Figure 3. Changes in total blood volume, mean circulatory filling pressure, venous compliance, and left ventricular end-diastolic volume. $N = 9$ for control and captopril, $n = 10$ for losartan; *$p < 0.05$ compared to control.

3. ANGIOTENSIN RECEPTOR BLOCKADE IN OTHER MODELS OF HEART FAILURE

3.1. Rapid Pacing Model

Recently, the rapid pacing model of heart failure has been used to examine neurohumoral activation in heart failure. The advantages of the pacing model are that it is technically easy to establish and results in a congestive state with activation of several neurohumoral systems, including the renin–angiotensin system, and can be applied in a variety of animal models. Although the pacing heart failure model pathophysiologically bears little resemblance to clinical heart failure, the data generated experimentally have helped explore some aspects of heart failure. For example, the acute effects of intravenous losartan and captopril were studied in ewes with heart failure induced by rapid ventricular pacing.[13] Both losartan and captopril resulted in increases in plasma renin activity, but only losartan significantly increased plasma ANG II concentration. In these short-term studies, glomerular filtration rate and urine sodium excretion were maintained despite a fall in renal perfusion with both agents.

In conscious dogs with heart failure produced with rapid ventricular pacing, the acute intravenous effects of losartan and captopril were compared in compensated (after 5 days of pacing) and in decompensated (after 15 days of pacing) heart failure.[14] In the compensated state both agents decreased mean arterial pressure and increased renal blood flow but did not change cardiac output. The hemodynamic changes were similar in decompensated heart failure except that after 15 days of pacing, captopril but not losartan increased cardiac output. Both drugs increased plasma renin activity and reduced plasma aldosterone, vasopressin, and norepinephrine concentrations. These data suggest that in the pacing model of heart failure, the acute hemodynamic response to both agents is similar with the exception that captopril increases cardiac output only in decompensated heart failure. Furthermore, both agents induce similar compensatory increases in neurohumoral activity.

3.2. Arterial–Venous Shunt Model

The effects of treatment for 1 week with captopril and losartan have been studied in rats with high-output failure induced by aortocaval shunts.[15] Both agents decreased mean arterial pressure and reduced left ventricular end-diastolic pressure to sham-operated levels. Regression of cardiac hypertrophy was seen with both treatments. Plasma COOH-terminal atrial natriuretic factor (ANF) levels were decreased and urinary volume and hematocrit were increased in the losartan-treated rats. The authors concluded that chronic angiotensin-converting enzyme inhibition and ANG II receptor antagonism improved hemodynamics, diminished water retention, reversed cardiac hypertrophy, and restored plasma and tissue ANF to more "normal" levels in rats with moderate high-output heart failure.

The effects of enalapril and losartan also were compared in a model of volume overload produced by an aortocaval shunts in rats.[16] Both agents lowered left ventricular end-diastolic pressure to the same extent, but only losartan blunted the hypertrophic response, that is, decreased left and right ventricular weights/body weight. Interestingly, the only neurohumoral change was an increase in the left ventricular renin activity in the shunt animals treated with enalapril. The authors concluded that the control of cardiac hypertrophy in the volume overload state was influenced more by neurohumoral than by hemodynamic mechanisms. One explanation for these findings may be that tissue levels of ANG II in the heart are a trophic factor in volume overload.

3.3. Aortic Regurgitation Model

Blockade of the renin–angiotensin system appears to be beneficial in volume overload states independent of the cause. While most investigations have focused on ischemic left ventricular dysfunction, there are experimental data in other volume overload states, for example, rats with experimentally induced aortic regurgitation. In this model, converting enzyme inhibition with captopril decreased aortic pulse pressure, left ventricular systolic pressure, and end-diastolic volume.[17] These beneficial effects were associated with decreases in left ventricular weight and a leftward shift of the left ventricular pressure–volume relationship.

4. TISSUE RENIN–ANGIOTENSIN SYSTEM IN THE HEART

The presence of an ANG II-forming system in the heart and blood vessels has been substantiated by documenting the presence of angiotensinogen, the converting enzyme, and ANG II receptors in the heart and the aorta.[6,18] Even though the system is present, it is not clear how important the local tissue system is in the pathophysiology of heart failure. In 3-day-old piglet hearts both enalapril and losartan decreased left ventricular weight/body weight, RNA content, and local RNA,[19] and the authors concluded that these data supported their hypothesis that the tissue ANG II system is required for rapid growth in the neonatal pig heart. In rats with aortic stenosis, cardiac angiotensin-converting enzyme and its mRNA were increased as left ventricular mass increased.[20] When these hearts were perfused with enalaprilat, intracardiac conversion from ANG I to ANG II was decreased, suggesting that in the hypertrophied heart the converting enzyme is important to maintain a functioning angiotensin system. Nevertheless, this technique does not allow estimation of the relative effects of circulating versus tissue angiotensin-converting enzyme systems.

5. IMPLICATIONS AND FUTURE DIRECTIONS

The data that have been generated with angiotensin-converting enzyme inhibition and angiotensin receptor blockade as pharmacological tools in experimental animal models have increased our understanding of the pathophysiology of heart failure. Although activation of several neurohumoral systems occur in patients with heart failure, it has been difficult to precisely define which of these systems are causative and which are epiphenomena. Most studies have suggested that activation of the sympathetic nervous system, vasopressin, and atrial natriuretic peptide systems are all responsible for alterations in vascular tone seen in heart failure. While it is still not possible to assign the specific contributions of each system, the comparable hemodynamic responses observed with converting enzyme inhibition and specific ANG II blockade in the rat coronary artery ligation model has strengthened the argument that activation of the renin–angiotensin system is in part responsible for the inappropriate systemic vasoconstriction seen in heart failure. This argument does not mean that activation of other systems are not part of the specific pathophysiology of heart failure.

The development of a specific nonpeptide antagonist of ANG II has made possible more precise evaluation of the renin–angiotensin system, since these antagonists have no known effects on the bradykinin or prostaglandin systems. As discussed above, our laboratory demonstrated that specific blockade of ANG II with losartan in rats with heart failure due to ischemic left ventricular dysfunction has peripheral hemodynamic effects that are indistinguishable from those of converting enzyme inhibitors.[10] Nevertheless, that study and others cited in this chapter[11,13,16] were not designed to test mechanisms of action of these treatments—that is, to what degree were these effects mediated by alterations in prostaglandins or bradykinins—but rather to furnish observations of overall hemodynamic effects.

There are other areas of investigations that need to be examined with reference to ANG II receptor blockade. The kidney is an important target organ of ANG II, bradykinin, and prostaglandins and plays a crucial adaptive role in salt and water homeostasis in heart failure. An evaluation of renal hemodynamics and sodium handling during blockade of the sequential steps of the renin–angiotensin system could, therefore, provide important information on the mechanisms of action of ANG II and the relative corollary effects of these other hormone systems. Therefore, future investigations should focus on the potentially contributory effects of enhanced bioavailability of bradykinin and prostaglandin due to blockade of the renin–angiotensin system by converting enzyme inhibitors versus receptor blockers. These studies could be expanded by examining renal hemodynamics and salt handling after specific blockade of bradykinin and prostaglandin.

ACKNOWLEDGMENT. This work is supported by grants from DuPont Merk Pharmaceutical, the Veterans Administration, the Arizona Disease Control Research Commission, the American Heart Association, and the National Institutes of Health (HL-20984) and (HL48163-01A1).

REFERENCES

1. Pfeffer JM, Pfeffer MA, Braunwald E: Influence of chronic captopril therapy on the infarcted left ventricle of the rat. *Circ Res* 57:84–95, 1985.
2. Pfeffer MA, Braunwald EB, Moye LA, et al: The effect of captopril on mortality and morbidity in patients with left ventricular dysfunction following myocardial infarction. *N Engl J Med* 327:669–677, 1992.
3. CONSENSUS Trial Study Group: Effects of enalapril on mortality in severe congestive heart failure: results of the Cooperative North Scandinavian Enalapril Survival Study (CONSENSUS). *N Engl J Med* 316:1429–1435, 1987.
4. Raya T, Gay RG, Lancaster L, et al: Serial changes in left ventricular relaxation and chamber stiffness after large myocardial infarction in rats. *Circulation* 77:1424–1431, 1988.
5. Cohn JN, Johnson G, Ziesche S, et al: A comparison of enalapril with hydralazine-isosorbide dinitrate in the treatment of chronic congestive heart failure. *N Engl J Med* 325:303–310, 1991.
6. Hirsch AT, Talsness CE, Smith AD, et al: Differential effects of captopril and enalapril on tissue renin–angiotensin systems in experimental heart failure. *Circulation* 86:1566–1574, 1992.
7. Gay R, Wool S, Paquin M, et al: Alterations in the systemic venous circulation in rats with heart failure. *Am J Physiol* 251:H483–H489, 1986.
8. Michel JB, Lattion AL, Salzman, JL, et al: Hormonal and cardiac effects of converting enzyme inhibition in rat myocardial infarction. *Circ Res* 62:641–650, 1988.
9. Raya TE, Gay RG, Aquirre M, et al: The importance of venodilatation in the prevention of left ventricular dilatation after chronic large myocardial infarction in rats: A comparison of captopril and hydralazine. *Circ Res* 64:330–338, 1989.
10. Raya TE, Fonken SJ, Lee RW, et al: Hemodynamic effects of direct angiotensin II blockade compared to converting enzyme inhibition in rat model of heart failure. *Am J Hypertens* 4:334S–340S, 1991.
11. Smits JFM, van Krimpen C, Schoemaker RG, et al: Angiotensin II receptor blockade after myocardial infarction in rats: Effects on hemodynamics, myocardial DNA synthesis, and interstitial collagen content. *J Cardiovasc Pharmacol* 20:772–778, 1992.
12. Lee RW, Lancaster LD, Buckley D, et al: Peripheral circulatory control of preload–afterload mismatch with angiotensin in dogs. *Am J Physiol* 253:H126–H132, 1987.
13. Fitzpatrick MA, Rademaker MT, Charles CJ, et al: Angiotensin II receptor antagonism in ovine heart failure: Acute hemodynamic, hormonal, and renal effects. *Am J Physiol* 263:H250–H256, 1992.
14. Murakami M, Suzusi H, Naitoh M, et al: Significance of action of angiotensin II in congestive heart failure in conscious dogs. *Circulation* 84:II-107, 1991.
15. Qing G, Garcia R: Chronic captopril and losartan (DuP 753) administration in rats with high output failure. *Am J Physiol* 32:H833–H840, 1992.
16. Ruzicka M, Yuan B, Harmsen E, et al: The renin–angiotensin system and volume overload-induced cardiac hypertrophy in rats: Effects of angiotensin converting enzyme inhibitor versus angiotensin II receptor blocker. *Circulation* 87:921–930, 1993.
17. Gay R: Captopril reduces left ventricular enlargement induced by chronic volume overload. *Am J Physiol* 259:H976–H803, 1990.
18. Lindpainter K, Ganten D: The cardiac renin–angiotensin system: An appraisal of present experimental and clinical evidence. *Circ Res* 68:905–921, 1991.
19. Beinlich CJ, White GJ, Baker KM, et al: Angiotensin II and left ventricular growth in newborn pig heart. *J Mol Cell Cardiol* 23:1031–1038, 1991.
20. Schunkert H, Jackson B, Tang SS, et al: Distribution and functional significance of cardiac angiotensin converting enzyme in hypertrophied rat hearts. *Circulation* 87:1328–1339, 1993.

REFERENCES

17

Angiotensin Antagonists in Models of Hypertension

Pancras C. Wong

1. INTRODUCTION

During the past 30 years, substantial progress has been made in the development of many antihypertensive agents with different sites and modes of action.[1] Newer antihypertensive drugs are effective in the control of high blood pressure with fewer adverse effects than the early agents. Among these new antihypertensive agents, the angiotensin-converting enzyme (ACE) inhibitors such as captopril, enalapril, and lisinopril have all been reported to be effective in the treatment of various forms of hypertension and congestive heart failure. Further, long-term administration of captopril was associated with an improvement in survival and reduced morbidity and mortality in patients with left ventricular dysfunction after a myocardial infarction.[2] These impressive clinical benefits of ACE inhibitors have established the renin–angiotensin system (RAS) as an important target for drug research.[3]

As angiotensin II (ANG II) is the primary molecule of the RAS, a receptor antagonist would provide a direct approach to block the system. The early ANG II receptor antagonists are peptide analogues of ANG II, and thus their therapeutic applications are limited because of their agonistic activity, short half-life, and lack of oral bioavailability.[4] Furukawa *et al.*[5,6] reported the first nonpeptide ANG II receptor antagonists, which we later characterized as weak but selective ANG II receptor antagonists and antihypertensive agents.[7,8] Based on these prototype molecules, highly potent and orally active nonpeptide ANG

Pancras C. Wong • The Du Pont Merck Pharmaceutical Company, Wilmington, Delaware 19880-0400.

Angiotensin Receptors, edited by Juan M. Saavedra and Pieter B.M.W.M. Timmermans. Plenum Press, New York, 1994.

II receptor antagonists without agonistic activity, including losartan (DuP 753; MK 954), have been synthesized.[9] Moreover, this series of antagonists has played a major role in the identification of type 1 (AT_1) and type 2 (AT_2) ANG II receptors.[10]

Losartan (Fig. 1) is the prototypic AT_1 receptor antagonist and inhibits most of the major physiological responses to ANG II.[10,11] The chemistry, pharmacology, pharmacokinetics/metabolism, and toxicology of losartan have been reviewed recently.[12] Losartan is the first nonpeptide ANG II receptor antagonist studied in man and is currently in the phase III clinical trial. It should be noted that losartan generates an active carboxylic acid metabolite EXP3174 (Fig. 1) in both rats and humans.[12] EXP3174 is about 33-fold more potent than losartan in inhibiting the contractile effect of ANG II in rabbit aorta. Unlike losartan, EXP3174 exhibits noncompetitive ANG II antagonism.[13] Because high levels of EXP3174 were found in rat and human plasma after the oral administration of losartan, the active metabolite may be responsible for part of the antihypertensive effect and the long duration of action of losartan in rats and humans.[12]

This chapter reviews primarily the effects of losartan and other nonpeptide ANG II receptor antagonists in various animal models of hypertension.

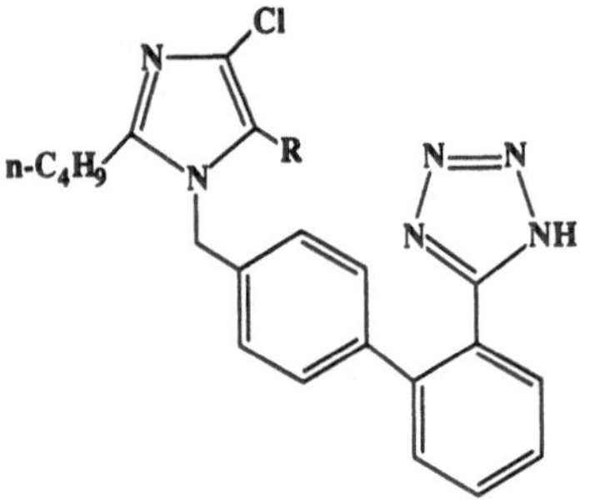

Figure 1. Chemical structures and pharmacological properties of losartan and EXP3174. [a]Losartan as a potassium salt. [b]Inhibition of specific binding of [^{125}I]-ANG II (0.05 nM) to rat adrenal cortical microsomes in the absence of bovine serum albumin. IC_{50}, concentration to reduce specific binding by 50%; [c]K_B, dissociation constant. Antagonism of ANG II-induced contraction of rabbit aorta. Antihypertensive potency in renal hypertensive rats after IV or PO administration. [d]ED_{30}, dose to decrease mean arterial pressure by 30 mm Hg. (From Wong *et al.*[10] Reprinted with permission.)

Parameter	Losartan[a]	EXP3174
R	-CH$_2$OH	-COOH
Binding (IC_{50}, M)[b]	5.5×10^{-9}	1.3×10^{-9}
K_B (M)[c]	3.3×10^{-9}	1.0×10^{-10}
AII Antagonism[c]	Competitive	Noncompetitive
IV ED30 (mg/kg)[d]	0.78	0.04
PO ED30 (mg/kg)[d]	0.59	0.66

2. ANIMAL MODELS OF HYPERTENSION

2.1. Renal Hypertensive Animals (Table I)

2.1.1. Renal Artery-Ligated Rats

Rats were made hypertensive by complete ligation of the left renal artery.[14] Only partially renal-infarcted rats developed hypertension, whereas rats with total renal infarction were not hypertensive. Plasma renin activity and blood pressure were elevated in this model 7 days after renal artery ligation. Twenty-eight days after ligation, plasma renin activity but not blood pressure was decreased to the control level.[14] We also observed that hypertension was fully developed and plasma renin activity was increased about eightfold 6 to 7 days after the ligation of the left renal artery.[15] Further, we demonstrated that the hypertension in these rats, 6 to 7 days after surgery, was maintained by ANG II since inhibitors of the RAS such as captopril, saralasin, and the ANG II monoclonal antibody KAA8 decreased blood pressure to a near-normotensive level.[15]

Losartan is a potent and long-acting antihypertensive agent in renal artery-ligated hypertensive rats with intravenously (IV) and orally (PO) ED_{30}s of 0.78 and 0.59 mg/kg, respectively, and has the same antihypertensive efficacy as

Table I. Effects of Losartan
on Blood Pressure of Renal Hypertensive Animals

| Animal model | Losartan | | | | |
	Dose (mg/kg)	Route	Duration	Comments	Reference
Renal artery-ligated rats	0.1–10	IV	Acute	IV ED_{30} = 0.78 mg/kg	16
	1–100	PO	Acute	PO ED_{30} = 0.59 mg/kg	16
	1; 3	IA	Acute	56–64 mm Hg decrease	18
	10	ID	Acute	28% decrease in BP	19
2-Kidney, 1-clip rats	10	IV	Acute	Anesth., 19 mm Hg decrease	25
	3	IV	Acute	Anesth., 36% decrease in BP	27
	2.5–3	IV	Acute	Anesth., 13–32 mm Hg	28
	125 mg/liter in drinking water		5 days	63 mm Hg decrease	26
2-Kidney, 1-clip monkeys	1–10	PO	Acute	5–24 mm Hg decrease	29
Aortic-coarctated rats	1	PO	Acute	60 mm Hg	32
	3/day	PO	6 weeks	23–35 mm Hg	34
Renal-ablated rats	3/day	IV	7 days	prevent hypertension	38
	30/day	PO	8 weeks	57 mm Hg decrease	37
	180 mg/liter in drinking water		10 weeks	60 mm Hg decrease	36

captopril.[16] Despite a decrease in blood pressure, losartan did not cause tachycardia. The antihypertensive effect of losartan in renal hypertensive rats is likely to be due to the blockade of the vasoconstrictor action of ANG II since losartan is a selective ANG II receptor antagonist and pretreatment of renal hypertensive rats with captopril or saralasin, but not prazosin or indomethacin, inhibited the antihypertensive effect of losartan.[16] Further, the IV antihypertensive potency of losartan correlates well with its *in vitro* potencies in inhibiting the effects of ANG II (i.e., the specific binding of ANG II and the contractile effects of ANG II).[17]

Although most studies demonstrated the selectivity of losartan for the ANG II receptors, other unexpected pharmacological effects of losartan have also been reported. Akers *et al.*[18] observed that losartan at 1 and 3 mg/kg intra-arterially (IA) in the renal artery-ligated rats inhibited the ANG II pressor maximally at 1–5 hr but decreased blood pressure maximally at 5–7 and at 3–7 hr, respectively. Further, there was a lack of correlation between the hypotensive effect and the ANG II antagonistic of losartan, suggesting that the hypotensive effect of losartan may not depend on its immediate blockade of the vasoconstrictor effect of exogenous ANG II. A possible explanation is that losartan may require a longer period to penetrate completely the vascular compartment to exert its maximal blockade of the vasoconstrictor effect of endogenous ANG II.

Ohlstein *et al.*[19] reported that losartan given at 10 mg/kg intraduodenally (ID) decreased blood pressure in this model for greater 48 hr. Although losartan blocked the pressor responses to ANG I and ANG II acutely, only the pressor response to ANG I was blocked at 24 hr postdose, suggesting that losartan may have the ACE inhibitorylike activity. However, this does not agree with the findings showing that losartan lacks effects on ACE and does not alter the contractile and depressor responses to bradykinin in guinea pig ileum and rats, respectively.[12,17] At 48 hr postdose, neither the pressor response to ANG I nor that to ANG II was blocked by losartan despite the fact that blood pressure was still decreased, suggesting additional mechanisms beyond ANG II receptor blockade accounting for the antihypertensive effect of losartan. Ohlstein *et al.*[19] also reported that in renal artery-ligated rats losartan (10 mg/kg ID) as well as prazosin (0.3 mg/kg ID) produced orthostatic hypotensive response as shown by their potentiation of the decrease in blood pressure induced by a 90° tilt, whereas enalapril (1 mg/kg ID) caused a lesser orthostatic response. However, no such effect was observed with SK&F 108566, (E)-α-[[2-butyl-1[(4-carboxyphenyl)methyl]-1*H*-imidazol-5-yl]methylene]-2-thiophenepropanoic acid, a competitive and selective AT_1 receptor antagonist.[20] This is not consistent with the data showing that losartan lacks α_1-adrenergic binding affinity and does not alter the contractile effect of norepinephrine.[12] Nevertheless, these unexpected findings of losartan require cautious interpretation and warrant further investigation.

The renal artery-ligated hypertensive rat model has also been used by other pharmaceutical companies in the evaluation of their ANG II receptor antagonists. A-81282, 4-{N-butyl-N-[(2′[1H-tetrazol-5-yl]biphenyl-4-yl)methyl]amino}pyrimidine-5-carboxylic acid, decreased blood pressure dose-dependently in renal-artery ligated rats with IV and PO ED_{30}s of 0.08 and 2.2 mg/kg, respectively. At 10 mg/kg PO, A-81282 decreased blood pressure to a normotensive level for greater than 24 hr and did not change heart rate.[21] SK&F 108566 at 1 to 10 mg/kg ID decreased blood pressure dose-dependently in renal artery-ligated rats.[22] At 10 mg/kg ID, the antihypertensive effect of SK&F 108566 lasted for at least 90 min. At this dose, it decreased blood pressure and total peripheral resistance and increased cardiac output and heart rate, suggesting systemic vasodilatation. GR117289, 1-[[3-bromo-2-[2-(1H-tetrazol-5-yl)phenyl]-5-benzofuranyl]methyl]-2-butyl-4-chloro-1H-imidazole-5-carboxylic acid, at 0.3, 1, and 3 mg/kg IA decreased diastolic blood pressure in renal artery-ligated rats by 35, 43, and 70 mm Hg, respectively, and at 1, 3, and 10 mg/kg PO decreased diastolic blood pressure by 20, 42, 59 mm Hg, respectively. The antihypertensive effect of GR117289 at 3 and 10 mg/kg PO was greater than 24 hr.[23]

2.1.2. Two-Kidney, One-Clip Renal Hypertensive Models

The two-kidney, one-clip renal hypertensive rat is a high-renin model and the high blood pressure was maintained by the RAS in both the initial and established phases for at least several weeks.[24] Braam *et al.*[25] reported that losartan at 10 mg/kg IV decreased blood pressure significantly in anesthetized two-kidney, one-clip Goldblatt hypertensive rats (5–9 days after clipping) and reduced tubuloglomerular feedback responses in the nonclipped kidney. Losartan given in drinking water (125 mg/liter) for 5 days normalized blood pressure and glomerular capillary pressure of the nonclipped kidney in two-kidney, one-clip renal hypertensive rats (1 month after clipping).[26]

In anesthetized two-kidney, one-clip renal hypertensive rats (1 month after clipping), losartan at 3 mg/kg IV decreased blood pressure by 36% and decreased glomerular filtration rate, renal blood flow, and diuresis in both kidneys with a greater effect on the clipped kidney.[27] By contrast, Lee and Blaufox[28] showed that losartan at 2.5 and 5 mg/kg IV effectively lowered blood pressure and increased renal blood flow and glomerular filtration rate in the nonclipped kidney, but decreased renal blood flow and glomerular filtration rate in the clipped kidney of the conscious two-kidney, one-clip renal hypertensive rats (7–10 days after clipping). Differences in the time after clipping and the condition of the rats may partly account for the discrepancies between these two studies. It appears that the renal function of the clipped kidney is very sensitive to a decrease in blood pressure.

In two-kidney, one-clip hypertensive cynomolgus monkeys, losartan at 1,

3, and 10 mg/kg PO decreased blood pressure by 5 ± 2, 19 ± 3, and 24 ± 3 mm Hg, respectively, with a duration of about 6 hr.[29] The maximal antihypertensive effect of losartan at 10 mg/kg PO is similar to that of saralasin at 20 µg/kg per min IV. It is possible that losartan does not generate enough of the active metabolite EXP3174 in monkeys, which may account for the shorter duration of its blood pressure-lowering effect.

In two-kidney one-clip hypertensive dogs (using an ameroid constrictor) SK&F 108566 at 10 mg/kg PO caused a similar decrease in blood pressure (from 150 to 120 mm Hg) as 1 mg/kg PO of enalapril, confirming the importance of ANG II in maintaining high blood pressure in this model.[22]

2.1.3. Aortic-Coarctated Models

Rats were made hypertensive by complete or partial occlusion of the aorta at a point 2 mm above the left renal artery.[30,31] Plasma renin activity was elevated within days of the coarctation.[30] Losartan at 1 mg/kg PO decreased blood pressure by 60 mm Hg maximally at 6 hr postdose with a duration of greater than 24 hr.[32] L-158,809,5,7-dimethyl-(2-ethyl-3-[[2'-(1H-tetrazol-5-yl)[1,1']-biphenyl-4-yl]methyl]3H-imidazo[4,5-b]pyridine, is a noncompetitive and selective AT$_1$ receptor antagonist[33] and decreased blood pressure significantly at 0.1 and 0.3 mg/kg PO for greater than 24 hr. The acute maximal hypotensive effect of L-158,809 was similar to that of enalapril in these rats.[31] Losartan given at 3 mg/kg per day for 6 weeks given either 1 day (prevention study) or 6 weeks after aortic coarctation (regression study) reduced blood pressure significantly.[34] Losartan was effective in reducing cardiac hypertrophy only in the regression study but not in the prevention study, whereas the ACE inhibitor ramipril at 10 µg/kg per day, a nonhypotensive dose, was effective in reducing cardiac hypertrophy in both studies. It is conceivable that potentiation of bradykinin by ACE inhibition may contribute to the prevention of myocardial hypertrophy. However, interpretation of the data has to be cautious since higher doses such as 10 mg/kg per day of losartan has been shown to reduce cardiac hypertrophy in spontaneously hypertensive rats[35] and have not been studied in this model.

2.1.4. Renal-Ablated Models

Reduction of renal mass in rats causes systemic and glomerular hypertension followed by progressive sclerosis of remnant glomeruli and proteinuria.[36] Partial renal infarct did not produce hypertension in rats treated with losartan (30 mg/day in drinking water for 8 weeks) but caused hypertension within 3 weeks following the discontinuation of losartan treatment, indicating that infarct hypertension does not develop during ANG II receptor blockade.[37] Losartan at 3 mg/kg per day IV prevented hypertension and reduced the increase in blood urea nitrogen in reduced renal mass rats during 7 days of high salt intake, suggesting ANG II acting at the AT_1 receptors contributes to high salt-induced hypertension in renal-ablated rats.[38]

Losartan given in drinking water at 180 mg/liter for 10 weeks reduced systolic blood pressure from 185 ± 4 mm Hg to 125 ± 2 mm Hg in renal-ablated rats.[36] Further, proteinuria and glomerular sclerosis were also reduced. Similar results were also obtained with enalapril but not with a combination of reserpine, hydralazine, and hydrochlorothiazide, which caused a similar decrease in blood pressure as losartan.[36] These results suggest that the blockade of the renal effects of ANG II but not hypotension per se retards the deterioration of renal failure in renal-ablated rats. Similarly, the AT_1 receptor antagonist L-158,809 given in drinking water at 3 to 100 mg/liter for 12 weeks prevents the development of hypertension in Munich Wistar rats with renal ablation.[39]

Kohara *et al.*[40] reported that TCV-116((±)-1-(cyclohexyloxycarbonyl-oxy)ethyl-2-ethoxy-1-[[2′(1*H*-tetrazol-5-yl)biphenyl-4-yl]methyl]benzimidazole-7-carboxylate) at 1 mg/kg per day PO for 4 weeks in 12-week-old spontaneously hypertensive rats with 5/6 nephrectomy (a severe renal failure rat model with hypertension) reduced proteinuria and glomerular sclerosis as well as improved survival rate, suggesting the involvement of ANG II via the activation of AT_1 receptors in the development of renal deterioration by hypertension.

2.2. Genetic Hypertensive Animals (Table II)

2.2.1. Spontaneously Hypertensive Rats

The spontaneously hypertensive rat (SHR) is a model of genetic hypertensive rats with a normal plasma renin activity.[41] Losartan given either intravenously or orally is an effective antihypertensive agent in conscious SHR.[42,43] Losartan has a minimal effect on blood pressure in volume-expanded SHR, which is probably due to the suppression of the influence of the RAS following volume expansion.[44] In our early study, we observed that the acute antihypertensive efficacy of IV bolus injection of captopril was less than that of IV bolus injection of losartan in conscious SHR.[42] These SHRs were surgically implanted with catheters 3 hr before the dosing of losartan and had plasma renin activity of ANG I 10.4 ng/ml per hr. In contrast, Bunkenburg *et al.*[45] reported that losartan infused intravenously at 10 or 30 mg/kg per day reduced blood pressure to the same extent as benazeprilat, an ACE inhibitor, infused intravenously at 3 or 10 mg/kg per day during 48-hr period in 11-week-old SHR that were operated 72 hr before the experiment, suggesting that losartan and benazeprilat have similar antihypertensive efficacies. Subsequently, we also observed that chronic oral administration of captopril and losartan at 10 mg/kg

Table II. Effects of Losartan
on Blood Pressure of Genetic Hypertensive Animals

| Animal model | Losartan | | | Comments | Reference |
	Dose (mg/kg)	Route	Duration		
SHR	3–30	IV	Acute	25–40 mm Hg decrease	42
	3; 10	PO	Acute	20–40 mm Hg decrease	42
	10/day	PO	15 days	Prevent hypertension	46
	10/day	IV	7 days	25 mm Hg decrease	35
	10/day	PO	2 weeks	34% decrease in SBP[b]	35,47
	60/day	PO	10 weeks	Prevent hypertension	48
	15/day	PO	10 weeks	Prevent hypertension in	49
	10	IV	Acute	~25 mm Hg decrease	19
	30	IV	Acute	33 mm Hg decrease	61
	1, 10	IV	Acute	12–28 mm Hg decrease	58
	10 µg	ICV	Acute	19 mm Hg increase	57
	1–1000 µg/kg	ICV	Acute	No effect on BP	58
	3/day	SC	13 days	30 mm Hg decrease in SBP	43
SHR-S	40 µg	IAH[a]	Acute	15 mm Hg decrease	59
spSHR	30/day	PO	12 weeks	13% decrease in SBP	64
Dahl S rats	30/day	PO	7–10 weeks	Transient decrease in BP	65–67
	10 µg	ICV	Acute	7% decrease in BP	68
TGR(mREN2)27	10/day	PO	4.5 weeks	Normalize BP	70
PH dogs	1–30	IV	Acute	10–15 mm Hg decrease	72

[a]IAH, injection into anterior hypothalamus.
[b]SBP, systolic blood pressure.

per day for 15 days produced a similar maximal decrease in blood pressure in conscious SHR.[46] The discrepancy between the study of Bunkenburg et al.[45] and our early study[46] may be related to differences in methods of drug administration (IV bolus vs IV infusion) and experimental conditions of animals (3 hr vs. 72 hr after surgery.)

Losartan given orally at 10 mg/kg per day for 2 weeks to 10- to 11-week-old SHR decreased blood pressure, plasma aldosterone, and left ventricular weight, but increased plasma renin activity, plasma ANG I, and ANG II.[35,47] Interestingly, the antihypertensive effect of losartan after 2 weeks treatment was greater than that after 1 week treatment, suggesting a relatively slow onset of antihypertensive activity of losartan.

Losartan given in drinking water at 60 mg/kg per day for 10 weeks to 4-week-old SHR prevented the development of hypertension, left ventricular hypertrophy, and abnormal vascular amplifier properties (as reflected by a reduction of maximum vasodilatation and an increase in maximum vasoconstriction in the perfused hindlimb). There was still a persistent long-term lowering of blood pressure following the termination of losartan treatment.[48] These data support that ANG II contributes to the development of hypertension, cardiovascular hypertrophy, and abnormal vascular amplifier properties in SHR. Losartan (15 mg/kg per day) and captopril (100 mg/kg per day) given in the drinking water for 10 weeks prevented the development of hypertension and vascular hypertrophy in young SHR. These effects of losartan and captopril were still evident even 17 weeks after treatment was stopped.[49] Similarly, a lesser increase in blood pressure was observed in 90-day-old SHR which were previously treated with losartan at 10 mg/kg per day subcutaneously (SC) on postnatal days 10 through 20.[50] A similar result was also observed with another ANG II receptor antagonist, TCV-116.[51] These results support the notion that ANG II may play an important role in the early onset of hypertension in SHR.

Tofovic *et al.*[52] demonstrated that losartan but not the AT$_2$-selective ligand PD 123177 blocked the pressor, aldosterone release, mesenteric vasoconstrictor responses to ANG II in SHR and Wistar–Kyoto rats (WKY). It appears that SHR is more sensitive than WKY to the blockade of ANG II-induced aldosterone release by losartan. Further, the *in vivo* pre- and postjunctional effects of ANG II on noradrenergic neurotransmission in the mesenteric vascular bed of WKY are mediated by AT$_1$ receptors. Losartan infused at 3 mg/kg SC for 13 days decreased systolic blood pressure by 30 mm Hg and increased sensitivity of baroreceptor reflex in 12 to 14-week-old SHR, suggesting a role of ANG II in hypertension and the impairment of baroreceptor reflex.[43]

Because of the effectiveness of ACE inhibitors in lowering blood pressure in SHR (which has a normal plasma renin activity), it has been suggested that vascular-generated ANG II rather than circulating ANG II maintains high blood pressure in SHR. This is consistent with our finding showing that ANG II monoclonal antibody, which neutralizes circulating ANG II, did not lower blood pressure in SHR.[53] Since losartan did not lower blood pressure in 24-hr bilateral-nephrectomized SHR,[42,54] plasma renin derived from the kidney may be an important source of vascular renin for the local vascular production of ANG II in SHR. This agrees well with the observation that a specific anti-rat renin antibody, which neutralizes plasma renin, lowered blood pressure in SHR.[54] Recently, Wood *et al.*[55] reported in a preliminary study that blood pressure of 12 to 14-week-old SHR was decreased by an intrarenal infusion of the AT$_1$ receptor antagonist CGP 48933, (S)-*N*-Valeryl-*N*-{[2'-(1*H*-tetrazol-5-yl)biphenyl-4-yl]-methyl}-valine, at 300 µg/kg per day for 3 days. The same dose of CGP 48933 infused intravenously did not alter blood pressure. This result suggests that ANG II generated locally in the kidney is important in the maintenance of high blood pressure in SHR, which supports the Guyton's theory that the kidney is important in the long-term control of blood pressure.[56]

Although it appears that the peripheral RAS may participate in the control of blood pressure in SHR, the role of the brain RAS has not been well defined. DePasquale *et al.*[57] reported that intracerebroventricular (ICV) injection of losartan at 10 µg, which inhibited the pressor response to ICV ANG II, did not lower blood pressure but caused a transient pressor response of 19 mm Hg in conscious SHR. This result suggests that the antihypertensive effect of systemic-administered losartan is due to the blockade of peripheral but not central ANG II receptors. By contrast, Kawano *et al.*[58] reported that ICV injections of losartan at 1–1000 µg/kg elicited no acute changes in blood pressure in either conscious SHR or WKY. However, in salt-sensitive SHR (SHR-S) but not in WKY, injection of losartan at 20 and 40 µg into the anterior hypothalamus but not posterior hypothalamus lowered blood pressure dose-dependently for at least 1 hr.[59] The selective AT_2 antagonist PD 123319 injected into anterior hypothalamus did not lower blood pressure in this model. Further, the hypotensive effect of ICV losartan was enhanced in SHR-S fed with high salt intake.[60] These data suggest that ANG II in the anterior hypothalamus regulates blood pressure via the activation of AT_1 receptors in the SHR-S but not in WKY. High salt intake in SHR-S caused a further increase in blood pressure which may be due to an enhanced activation of AT_1 receptors in anterior hypothalmus.[60]

Ohlstein *et al.*[9] showed that losartan at 10 mg/kg IV lowered blood pressure with a duration of greater than 24 hr in SHR that were surgically implanted with catheters 3 to 4 days before the experiment. By contrast, enalapril at 1 mg/kg IV and an AT_1 receptor antagonist SK&F 108566 at 10 mg/kg IV did not lower blood pressure in SHR. These results imply that additional mechanisms beyond ANG II receptor blockade may account for the antihypertensive effect of losartan. However, interpretation of these data has to be made cautiously. First, enalapril is a prodrug, and thus enalapril given at 1 mg/kg IV conceivably may not generate enough of the active metabolite enalaprilat to lower blood pressure. As discussed above, losartan generates an active metabolite EXP3174, which is a noncompetitive ANG II receptor antagonist,[13] whereas SK&F 108566 is a competitive ANG II receptor antagonist.[20] Because ANG II receptor antagonists stimulate a compensatory rise in renin secretion, it is tempting to speculate that the resulting increase in ANG II may counteract the receptor blockade induced by a competitive ANG II receptor antagonist more than that by a noncompetitive receptor antagonist. Thus, higher doses of SK&F 108566 may be needed to lower blood pressure in SHR. Whether this factor and/or others factor accounting for the lack of blood pressure lowering effect of SK&F 108566 at 10 mg/kg IV in SHR remain to be studied.

Cachofeiro *et al.*[61] reported that the kinin receptor antagonist D-Arg-Arg-Pro-Hyp-Gly-Thi-Ser-D-Phe-Thi-Arg-trifluoroacetic acid did not alter blood pressure in SHR but attenuated the hypotensive effect of captopril and ramiprilat but not losartan, suggesting that kinins contribute to the hypotensive effects of ACE inhibitors but not losartan. Interestingly, N^w-nitro-L-arginine-methyl ester (L-NAME), a nitric oxide synthase inhibitor (15 mg/kg + 10 mg/kg per hr IV), increased blood pressure in conscious SHR and reduced the blood pressure-lowering effect of captopril, ramiprilat, and losartan, suggesting a contribution of nitric oxide to the hypotensive effects of ACE inhibitors and losartan.

2.2.2. Stroke-Prone Spontaneously Hypertensive Rats

In stroke-prone SHR (spSHR) fed a high sodium diet, there was a paradoxical rise in plasma renin activity and the elevated ANG II may be responsible for the increase in mortality and end-organ damage.[62] Both enalapril (15 mg/kg per day PO) and losartan (10 mg/kg per day PO) prevented stroke but not hypertension and protected against renal damage through 28 weeks of age in salt-loaded spSHR.[63] At a higher dose, losartan (30 mg/kg per day in drinking water for 12 weeks) decreased blood pressure and end-organ damage and improved survival in spSHR fed a high sodium diet.[64]

2.2.3. Dahl Salt-Sensitive Hypertensive Rats

In 6-week-old Dahl salt-sensitive (Dahl-S) rats fed high sodium diet (8% NaCl), losartan given at 30 mg/kg per day in drinking water for 10 weeks reduced blood pressure only transiently but delayed the progression of renal damage and stroke and increased survival, suggesting that an activated RAS may be responsible for the morbidity and mortality in salt-loaded Dahl-S rats.[65,66] Similarly, Sugimoto *et al.*[67] reported that losartan (30 mg/kg per day PO for 7 weeks) only caused a transient decrease in blood pressure at week 6 of losartan treatment but reduced cardiac hypertrophy in 6-week-old Dahl-S rats fed a high sodium diet (4% NaCl). Thus, ANG II may contribute to the development of cardiac hypertrophy in this low-renin-volume-dependent hypertensive model.

A moderate hypotensive effect of 7% and a decrease in heart rate of 21% were noted when losartan was injected ICV at 10 μg in conscious Dahl-S rats, suggesting a moderate role of ANG II in the brain for regulating blood pressure and heart rate in Dahl-S rats.[68]

2.2.4. Transgenic Hypertensive Rats

Mullins *et al.*[69] established a transgenic hypertensive rat strain, TGR(mREN2)27, by introducing an additional renin gene, the murine *Ren2* gene, into the germ line of rats. These rats develop severe hypertension with an overexpression of renin in the adrenal but not in the kidney.[70] Although the plasma renin and ANG II levels were normal in these animals, vascular ANG production was enhanced.[71] Losartan given in drinking water at 10 mg/kg per day for 4.5 weeks normalized blood pressure and increased plasma renin and ANG II.[70] These results suggest a major role of ANG II in hypertension of

TGR(mREN2)27 rats and that this model may be useful for studying roles of local adrenal RAS in hypertension.

2.2.5. Spontaneously Hypertensive PH Dogs

The genetic Penn Hypertensive (PH) dogs represent a canine model of essential hypertension with a normal plasma renin level.[72] Losartan at 1–30 mg/kg IV caused a moderate but dose-dependent decrease in blood pressure and increased renal blood flow as well as glomerular filtration rate, suggesting a beneficial renal hemodynamic effect of losartan in this model.[72]

2.3. Deoxycorticosterone Acetate-Salt Hypertensive Animals

The deoxycorticosterone acetate (DOCA)-salt hypertensive rat is a low renin model, and the pathogenesis of DOCA-salt hypertension has been recently reviewed.[73] Losartan at 10 mg/kg IV did not alter blood pressure in DOCA-salt hypertensive rats with a low plasma renin activity.[16] This dose of losartan was effective to decrease blood pressure to a normotensive level in renal artery-ligated hypertensive rats (see above). Thus, the acute antihypertensive effect of losartan in rats appears to depend on the pretreatment level of plasma renin activity. A new ANG II receptor antagonist TCV-116 at 10 mg/kg PO also did not lower blood pressure in DOCA-salt hypertensive rats, whereas it lowered blood pressure in two-kidney one-clip renal hypertensive rats with an oral ED_{25} of 0.03 mg/kg.[74]

2.4. Miscellaneous Hypertensive Models (Table III)

Losartan reversed the increase in blood pressure induced by ANG II infusion in rats fed on a high sodium diet.[75] Losartan significantly decreased blood pressure within 5 min of administration, probably due to blockade of the direct vasoconstrictor effect of ANG II. Interestingly, the time required to reach the maximum depressor effect increased progressively with increased duration

Table III. Effects of Losartan on Blood Pressure
of Miscellaneous Models of Hypertensive Animals

| Animal model | Losartan | | | | |
	Dose (mg/kg)	Route	Duration	Comments	Reference
ANG II-infused rats	3	IA	Acute	Decrease BP	75
Endothelin-infused rats	3	IV	Acute	Decrease BP	77
L-NAME-treated rats	30/day	PO	4 weeks	39 mm Hg decrease	80
	30/day	PO	24 days	Prevent hypertension	81
Pulmonary hypertensive rats	10	SC	21 days	No effect on BP	82

of ANG II infusion, probably related to the slow pressor effect of ANG II,[75] which may operate through a central nervous system mechanism.[76]

Losartan decreased blood pressure in endothelin-infused hypertensive rats.[77] Since endothelin elevates plasma renin activity *in vivo* and stimulates ACE activity *in vitro*,[78] the increased ANG II may contribute to the high blood pressure in endothelin-infused rats, which may account for the observed anti-hypertensive effect of losartan in this model. Further, Yoshida *et al.*[79] demonstrated that blood pressure was increased in rats by subpressor doses of endothelin-1 and ANG II, suggesting that endothelin-1 may increase the pressor sensitivity to ANG II and vice versa.

Ribeiro *et al.*[80] reported chronic inhibition of nitric oxide with L-NAME at about 60 mg/kg per day in drinking water for 4 weeks in rats, increased blood pressure as well in plasma renin activity, depressed renal function, and widespread renal morphological changes. Co-administration of losartan at 30 mg/kg per day and L-NAME in drinking water caused a lesser increase in blood pressure (125 ± 6 mm Hg vs. control 164 ± 6 mm Hg) and renal vascular resistance and prevented the decrease in glomerular filtration rate and changes in renal morphology. This study suggests that chronic inhibition of nitric oxide synthesis with L-NAME activates the RAS, which may account partly for the hypertension. A similar finding was also observed by Jover *et al.*,[81] showing that hypertension (159 ± 3 mm Hg) and cardiac hypertrophy induced by L-NAME (10 mg/kg PO bid) for 24 days in rats was prevented by co-administration of losartan (30 mg/kg per day PO).

A role of ANG II in the development of pulmonary hypertension has been suggested in several animal models of pulmonary hypertension.[82] In the monocrotaline-induced pulmonary hypertension rat model, losartan at 10 mg/kg subcutaneously for 21 days did not prevent the development of hypertensive pulmonary vascular disease.[82] Whether ANG II may play a role in other models such as the hypoxia-induced pulmonary hypertension remains to be investigated.

3. SUMMARY

Losartan is an effective antihypertensive agent in several animal models of hypertension especially in renin-dependent models but is not active in the very low renin DOCA-salt hypertensive model. The antihypertensive efficacy of losartan in human essential hypertension has also been demonstrated recently.[83] It is likely that the blockade of the vasoconstrictor effect of ANG II is involved in the blood pressure-lowering effect of losartan, but additional mechanisms beyond ANG II blockade have also been suggested. In most cases, losartan appears to be as effective as ACE inhibitors in lowering blood pressure in hypertensive animals and humans. However, ANG II may be produced in the vasculature from other pathways that are insensitive to the inhibition of ACE inhibitors.[84] Thus, it is conceivable that in some cases losartan may be more efficacious than ACE inhibitors in blocking the effects of endogenous ANG II. Similar to ACE inhibitors, losartan reduces end-organ damage and improves survival in spSHR and salt-loaded Dahl-S rats.

As discussed above, losartan is a selective receptor antagonist for the subtype AT_1 receptors. All the known physiological effects of ANG II are mediated by the AT_1 receptors. Few functions of the AT_2 receptors have been reported but the physiological relevance of these findings is not clear.[10] It is noteworthy that losartan stimulates a compensatory rise in plasma ANG II *in vivo* and the increased levels of ANG II may activate the AT_2 receptor, which is not blocked by losartan. However, no unexpected effects of losartan have been reported in animals and humans treated with losartan. Whether a nonpeptide ANG II receptor antagonist that blocks both AT_1 and AT_2 receptors may have a better clinical efficacy over losartan remains to be studied.

Because of the multisubstrate specificity of ACE, ACE inhibitors block the formation of ANG II from ANG I, but also prevent the degradation of bradykinin and other peptides. These additional activities of ACE inhibitors may account for some of their side effects such as dry cough and proinflammatory responses.[85] Inasmuch as losartan does not affect the metabolism of bradykinin, it may not cause these side effects of ACE inhibitors. Whether losartan would offer additional therapeutic benefits over the ACE inhibitors by blocking selectively at the ANG II receptors remains to be determined in clinical trials.

REFERENCES

1. Robson RD: The evolution of hypertensive therapy. *Clin Exp Pharmacol Physiol* 19:41–46, 1992.
2. Pfeffer MA, Braunwald E, Moyé LA, et al: Effect of captopril on mortality and morbidity in patients with left ventricular dysfunction after myocardial infarction. *N Engl J Med* 327:669–677, 1992.
3. Greenlee WJ, Siegl PKS: Angiotensin/renin modulators. *Annu Rep Med Chem* 27:59–68, 1992.
4. Garrison JC, Peach MJ: Renin and angiotensin. In Gilman AG, Rall TW, Nies AS, Taylor P (eds): *Goodman and Gilman's The Pharmacological Basis of Therapeutics,* 8th ed. Pergamon Press, New York, 1990, pp. 749–763.
5. Furukawa Y, Kishimoto S, Nishikawa K: Hypotensive imidazole derivatives. U.S. Patents 4,340,598. Issued to Takeda Chemical Industries, Ltd. (Osaka, Japan), 1982.
6. Furukawa Y, Kishimoto S, Nishikawa K: Hypotensive imidazole-5-acetic acid derivatives. U.S. Patents 4,355,040. Issued to Takeda Chemical Industries, Ltd. (Osaka, Japan), 1982.
7. Wong PC, Chiu AT, Price WA, et al: Nonpeptide angiotensin II receptor antagonists. I.

Pharmacological characterization of 2-*n*-butyl-4-chloro-1-(2-chlorobenzyl)imidazole-5-acetic acid, sodium salt (S-8307). *J Pharmacol Exp Ther* 247:1–7, 1988.

8. Chiu AT, Carini DJ, Johnson AL, et al: Nonpeptide angiotensin II receptor antagonists. II. Pharmacology of S-8308. *Eur J Pharmacol* 157:13–21, 1988.

9. Duncia JV, Carini DJ, Chiu AT, et al: The discovery of DuP 753, a potent, orally nonpeptide angiotensin II receptor antagonist. *Med Res Rev* 12:149–191, 1992.

10. Wong PC, Chiu AT, Duncia JV, et al: Angiotensin II receptor antagonists and receptor subtypes. *Trends Endocrinol Metab* 3:211–217, 1992.

11. Smith RD, Chiu AT, Wong PC, et al: Pharmacology of nonpeptide angiotensin II receptor antagonists. *Annu Rev Pharmacol Toxicol* 32:135–165, 1992.

12. Wong PC, Barnes TB, Chiu AT, et al: Losartan (DuP 753), an orally active nonpeptide angiotensin II receptor antagonist. *Cardiovasc Drug Rev* 9:317–339, 1991.

13. Wong PC, Price WA, Chiu AT, et al: Nonpeptide angiotensin II receptor antagonists. XI. Pharmacology of EXP3174, an active metabolite of DuP 753—an orally active antihypertensive agent. *J Pharmacol Exp Ther* 255:211–217, 1990.

14. Cangiano JL, Rodriguez-Sargent C, Martinez-Maldonada M: Effects of antihypertensive treatment on systolic blood pressure and renin in experimental hypertension in rats. *J Pharmacol Exp Ther* 208:310–313, 1979.

15. Reilly TM, Wong PC, Price WA, et al: Characterization of the functional antagonism and antihypertensive activity displayed by a monoclonal antibody to angiotensin II. *J Pharmacol Exp Ther* 244:160–165, 1988.

16. Wong PC, Price WA, Chiu AT, et al: Nonpeptide angiotensin II receptor antagonists. IX. Antihypertensive activity in rats of DuP 753, an orally active antihypertensive agent. *J Pharmacol Exp Ther* 252:726–732, 1990.

17. Wong PC, Price WA, Chiu AT, et al: Nonpeptide angiotensin II receptor antagonists: Studies with EXP9270 and DuP 753. *Hypertension* 15:823–834, 1990.

18. Akers JS, Hilditch A, Robertson MJ, et al: Does a relationship exist between the antihypertensive action and the angiotensin receptor antagonist activity of DuP 753 in conscious rats? (abstract). *Br J Pharmacol* 104(Oct Suppl):126P, 1991.

19. Ohlstein EH, Gellai M, Brooks DP, et al: The antihypertensive effect of the angiotensin II receptor antagonist DuP 753 may not be due solely to angiotensin II receptor antagonism. *J Pharmacol Exp Ther* 262:595–601, 1992.

20. Edwards RM, Aiyar N, Ohlstein EH, et al: Pharmacological characterization of the nonpeptide angiotensin II receptor antagonist, SK&F 108566. *J Pharmacol Exp Ther* 260:175–181, 1992.

21. Lee JY, Brune M, Warner R, et al: Antihypertensive activity of Abbott (A)-81282, a nonpeptide angiotensin II antagonist in the renal artery-ligated hypertensive rat (abstract). *Pharmacologist* 34:165, 1992.

22. Brooks DP, Fredrickson TA, Weinstock, et al: Antihypertensive activity of the nonpeptide angiotensin II receptor antagonist. SK&F 108566, in rats and dogs. *Naunyn Schmiedebergs Arch Pharmacol* 345:673–678, 1992.

23. Hilditch A, Akers JS, Travers A, et al: Cardiovascular effects of the angiotensin receptor antagonist, GR117289, in conscious renal hypertensive and normotensive rats (Abstract). *Br J Pharmacol* 104(Suppl):423P, 1991.

24. Goldblatt H, Lynch J, Hanzal RF, et al: Studies in experimental hypertension I. The production of persistent elevation of systolic blood pressure by means of renal ischemia. *J Exp Med* 59:347–379, 1934.

25. Braam B, Mitchell KD, Navar LG: Attenuation of tubuloglomerular feedback response by the AT₁ receptor antagonist, DuP 753, in Goldblatt hypertension (abstract). *FASEB J* 6(4):A980, 1992.

26. DeNicola L, Keiser JA, Blantz RC, et al: Angiotensin II and renal functional reserve in rats with Goldblatt hypertension. *Hypertension* 19:790–794, 1992.

27. El Amrani AIK, Philippe M, Michel JB: Bilateral renal responses to the angiotensin II receptor antagonist, losartan, in 2K-1C Goldblatt hypertensive rats (abstract). *J Hypertens* 19(Suppl 4):206, 1992.

28. Lee B, Blaufox MD: Renal effects of DuP 753 in renovascular hypertension (Abstract). *Am J Hypertens* 4(5, Part 2):84A, 1991.

29. Keiser JA, Painchaud CA, Hicks GW, et al: Effects of oral DuP 753 in renal hypertensive primates (abstract). *Am J Hypertens* 4(5, Part 2):32A, 1991.

30. Sweet CS, Columbo JM, Gaul SL: Central antihypertensive effects of inhibitors of the renin–angiotensin system in rats. *Am J Physiol* 231:1794–1799, 1976.

31. Siegl PKS, Chang RSL, Mantlo BN, et al: In vivo pharmacology of L-158,809, a new highly potent and selective nonpeptide angiotensin II receptor antagonist. *J Pharmacol Exp Ther* 262:139–144, 1992.

32. Mantlo NB, Chakravarty PK, Ondeyka DL, et al: Potent, orally active imidazo[4,5-*b*]pyridine-based angiotensin II receptor antagonists. *J Med Chem* 34:2919–2922, 1991.

33. Chang RSL, Siegl PKS, Clineschmidt BV, et al: In vitro pharmacology of L-158,809, a new highly potent and selective angiotensin II receptor antagonist. *J Pharmacol Exp Ther* 262:133–138, 1992.

34. Linz W, Henning R, Schölkens BA: Role of angiotensin II receptor antagonism and converting enzyme inhibition in the progression and regression of cardiac hypertrophy in rats. *J Hypertens* 9(Suppl 6):S400–S401, 1991.

35. Mizuno K, Tani M, Hashimoto S, et al: Effects of losartan, a nonpeptide angiotensin II receptor antagonist, on cardiac hypertrophy and the tissue angiotensin II content in spontaneously hypertensive rats. *Life Sci* 51:367–374, 1992.

36. Lafayette RA, Mayer G, Park SK, et al: Angiotensin II receptor blockade limits glomerular injury in rats with reduced renal mass. *J Clin Invest* 90:766–771, 1992.

37. Dzielak DJ, Doherty MC, Callahan J, et al: Partial renal infarct hypertension does not develop in animals treated with DuP 753 (abstract). *FASEB J* 6(4):A945, 1992.

38. Kanagy NL, Fink GD: Losartan (DuP 753) prevents salt-induced hypertension in reduced renal mass rats (Abstract). *FASEB J* 6(5):A1810, 1992.

39. Gabel RA, Kivlighn SD, Siegl PKS: The effect of chronically administered L-158,809 on the development of hypertension in subtotally nephrectomized Munich Wistar rats (abstract). *FASEB J* 6(4):A982, 1992.

40. Kohara K, Mikami H, Higaki J, et al: Selective angiotensin type 1 receptor blockade ameliorates proteinuria and survival rate in spontaneously hypersensitive rats with reduced renal mass (Abstract). *Hypertension* 20:427, 1992.

41. Okamoto K, Aoki K: Development of a strain of spontaneously hypertensive rats. *Jpn Circ J* 27:282–293, 1963.

42. Wong PC, Price WA, Chiu AT, et al: Hypotensive action of DuP 753, an angiotensin II antagonist, in spontaneously hypertensive rats. Nonpeptide angiotensin II receptor antagonists. X. *Hypertension* 15:459–468, 1990.

43. Kawano Y, Yoshida K, Kuramochi M, et al: Chronic effects of losartan, an angiotensin AT₁ receptor antagonist, on blood pressure and baroreceptor reflex in spontaneously hypertensive rats (Abstract). *J Hypertens* 19(Suppl 4):244, 1992.

44. Fenoy FJ, Milicic I, Smith RD, Wong PC, et al: Effects of DuP 753 on renal function of normotensive and spontaneously hypertensive rats. *Am J Hypertens* 4(4, Part 2):321S–326S, 1991.

45. Bunkenburg B, Schnell C, Baum HP, et al: Prolonged angiotensin II antagonism in spon-

taneously hypertensive rats: Hemodynamic and biochemical consequences. *Hypertension* 18:278–288, 1991.

46. Wong PC, Price WA, Chiu AT, et al: In vivo pharmacology of DuP 753. *Am J Hypertens* 4(4, Part 2):288S–298S, 1991.

47. Mizuno K, Niimura S, Tani M, et al: Antihypertensive and hormonal activity of MK 954 in spontaneously hypertensive rats. *Eur J Pharmacol* 215:305–308, 1992.

48. Oddie CJ, Dilley RJ, Bobik A: Long-term angiotensin II antagonism in spontaneously hypertensive rats: Effects on blood pressure and cardiovascular amplifiers. *Clin Exp Pharmacol Physiol* 19:392–395, 1992.

49. Morton JJ, Beattie EC, McPherson F: Treatment of young spontaneously hypertensive rats with angiotensin receptor antagonist reduces hypertension and vascular hypertrophy in adulthood (abstract). *J Hypertens* 10(Suppl 4):S151, 1992.

50. Kirby RF, Nanda A, Henry M, et al: Preweanling losartan treatment reduces adult blood pressure in the spontaneously hypertensive rat (abstract). *FASEB J* 6(5):A1872, 1992.

51. Ogihara T, Inada Y, Nishikawa K: Long term cardiovascular effects after a brief treatment with angiotensin II receptor antagonist, TCV-116, in young spontaneously hypertensive rats (Abstract). *J Hypertens* 10(Suppl 4):291, 1992.

52. Tofovic SP, Pong AS, Jackson EK: Role of angiotensin subtype 1 and subtype 2 receptor antagonists in normotensive versus hypertensive rats. *Hypertension* 18:774–782, 1991.

53. Wong PC, Reilly TM, Timmermans PBMWM: Angiotensin II monoclonal antibody: Blood pressure effects in normotensive and spontaneously hypertensive rats. *Eur J Pharmacol* 186:353–356, 1990.

54. Inagami T, Murakami T, Higuchi K, et al: Roles of renal and vascular renin in spontaneous hypertension and switching of mechanism upon nephrectomy: Lack of hypotensive effects of inhibition of renin, converting enzyme, angiotensin II receptor blocker after bilateral nephrectomy. *Am J Hypertens* 4(1, Part 2):15S–22S, 1991.

55. Wood JM, Schnell Cr, Levens NR: The kidney is an important target for the antihypertensive action of an angiotensin II receptor antagonist in spontaneously hypertensive rats (abstract). *Hypertension* 20:436, 1922.

56. Guyton AC, Hall JE, Lohmeier TE, et al: Blood pressure regulation regulation: Basic concepts. *Fed Proc* 40:2252–2256, 1981.

57. DePasquale MJ, Fossa AA, Holt WF, et al: Central DuP 753 does not lower blood pressure in spontaneously hypertensive rats. *Hypertension* 19:668–671, 1992.

58. Kawano Y, Yoshide K, Yoshimi H, et al: Central and peripheral effects of DuP 753, an angiotensin II receptor antagonist, on blood pressure in hypertensive and normotensive rats. *Ther Res* 13:333–338, 1992.

59. Yang RH, Jin HK, Wyss JM, et al: Depressor effect of blocking angiotensin subtype 1 receptors in anterior hypothalamus. *Hypertension* 19:475–481, 1992.

60. Yang RH, Jin H, Wyss JM, et al: High NaCl intake enhances the depressor effect of angiotensin II receptor blockade in anterior hypothalamic area of sodium chloride sensitive spontaneously hypertensive rats (abstract). *Hypertension* 18:421, 1991.

61. Cachofeiro V, Sakakibara T, Nasjletti A: Kinins, nitric oxide, and the hypotensive effect of captopril and ramiprilat in hypertension. *Hypertension* 19:138–145, 1992.

62. Volpe M, Camargo MJF, Mueller FB, et al: Relation of plasma renin to end organ damage and to protection of K feeding in stroke-prone hypertensive rats. *Hypertension* 15:318–326, 1990.

63. Stier CT Jr, Mahboubi K, Chander P, et al: Role of the renin–angiotensin system in the pathology of salt-loaded stroke-prone SHR (abstract). *Hypertension* 18:410, 1991.

64. Camargo MJF, von Lutterotti N, Pecker MS, et al: DuP 753 increases survival in spontaneously hypertensive stroke-prone rats fed a high sodium diet. *Am J Hypertens* 4(4, Part 2):341S–345S, 1991.

65. Von Lutterotti N, Camargo MJF, Mueller FB, et al: Angiotensin II receptor antagonist markedly reduces morbidity in salt-loaded Dahl S rats. *Am J Hypertens* 4(4, Part 2):346S–349S, 1991.

66. Von Lutterotti N, Camargo MJF, Campbell WG, et al: Angiotensin II receptor antagonist delays renal damage and stroke in salt-loaded Dahl salt-sensitive rats. *J Hypertens* 10:949–957, 1992.

67. Sugimoto K, Gotoh E, Takasaki I, et al: Effect of angiotensin II receptor antagonist on cardiac hypertrophy in Dahl salt-sensitive rats (abstract). *Hypertension* 20:419, 1992.

68. Lark LA, Hess EJ, Weyhenmeyer JA: Cardiovascular effects of centrally administered peptide and nonpeptide angiotensin II receptor antagonists in Dahl salt-sensitive rats (abstract). *FASEB J* 5(4):A852, 1992.

69. Mullins JJ, Peters J, Ganten D: Fulminant hypertension in rats harbouring the mouse *ren-2* gene. *Nature* 344:541–544, 1990.

70. Bader M, Zhao Y, Sander M, et al: The transgenic rats TGR(mREN2)27: Role of tissue renin in the pathophysiology of hypertension. *Hypertension* 19:681–686, 1992.

71. Hilgers KF, Peters J, Veelken R, et al: Increased vascular angiotensin formation in female rats harboring the mouse *Ren-2* gene. *Hypertension* 19:687–691, 1992.

72. Bovee KC, Wong PC, Timmermans PBMWM, et al: Effects of the nonpeptide angiotensin II receptor antagonist DuP 753 on blood pressure and renal functions in spontaneously hypertensive PH dogs. *Am J Hypertens* 4(4, Part 2):327S–333S, 1991.

73. Schenk J, McNeill JH: The pathogenesis of DOCA-salt hypertension. *J Pharmacol Toxicol Methods* 27:161–170, 1992.

74. Wada T, Inada Y, Shibouta Y, et al: Antihypertensive action of a nonpeptide angiotensin II antagonist, TCV-116, in various hypertensive rats (abstract). *J Hypertens* 10(Suppl 4):S144, 1992.

75. Gorbea-Oppliger VJ, Kanagy NL, Fink GD: Losartan (DuP 753) reverses angiotensin-induced hypertension in conscious rats (abstract). *FASEB J* 6(5):A1810, 1992.

76. Smits GJ, Koepke JP, Blaine EH: Reversal of low dose angiotensin hypertension by angiotensin receptor antagonists. *Hypertension* 18:17–21, 1991.

77. Mortensen LH, Fink GD: Losartan (DuP 753) acutely attenuates endothelin-hypertension in conscious rats (abstract). *FASEB J* 6(4):A945, 1992.

78. Rubanyi GM, Parker Botelho LH: Endothelins. *FASEB J* 5:2713–2720, 1991.

79. Yoshida K, Yasujima M, Kohzuki M, et al: Endothelin-1 augments pressor response to angiotensin II infusion in rats. *Hypertension* 20:292–297, 1992.

80. Ribeiro MO, Antunes E, de Nucci G, et al: Chronic inhibition of nitric oxide synthesis. A new model of arterial hypertension. *Hypertension* 20:298–303, 1992.

81. Jover B, Ventre F, Herizi A, et al: Sodium and angiotensin in hypertension induced by chronic NO synthase inhibition in the rat (abstract). *Hypertension* 20:399, 1992.

82. Cassis LA, Rippetoe PE, Soltis EE, et al: Angiotensin II and monocrotaline-induced pulmonary hypertension: Effect of losartan (DuP 753), a nonpeptide angiotensin type 1 receptor antagonist. *J Pharmacol Exp Ther* 262:1168–1172, 1992.

83. Nelson E, Merrill D, Sweet CS, et al: Efficacy and safety of oral MK-954 (DuP 753), an angiotensin receptor antagonist, in essential hypertension. *J Hypertens* 9(Suppl 6):S468–S469, 1991.

84. Okamaur T, Okunish H, Ayajika K, et al: Conversion of angiotensin I to angiotensin II in dog isolated renal artery: Role of two different angiotensin II-generating enzymes. *J Cardiovasc Pharmacol* 15:353–359, 1990.

85. Lindgren BR, Andersson, RGG: Angiotensin-converting enzyme inhibitors and their influence on inflammation, bronchial reactivity and cough. *Med Toxicol Adverse Drug Exp* 4:369–380, 1989.

18

Angiotensin II Antagonism and Renal Function

Keshwar Baboolal and Timothy W. Meyer

1. ANGIOTENSIN II RECEPTOR ANTAGONISM IN THE NORMAL KIDNEY

1.1. Angiotensin II Receptors in the Kidney

The kidney is abundantly supplied with angiotensin II (ANG II) receptors. In the tubules, ANG II receptors are found in greatest density in the first part of the proximal convolution.[1] In the vasculature, ANG II receptors are found in both afferent (preglomerular) and efferent (postglomerular) arteries and arterioles, in the glomerular capillary tuft, and in the vasa recta that supply blood to the renal medulla.[2]

1.2. Effects of Exogenous ANG II

The contribution of this complex receptor distribution to the control of renal function is not fully understood. However, the predominant effect of ANG II appears to be to reduce renal excretion of sodium. Tubule microperfusion studies have shown that stimulation of ANG II receptors in the proximal tubule augments sodium transport by epithelial cells of this nephron segment.[3-5] Glomerular micropuncture studies suggest that stimulation of ANG II receptors in the renal cortical vasculature may also promote sodium reabsorption. These studies have shown that low doses of ANG II cause more constriction of efferent than afferent arterioles.[6,7] The effect of predominant constriction of

Keshwar Baboolal and Timothy W. Meyer • Department of Medicine, Stanford University, Stanford, California 94304.

Angiotensin Receptors, edited by Juan M. Saavedra and Pieter B.M.W.M. Timmermans. Plenum Press, New York, 1994.

efferent arterioles is to increase glomerular capillary pressure so that the glomerular filtration rate is maintained while renal blood flow is reduced. Increasing the ratio of glomerular filtration to glomerular perfusion in turn increases the protein concentration and oncotic pressure of blood flowing into the peritubular capillary network. Stimulation of arteriolar ANG II receptors may thus promote sodium reabsorption by increasing Starling forces, drawing fluid into the peritubular capillaries, while stimulation of tubular ANG II receptors promotes sodium reabsorption by increasing epithelial sodium transport.

The effects of stimulation of ANG II receptors in the specialized blood vessels of the glomerulus and vasae rectae are less clearly defined. Within the glomerulus, ANG II receptors are found on both mesangial cells and visceral epithelial cells, or podocytes.[8] It is often assumed that stimulation of these receptors alters glomerular permeability to water and macromolecules. Direct evidence for this assumption, however, has not been obtained. Infusion of ANG II in low doses has been shown to reduce the hydraulic permeability of the glomerulus.[6,7] This effect, along with the reduction of blood flow mediated by ANG II, keeps the glomerular filtration rate stable when glomerular capillary pressure rises during ANG II infusion. The observation that ANG II causes contraction of mesangial cells *in vitro* prompted the suggestion that ANG II reduces glomerular hydraulic permeability because contraction of mesangial cells shuts off perfusion of part of the glomerular capillary network.[9] Morphological studies, however, have repeatedly failed to document closure of glomerular capillary loops or reduction of glomerular capillary surface area in animals infused with ANG II.[10,11] While infusion of ANG II in low doses reduces glomerular hydraulic permeability, infusion of ANG II in high doses has been shown to increase glomerular permeability to macromolecules, leading to the appearance of protein in the urine.[12,13] Again, it has not been established that this functional change is mediated by stimulation of glomerular ANG II receptors. It should be noted that ANG II receptors may have effects on glomerular cells which are not directly related to the control of glomerular permeability.[14-16] The role of ANG II receptors found on the vasae rectae is still less well defined. Stimulation of these receptors appears not to cause vasoconstriction, as infusion of ANG II has been shown to increase vasa recta blood flow.[17]

1.3. Effects of Endogenous ANG II Revealed by Receptor Antagonism

It is important to emphasize that the physiological effects of ANG II on renal function are probably not exactly replicated by infusion of exogenous ANG II. This is particularly the case since ANG II is both produced and degraded within the kidney, and concentrations of ANG II may vary at different locations in the kidney.[18,19] The best available means to measure the effects of endogenous ANG II is probably to assess the effects of ANG II blockade on kidney function. Until recently, ANG II blockade has been accomplished with peptide ANG II receptor blockers or converting enzyme inhibitors. Studies with these agents have shown that endogenous ANG II, like exogenous ANG II, increases blood pressure while reducing renal blood flow and sodium excretion. These effects of ANG II are most prominent when extracellular fluid volume is low and circulating renin and ANG II levels are high.[20,21] Because both peptide receptor blockers and converting enzyme inhibitors have effects other than ANG II receptor blockade, it has been harder to assess the effect of endogenous ANG II on renal function when extracellular fluid volume is nearer normal. Antagonist actions of peptide ANG II receptor blockers may have obscured the effects of blocking low levels of ANG II in some studies. Conversely, potentiation of kinin and prostanoid activities by converting enzyme inhibitors may have augmented the effects of blocking low levels of ANG II in other studies. Both these problems can now be avoided by using nonpeptide receptor antagonists to block endogenous ANG II activity. Future studies with these agents should settle the question of whether ANG II influences renal function when sodium intake is liberal and circulating renin and ANG II levels are low. Studies with nonpeptide ANG II receptor blockers should also reveal whether ANG II has different effects as the renin–angiotensin system is progressively stimulated, so that proximal sodium reabsorption, for instance, is stimulated when ANG II levels are low, while intrarenal vasoconstriction is stimulated only when ANG II levels are higher.

At present, practically all renal ANG II receptors have been classified as AT_1, because losartan displaces bound ANG II throughout the kidney and blocks all measurable effects of ANG II on renal function.[22–26] A few exceptions to this generalization have been reported. AT_2 receptors have been found in the renal capsule; it is difficult to conceive of any physiological function for these receptors.[22] Binding studies have also detected AT_2 receptors in the renal vasculature of the rhesus monkey but not of other species.[25] Finally, functional studies have shown that PD 123177 as well as losartan inhibits ANG II-mediated sodium reabsorption in the proximal tubule.[6] This finding is provocative, but could reflect only blockade of AT_1 receptors by a high level of PD 123177. However, since tubular and vascular ANG II receptors have not so far been shown to be structurally identical, it can still be hoped that future developments in ANG II receptor pharmacology will make it possible to selectively block the vascular and tubular actions of ANG II. Knowledge that the vascular vasopressin receptor, which causes vasoconstriction by activating protein kinase C and increasing cell calcium, is distinct from the tubule vasopressin receptor, which promotes epithelial water transport by activating adenylyl cyclase, encourages hope that vascular and tubular ANG II receptors may prove to be distinct. Agents that blocked tubular but not vascular ANG II receptors would have potential clinical as well as physiological value. Experimentally, such agents could be used to distinguish the contributions of epithelial sodium reabsorption and arteriolar vasoconstriction of ANG II-mediated hypertension. Clinically, such agents might promote natriuresis with a minimal risk of orthostatic hypotension.

2. RECEPTOR ANTAGONISM IN EXPERIMENTAL RENAL DISEASE

2.1. Renal Ablation

Nonpeptide ANG II receptor blockers have facilitated study of the contribution of ANG II to the progression of renal disease as well as study of the influence of ANG II on normal renal function. Clinical reports have emphasized that progressive loss of renal function appears inevitable in patients with serious renal insufficiency. These findings suggest that after a certain point reduction of nephron number cause eventual failure of the remaining nephron units.[27,28] Experimentally, reduction in nephron number is most often induced by surgical ablation of renal tissue in the rat. Hypertension and progressive sclerosis of initially normal remnant glomeruli follow ablation of major portions of the renal mass in this species. Numerous studies have therefore employed the rat renal ablation model to identify factors potentially responsible for the progression of human renal insufficiency.[27,28]

One line of evidence suggests that ANG II contributes to remnant glomerular injury in the rat renal ablation model. Early studies showed that remnant glomerular sclerosis was proceeded by increases in glomerular capillary pressure and flow.[29] These "adaptive" increases in glomerular pressure and flow elevate the glomerular filtration rate (GFR) of single-remnant nephrons and thus partially offset the fall in total GFR caused by reduction in nephron number. Restriction of dietary protein was shown to lower remnant glomerular pressure and flow and retard the progression of remnant glomerular injury.[29] Because the increase in glomerular pressure observed in rats subjected to renal ablation was similar to the increase in glomerular pressure observed in normal rats infused with ANG II, the effect of converting enzyme inhibitors on remnant glomerular injury was also assessed. Doses of converting enzyme inhibitors that normalized systemic blood pressure were shown also to normalize glomerular capillary pressure and largely to prevent the development of remnant glomerular injury following renal ablation.[30,31]

The studies described above showed that converting enzyme inhibition prevents glomerular injury in renal-ablated rats even though these animals have low circulating renin levels. The availability of nonpeptide ANG II receptor blockers has made it possible to determine whether the beneficial effects of converting enzyme inhibition in renal-ablated rats are attributable to reduction in ANG II activity or to some other pharmacological effect of converting enzyme inhibition. A recent study from our laboratory employed losartan to address this question in male Munich Wistar rats subjected to five-sixth renal ablation.[32] Two weeks after ablation rats were divided into four groups that were then followed for a further period of 10 weeks prior to micropuncture and morphologic studies. Group 1 received no treatment and served as controls. Group 2 received losartan, 180 mg/liter, added to the drinking water. Group 3 received enalapril, 25 mg/liter, added to the drinking water. Group 4 received a combination of reserpine 3 mg/liter, hydralazine 40 mg/liter, and hydrochlorothiazide 13 mg/liter, added to the drinking water.

Results of systolic blood pressure measurements in this study are depicted in Fig. 1. The four groups were matched for blood pressure at 2 weeks after renal ablation. Untreated group 1 rats exhibited sustained hypertension with systolic blood pressure averaging 185 ± 4 mm Hg over the remainder of the study period. In contrast, average systolic blood pressure over the 10-week treatment period was reduced to 125 ± 2 mm Hg in group 2 rats, 127 ± 2 mm Hg in group 3 rats, and 117 ± 4 mm Hg in group 4 rats. Average blood pressure values for the three treated groups were not statistically different from

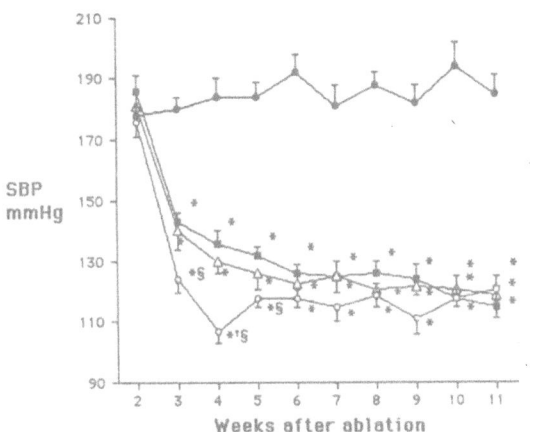

Figure 1. Systolic blood pressure (SBP) after renal ablation. Groups were matched for blood pressure at 2 weeks. Untreated group 1 rats (●) exhibited sustained hypertension over the remainder of the study. Blood pressure was reduced near to normal values by ANG II blockade in group 2 rats (△), by converting enzyme inhibition in group 3 rats (■), and by a combination of reserpine, hydralazine, and hydrochlorothiazide in group 4 rats (○). *,$p < 0.05$ group 2, 3, or 4 vs. group 1; †, $p < 0.05$ group 4 vs group 2; §, $p < 0.05$ group 4 vs. group 3. (From Lafayette *et al.*[32] Reprinted with permission.)

one another and were similar to blood pressure values observed in intact rats in our laboratory.

Mean values for arterial pressure under anesthesia ($\overline{AP}$), whole kidney GFR, single nephron GFR, glomerular pressure, and the glomerular ultrafiltration coefficient, K_f, are summarized in Table I. Values for $\overline{AP}$ paralleled values for systolic blood pressure in awake animals. Untreated group 1 rats exhibited marked hypertension with $\overline{AP}$ averaging 140 ± 5 mm Hg. ANG II receptor blockade reduced $\overline{AP}$ to 98 ± 4 mm Hg in group 2 rats, converting enzyme inhibition reduced $\overline{AP}$ to 99 ± 2 mm Hg in group 3 rats, and the combination regimen reduced $\overline{AP}$ to 101 ± 2 mm Hg in group 4 rats. Systemic hypertension was associated with elevation of glomerular transcapillary hydraulic pressure (ΔP) in untreated group 1 rats. The value of 49 ± 1 mm Hg for ΔP in these animals was similar to values observed in previous studies of rats subjected to renal ablation.[30,31] In contrast, ΔP averaged 42 ± 2 mm Hg in group 2 rats and 40 ± 2 mm Hg in group 3 rats. Thus control of systemic hypertension by ANG II receptor blockade and by converting enzyme inhibition was associated with the reduction of ΔP to values close to those observed in normal rats. Control of systemic hypertension by the combination regimen, however, was not associated with a similar reduction in ΔP. The average value for ΔP in group 4 rats was 47 ± 1 mm Hg. This value was significantly greater than the values for ΔP in groups 2 and 3 and only slightly and not significantly less than the value for ΔP in group 1.

Despite the lower systemic blood pressures, values for remnant kidney GFR were not reduced in the treated groups, averaging 0.91 ± 0.07 ml/min in group 2, 0.96 ± 0.06 ml/min in group 3, and 0.72 ± 0.05 ml/min in group 4, as compared to a value of 0.85 ± 0.06 ml/min in group 1. In accord with results

Table I. Summary of Renal Cortical Microcirculation Studies[a]

	$\overline{AP}$[b] (mm Hg)	P_{GC} (mm Hg)	ΔP (mm Hg)	GFR (ml/min)	SNGFR (nl/min)	FF (%)	Q_A (nl/min)	K_f nl/(s•mm Hg)
Group 1 (n = 11)	140 ± 5	65 ± 1	49 ± 1	$0.85 \pm .06$	118 ± 4	$0.29 \pm .02$	435 ± 38	83 ± 4
Group 2 (n = 7)	$98 \pm 4^*$	$56 \pm 2^*$	$42 \pm 2^*$	$0.91 \pm .07$	107 ± 6	$0.30 \pm .01$	358 ± 22	109 ± 11
Group 3 (n = 7)	$99 \pm 2^*$	$54 \pm 2^*$	$40 \pm 2^*$	$0.96 \pm .06$	112 ± 6	$0.28 \pm .01$	410 ± 31	$137 \pm 25^*$
Group 4 (n = 9)	$101 \pm 2^*$	$60 \pm 1^*\S$	$47 \pm 2\dagger\S$	$0.72 \pm .05$	111 ± 6	$0.27 \pm .01$	416 ± 32	$80 \pm 4\S$

[a]Mean values $\pm$ SEM.
[b]Abbreviations: $\overline{AP}$, mean arterial pressure; P_{GC}, mean glomerular capillary hydraulic pressure; ΔP, mean glomerular transcapillary hydraulic pressure difference; GFR, glomerular filtration rate; SNGFR, single nephron glomerular filtration rate; FF, filtration fraction; Q_A, glomerular plasma flow rate; K_f, glomerular capillary ultrafiltration coefficient. $^*p < 0.005$ group 2, 3, or 4 vs. group 1; $\dagger$group 4 vs. group 2; $\S$group 4 vs. group 3.

of prior studies, rats subjected to renal ablation exhibited prominent remnant nephron hyperfiltration with values for SNGFR increased to more than twice those observed in normal rats. Persistent remnant nephron hyperperfusion, attributable largely to reductions in afferent arteriolar resistance contributed to the maintenance of hyperfiltration when blood pressure was reduced by drug treatment. Increased values for the ultrafiltration coefficient, K_f, also contributed to maintenance of hyperfiltration in groups 2 and 3 treated with ANG II blockade and converting enzyme inhibition, but the increase in K_f attained statistical significance only in group 3. The mean value for K_f in group 4 rats receiving the combination regimen was practically identical to the mean value for K_f in group 1 rats receiving no treatment.

As expected, untreated group 1 rats with systemic and glomerular hypertension developed progressive glomerular injury manifested by increasing proteinuria (Fig. 2). Protein excretion in this group averaged 145 ± 22 mg/day at 6 weeks after ablation and 251 ± 23 mg/day at 11 weeks after ablation. ANG II blockade and converting enzyme inhibition afforded equal protection against the development of proteinuria. Thus protein excretion in group 2 rats averaged 52 ± 7 mg/day at 6 weeks after ablation and rose only to 85 ± 16 mg/day at 11 weeks after ablation. Similarly, protein excretion in group 3 rats averaged 39 ± 8 mg/day at 6 weeks after ablation and rose only to 67 ± 13 mg/day at 11

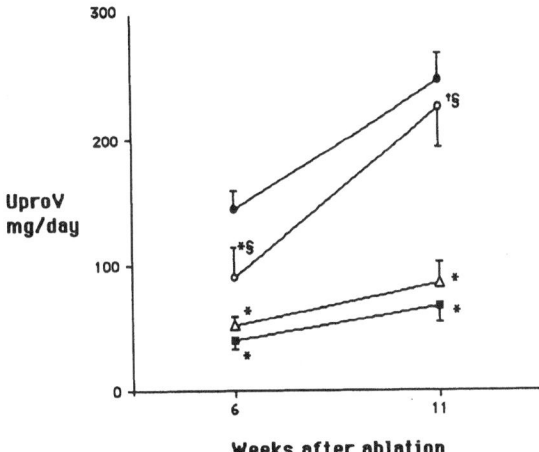

Figure 2. Twenty-four-hour urine protein excretion (UproV) at 6 and 11 weeks after renal ablation. Untreated rats (group 1, ●) developed heavy proteinuria. ANG II blockade (group 2, △) and converting enzyme inhibition (group 3, ■) were equally effective in limiting development of proteinuria. The combination of reserpine, hydralazine, and hydrochlorothiazide (group 4, ○) was less effective in limiting proteinuria at 6 weeks after ablation and ineffective in limiting proteinuria at 11 weeks after ablation. *$p < 0.05$ group 2, 3, or 4 vs. group 1; §$p < 0.05$ group 4 vs. group 3. (From Lafayette *et al.*[32] Reprinted with permission.)

weeks after ablation. In contrast, reduction of blood pressure with the combination regimen did not afford effective protection against the development of proteinuria. At 6 weeks after ablation, protein excretion in group 4 rats averaged 90 ± 26 mg/day, a value intermediate between the values observed in groups 2 and 3 and the value observed in group 1. By 11 weeks after ablation, however, protein excretion in group 4 rats averaged 247 ± 34 mg/day, a value close to that observed in group 1 rats receiving no treatment.

Morphological studies disclosed that the prevalence of segmental glomerular lesions was closely associated with the degree of proteinuria. Group 1 rats receiving no treatment exhibited segmental lesions in $41 \pm 3\%$ of glomeruli. The prevalence of segmental glomerular lesions was $9 \pm 1\%$ both in group 2 rats receiving losartan and in group 3 rats receiving enalapril, indicating that ANG II blockade and converting enzyme inhibition provided equivalent protection against glomerular structural injury. In contrast, the prevalence of segmental glomerular lesions was $33 \pm 6\%$ in group 4 rats, indicating that reduction of blood pressure with the combination regimen provided significantly less, if any, protection against glomerular injury.

Overall, in the study described above the effects of ANG II receptor blockade were not distinguishable from those of converting enzyme inhibition. This study thus strongly suggests that converting enzyme inhibition prevents glomerular injury by reducing ANG II activity in rats subjected to renal ablation. Results of this study do not, however, prove that ANG II activity is increased following renal ablation. It is possible that renal ablation induces a primary disturbance in some other vasomotor control system, but that the effects of this disturbance on blood pressure and glomerular structure do not develop when ANG II activity is reduced. Alternatively, hypertension and glomerular injury in rats subjected to renal ablation could be caused by activation of the renin angiotensin system. Early investigators showed that circulating renin levels are not elevated in rats with reduced nephron number. The hypothesis that increased renin release causes hypertension in these animals was therefore largely abandoned. It has since been recognized, however, that maintenance of "normal" values for circulating renin activity in the setting of increased blood pressure or sodium excretory load may reflect activation of the renin–angiotensin system.[33,34] Thus failure of plasma renin levels to decrease despite a prominent increase in blood pressure in rats subjected to renal ablation could reflect an increase in the activity of the circulating renin–angiotensin system in these animals. It is notable that plasma renin levels are markedly increased when antihypertensive agents that have no direct effect on the renin angiotensin system are used to lower blood pressure in renal ablated rats.[35] These findings suggest that enalapril and losartan could maintain normal blood pressure following renal ablation by blocking the effect of circulating renin activity.

An additional possibility is that converting enzyme inhibition and ANG II receptor blockade prevent an increase in the intrarenal activity of the renin–angiotensin system following renal ablation. It has recently been recognized that local activity of the renin–angiotensin system may be increased without alteration of circulating renin levels.[36,37] Within the kidney, renin is released into the perivascular interstitium, and renin concentrations in renal lymph may exceed those in plasma.[38] Angiotensinogen is also produced within the kidney, and intrarenal angiotensinogen synthesis may modulate intrarenal ANG II production independent of circulating renin activity.[37] These observations have encouraged speculation that intrarenal ANG II activity promotes hypertension and progressive glomerular injury in renal disease.[39,40] Proof that the activity of the intrarenal renin angiotensin system is increased in rats subjected to renal ablation has not yet been obtained. Recent studies suggest, however, that systemic and glomerular hypertension in rats subjected to renal ablation is caused by hypoperfusion of nephrons adjacent to areas of renal scarring.[41] Hypoperfusion of this nephron population could presumably stimulate release of renin into the renal interstitium and thereby cause an increase in intrarenal ANG II activity. We have recently found that administration of losartan acutely increases sodium excretion without lowering glomerular pressure in rats subjected to renal ablation.[42]

In the study described above, glomerular capillary pressure was reduced along with systemic blood pressure in renal-ablated rats receiving enalapril and losartan. Glomerular transcapillary pressure remained elevated, however, when blood pressure was reduced with a combination of reserpine, hydralazine, and hydrochlorothiazide. Findings in rats treated with enalapril and with the combination regimen were similar to those originally reported by Anderson *et al.*[35] The reason that these antihypertensive regimens have different effects on glomerular function remains to be elucidated. It has recently been shown that acute reduction of renal perfusion pressure by aortic constriction reduces glomerular capillary pressure in rats subjected to renal ablation.[43] Anderson *et al.*[35] showed that sustained reduction of systemic pressure by administration of reserpine hydralazine, and hydrochlorothiazide is associated with elevation of plasma renin levels as well as with maintenance of glomerular hypertension in this disease model. Together, these results suggest that glomerular transcapillary pressure might initially be reduced by any therapy that reduces systemic blood pressure but that blockade of ANG II activity may be necessary for sustained reduction of glomerular transcapillary pressure along with systemic blood pressure in rats subjected to renal ablation.

Finally, it should be emphasized that the current study does not identify the mechanism by which reducing ANG II activity preserves remnant glomerular structure. Protection against glomerular sclerosis was associated with reduction of glomerular transcapillary pressure both in rats receiving losartan and in rats receiving enalapril. These findings are consistent with the hypothesis that glomerular injury in rats subjected to renal ablation is caused by capillary hypertension.[30,31] The effects of ANG II on the glomerulus, however, are not limited to increasing transcapillary pressure. *In vitro* studies have shown that ANG II causes growth of smooth muscle and mesangial cells and may cause proliferation of these cells.[14–16] A sustained increase in intrarenal ANG II activity could thus cause progressive mesangial expansion. The finding that glomerular epithelial cells have ANG II receptors suggests that ANG II could also have a direct effect on glomerular barrier function.[8] At present, it is not possible to selectively block any one of these actions of ANG II without also blocking the others. It is thus not possible to determine whether ANG II receptor blockade protects remnant glomeruli by lowering glomerular capillary pressure or by preventing some other action of ANG II in the remnant kidney.

2.2. Diabetes Mellitus

Diabetic nephropathy is the most common, and therefore the most thoroughly studied, cause of renal failure in the Western world. The course of diabetic nephropathy is characterized by an interval of many years between the onset of diabetes and the appearance of glomerular injury characterized by proteinuria. Renal function and structure, however, are not normal during this interval. There is an early increase in the GFR, accompanied by an increase in kidney size.[44,45] Similar increases in GFR and kidney size have been observed in rats with diabetes induced by the β-cell toxin streptozotocin. Micropuncture studies have shown that elevation of GFR in these animals is attributable to increases in glomerular pressure and flow similar to those observed in rats subjected to renal ablation.[46] Moreover, early increases in capillary pressure and flow are followed by development of proteinuria and glomerular sclerosis in diabetic rats since they are in renal-ablated rats. In diabetes, however, the early increase in capillary pressure is not attended by systemic hypertension.

The finding that glomerular injury in diabetic rats is associated with an increase in capillary pressure similar to that observed after renal ablation prompted examination of the effects of converting enzyme inhibition in the diabetic model. Converting enzyme inhibition was shown to normalize glomerular pressure and largely to prevent development of proteinuria and glomerular sclerosis in rats with streptozotocin diabetes.[47] Subsequent studies showed that blood pressure-lowering agents that do not reduce capillary pressure are less effective than converting enzyme inhibitors in preventing glomerular injury in rats with diabetes as well as in rats subjected to renal ablation.[48,49]

Anderson *et al.*[50,51] have recently used losartan to examine whether converting enzyme inhibitors lower glomerular pressure in experimental diabetes by blocking ANG II activity in the diabetic kidney. They showed that glomerular pressure in diabetic rats was reduced to normal by both acute and chronic administration of losartan. Reduction of glomerular pressure was accompanied by an increase in glomerular hydraulic permeability and by maintenance of renal blood flow, so that GFR remained stable during ANG II receptor blockade. These results form an interesting contrast to those obtained with losartan in the renal ablation model in which acute ANG II receptor blockade causes an increase in sodium excretion but no decrease in glomerular capillary pressure.[42] They suggest that intrarenal ANG II activity is high enough to cause a sustained increase in efferent arteriolar tone in experimental diabetes. The mechanism by which intrarenal ANG II activity is increased in diabetes remains to be elucidated. Circulating renin levels are normal or low early in the course of both experimental and clinical diabetes.[52,53] Glomerular ANG II receptor number is also normal or low in experimental diabetes, and the majority of studies have failed to reveal increases in mRNA for renin or angiotensinogen in the kidneys of diabetic rats.[52,54] It remains possible that a highly localized increase in ANG II production causes increased efferent arteriolar tone in diabetes or that reduced ANG II degradation increases intrarenal ANG II activity in this disorder.

It should be noted that the diabetic rat does not provide an exact model of human diabetic nephropathy. In particular, though rats maintained diabetic for 1 year develop segmental glomerular sclerosis, they do not exhibit the prominent mesangial expansion observed in patients who develop renal disease after 20 years of diabetes. However, converting enzyme inhibitors have been shown to reduce proteinuria to slow disease progression in patients with diabetic nephropathy,[55,56] suggesting that ANG II may contribute to glomerular injury in clinical as well as in experimental diabetes. The observations will undoubtedly stimulate further studies with ANG II receptor blockers designed to elucidate the effects of diabetes on intrarenal ANG II activity.

2.3. Nephrosis

An important additional group of renal diseases are those characterized by heavy proteinuria, or nephrosis. A common finding in these diseases is that if proteinuria is sustained, the risk of renal failure increases with the amount of protein in the urine. It is therefore widely assumed that therapies that reduce proteinuria will prevent progressive loss of renal function in nephrotic patients, and extensive efforts have been made to identify such therapies. It should be emphasized that heavy proteinuria may be caused by a variety of mechanisms, so that not all nephrotic diseases should be expected to respond equally to individual therapies. Among the best characterized models of proteinuric renal disease is the Heymann nephritis model, which exhibits morphological features similar to those observed in membranous nephropathy in humans. Prompted by reports that converting enzyme inhibition reduces proteinuria in nephrotic humans, Kaysen and associates[57] showed that converting enzyme inhibition markedly reduces proteinuria in rats with passive Heymann nephritis. Hutchison and co-workers[58,59] have subsequently used losartan to examine whether converting enzyme inhibition reduces proteinuria in Heymann nephritis by blocking ANG II activity. They found that losartan lowers fractional protein clearance in Heymann nephritis, but that it is less effective than enalapril, which not only reduces ANG II activity but also increases kinin activity.[58] Of note, losartan caused an increase in the GFR of nephrotic rats, which was not seen with enalapril. Further studies showed that other agents that increase kinin activity also reduce proteinuria in Heymann nephritis.[59] Together these findings indicate that ANG II receptor blockade reduces glomerular protein leakage in Heymann nephritis and that the effects of blocking ANG II activity are additive to those of increasing kinin activity.

The effects of ANG II receptor blockade on proteinuria have thus far been examined in only a limited number of disease models other than Heymann nephritis. Yoshioka *et al.*[60] showed that saralasin reduced the protein excretion rate in rats with unilateral renal vein constriction. Studies from our laboratory have shown that losartan reduces the protein excretion rate in rats subjected to five-sixths renal ablation and then left untreated for the first 4 weeks after operation.[61] In both these studies, ANG II receptor blockade reduced glomerular protein permeability and also reduced glomerular capillary pressure. These results do not establish, however, that reduction in glomerular pressure causes improvement in glomerular permselective function when ANG II activity is blocked. As noted above, there are AT_1 receptors on glomerular epithelial cells, and it has been suggested that ANG II alters glomerular macromolecule permeability by a direct effect on glomerular capillary wall structure. Studies in diabetic nephropathy have shown that converting enzyme inhibitors can reduce proteinuria without markedly altering blood pressure.[56,62] These studies have been taken to suggest that reduction in ANG II activity can improve glomerular permselective function without altering glomerular hemodynamic function. Recent studies by Heeg *et al.*[63] have shown that infusion of exogenous ANG II increases blood pressure and filtration fraction without reversing the antiproteinuric effect of converting enzyme inhibition in nephrotic humans. Taken together, these findings suggest either that the effects of ANG II activity on glomerular permselectivity are not mediated by effects of ANG II on glomerular pressure, or that sustained changes in glomerular pressure are required to alter glomerular permselectivity. As discussed by Heeg *et al.*,[63] further studies are required to establish whether changes in glomerular hemodynamic functions cause changes in glomerular permselectivity when ANG II activity is blocked.

2.4. Other Conditions

The effects of ANG II receptor blockade have been examined in a limited number of other renal diseases. Two recent studies have examined the effects of losartan in rat models of hypertension characterized by glomerular and vascular injury. Ribeiro et al.[64] have shown that losartan normalizes GFR and returns blood pressure part way to normal in rats with hypertension caused by inhibition of nitric oxide synthesis. Imamura et al.[65] have shown that the losartan normalizes blood pressure and increases GFR while reducing protein excretion from the unclipped kidney of rats with two-kidney one-clip hypertension. These reports add to the number of renal diseases in which ANG II receptor blockade may prove to have a therapeutic effect. The report of Ribeiro et al.[64] is particularly interesting in that it suggests that intrarenal ANG II antagonizes the effects of intrarenal nitric oxide and that ANG II receptor blockade could prove beneficial in the variety of disorders now being considered as possibly due to EDRF deficiency.

3. ANG II RECEPTOR ANTAGONISM IN CLINICAL RENAL DISEASE

Studies of the effects of losartan in clinical renal disease have only recently been initiated and results are not yet available. Converting enzyme inhibition, however, has been shown to reduce proteinuria in a variety of renal diseases,[56,63] and it seems likely that ANG II receptor antagonism will also have a beneficial effect in many of these disorders. A major clinical question will be whether ANG II receptor blockers offer any therapeutic advantages compared to converting enzyme inhibitors. One potential advantage of receptor blockers would be the capacity to selectively block new receptor subtypes, if such receptor subtypes are in the future shown to play different roles in the regulation of renal function.

ACKNOWLEDGMENTS. We are grateful to Brittmarie Anderssen and to Helen Kwan for expert technical assistance. The work has been supported by grants from the Research Service of the Veterans Administration, the NIH (DK 42093), and Merck Inc. Dr. Baboolal is the recipient of a Fellowship award from the National Kidney Foundation.

REFERENCES

1. Mujais SK, Kauffman S, Katz AI: Angiotensin II binding sites in individual segments of the rat nephron. J Clin Invest 77:315–318, 1986.
2. Mendelsohn FAO, Dunbar M, Allen A, et al: Angiotensin II receptors in the kidney. Fed Proc 45:1420–1425, 1986.

3. Liu F-Y, Gocan MC: Angiotensin II stimulation of hydrogen ion secretion in the rat early proximal tubule. Modes of action, mechanism, and kinetics. *J Clin Invest* 82:601–607, 1988.

4. Xie M-H, Liu F-Y, Wong PC, et al: Proximal nephron and renal effects of DuP 753, a nonpeptide angiotensin II receptor antagonist. *Kidney Int* 38:473–479, 1990.

5. Cogan MG, Liu F-Y, Wong PC, et al: Comparison of inhibitory potency by nonpeptide angiotensin II receptor antagonists PD 123177 and DuP 753 on proximal nephron and renal transport. *J Pharmacol Exp Ther* 259:687–691, 1991.

6. Blantz RC, Konnen KS, Tucker BJ: Angiotensin II effects upon the glomerular microcirculation and ultrafiltration coefficient of the rat. *J Clin Invest* 57:419–434, 1976.

7. Steinhausen M, Endlich K, Wiegman D: Glomerular blood flow. *Kidney Int* 38:769–784, 1976.

8. Bianchi C, Gutkowska J, Thibault G, et al: Distinct localization of atrial natriuretic factor and angiotensin II binding sites in the glomerulus. *Am J Physiol* 251:F594–F602, 1986.

9. Ausiello DA, Kreisberg JI, Roy C, et al: Contraction of cultured rat glomerular cells of apparent mesangial origin after stimulation with angiotensin II and arginine vasopressin. *J Clin Invest* 65:754–760, 1980.

10. Haley DP, Sarrafian M, Bulger RE, et al: Structural and functional correlates of effects of angiotensin-induced changes in rat glomerulus. *Am J Physiol* 253:F111–F119, 1987.

11. Denton KM, Fennessy PA, Alcorn D, et al: Morphometric analysis of the actions of angiotensin II on renal arterioles and glomeruli. *Am J Physiol* 262:F367–F372, 1992.

12. Eisenbach GM, Van Liew JB, Boylan JW: Effect of angiotensin on the filtration of protein in the rat kidney: a micropuncture study. *Kidney Int* 8:80–87, 1975.

13. Olivetti G, Kithier K, Giacomelli F, et al: Characterization of glomerular permeability and proteinuria in acute hypertension in the rat. *Kidney Int* 25:599–607, 1984.

14. Fuzibayashi M, Fujiwara Y, Fukunaga M, et al: Angiotensin II stimulates interleukin-6 release from cultured mouse mesangial cells. *J Am Soc Nephrol* 1:454, 1990.

15. Fujiwara Y, Takama T, Shin S, et al: Angiotensin II stimulates mesangial cell growth through phosphoinositide cascade. *Kidney Int* 35:172, 1989.

16. Naftilan AJ, Pratt RE, Dzau VJ: Induction of platelet-derived growth factor A-chain and *c-myc* gene expressions by angiotensin II in cultured rat vascular smooth muscle cells. *J Clin Invest* 83:1419–1424, 1989.

17. Nobes MS, Harris PJ, Yamada H, et al: Effects of angiotensin on renal cortical and papillary blood flows measured by laser-Doppler flowmetry. *Am J Physiol* 261:F998–F1006, 1991.

18. Rosivall L, Narkates AJ, Oparil S, et al: De novo intrarenal formation of angiotensin II during control and enhanced renin secretion. *Am J Physiol* 252:F1118–F1123, 1987.

19. Seikaly MG, Arant Jr BS, Seney Jr FD: Endogenous angiotensin concentrations in specific intrarenal fluid compartments of the rat. *J Clin Invest* 86:1352–1357, 1990.

20. Levens NR, Peach MJ, Carey RM: Role of the intrarenal renin–angiotensin system in the control of renal function. *Circ Res* 48:157–167, 1981.

21. Wang YX, Gavras I, Wierzba T, et al: Comparison of systemic and regional hemodynamic effects of a diuretic, and angiotensin II receptor antagonist, and an angiotensin-converting enzyme inhibitor in conscious renovascular hypertensive rats. *J Lab Clin Med* 119:267–272, 1992.

22. Herblin WF, Chiu AT, McCall DE, et al: Angiotensin II receptor heterogeneity. *Am J Hypertens* 4:299S–302S, 1991.

23. Edwards RM, Stack EJ, Weidley EF, et al: Characterization of renal angiotensin II receptors using subtype selective antagonists. *J Pharmacol Exp Ther* 260:933–938, 1992.

24. Chansel D, Czekalski S, Pham P, et al: Characterization of angiotensin II receptor subtypes in human glomeruli and mesangial cells. *Am J Physiol* 262:F432–F441, 1992.

25. Gibson RE, Thorpe HH, Cartwright ME, et al: Angiotensin II receptor subtypes in renal cortex of rats and rhesus monkeys. *Am J Physiol* 261:F512–F518, 1991.

26. Sechi LA, Grady EF, Griffin CA, et al: Distribution of angiotensin II receptor subtypes in rat and human kidney. *Am J Physiol* 262:F236–F240, 1992.

27. Brenner BM, Meyer TW, Hostetter TH: Dietary protein intake and the progressive nature of kidney disease. *N Engl J Med* 307:652–659, 1982.

28. Olson JL, Heptinstall RH: Biology of disease. Nonimmunologic mechanisms of glomerular injury. *Lab Invest* 59:564–578, 1988.

29. Hostetter TH, Olson JL, Rennke HG, et al: Hyperfiltration in remnant nephrons: A potentially adverse response to renal ablation. *Am J Physiol* 241:F85–F93, 1981.

30. Anderson S, Meyer TW, Rennke HG, et al: Control of glomerular hypertension limits glomerular injury in rats with reduced renal mass. *J Clin Invest* 76:612–619, 1985.

31. Meyer TW, Anderson S, Rennke HG, et al: Reversing glomerular hypertension stabilizes established glomerular injury. *Kidney Int* 31:752–759, 1987.

32. Lafayette RA, Mayer G, Park SK, et al: Angiotensin II receptor blockade limits glomerular injury in rats with reduced renal mass. *J Clin Invest* 90(3):766–771, 1992.

33. Warren DJ, Ferris TF: Renin secretion in renal hypertension. *Lancet* 1:159–162, 1970.

34. Chapman AB, Johnson A, Gabow PA, et al: The renin–angiotensin–aldosterone system and autosomal dominant polycystic kidney disease. *N Engl J Med* 323:1091–1095, 1990.

35. Anderson S, Rennke HG, Brenner BM: Therapeutic advantage of converting enzyme inhibitors in arresting progressive renal disease associated with systemic hypertension in the rat. *J Clin Invest* 77:1993–2000, 1986.

36. Navar GN, Rosivall L: Contribution of the renin–angiotensin system to the control of intrarenal hemodynamics. *Kidney Int* 25:857–868, 1984.

37. Ingelfinger JR, Pratt RE, Ellison K, et al: Sodium regulation of angiotensinogen mRNA expression in rat kidney cortex and medulla. *J Clin Invest* 78:1311–1315, 1986.

38. Kriz W: A periarterial pathway for intrarenal distribution of renin. *Kidney Int* 31:S51–S56, 1987.

39. Sealey JE, Blumenfeld JD, Bell GM, et al: On the renal basis for essential hypertension: Nephron heterogeneity with discordant renin secretion and sodium excretion causing a hypertensive vasoconstriction-volume relationship. *J Hypertens* 6:763–777, 1988.

40. Rosenberg ME, Kren SM, Hostetter TH: Effect of dietary protein on the renin–angiotensin system in subtotally nephrectomized rats. *Kidney Int* 38:240–248, 1990.

41. Meyer TW, Rennke HG: Progressive glomerular injury after limited renal infarction in the rat. *Am J Physiol* 254:F856–F862, 1988.

42. Baboolal K, Meyer TW: Acute angiotensin II (AII) receptor blockade does not reverse systemic or glomerular capillary hypertension in remnant kidney rats (abstract). *J Am Soc Nephrol* 3(3):557, 1992.

43. Pelayo JC, Westcott JY: Impaired autoregulation of glomerular capillary hydrostatic pressure in the rat remnant nephron. *J Clin Invest* 88:101–105, 1991.

44. Mogensen CE, Østerby R, Gundersen HJG: Early functional and morphologic vascular renal consequences of the diabetic state. *Diabetologia* 17:71–76, 1979.

45. Mogensen CE, Schmitz O: The diabetic kidney: From hyperfiltration and microalbuminuria to end-stage renal failure. *Med Clin N Am* 72:1465–1492, 1988.

46. Hostetter TH, Troy JL, Brenner BM: Glomerular hemodynamics in experimental diabetes mellitus. *Kidney Int* 19:410–415, 1981.

47. Zatz R, Dunn R, Meyer TW, et al: Prevention of diabetic glomerulopathy by pharmacological amelioration of glomerular capillary hypertension. *J Clin Invest* 77:1925–1930, 1986.

48. Anderson S, Rennke HG, Garcia DL, et al: Short and long term effects of antihypertensive therapy in the diabetic rat. *Kidney Int* 36:526–536, 1989.

49. Fujihara CK, Padilha RM, Zatz R: Glomerular abnormalities in long-term experimental diabetes. Role of hemodynamic and nonhemodynamic factors and effects of antihypertensive therapy. *Diabetes* 41:286–293, 1992.

50. Anderson S, Bouyounes B, Clarey LE, et al: Intrarenal renin–angiotensin system (RAS) in experimental diabetes (abstract). *J Am Soc Nephrol* 1:621, 1990.

51. Anderson S, Wu Z, Reams GP, et al: Discordant renin–angiotensin II (ANG II) axis in diabetic rats (abstract). *J Am Soc Nephrol* 2:289, 1991.

52. Ballermann BJ, Skorecki KL, Brenner BM: Reduced glomerular angiotensin II receptor density in early untreated diabetes mellitus in the rat. *Am J Physiol* 247:F110–F116, 1984.

53. Bjorek S, Aurell M: Diabetes mellitus, the renin angiotensin system, and angiotensin converting enzyme inhibition. *Nephron* 55(Suppl 1):10–20, 1990.

54. Correa-Rotter R, Hostetter TH, Rosenberg ME: Renin and angiotensinogen gene expression in experimental diabetes mellitus. *Kidney Int* 41:796–804, 1992.

55. Parving H-H, Hommel E, Damkjær Nielsen M, et al: Effect of captopril on blood pressure and kidney function in normotensive insulin dependent diabetics with nephropathy. *Br Med J* 299:533–536, 1989.

56. Mathiesen ER, Hommel E, Giese J, et al: Efficacy of captopril in postponing nephropathy in normotensive insulin dependent diabetic patients with microalbuminuria. *Br Med J* 303:81–87, 1991.

57. Hutchison FN, Schambelan M, Kaysen GA: Modulation of albuminuria by dietary protein and converting enzyme-inhibition. *Am J Physiol* 253:F719–F725, 1987.

58. Hutchison FN, Webster SK: Effect of ANG II receptor antagonist on albuminuria and renal function in passive Heymann nephritis. *Am J Physiol* 263:F311–318, 1992.

59. Hutchison FN, Martin VI, Jones H, et al: Differing actions of dietary protein and enalapril on renal function and proteinuria. *Am J Physiol* 258:F126–F132, 1990.

60. Yoshioka T, Miyarai T, Kon V, et al: Role for angiotensin II in an overt functional proteinuria. *Kidney Int* 30:538–545, 1986.

61. Mayer G, Lafayette RA, Oliver J, et al: Effects of angiotensin II receptor blockade on remnant glomerular permselectivity. *Kidney Int* 43:346–353, 1993.

62. Morelli E, Loon N, Meyer T, et al: Effects of converting-enzyme inhibition on barrier function in diabetic glomerulopathy. *Diabetes* 39:76–82, 1990.

63. Heeg JE, De Jong PE, Van der Hem GK, et al: Angiotensin II does not acutely reverse the reduction of proteinuria by long-term ACE inhibition. *Kidney Int* 40:734–741, 1991.

64. Ribeiro MO, Antunes E, de Nucci G, et al: Chronic inhibition of nitric oxide synthesis. A new model of arterial hypertension. *Hypertension* 20:298–303, 1992.

65. Imamura A, Mackenzie HS, Hutchison FN, et al: Albuminuria and the effects of chronic renin–angiotensin system inhibition in 2-kidney, 1 clip hypertensive rats (abstract). *J Am Soc Nephrol* 3(3):521, 1992.

19

Angiotensin Receptors and Renal Function

David P. Chan and John C. Burnett, Jr.

1. INTRODUCTION

Angiotensin II (ANG II) is an important and potent regulator of renal function. With the discovery of specific types of angiotensin receptors and the availability of specific receptor antagonists, a body of information is evolving to describe the specific function of these important receptors. Along with insights in the physiological actions of ANG II on these receptors, variability in the type and localization of ANG II receptors within the renal system has been described. In addition, a maturational difference in the type of receptors exists in the development of the kidney.

Many reviews are available describing the actions of ANG II on the kidney.[1-5] The purpose of this chapter is to discuss the recent reports regarding the recently identified ANG II receptors in the kidney. Specifically, the distribution and physiological actions of ANG II receptors at the various sites within the renal system will be discussed.

2. RECEPTOR TYPES

Heterogeneity of ANG II receptors have been suggested by Douglas.[3] Specific identification of these receptors was achieved by binding studies using

David P. Chan • Division of Cardiology, Children's Hospital, Ohio State University, Columbus, Ohio 43205.   *John C. Burnett, Jr.* • Department of Cardiovascular Diseases, Mayo Clinic, Rochester, Minnesota 55905.

Angiotensin Receptors, edited by Juan M. Saavedra and Pieter B.M.W.M. Timmermans. Plenum Press, New York, 1994.

rat adrenal glands.[6] The two types of receptors described to date are AT_1 and AT_2. Subtypes of AT_1 receptors have also been described, AT_{1a} and AT_{1b}. This chapter will discuss the function of these receptors as a common type, i.e., AT_1. The receptors were characterized primarily by using the specific binding properties of two nonpeptide antagonists, DuP 753 (losartan) and PD 123177. Both nonpeptides are potent antagonists, with high affinity, specificity, and selectivity for the AT_1 and AT_2 receptors respectively.[7–11] Clones of the AT_1 receptors have been isolated and the cDNA sequenced.[12] The peptide is made up of 359 amino acids.

Both AT_1 and AT_2 receptors exist within the renal system (Table I). Species variability have been reported. The concentration of receptor sites within the kidney are not uniformly distributed. Indeed, localization of receptors are consistent with the described renal actions of ANG II. In rat and human adult kidneys, binding of ANG II to renal receptors was almost completely inhibited by losartan.[13–15] In contrast, no significant change in peptide binding was noted in the presence of PD 123177. This would suggest that the predominant ANG II receptor in the rat and human adult kidney is the AT_1 type. Subsequent reports have suggested that a small population of AT_2 receptors may exist on large preglomerular vessels in adult human kidneys.[15] Studies using rhesus monkey's adult kidney cortex have identified both AT_1 and AT_2 receptors.[15] Specifically, AT_2 receptors have been localized to the interlobar arteries and portions of the juxtaglomerular apparatus. The significance of this variability among species remains unclear.

Using rat renal tissue from fetuses and neonatal pups at various stages of development, Grady et al.[16] reported a marked increase in ANG II binding to AT_2 type receptors with increasing gestation. The binding affinity of these receptors declined at 12 hr postparturition. In contrast to adult renal tissue, very

Table I. Distribution and Function of Renal ANG II Receptor Subtypes

Location	Subtype	Function
Glomerulus	AT_1	GFR, TGF
Mesangial cells	AT_1	Kf
Vasculature		
Large vessels	AT_2	?
Efferent and afferent arterioles	AT_1	Vasoconstriction
Vasa recta	AT_1	?Vasoconstriction
Proximal tubules	AT_1, AT_2	Chloride, bicarbonate, and water reabsorption
Juxtaglomerular apparatus	AT_1	Renin release
Adrenal glands	AT_1	Mineralcorticoid release
Renal pelves	AT_1	?
Fetal kidney	AT_2	?Fetal renal development

Adapted from Edwards and Aiyar,[28] with permission.

little effect on ANG II binding was noted in the presence of losartan. Binding was reduced with PD 123177. This would suggest the relative absence of AT_1 type receptors during renal development. Gröne et al.,[17] analyzing kidneys of human fetuses of 17 weeks gestation or older, also found a predominance of AT_2 type receptors with high binding affinity. The receptors were diffusely located throughout the cortex and the medulla. The predominance and the binding characteristics of these receptors suggest that AT_2 receptors may play an important role in renal development. Future studies with long-term in utero AT_2 receptors blockade may aid in addressing this question.

ANG II also regulates renal function via its actions on the adrenal gland. Adult rabbit adrenal glands have been reported to have predominately AT_1 receptors.[18] Subsequent studies by Herblin et al.[6] identified a low population of AT_2 receptors. Interestingly, the regulation of AT_1 receptor mRNA expression in the adrenal gland of rats are dependent on the renin–angiotensin system.[19] This amplification of AT_1 receptors in the presence of ANG II in the adrenal gland would potentiate the renal actions of this peptide by increasing aldosterone secretion.

3. LOCALIZATION OF RECEPTORS

Using autoradiographic labeling of kidney tissue, localization of ANG II receptors have been achieved. Binding of radiolabeled ANG II was demonstrated in both the cortex and medulla of the kidney.[6,20] In the cortex, high concentrations of binding sites were noted in the glomerular tufts in various mammalian tissue. Isolating human glomeruli and mesangial cells, saturation kinetic studies found nearly the same response in both preparations.[21] This would suggest that the binding sites in the glomerulus are almost exclusively on mesangial cells. However, the concentration of the receptors per milligram of tissue was tenfold higher in the isolated glomeruli than in mesangial cells. This difference was proposed to be secondary to modification of the phenotypic expression of ANG II binding sites. It is unclear if this difference may also be due to the binding sites on the juxtaglomerular apparatus and proximal tubules. Gibson et al.[15] reported in rhesus monkey kidneys, high-density ANG II binding to the juxtaglomerular apparatus, interlobular arterial smooth muscle, and diffuse binding to the renal tubules.[15]

High concentrations of ANG II receptors are located on the tubules.[22,23] The observed biological actions with both AT_1 and AT_2 receptor antagonism have suggested both types of receptors are present. However, autoradiographic labeling have detected only the presence of AT_1 receptors on renal tubules.[24]

Binding of ANG II in the medulla was most concentrated in the outer medulla in longitudinal bands of the inner stripe that represent the medullary vascular bundles.[25] This would correspond specifically to the vasa recta. Very little binding was observed in the remaining portion of the medulla or the papilla.

Additional ANG II receptors were found in the medial layer of large renal vessels (AT_2)[17] and the muscular layer of renal pelves(AT_1).[25] Nonspecific binding of ANG II was also demonstrated in the outer stripe of the outer medulla.[25] The biological significance of these binding sites have yet to be established.

4. FUNCTION OF AT_1 RECEPTORS

ANG II is involved in the basal regulation of renal function.[26] Using normal dogs, specific AT_1 receptor antagonism was associated with marked renal hemodynamic and excretory changes (Table II). Clark et al.[27] have found similar results. An increase in glomerular filtration rate (GFR) and renal blood flow (RBF) was measured. A diuretic and natriuretic effect was also induced. In part, this was due to a decrease in the reabsorption of sodium both at the proximal and distal tubules. This would suggest that ANG II regulates renal function primarily via the AT_1 receptors.[28]

4.1. Glomerulus

Exogenous ANG II administration results in marked decreases in GFR. Using the specific actions of losartan, AT_1 antagonism was associated with significant increases in GFR in the anesthetized animal. The actions of the nonpeptide to influence glomerular filtering may be due to several mechanisms. Indirectly, an increase in RBF and transglomerular hydrostatic pressure due to the presence of AT_1 receptors on renal vessels may have contributed to the

Table II. Summary of Renal Function with ANG II Antagonism[a]

	Baseline	C1	C2	C3
GFR (ml/min)[b]	19.7 ± 0.9	28.7 ± 3.3*	25.6 ± 1.1*	24.0 ± 1.8*
RBF (ml/min)	151 ± 20	184 ± 27*	191 ± 25*	192 ± 27*
RVR (mmHg/ml/	0.85 ± 0.10	0.67 ± 0.06*	0.64 ± 0.06*	0.65 ± 0.05*
UV (ml/min)	0.16 ± 0.05	0.59 ± 0.28*	0.57 ± 0.20*	0.54 ± 0.16*
UNaV (μEq/min)	31.2 ± 7.0	92.1 ± 21.1*	103.0 ± 23.8*	106.5 ± 24.1*
FENa (%)	1.10 ± 0.26	2.23 ± 0.59*	2.65 ± 0.61*	2.95 ± 0.69*
FELi (%)	31.2 ± 2.2	38.3 ± 3.5	40.1 ± 4.5*	44.0 ± 4.9*

[a]Values expressed as mean and sem; $n = 6$; C1, C2, and C3: with Losartan at 35, 50, 65 min.
[b]Abbreviations: GFR, glomerular filtration rate; RBF, renal blood flow; RVR, renal vascular resistance; UV, urine volume; UNaV, urinary sodium excretion; FENa, fractional excretion of sodium; FELi, fractional excretion of lithium. *$p < 0.05$ by ANOVA and Fisher's test for least significant difference.
Adapted from Chan et al.,[26] with permission.

observed effects of losartan. This will be discussed further below. An additional site for ANG II-related modulation of GFR is the attenuation of the tubulo-glomerular feedback system. Schnermann and Briggs[29] have reported ANG II directly influences the responsiveness of the tubuloglomerular feedback (TGF) system in normal rats. Therefore, despite the initial increases in the solute load to the distal tubules, the feedback mechanism that normally would have limited glomerular filtration was blocked in the presence of AT_1 receptor antagonism.

A more direct effect of ANG II on glomerular function is modulating the ultrafiltration coefficient (K_f). This is presumably due to the action of the peptide on mesangial cells. *In vitro* studies with exogenous administration of ANG II have been noted to result in marked contraction of mesangial cells.[2] This would be consistent with the heavy concentration of ANG II binding to the glomerulus with autoradiographic studies as described above.

4.2. Renal Vasculature

The potent action of ANG II on the renal vasculature has long been recognized. The proposed vascular sites of action by the peptide have included the afferent and efferent arterioles. The beneficial enhancement of renal function to low-dose ANG II have suggested that there is a difference in the sensitivity between the pre- and postglomerular vessels. Under nonpathological conditions, specific AT_1 antagonism resulted in a significant increase in renal blood flow and a decrease in whole kidney vascular resistance.[26] In contrast, blockade of AT_2 receptors with PD 123319 had no effect on renal hemodynamics.[30] This would suggest that the actions of ANG II on the renal vasculature is primarily via AT_1 receptors. Along with the increase in RBF, an increase in GFR with AT_1 antagonism could be due to an increase in the transglomerular hydrostatic pressure. This is consistent with a relative difference in the sensitivity in the afferent and efferent ANG II receptors to competitive anatagonism. Likewise, exogenous ANG II administration has resulted in greater vascular constriction in the efferent than the afferent vessels.[4,31] This is compatible with the concept that there is a greater population of occupied receptors by ANG II on the afferent arterioles in normal homeostasis.

The vasa recta plays an important role in reabsorbing the large volume of filtrate reclaimed from the loop of Henle. The flow through the vasa recta is relatively slow and low pressure. The primary force to reabsorption of the fluid is driven by the high colloidal pressure difference between the peritubular vessels and the interstitium. As discussed above, there is a significant population of ANG II receptors along these structures in the kidney. Presumably this will influence the ability of the nephron to concentrate urine. To date, the function of these receptors and how they regulate renal function at the vasa recta remain unclear. Studies have demonstrated a decrease in papillary blood flow with exogenous ANG II administration.[32] This may be secondary to contraction of pericytes that surround the vasa recta.

4.3. Juxtaglomerular Apparatus and Adrenal Gland

The juxtaglomerular apparatus is responsible for a large proportion of renal renin production. Identification of ANG II receptors would indicate a role for the peptide to stimulate renin release. *In vivo* studies with AT_1 antagonism resulted in significant changes in circulating plasma renin activity.

AT_1 receptors have been localized in the adrenal gland. These receptors are involved in the regulation of adrenal mineral corticoid secretion. This is consistent with the observed decrease in circulating levels of aldosterone with AT_1 receptor antagonism.[26,27]

4.4. Renal Tubules

The renal tubules are important in reclaiming the vital salts and water filtered by the glomerulus. The proximal tubules is responsible for reabsorbing 65% of the solute and water filtered by the glomerulus. This occurs primarily at the S_1 segment of the proximal tubules. The identification of high concentrations of ANG II receptors on the tubules would suggest a marked response to both exogenous ANG II administration and specific ANG II receptor antagonism. Indeed, a dose-response modulation of sodium and water excretion is evident with ANG II infusion. Likewise, specific blockade of AT_1 receptors resulted in marked increases in the urine flow and a decrease in reabsorption of sodium, chloride, and bicarbonate.[26] An increase in fractional excretion of sodium (FENa) would indicate the importance of AT_1 receptors in modulating whole kidney tubular handling of sodium. Associated with the increase in FENa, a significant increase in the fractional excretion of lithium was measured, suggesting specifically that the proximal tubules are in part involved in the observed modulation of sodium excretion. Cogan *et al.*[23] have found a similar dramatic response of the proximal cortical tubular cells to both AT_1 and AT_2 receptor antagonism in solute reabsorption. This would suggest a dual set of receptors that are ANG II responsive in modulating tubular handling of sodium.

5. FUNCTION OF AT₂ RECEPTORS

The renal action of AT_2 receptors remains unclear. To date, the primary function of AT_2 has been suggested to involve the renal tubules and free water handling. Cogan *et al.*,[23] using *in vivo* microperfusion techniques in the rat, demonstrated a reduction in bicarbonate, chloride, and water absorption by the proximal tubules in the presence of AT_2 antagonism (Fig. 1).

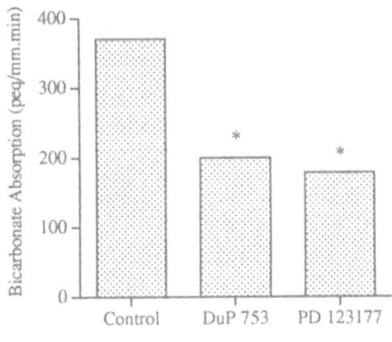

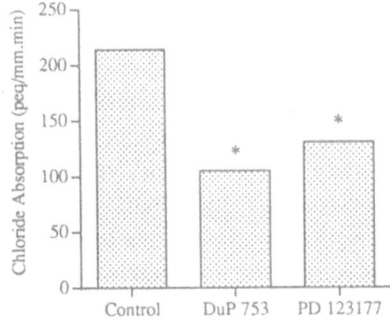

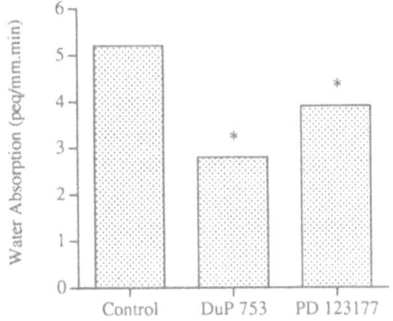

Figure 1. Proximal tubular absorption of bicarbonate, chloride, and water in the presence of vehicle (control), losartan (10mg/kg), and PD123177 (60 or 120 mg/kg). (Adapted from Cogan *et al.*,[23] with permission.)

A recent report by Keiser *et al.*,[30] using another specific AT_2 receptor antagonist, PD 123319, demonstrated a dose-dependent increase in urine volume and free water clearance in the anesthetized dog. Presumably this was due to the actions of the antagonist on AT_2 receptors located at the renal tubules. Concurrently, AT_2 antagonism was not associated with significant changes in renal hemodynamic function.

The predominance of AT_2 receptors in the fetal kidney poses many intriguing questions. Indeed, the physiological and developmental actions of ANG II on AT_2 receptors in the fetal kidney have not been elucidated. One could propose that the exclusive presence of AT_2 receptors could represent a role for ANG II in normal renal maturation.

6. SUMMARY

ANG II is a potent regulator of renal function. With the availability of specific antagonists, it has been demonstrated that the physiological actions of ANG II is predominately via AT_1 receptors. These receptors have been localized to the mesangial cells, afferent and efferent arterioles, vasa recta, renal tubules with majority of the receptors at the proximal tubules, juxtaglomerular apparatus, and the adrenal gland. AT_1 receptor antagonism is associated with an increase in glomerular filtration rate, renal blood flow, urine flow, total urinary sodium excretion, fractional excretion of sodium, and fractional excretion of lithium. In addition, modulation of plasma renin activity and aldosterone secretion was measured.

AT_2 receptors have been demonstrated to be involved in the physiological actions of the S_1 segment of the proximal tubules. Antagonism of these receptors resulted in a decrease in bicarbonate, chloride, and water reabsorption at the proximal tubules. No hemodynamic actions have been documented in the presence of AT_2 receptor antagonists.

REFERENCES

1. Blantz RC: The glomerular and tubular actions of angiotensin II. *Am J Kidney Dis* 10(1 Suppl 1):2–6, 1987.
2. Dworkin LD, Brenner BM: Hormonal modulation of glomerular function. *Am J Physiol* 244:F95–F104, 1983.
3. Douglas JG: Angiotensin receptor subtypes of the kidney cortex. *Am J Physiol* 253:F1–F7, 1987.
4. Hall JE: Control of sodium excretion by angiotensin II: Intrarenal mechanisms and blood pressure regulation. *Am J Physiol* 250:R960–R972, 1986.
5. Levens NR, Peach MJ, Carey RM: Role of the intrarenal renin–angiotensin system in the control of renal function. *Circulation* 48(2):157–167, 1981.
6. Herblin WF, Chiu AT, McCall DE, et al: Angiotensin II receptor heterogeneity. *Am J Hypertens* 4:299S–302S, 1991.
7. Chiu AT, Herblin WF, McCall DE, et al: Identification of angiotensin II receptor subtypes. *Biochem Biophys Res Commun* 165(1):196–203, 1989.
8. Chiu AT, McCall DE, Price WA, et al: Nonpeptide angiotensin II receptor antagonists. VII.

Cellular and biochemical pharmacology of DuP 753, an orally active antihypertensive agent. *J Pharmacol Exp Ther* 252(2):711–718, 1990.

9. Rhaleb N-E, Rouissi N, Nantel F, et al: DuP 753 is a specific antagonist for the angiotensin receptor. *Hypertension* 17:480–484, 1991.

10. Wong PC, Price WAJ, Chiu AT, et al: In vivo pharmacology DuP 753. *Am J Hypertens* 4(4):288S–298S, 1991.

11. Wong PC, Price WA, Chiu AT, et al: Nonpeptide angiotensin II receptor antagonists. VII. Characterization of functional antagonism displayed by DuP 753, an orally active antihypertensive agent. *J Pharmacol Exp Ther* 252(2):719–725, 1990.

12. Burns KD, Inagami T, Harris RC: Cloning of a rabbit kidney cortex AT_1 angiotensin II receptor that is present in proximal tubule epithelium. *Am J Physiol* 264(4 Pt 2):F645–654, 1993.

13. Sechi LA, Griffin CA, Grady EF, et al: Characterization of angiotensin II receptor subtypes in rat heart. *Circ Res* 71(6):1482–1489, 1992.

14. Chang RS, Lotti VJ: Angiotensin receptor subtypes in rat, rabbit and monkey tissues: Relative distribution and species dependency. *Life Sci* 49(20):1485–1490, 1991.

15. Gibson RE, Thorpe HH, Cartwright ME, et al: Angiotensin II receptor subtypes in renal cortex of rats and rhesus monkeys. *Am J Physiol* 261(3 Pt 2):F512–F518, 1991.

16. Grady EF, Sechi LA, Griffin CA, et al: Expression of AT_2 receptors in the developing rat fetus. *J Clin Invest* 88(3):921–933, 1991.

17. Gröne HJ, Simon M, Fuchs E: Autoradiographic characterization of angiotensin receptor subtypes in fetal and adult human kidney. *Am J Physiol* 262(2 Pt 2): F326–F331, 1992.

18. Dudley DT, Panek RL, Major TC, et al: Subclasses of angiotensin II binding sites and their functional significance. *Mol Pharmacol* 38(3):370–377, 1990.

19. Iwai N, Inagami T: Regulation of the expression of the rat angiotensin II receptor mRNA. *Biochem Biophys Res Commun* 182(3):1094–1099, 1992.

20. Yamada H, Sexton PM, Chai SY, et al: Angiotensin II receptors in the kidney. Localization and physiological significance. *Am J Hypertens* 3(3):250–255, 1990.

21. Chansel D, Czekalski S, Pham P, et al: Characterization of angiotensin II receptor subtypes in human glomeruli and mesangial cells. *Am J Physiol* 262(3 Pt 2):F432–441, 1992.

22. Cogan MG, Xie MH, Liu FY, et al: Effects of DuP 753 on proximal nephron and renal transport. *Am J Hypertens* 4(4 Pt 2):315S–320S, 1991.

23. Cogan MG, Liu FY, Wong PC, et al: Comparison of inhibitory potency by nonpeptide angiotensin II receptor antagonists PD 123177 and DuP 753 on proximal nephron and renal transport. *J Pharmacol Exp Ther* 259(2):687–691, 1991.

24. Edwards RM, Stack EJ, Weidley EF, et al: Characterization of renal angiotensin II receptors using subtype selective antagonists. *J Pharmacol Exp Ther* 260(3):933–938, 1992.

25. Sechi LA, Grady EF, Griffin CA, et al: Distribution of angiotensin II receptor subtypes in rat and human kidney. *Am J Physiol* 262(2 pt 2):F236–240, 1992.

26. Chan DP, Sandok EK, Aarhus LL, et al: Renal specific actions of angiotensin II receptor antagonism in the anesthetized dog. *Am J Hypertens* 5:354–360, 1992.

27. Clark KL, Robertson MJ, Drew GM: Role of angiotensin AT_1 and AT_2 receptors in mediating the renal effects of angiotensin II in the anaesthetized dog. *Br J Pharmacol* 109(1):148–156, 1993.

28. Edwards RM, Aiyar N: Angiotensin II receptor subtypes in the kidney. *J Am Soc Nephrol* 3:1643–1652, 1993.

29. Schnermann J, Briggs JP: Effect of angiotensin and other pressor agents on tubuloglomerular feedback responses. *Kidney Int Suppl* 30:S77–80, 1990.

30. Keiser JA, Bjork FA, Hodges JC, et al: Renal hemodynamic and excretory responses to

PD 123319 and losartan, nonpeptide AT_1 and AT_2 subtype-specific angiotensin II ligands. *J Pharmacol Exp Ther* 262(3):1154–1160, 1992.

31. Myers BD, Deen WM, Brenner, BM: Effects of norepinephrine and angiotensin II on the determinants of glomerular ultrafiltration and proximal tubule fluid reabsorption in the rat. *Circ Res* 37:101–110, 1975.

32. Nobes MS, Harris PJ, Yamada H, et al: Effects of angiotensin on renal cortical and papillary blood flows measured by laser-Doppler flowmetry. *Am J Physiol* 261(6 Pt 2):F998–1006, 1991.

20

Clinical Application
of Treatment with Angiotensin
Receptor Antagonists

Michel Burnier, Bernard Waeber, and Hans R. Brunner

1. INTRODUCTION

Today, interruption of the renin–angiotensin system with angiotensin-converting enzyme (ACE) inhibitors represents a well-accepted therapeutic approach to control high blood pressure and to treat patients with congestive heart failure. After more than 10 years of a wide clinical use, ACE inhibitors have not only turned out to be effective in reducing blood pressure of hypertensive patients and in improving survival of patients with congestive failure, they have also produced some interesting metabolic effects, such as raising potassium levels and restoring normal insulin sensitivity. In addition, they can alter the distribution of cardiac output, resulting in marked changes in cardiac, cerebral, and renal tissue perfusion with potential long-term beneficial effects that may go beyond those expected through the reduction of blood pressure. Thus, blood pressure reduction with ACE inhibitors has been shown to provide cardiac protection following myocardial infarction as well as renal protection in patients with diabetic or nondiabetic chronic renal failure. These favorable metabolic and tissue-protecting effects of ACE inhibitors have enlarged considerably their clinical applications.[1]

It is generally assumed that the main cardiovascular effects of ACE inhibitors can be attributed to blockade of angiotensin II (ANG II) generation.

Michel Burnier, Bernard Waeber, and Hans R. Brunner • Division of Hypertension and Cardiovascular Research Group, University Hospital, 1010 Lausanne, Switzerland.

Angiotensin Receptors, edited by Juan M. Saavedra and Pieter B.M.W.M. Timmermans. Plenum Press, New York, 1994.

However, since the converting enzyme is quite promiscuous and many other peptides are processed by this same enzyme, some of the therapeutic efficacy and perhaps also some of the side effects of ACE inhibitors might be attributed to angiotensin-independent mechanisms. The recent development of specific, orally active, nonpeptide ANG II receptor antagonists now offers the possibility to block the activity of the renin–angiotensin system by inhibiting the hormone from binding to its receptor.[2] Interestingly, blockade at the receptor level not only provides the most specific access to inhibition of the renin–angiotensin system, it also offers an opportunity to evaluate the real benefits linked to the sole blockade of the renin angiotensin system. Thus, if all properties of ACE inhibitors are indeed due to blockade of ANG II generation, ANG II receptor antagonists should a priori have similar hemodynamic, metabolic, and tissue protecting properties. In the following, the clinical experience acquired so far with the new ANG II receptor antagonist losartan in normotensive subjects as well as in hypertensive patients will be discussed in order to have a first insight into the potential clinical applications of this new class of agents.

2. ANGIOTENSIN II RECEPTOR BLOCKADE IN NORMOTENSIVE VOLUNTEERS

2.1. Pharmacodynamic–Pharmacokinetic and Dose-Finding Studies

The initial studies in healthy normotensive volunteers assessed the tolerance and inhibitory effect of losartan (Dup 753) on the pressor action of exogenous ANG I or ANG II. These studies were designed to determine the range of the effective doses of losartan and to characterize the pharmacodynamic–pharmacokinetic profile of losartan after single and repeated oral administration.

In the first protocol, single oral doses of losartan (2.5, 5, 10, 20, and 40 mg) were studied versus placebo.[3] A dose–response relationship was established with regard to systolic blood pressure responses to increasing intravenous (IV) doses of ANG I in each subject. The test dose of ANG I selected was one that increased systolic blood pressure by about 30 mm Hg. This dose was then used to define the baseline systolic blood pressure response to angiotensin and became the challenge dose for that subject. The effect of losartan was established by bolus injections of the challenge dose of ANG I given from 15 min to 27 hr after ingestion of losartan. Plasma drug levels were determined throughout the experiment.

The results of this evaluation revealed a dose-dependent increase in the degree of ANG II inhibition, with the 40-mg dose producing a peak inhibition of approximately 70%. Furthermore, losartan had a long duration of action since blood pressure response to ANG I was still reduced 24 hr after intake of the 20- and 40-mg doses (Fig. 1). Peak plasma drug levels were obtained 30 min after drug intake for all doses tested (10, 20, and 40 mg), although the highest inhibitory effects were seen 7, 5, and 3 hr after administration of the 10-, 20-, and 40-mg doses, respectively.

The second study of losartan in human volunteers was a multiple-dose tolerance study in which 5, 10, 20, or 40 mg losartan or a placebo was administered once a day for 8 days, again to male normotensive volunteers.[3] As for the first study, a dose–response curve was established before the losartan was administered, using ANG II this time. On days 1, 4, and 8 of the losartan administration, the volunteers were challenged with the preestablished doses of ANG II, before and 6 and 12 hr after drug intake; further doses of ANG II were injected 24 hr after the drug on day 1, and 24, 30, and 36 hr after the last dose on day 8.

Again, clear evidence of dose-dependent ANG II inhibition throughout the treatment period was obtained, although the pressor response to the exogenous ANG II was not completely blocked. On days 1, 4, and 8, the 6-hr postdrug effect was similar with both the 20- and the 40-mg doses, about 70% inhibition of the blood pressure increase. However, with the 40-mg dose, the trough (predrug) inhibition of ANG II increased continuously from day 1 to day 8, so that on day 8 the pressor response to ANG II was only 75% of the comparable response on day 1 (Fig. 2). This suggested that losartan had a carryover effect lasting more than 24 hr.

The results of these first two studies clearly suggested that ANG II inhibition continued long after plasma losartan levels had fallen to undetectable levels (5–13 hr after drug intake). The drug half-life was estimated at 1.5–3.3 hr for all subjects (mean ±SD, 2.1 ± 0.6). Meanwhile, it was demonstrated in animals that the drug produced an active metabolite, EXP3174, which exhibits an at least tenfold higher affinity to the ANG II receptor subtype 1 and a

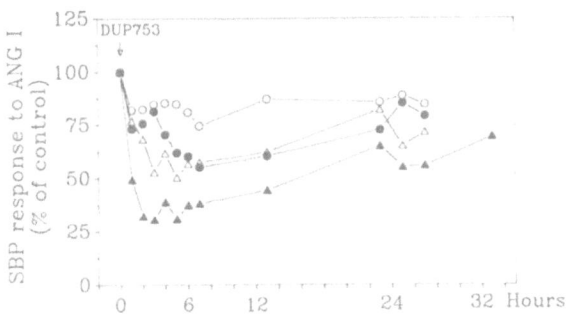

Figure 1. Effects of single doses of oral losartan [10 mg (●, $n = 6$); 20 mg (△, $n = 8$); 40 mg (▲, $n = 7$)] or placebo (○, $n = 6$) on systolic blood pressure (SBP) response to test doses of ANG I in healthy volunteers. (From Christen *et al.*[3] Reprinted with permission.)

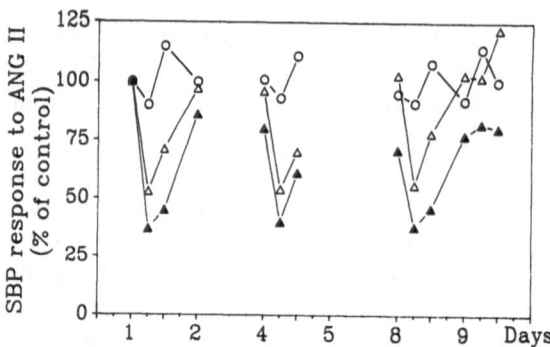

Figure 2. Effects of 8 consecutive days of treatment with two oral doses of losartan [20 mg (△, $n = 5$); 40 mg (▲, $n = 6$)] or placebo (○, $n = 6$) on systolic blood pressure (SBP) response to test doses of ANG II in healthy volunteers. (From Christen *et al.*[3] Reprinted with permission.)

longer half-life than the mother compound losartan.[4] For these reasons, a third study was designed in which higher doses of losartan (40, 80, and 120 mg) were administered and plasma levels of the drug and its active metabolite EXP3174 were determined.[5] ANG II blockade was assessed as described previously. A maximal blockade of the systolic blood pressure response to this peptide (> 80%) was produced with the 80-mg dose and no real gain was obtained with the 120-mg dose. Moreover, the duration of inhibition was not prolonged with the highest dose. The metabolite EXP3174 reached higher plasma concentrations and was eliminated more slowly than losartan. More interestingly, the plasma EXP3174 concentrations paralleled the profile of ANG II blockade more closely than did plasma levels of losartan.

Taken together, the results of these three studies have allowed us to estimate that the optimum dose of losartan probably ranges between 40 and 80 mg and that no real benefit is to be expected from higher doses. In addition, the ability of losartan to block the pressor response to exogenous ANG I or II for a prolonged period of time is due mainly to the sustained activity of its long-acting metabolite, EXP3174.

The pharmacokinetics of losartan were also evaluated in healthy male volunteers after single and multiple oral administration.[6] Thirty-three men received orally losartan in single rising doses (25, 50, 100, and 200 mg) or 100 mg/day for 7 days. Plasma and urinary concentrations of losartan and its active metabolite, EXP3174, were determined by high-pressure liquid chromatography (HPLC). Plasma concentrations of losartan were proportional to the dose over the range of 25 to 200 mg. The terminal half-lives were dose-independent and ranged from 1.5 to 2.5 hr. The plasma concentrations of EXP3174 were higher than those of losartan at all doses and the terminal half-life of the metabolite was two times longer than that of losartan. The pharmacokinetics of losartan and EXP3174 did not change after multiple dosing for 7 days.

2.2. Effects on the Renin–Angiotensin System

ANG II receptor blockade in human subjects is associated with a significant increase in plasma renin activity and a rise in plasma ANG II levels.[3,5–9] In one study, there was a relationship between the compensatory rise in circulating renin, and hence ANG II levels and the plasma concentration of the active metabolite.[5] In the studies cited above, the increases in plasma renin activity and in plasma ANG II were clearly dose-dependent and were greater after 8 days of treatment than on the first day. However, the changes in plasma renin activity and ANG II in these normotensive subjects were highly variable and could not be used to predict the inhibition in individual subjects. Plasma aldosterone levels did not decrease significantly with the small doses of losartan (10 to 40 mg/day) but with the higher doses of 80 and 120 mg a significant fall in aldosterone was found.[3,5]

So far, the clinical impact of the marked increase in plasma ANG II levels occurring during ANG II receptor blockade is not known. Theoretically, two consequences could be anticipated. First, because of the high specificity of losartan and its metabolite for the AT_1 subtype of ANG II receptors, the reactive increase in plasma ANG II may lead to an excessive stimulation of the unprotected AT_2 subtype of receptor. However, the function of this receptor subtype is not known. Another potential consequence might be a rebound effect with an increase in blood pressure upon drug withdrawal. So far no such phenomenon has been reported in normotensive or hypertensive subjects treated with losartan.

2.3. Effect on Blood Pressure

In normotensive volunteers maintained on a regular sodium diet and water intake, a single or repeated administration of losartan induced either no change or slight decreases in systolic and diastolic blood pressure.[3,5–7] Heart rate did not vary during angiotensin receptor blockade. In water-loaded normotensive subjects maintained on high (200 mmole/day) or low (50 mmole/day) sodium diet, losartan also had no effect on blood pressure and heart rate.[9] However, in normotensive subjects with a renin–angiotensin system activated by a combined salt and water depletion with a low sodium diet (40 mmole/day) and furosemide (40 mg twice daily) given for 3 days before evaluation, losartan induced dose-dependent decreases in supine and erect blood pressure.[8] The falls in blood pressure obtained with the higher doses of 50 and 100 g were statistically significant and were accompanied by a slight increase in heart rate. The results of these two last studies further emphasize the important role of volume and/or salt depletion in accentuating the hemodynamic effects of angiotensin receptor blockade.

2.4. Renal Effects of Losartan

Because ANG II exerts several regulatory functions in the kidney, including modulation of renal blood flow, glomerular filtration rate (GFR), glomerular permselectivity, tubular sodium reabsorption, and inhibition of renin, the renal effects of the ANG II antagonist losartan were also examined in normotensive subjects.[9] For this purpose, the renal hemodynamic and tubular effects of a single dose of losartan (100 mg) have been evaluated in normotensive volunteers maintained for 6 days on a low (50 mmole/day) and 6 days on a high sodium (200 mmole/day) diet according to a crossover design. As mentioned earlier, the subjects were studied in water-loaded conditions to ensure an adequate and spontaneous urine output. In these experimental conditions, losartan had no effect on renal blood flow or glomerular filtration rate whatever the sodium intake. In contrast to these results, an increase in renal blood flow was observed 90 min after the administration of losartan in another study involving normotensive volunteers.[10]

In salt-depleted subjects, the tubular effects of losartan were characterized by a slight increase in urine output but a marked increase in urinary sodium excretion. Fractional excretion of lithium did not change under losartan, suggesting no effect on proximal reabsorption in these experimental conditions. This natriuretic effect of losartan was accompanied by a transient increase in potassium, phosphate, and magnesium excretion. During salt repletion, the effects of losartan were similar but they were generally attenuated.

In a pilot study, Nakashima *et al.*[11] demonstrated a significant uricosuric effect of losartan in healthy male volunteers. Thus, in a single-dose study, administration of increasing doses of losartan (25, 50, 100, or 200 mg) produced a dose-dependent decrease in serum uric acid and increase in urinary uric acid excretion. The uricosuric effect was still present and of the same magnitude after 7 days of losartan administration. In volunteers studied on a high- and low-sodium diet, urinary uric acid excretion increased by 300% (Fig. 3). The uricosuric effect appeared to be independent of the degree of activity of the renin–angiotensin system and was comparable on both sodium diets. Recent results suggest that the ability of losartan to increase uric acid excretion is a property of the mother compound only. Indeed, infusion of the active metabolite EXP3174 did not increase uric acid excretion.

3. ANG II RECEPTOR BLOCKADE IN HYPERTENSION

Hypertension represents the main clinical application of drugs that inhibit the activity of the renin–angiotensin system. As mentioned earlier, ANG II

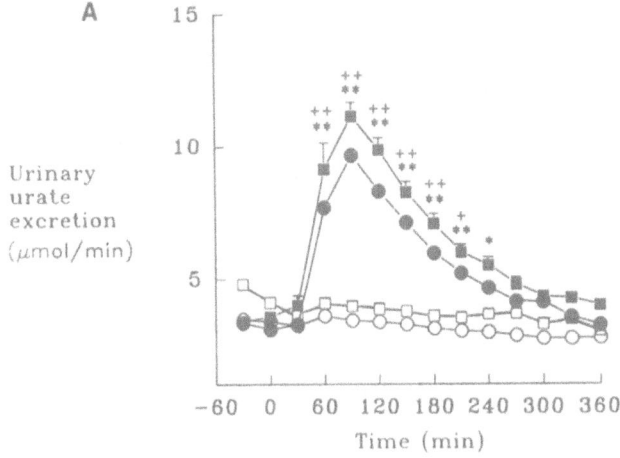

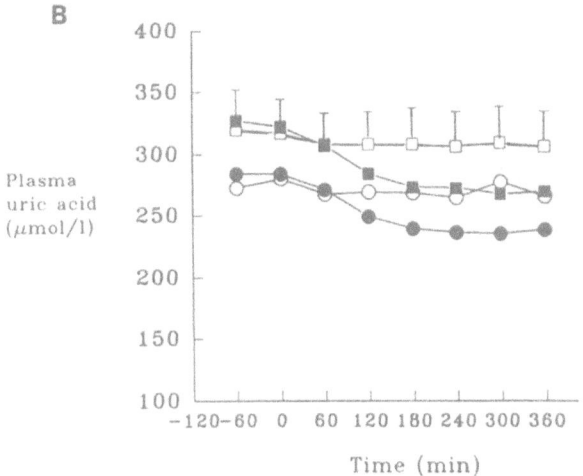

Figure 3. Effect of losartan (100 mg) or a placebo on (A) urinary uric acid excretion and (B) plasma uric acid levels in water-loaded healthy volunteers on a high- and low-sodium diet; low salt diet/placebo (□, $n = 7$); low salt diet/losartan (■, $n = 16$); high salt diet/placebo (○, $n = 7$); high salt diet/losartan (●, $n = 16$); $*p < 0.05$; $**p < 0.01$ vs. time 0; $+p < 0.05$; $++p < 0.01$ vs. placebo. (From Burnier *et al.*[9] Reprinted with permission.)

receptor antagonists are more selective than ACE inhibitors, which are also involved in the breakdown of bradykinin and may therefore lead to a potentiation of the vasodilating action of bradykinin. Theoretically therefore the higher specificity of ANG II antagonists might be linked with a decreased therapeutic efficacy.

The antihypertensive efficacy of the ANG II antagonist losartan has been evaluated and compared to that of an ACE inhibitor in two large studies of which only preliminary results have been published so far.[12,13] The first trial has been conducted in hospitalized patients with in-hospital diastolic blood pressures of at least 95 mm Hg immediately before treatment. The patients were randomly assigned into five treatment groups, each with about 20 patients. Three of these groups were given losartan in once-daily doses of either 50, 100, or 150 mg. One group received a placebo and the fifth group 10 mg enalapril. The patients were treated for 5 days.

Compared with the placebo, each of the four treatments induced significant decreases in systolic and diastolic blood pressures both at peak and at trough. Interestingly, there was no dose–response relationship with the three doses of losartan, suggesting that 50 mg dose of losartan is as effective as higher doses. By day 5, all three losartan doses and enalapril had reduced diastolic pressure to a similar degree (Fig. 4).

The second study was a large-scale multicenter outpatient trial conducted in hypertensive patients, free of treatment for at least 3 weeks, whose supine diastolic blood pressure was ≥ 100 mm Hg at the end of a placebo period. Five doses of losartan (10, 25, 50, 100, and 150 mg) were compared to a placebo and to 20 mg enalapril The groups were constituted of 68 to 82 patients. All medications were administered as a single daily dose for 8 weeks. The 10- and 25-mg doses of losartan had no significant effect on blood pressure as compared to the placebo group. In contrast, significant decreases in systolic and diastolic blood pressures were obtained with the 50-mg and higher doses of

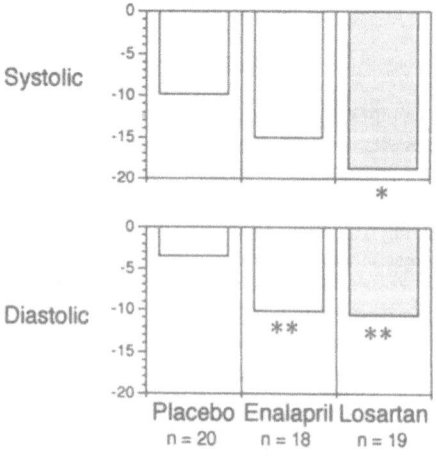

Figure 4. Decreases in systolic and diastolic blood pressures (in mm Hg) from baseline measured 24 hr after the fifth dose of losartan (150 mg/day, $n = 19$), enalapril (10 mg/day, $n = 18$), or a placebo ($n = 20$) in hospitalized white hypertensive patients (diastolic blood pressure ≥ 95 mm Hg); *$p < 0.05$; **$p < 0.01$ vs. placebo. (From Nelson *et al*.[12] Reprinted with permission.)

losartan. Again, no difference was observed between the 50-, 100-, and 150-mg doses of losartan. Moreover, the antihypertensive effect of these three higher doses of losartan was comparable to that of 20 mg enalapril.

The chronic effects of losartan on blood pressure were also examined in eight hospitalized patients with essential hypertension.[14] After a 1-week washout period, losartan was administered orally once a day for 2 to 4 weeks in increasing doses of 12.5, 25, 50, and 100 mg. The patients were maintained for 3 to 5 days at each dose level and the losartan doses were increased until the blood pressure in supine position had decreased more than 20 mm Hg (systolic) and 10 mm Hg (diastolic) or 13 mm Hg in mean blood pressure. Blood pressure and heart rate were monitored daily and ambulatory blood pressure recordings were performed before and after losartan treatment. In this small group of patients, the average dose of losartan needed to achieve the therapeutic goal was 59.4 ± 43.7 mg/day (mean $\pm$SD). With this average dose, significant falls in systolic and diastolic blood pressure were obtained with no significant increase in heart rate. There was no change in the circadian rhythm or variability of blood pressure under losartan. The influence of losartan on renal function was also studied in these patients. Losartan had no effect on serum electrolytes, creatinine clearance, urine volume, or urinary sodium excretion. However, losartan induced a significant decrease in serum uric acid concentration. This effect was associated with a significant increase in urinary uric acid excretion.

Since the main contribution to the ANG II receptor-blocking effect of losartan is due to the active metabolite EXP3174, the antihypertensive effect EXP3174 was recently investigated in 12 patients with essential hypertension.[15] Twenty milligrams of EXP3174 were administered intravenously for 4 hr and were compared to the infusion of a placebo in a double-blind parallel study. EXP3174 causes a significant reduction in supine diastolic blood pressure with a peak fall in blood pressure being observed 4 hr after termination of the infusion.

Thus, losartan given at doses between 50 and 100 mg appears to be an effective antihypertensive agent that is relatively comparable to an ACE inhibitor. According to the results available so far, hypertension should be one of the main clinical application of treatment with ANG II receptor antagonists.

4. ANG II RECEPTOR BLOCKADE IN PATIENTS WITH CHRONIC RENAL FAILURE OR CONGESTIVE HEART FAILURE

Inhibition of the renin–angiotensin system in hypertensive patients with diabetic or nondiabetic nephropathy have not only turned out to be very effective in controlling blood pressure. In addition, numerous clinical and experimental studies have shown that ACE inhibitors reduce proteinuria and afford renal protection. Thus, if ANG II receptor antagonists have the same properties as ACE inhibitors, hypertension with chronic renal failure may represent another large field of application of ANG II antagonists. In animals, losartan appears to have the same beneficial effects on blood pressure, protein-uria, and intraglomerular pressure.[16] Today, the clinical experience with ANG II antagonists in patients with chronic renal failure is relatively limited.

The efficacy and safety of losartan has been examined in 24 hypertensive patients with various levels of renal function.[17] The patients were treated for 7 days with losartan 100 mg once daily. They were distributed into four groups (6 patients/group) according to their creatinine clearance (group 1: 75 ml/min and above; group 2: 30–74 ml/min; group 3: 10–29 ml/min; and group 4: patients on hemodialysis). Losartan reduced blood pressure in all groups of patients on day 1 and the hypotensive effect was still present on day 7. In nonhemodialyzed patients, the decrease in blood pressure was accompanied by a decrease in urinary protein excretion. No significant changes in creatinine clearance, serum creatinine, or serum potassium were found in this small group of patients.

Gansevoort *et al.*[18] have compared the systemic and renal effects of losartan to those of enalapril in another small group of 7 patients with hyper-tension (diastolic blood pressure > 90 mm Hg), proteinuria (> 2 g/day), and moderate nondiabetic renal failure (mean GFR 75 ml/min). The protocol con-sisted of seven periods, each lasting 1 month, in which patients received once daily placebo, 50 mg losartan, 100 mg losartan, placebo, 10 mg enalapril, 20 mg enalapril, and placebo, respectively. Patients used no other medication and were maintained on a 100 mmole sodium-restricted diet. Dose-dependent decreases in diastolic blood pressure and proteinuria were obtained with the two doses of losartan as well as with the two doses of enalapril. With both drugs there was no change in glomerular filtration rate and a slight increase in renal plasma flow. The results of this study demonstrate that ANG II receptor blockade and ACE inhibition induce similar changes in blood pressure, renal hemodynamics, and proteinuria. Moreover, these data do not support the idea that the renal effects of ACE inhibitors are mediated by other hormonal system than the renin–angiotensin system.

Finally, Lo *et al.*[19] have studied the pharmacokinetics of losartan and its biotransformation into the active metabolite EXP3174 in 18 patients with varying degrees of renal insufficiency according to their creatinine clearance. Three groups of patients were analyzed: group 1 with a clearance ≥ 75 ml/min, group 2 with a clearance of 30–74 ml/min, and group 3 with a creatinine clearance between 10 and 29 ml/min. Losartan was given orally for 7 days at the dose of 100 mg once daily. Plasma and urinary concentrations of losartan and EXP3174 were measured for up to 60 hr postdose on day 7. Renal insuf-ficiency had only a minimal effect on the area under the concentration curve

(AUC 0–14 hr) of losartan or its metabolite. The half-life of losartan ranged between 2.1 ± 0.7 hr in group 1 and 3.2 ± 2.2 hr in group 3. The half-life of the active metabolite was 10 ± 3 hr in group 1 and 13 ± 6 hr in group 3. For both losartan and the metabolite, the percentage of the dose excreted in the urine at steady-state decreased with increasing degree of renal failure (from 3% in group 1 to 0.3% in group 3 for losartan and from 4.4 to 0.2% for the metabolite).

These preliminary results in patients with chronic renal failure therefore suggest that losartan is safe and effective in hypertensive patients with a limited renal function. So far, losartan appears to have the same beneficial effects to reduce blood pressure and proteinuria as ACE inhibitors.

Congestive heart failure is another clinical indication in which inhibition of the renin–angiotensin system was found to be very useful. Thus, the benefits of ACE inhibitors in patients with heart failure include symptomatic relief, improvement in left ventricular function, improved survival, and decreased incidence of myocardial infarction and unstable angina. Theoretically, ANG II antagonists are again expected to produce similar effects to those of ACE inhibitors. Unfortunately, there is yet no published information on the clinical efficacy of ANG II receptor antagonists in this indication. Preliminary results presented at recent meetings suggest that ANG II antagonists induce favorable systemic hemodynamic changes in heart failure.[20]

5. SIDE EFFECTS

The clinical application of a treatment depends not only on its efficacy in a given indication but also on the incidence of side effects. In all clinical studies cited above, losartan has been safe and no relevant side effect has been reported. However, the clinical experience is too limited to draw firm conclusions on the incidence of undesirable side effects. Some side effects observed during ACE inhibition might well be observed during ANG II receptor blockade because they are related directly to the blockade of the renin–angiotensin system and hence to the decrease in plasma ANG II and to some extent in plasma aldosterone levels.

Hypotension, hyperkalemia, and acute renal failure are typical side effects caused by blockade of ANG II production.[1] Dehydration and/or salt-depletion clearly increase the risk of developing hypertension during ACE inhibition. As far as acute renal failure is concerned, the risk exists mainly in patients with bilateral renal artery stenosis or with a stenosis of the artery to a solitary kidney. It may also occur if widespread intrarenal vascular lesions exist. Whether this kind of problem will occur under chronic ANG II receptor blockade is not known, but unless the postulated mechanism leading to this unwanted effect is wrong, there is no evident reason why these side effects should not be observed also with ANG II antagonists.

Whether plasma potassium levels will increase during long-term treatment with ANG II antagonists remains to be demonstrated. So far, no significant increase in plasma potassium has been reported in patients with essential hypertension and normal renal function or in those with some degree of renal insufficiency. In this context, it is important to note that in normal volunteers, losartan transiently increased rather than decreased urinary potassium excretion.[9] Moreover, the changes in plasma aldosterone observed so far under losartan therapy have been relatively small.

The major advantage of ANG II receptor antagonists should be the absence of cough, a side effect typically related to the inhibition of ACE per se. A large study examining the issue of cough during chronic losartan therapy is being conducted. Preliminary results presented at the last meeting of the European Society of Hypertension suggest that losartan compared to lisinopril and a diuretic indeed offers the advantage of not causing cough.[21] In this case, receptor antagonists may represent an ideal alternative to ACE inhibitors in patients whose blood pressure is well controlled with an ACE inhibitor but in whom cough has developed.

The side effects of drugs are sometimes related to the compound itself. With losartan, one effect appears to be a unique property of the mother compound, i.e., the uricosuric effect. Indeed, infusion of the active metabolite EXP3174 does not increase uric acid excretion (CS Sweet, personal communication). The uricosuric effect of losartan could be interpreted as an advantage since many hypertensive patients have increased plasma uric acid levels.[22] Nevertheless, it may also represent a limitation in some clinical circumstances such as acute dehydration, as it might enhance the risk of developing acute uric acid nephropathy if a state of supersaturation of undissociated uric acid is suddenly achieved. A marked decrease in plasma uric acid levels might also be associated with the occurrence of acute gout in patients prone to this disease.

6. CONCLUSIONS

The clinical experience accumulated so far with losartan has established that it is an effective antihypertensive agent that compares favorably with ACE inhibitors. It appears to be well tolerated, and besides its uricosuric effect, which may be a potential undesirable effect, no significant side effect has been reported. Like ACE inhibitors, ANG II receptor antagonists might play an important role in the treatment of congestive heart failure and in the management of hypertensive patients with a chronic nephropathy. Indeed, early reports in animal have suggested that losartan has the same beneficial effects on renal function, cardiac function, and animal survival as ACE inhibitors. The preliminary results of the first clinical studies tend to support this view but large trials are still needed to define all the clinical applications of ANG II receptor antagonists.

REFERENCES

1. Waeber B, Nussberger J, Brunner HR: Angiotensin-converting enzyme inhibitors in hypertension. In Laragh JH, Brenner BM (eds): *Hypertension: Pathophysiology, Diagnosis, and Management*. Raven Press, New York, 1990, pp. 2209–2232.
2. Timmermans PBMWM, Carini DJ, Chin AT, et al: The discovery of a new class of highly specific nonpeptide angiotensin II receptor antagonists. *Am J Hypertens* 4:275S–281S, 1991.
3. Christen Y, Waeber B, Nussberger J, et al: Oral administration of DuP 753, a specific angiotensin II receptor antagonist, to normal male volunteers. Inhibition of pressor response to exogenous angiotensin I and II. *Circulation* 83:1333–1342, 1991.
4. Wong PC, Price WA, Chiu AT, et al: Nonpeptide angiotensin II receptor antagonists. XI. Pharmacology of EXP3174: An active metabolite of DuP 753, an orally active antihypertensive agent. *J Pharmacol Exp Ther* 255:211–217, 1990.
5. Munafo H, Christen Y, Nussberger J, et al: Drug concentration response relationships in normal volunteers after oral administration of losartan, an angiotensin II receptor antagonist. *Clin Pharmacol Ther* 51(5):513–521, 1992.
6. Ohtawa M, Takayama F, Saithoh K, et al: Pharmacokinetics and biochemical efficacy after single and multiple oral administration of losartan, an orally active nonpeptide angiotensin II receptor antagonist, in humans. *Br J Clin Pharmacol* 35(3):290–297, 1993.
7. Goldberg MR, Tanaka W, Barchowsky A, et al: Effects of losartan on blood pressure, plasma renin activity, and angiotensin II in volunteers. *Hypertension* 21:704–713, 1993.
8. Doig JK, MacFayden RJ, Sweet CS, et al: Dose-ranging study of the angiotensin type I receptor antagonist losartan (DuP753/MD954), in salt-deplete normal man. *J Cardiovasc Pharmacol* 21(5):732–738, 1993.
9. Burnier M, Rutschmann B, Nussberger J, et al: Salt dependent renal effects of an angiotensin II antagonist in healthy subjects. *Hypertension* 22:339–347, 1993.
10. Martinez F, Schmitt F, Natov S, et al: Renal effects of losartan and enalapril alone and in combination in healthy volunteers. In Eliahou, HE, Iaina A, Bar-Khazim Y (eds): *Proceedings of the International Society of Nephrology*. Jerusalem, 1993, p. 228.
11. Nakashima M, Uematsu T, Kosuge K, et al: Pilot study of the uricosuric effect of DuP 753, a new angiotensin II receptor antagonist, in healthy subjects. *Eur J Clin Pharmacol* 42:333–335, 1992.
12. Nelson E, Arcuri K, Ikeda L, et al: Efficacy and safety of losartan in patients with essential hypertension. *Am J Hypertens* 5:19A, 1992.
13. Weber MA: Clinical experience with the angiotensin II receptor antagonist losartan. A preliminary report. *Am J Hypertens* 5:247S–251S, 1992.
14. Tsunoda K, Abe K, Hagino T, et al: Hypotensive effect of losartan a nonpeptide angiotensin II receptor antagonist in essential hypertension. *Am J Hypertens* 6(1):28–32, 1993.
15. Sweet CS, Bradstreet D, Nelson EB, et al: Pharmacodynamic activity of intravenous E-3174, the metabolite of losartan in hypertensive patients. *Am J Hypertens* 6(5):93A, 1993.
16. Lafayette RA, Mayer G, Park SK, et al: Angiotensin II receptor blockade limits glomerular injury in rats with reduced renal mass. *J Clin Invest* 90:766–771, 1992.
17. Shaw W, Keane W, Sica D, et al: Safety and antihypertensive effects of losartan MK-954 or

DuP 753 a new angiotensin II receptor antagonist in patients with hypertension and renal disease. *Clin Pharmacol Ther* 53(2):140, 1993.

18. Gansevoort T, de Zeeuw D, de Jong P: Similar systemic and renal effects of angiotensin-II (A-II) receptor antagonism and ACE inhibition (ACEi) in non-diabetic renal disease. *Kidney Int* 43:967, 1993.

19. Lo MW, Shahinfar S, Furtek CI, et al: Pharmacokinetics of losartan MK-954 or DuP 753 in patients with renal insufficiency. *Clin Pharmacol Ther* 53(2):160, 1993.

20. Gottlieb SS, Dickstein K, Fleck E, et al: Hemodynamic and neurohormonal effects of the angiotensin II antagonist losartan in patients with congestive heart failure. *Circulation* 86(4 suppl 1):I120, 1992.

21. Lacourciere Y, Snavely DB, Faison EP, et al: Effects of modulators of the renin angiotensin system in cough. Abstract book of the satellite symposium to the Sixth European Meeting on Hypertension, "Angiotensin II antagonism: A new advance in the treatment of hypertension," Milan 1993, pp. 22–23.

22. Cannon PJ, Stason WB, Demartini FE, et al: Hyperuricemia in primary and renal hypertension. *N Engl J Med* 275:457–464, 1966.

21

Angiotensin Receptor Stimulation of Transforming Growth Factor-β in Rat Skin and Wound Healing

M. Ian Phillips, Birgitta Kimura, and Robert Gyurko

1. INTRODUCTION

Numerous studies have implicated the octapeptide angiotensin II (ANG II) as a growth factor. There is a link between angiotensin and cell growth in cardiovascular hypertrophy, in proto-oncogene regulation[1] and in the excessive smooth muscle cell proliferation that occurs after balloon catheter injury.[2] The neointimal regrowth has been shown to be decreased by treatment with angiotensin-converting enzyme (ACE) inhibitors and losartan, an AT_1 receptor antagonist.[3] Although ACE inhibitors have many actions, further work by Powell *et al.*[4] support the role of ANG II as the growth promotor of vascular smooth muscle cells and matrix protein synthesis. Further evidence is that in aortic smooth muscle cells in culture, ANG II stimulates [^{3}H]thymidine incorporation but the effect depends on the types of cell cultures used and the presence of serum or platelet-derived growth factor or epidermal growth factor.[5] Thus the growth action of ANG II may depend on its ability to stimulate other growth factors. Majesky *et al.*[6] showed that transforming growth factor-β (TGF-β) mRNA expression in the blood vessel wall was increased after vascular injury. The neointimal smooth muscle cells had positive staining for TGF-β, and taken together these studies suggest that angiotensin in the vascular wall stimulates paracrine production of TGF-β and other growth factors.

M. Ian Phillips, Birgitta Kimura, and Robert Gyurko • Department of Physiology, College of Medicine, University of Florida, Gainesville, Florida 32610.

Angiotensin Receptors, edited by Juan M. Saavedra and Pieter B.M.W.M. Timmermans. Plenum Press, New York, 1994.

The actions of ANG II depend on the presence of functional receptors in tissue. Angiotensin receptors have been found in vascular smooth muscle, as well as many other tissues, notably adrenal gland, liver, kidney, pituitary, and brain. With the discovery of at least two major subtypes of angiotensin receptors[7] it was discovered that in fetal development the dominant ANG II receptor subtype is the AT_2 form. This is particularly apparent in skin, connective tissue, and skeletal muscle, all of which are rapidly growing tissues.[8] This receptor expression changes after birth and the AT_1 receptors become more predominant. This was established principally by autoradiography of whole fetuses and newborns and displacement of the radiolabeled ANG II with PD 123177, an AT_2 antagonist, or losartan, the AT_1 antagonist.[8,9] In rat fetal skin fibroblasts there is a progressive increase in AT_1 receptors during culture but the transition from AT_2 receptors to AT_1 can be inhibited in the primary cultures with actinomycin D.[10] Skin, which is one of the largest organs in the body, offers a convenient tissue to study angiotensin receptors and their role in growth. Growth and tissue remodeling are essential processes in wound healing of skin lesions and we have used this model to study angiotensin growth-promoting mechanisms in skin and fibroblasts. Based on the evidence for ANG II in growth facilitation, we hypothesized that ANG II would be involved in skin growth and wound healing.[11] It was attractive to hypothesize that if ontogeny recapitulates phylogeny, then the change of AT_2 to AT_1 receptors that occurs in fetal development would be repeated in reverse during wound healing. AT_2 receptors would reappear during the redevelopment and restructuring of tissue. Unfortunately this hypothesis has not been supportable in adult tissue,[12] although in young rats (2-week-old) where AT_2 receptors are still present in skin they become significantly enhanced 3 days after wounding.[13] Nevertheless, our studies of angiotensin receptors in skin indicate that angiotensin may act as a growth promotor and enhance wound healing. In this chapter we present evidence for a tissue renin–angiotensin system (RAS) in the skin, the changes that occur in ANG II and receptors after skin damage and the mechanism by which the angiotensin could exert its growth effects through the growth factor TGF-β.

2. TISSUE RAS IN SKIN

Although the renin–angiotensin system can produce high circulating quantities of blood-borne ANG II, there is strong evidence for the presence of independent, endogenous, RAS in many tissues. The genes for the various components of the RAS, including renin, angiotensinogen, ACE, and AT_1 receptors, have been shown to be expressed in the same tissues. Measurements of mRNA expression have shown the overlapping presence of RAS genes in adrenal, brain, testes, ovary, kidney, vascular tissue, and heart.[14,15] Direct measurements of gene product secondary proteins have also demonstrated measurable levels in the tissue.[16] It is hypothesized that the angiotensin-producing cells act in a paracrine fashion by secreting angiotensin to stimulate neighboring cells. In skin we hypothesize that there is an angiotensin production independent of the circulating RAS and that the cells that produce angiotensin (either directly through intracellular synthesis or indirectly through extracellular synthesis[16]) stimulate cells synthesizing growth factors that are involved in the normal regulation of epidermal and subepidermal growth and repair. During wound healing the tissue RAS of skin becomes more active and stimulates more paracrine cells to secrete growth factors to bring about the granulation tissue formation and remodeling that occurs in the wound healing process.

To establish that there is angiotensin in skin, skin samples were removed from rats and frozen at $-80°C$. Angiotensin was extracted and measured by radioimmunoassay and characterized by high-pressure liquid chromatography (HPLC).[17,18] Frozen tissue was boiled in 1 M acetic acid to prevent the possibility of *in vitro* production of ANG II. The tissue was homogenized and centrifuged. The supernatant was purified on SepPak C-18 cartridges and ANG II eluted with methanol/H_2O/TFA 55/44/1. The eluate was dried and redissolved in buffer and measured by radioimmunoassay (RIA). HPLC analysis was carried out on Pharmacia SMART System with a URPC C2/C18 column. A linear gradient of 17–40% 0.1 M morpholine/acetonitrile/methanol in 12 min gave separation of the pentapeptide, hexapeptide, ANG III, ANG II, and ANG I.[19] The levels of angiotensin found in skin is 104 ± 9 pg ANG II/g ($n = 18$). HPLC analysis of the angiotensin shows a single peak corresponding to the fractions that comigrate with ANG II (Fig. 1). There is no peak for ANG III or the pentapeptide in the tissue. In cut skin, however, there appears to be more metabolism of the angiotensins. Figure 1 shows the peak for ANG II and the higher peak that comigrates with the pentapeptide. There is also indication of a peak for ANG III being present. The level of angiotensin in skin per gram of tissue, is comparable to that found in testes, kidney, blood vessels, and brain. Further evidence for an endogenous skin RAS is the expression of angiotensinogen mRNA. Northern blot analysis was carried out with a ^{32}P-labeled RNA probe made from a full-length angiotensinogen cDNA. Figure 2 shows examples of Northern blots of angiotensinogen mRNA in skin taken from Wistar–Kyoto (WKY) and spontaneously hypertensive rats (SHR). The figure illustrates the clear presence of angiotensinogen mRNA in skin of adult animals. In hypertensive rats the levels are slightly lower than in the normotensive controls. In developing skin of young animals from the 18th day of gestation newborn and 5 days to 2 weeks old, we noted an increasing abundance of angiotensinogen mRNA in the normotensive WKY rat and an increase followed by a decrease at 2 weeks in SHR. ACE mRNA has also been detected

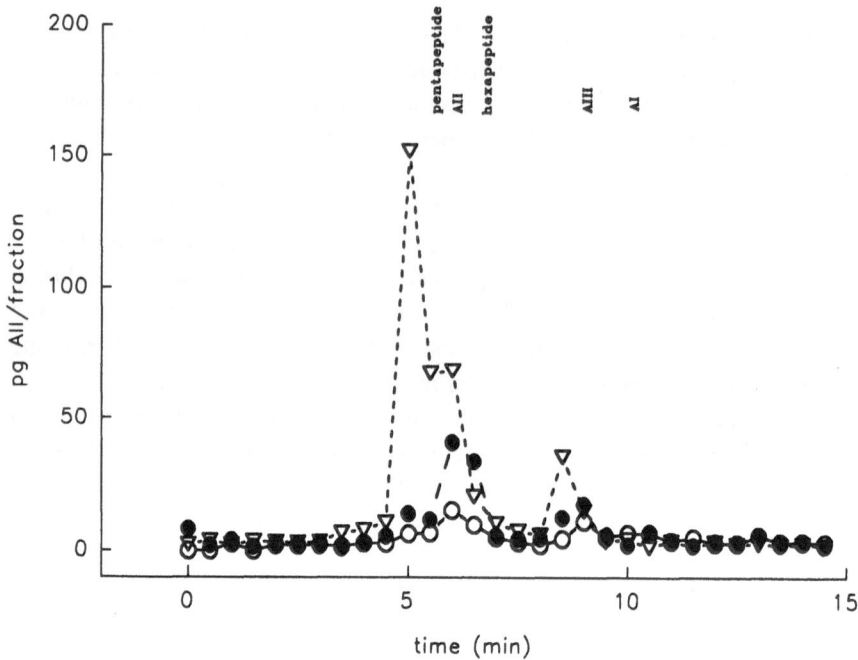

Figure 1. HPLC analysis of skin angiotensin. The control skin ANG II sample (●) has one peak for ANG II. The healing skin (▽) shows more variation and an increase in angiotensins. There is still a peak for ANG II, but in addition there is a large peak before ANG II, corresponding to the angiotensin pentapeptide. In addition, a peak for ANG III can be detected. Blank samples (○) provide a control baseline with no peaks. Each fraction collected from HPLC was measured by RIA for ANG II. A separate analysis with ANG I RIA also showed the presnece of ANG I in skin (not shown).

in skin (unpublished observation), but renin mRNA has not yet been determined in skin.

Specific binding of [^{125}I]-ANG II to skin membranes was established by homogenizing the tissue and purifying membranes that were incubated for 45 min at room temperature with [^{125}I]-ANG II and increasing concentration from 100 pM to 1 µm of ANG II or ANG II receptor antagonists. There was a linear relationship between moles [^{125}I]-ANG II bound and microgram protein. The dissociation of [^{125}I]-ANG II after the addition of excessive buffer showed a reversible dissociation of 50% in 10 min. Displacement curves with ANG II gave IC$_{50}$ values of 8.19 nM (Table I). Scatchard analysis revealed both a high-affinity and a low-affinity site which was also indicated by the hill slope coefficient (Fig. 3). The K$_d$ for the high-affinity site was 6.28 nM and for the low-affinity site 88.3 nM. The receptor binding was displaceable by losartan (1 µM), which displaced 94% of the [^{125}I]-ANG II specific binding (Fig. 4). Neither 1 µm PD 123177 nor 1 µm PD 123319 (the AT$_2$-specific

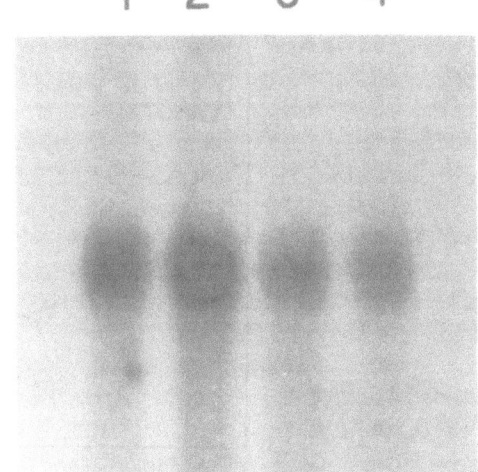

Figure 2. Northern blot analysis of angiotensin mRNA in skin. RNA was extracted from skin according to the method of Chomczynski and Sacchi.[32] Total RNA was separated by agarose gel electrophoresis, transferred to a nylon membrane and probed with a [32P]-RNA probe for angiotensinogen. The figure shows presence of angiotensinogen mRNA in skin. Lane 1: WKY control; lane 2: WKY DUP; lane 3: SHR control; lane 4: SHR DUP.

antagonists) displaced [125I]-ANG II binding. Thus there was no evidence of AT_2 receptors in the normal skin (Table I). This is in contrast to fetal rat skin where 97% of the receptors are of the AT_2 type. It is not known whether the AT_1 receptors develop from the AT_2 receptors by conversion or whether the AT_2 receptors are not expressed. Only a few tissues have residual AT_2 receptor

Table I. Kinetics of Angiotensin Receptor Binding in Control and Healing Skin (1 Day after Wounding)

	Control skin	Healing skin
High affinity		
B_{max} fm/mg	91.4 ± 16.0	35.9 ± 10.17[a]
K_d nM	6.284 ± .246	4.960 ± 1.095
Low affinity		
B_{max} fm/mg	606 ± 417	237 ± 47
K_d nM	88.3 ± 52.9	94.0 ± 25.6
IC_{50}		
All	8.19 nM	3.26 nM
(Hill slope)	(0.568)	(0.528)
DUP753	17.4 nM	17.5 nM
(Hill slope)	(0.664)	(0.506)
DP123319	> 10 μM	> 10 μM

[a]Significantly different from control $p < 0.05$.

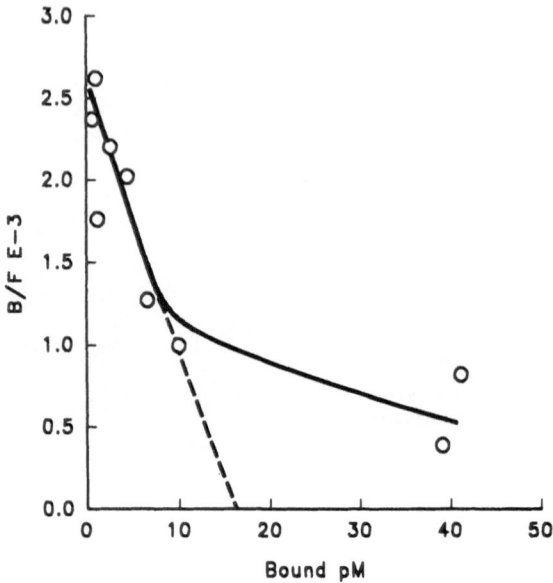

Figure 3. Scratchard analysis of ANG II receptor binding in skin show the presence of a high-
and a low-affinity site. (From Kimura *et al.*[12] Reprinted with permission.)

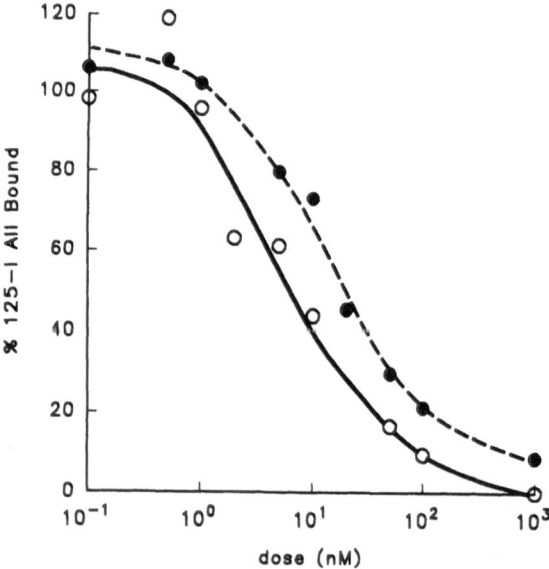

Figure 4. Displacement of [125I]-ANG II from the ANG II receptor. Both ANG II (○) and losartan
(DuP 753) (●) displaced the radioactive ligand showing the receptor to be of the AT₁ subtype.
(From Kimura *et al.*[12] Reprinted with permission.)

in the adult populations. In the brain, the inferior olivary nucleus is one example. In the adrenal medulla the AT_2 receptors are predominant.

3. ANGIOTENSIN RECEPTORS DURING WOUND HEALING

Adult rats (200–250 g body weight) were anesthetized and the abdominal area depilated. Surgical incisions were made with a scalpel through the epidermis, dermis, and subcutaneous fat. At different times after the wounds were made the rats were euthanized and the tissue from the healing wound biopsied. Control tissue was taken from an uncut tissue of the abdominal skin. Measurements were made by homogenizing the skin samples and carrying out binding studies with $[^{125}I]$-ANG II and competition studies with 1 μm losartan and 1 μm PD 123319 in the presence or absence of 5 mM dithiothreitol (DTT). Samples of tissue were taken at 3, 6, 12, 18, 24, 48, and 72 hr after wounding. The result is shown in Fig. 5 where it can be seen that at 12, 18, and 24 hr after wounding, there is a significant decrease in angiotensin binding ($p < 0.05$). In cut skin in the absence of DTT, the AT_2 receptor antagonist did not displace the $[^{125}I]$-ANG II binding significantly. However, in skin cut at 6, 12, and 18 hr and incubated with 5 mM DTT and PD 123177, there was a displacement of 21.7% of the $[^{125}I]$-ANG II binding ($n = 10$). This is suggestive of a brief

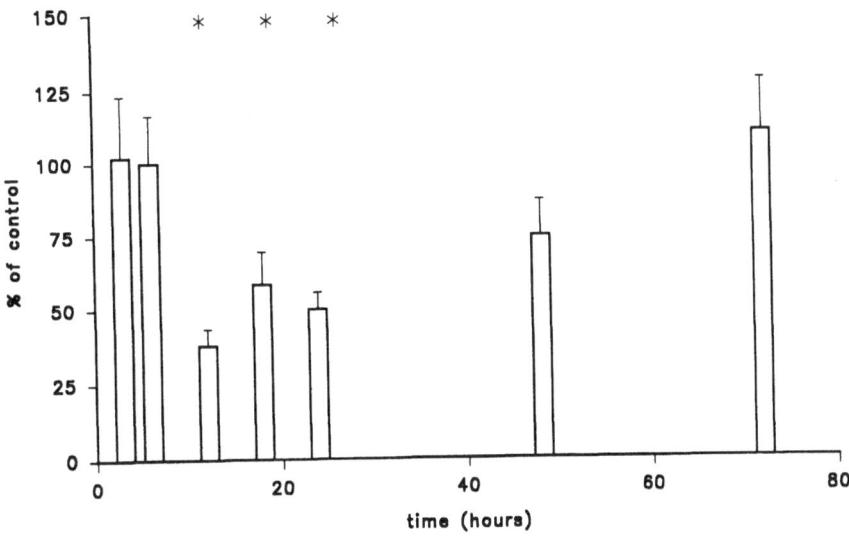

Figure 5. Specific $[^{125}I]$-ANG II binding in healing skin expressed as percent of control skin. There was a significant decrease ($p < 0.05$) at 12, 18, and 24 hr. (From Kimura *et al.*[12] Reprinted with permission.)

appearance of AT_2 receptors in cut skin but was variable and specific binding was low. Also, losartan displaced 95% ($n = 8$) of the [^{125}I]-ANG II binding. Thus the evidence for AT_2 receptors appearing in cut skin in the adult is not significant, whereas the evidence for AT_1 receptors is highly significant. The only difference between the AT_1 binding parameters in control skin and healing skin was in the number of binding sites (B_{max} control 91.4 ± 16, healing skin 35.9 ± 10.1 fm/mg, $p < 0.05$). The other measures of the high-affinity K_d and the low-affinity K_d were not different. Since these kinetic studies were carried out in homogenized skin, it is possible that the nonsignificant appearance of AT_2 binding may be due to significant levels in localized compartments of the skin. However, the AT_2 receptors, if they play a role in wound healing, do not appear to play a role in the second-messenger stimulation due to a discoupled G-protein linkage.

4. ANGIOTENSIN DURING WOUND HEALING

Adult rats (200–250 g), maintained in a light-controlled room with free access to food and water, were anesthetized on the day of the experiment with methafane and the hair removed from the abdominal area and a surgical incision made with a scalpel. Animals recovered from the surgery without incident. At different time intervals after the wounding, rats were euthanized with sodium pentobarbital and the tissue from the healing wound biopsied. Control tissue was taken from an uncut area of the abdominal skin. Skin samples were taken at 3, 6, 18 hr, 1, 3, 7, and 11 days after the wounding. The tissue was boiled in 1 M acetic acid, homogenized, and extracted as described previously for ANG II radioimmunoassay. The data were plotted on a time graph (Fig. 6) from which it can be seen that there is a rapid rise in the levels of skin ANG II beginning at 3 hr and rising to a peak at 24 hr, with a gradual decrease back to normal levels on day 11 when the wounded skin is fully healed. The dramatic increase in angiotensin is worth comparing to the changes that occur in [^{125}I]-ANG II binding (Fig. 5). Although binding remained unchanged, as 100% of control at 3 and 6 hr, by 12 hr it had significantly decreased and binding remained below control levels during the peak times of angiotensin increases at 18 and 24 hr. Thus the increase in angiotensin mirrors the decrease in angiotensin receptor binding with a slight delay. Taken together, these data suggest that the rapid rise in angiotensin in skin after wounding causes a downregulation of receptors during the first 24 hr. Even with the significant decreases in ANG II binding, the elevation of ANG II is comparatively higher than the amount of decrease in ANG II binding. Therefore, although there is downregulation, there is sufficient angiotensin present to provide for continuous effects of ANG II during this time period.

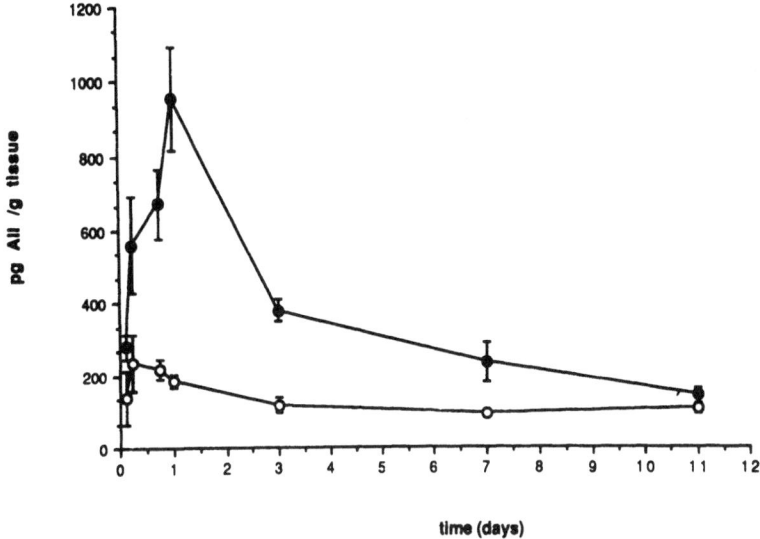

Figure 6. ANG II levels in skin during healing. The difference between healing skin (●) and control (○) was significant at 18, 24, and 72 hr ($p < 0.05$). The peak ANG II concentration is approximately 10 times higher than baseline concentrations.

The source of ANG II in the skin is not known. Immunocytochemical staining with an angiotensin antibody in our lab has not conclusively demonstrated the source of the increased angiotensin. There are many possibilities. Angiotensin is present (as shown by ranging degrees of staining) in vascular smooth muscle of blood vessels in the skin, in the hair follicles, dermis, and subcutaneous fat. It is also present in lymphocytes and since platelets and fibroblasts have binding sites for ANG II, they may sequester ANG II in these tissues after binding to these sites. A significant finding is that during this increase in local skin levels of ANG II, the circulating levels of angiotensin are not increased. Also in skin samples from the same rat, uncut skin did not have elevated ANG II levels. Therefore, unless angiotensin is being sequestered from blood, the data suggest that the increase in angiotensin is due to endogenous synthesis of local skin angiotensin and not to circulating ANG II. Skin fibroblasts do not appear to be a likely source of angiotensin synthesis because we have been unable to detect angiotensinogen mRNA in adult skin fibroblasts.

The process of wound healing follows a sequence of phases. The early phase is the inflammatory phase, leading to the formation of fibrin and fibronectin. During inflammation an increase in phagocytic cells diminishes the dead cells and the further phases of contraction, repair, and regeneration can then begin. During the contraction phase myofibroblasts appear, apparently from pericytes in the vascular wall. They migrate to the wound and have features intermediate between smooth muscle cells and fibroblasts. Based on the general principal that angiotensin is an activator of contractile tissue, the increase in angiotensin may play a role in activating the myofibroblasts. In the next stage of repair, active fibroblasts arise from the stem cells in the dermis of the skin and synthesize and secrete extracellular matrix compounds including fibronectin, proteoglycans, and collagens type I and II.[20] Platelet-derived growth factor (PDGF) are mitogenic for fibroblasts and stimulate chemotactically monocytes, neutrophils, and smooth muscle cells. TGF-β is present in very high concentration in platelets and may be important in wound healing by stimulating the expression of fibronectin and collagen from fibroblasts.[21] TGF-β has been shown to accelerate healing in incisional wounds in rats.[22] Thus if angiotensin stimulates fibroblasts or other cell types in skin, it could release growth factors and thereby contribute to the healing process.

5. ANGIOTENSIN RECEPTOR REGULATION OF PHOSPHATIDYLINOSITOL HYDROLYSIS

We showed above that the predominate receptor for ANG II in skin is the AT_1 receptor subtype. In other tissues, ANG II stimulates phosphatidylinositol (PI) hydrolysis through AT_1 receptor activation and mobilizes calcium.[10,23] In adrenal cortex, heart, liver, and smooth muscle ANG II binds to receptors that activate phospholipase C and increase inositol 1,4,5 triphosphate (IP_3). Ca^{2+} is activated by both influx through receptor-operated channel opening and by modulation of voltage-sensitive Ca^{2+} channels and by the release of intracellular Ca^{2+} by the PI hydrolysis product IP_3. To study the receptor coupling to signal transduction systems, we carried out experiments on adult rats using skin slices.[24] Skin samples were cut and placed in buffer. After five washes the skin slices were labeled with 30 μCi of myo-[^{3}H]inositol for 1 hr at 37°C in 6 ml buffer. PI hydrolysis assays were carried out after stimulating the skin slices with increasing concentrations of ANG II in the presence of 8 mM LiCl, 1 M losartan, or PD 123319 for 60 min. Control slices were treated with buffer only. After stopping the reactions with chloroform : ethanol : HCL (100 : 200 : 1), the aqueous phase was run on Dowex anion exchange columns and the [^{3}H]inositol phosphates were eluted with ammonium formate/formic acid (1 M/0.1 M). Radioactivity was determined and [^{3}H]inositol phosphates were also separated by HPLC. ANG II stimulated PI hydrolysis in skin. The stimulation was dose-dependent on the concentration of ANG II and there was significant increase in PI hydrolysis at 10^{-6}M concentration (Fig. 7). To determine which inositol phosphate isomers were formed during the ANG II stimulation, the samples were analyzed by HPLC (Fig. 8). HPLC analysis revealed 3 isomers of the Ins P_3: Ins(1,4,5)P_3, Ins(1,3,4)P_3, and an unidentified peak, possibly

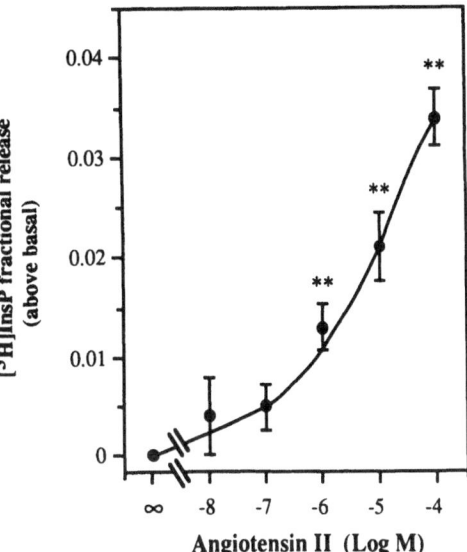

Figure 7. Dose-dependent increase in the release of [³H]inositol phosphates to angiotensin stimulation. Fresh skin slices were preincubated with [³H]inositol, then incubated with different concentrations of ANG II. After disrupting the cell membranes with a mixture of chloroform/methanol/HCl, inositol phosphates were separated on Dowex anion exchange resin, and [³H]inositol phosphates were detected in a liquid scintillation counter. (From Gyurko *et al.*[24] Reprinted with permission.)

Ins(2,4,5)P$_3$ (Fig. 6). Ins(1,3,4,5)P$_4$ was not detected in the assay. Ins P$_2$ consisted of Ins(1,4)P$_2$ and Ins(1,3)P$_2$. There were also two isomers of Ins P$_1$, namely, Ins(1)P$_1$ and Ins(4)P$_1$ formed during the angiotensin stimulation (Fig. 8).

The subtype of the ANG II receptor in the skin slices was determined by competition with binding ANG II, losartan, the AT$_1$ receptor, and PD 123319, the AT$_2$ receptor antagonist. 10^{-4} M losartan shifted the dose–response curve of inositol phosphates to the right, revealing a significant inhibition of the effects of 10^{-7} and 10^{-5} M ANG II ($p < 0.005$) (Fig. 9). The same dose of PD 123319 was not effective. A high dose of PD 123319 (10^{-3}M), however, had the opposite effect of losartan. 10^{-3} M of PD 123319 greatly potentiated instead of inhibiting the effect of ANG II on PI hydrolysis at concentrations of 10^{-7}M, 10^{-6}, and 10^{-4} ($p < 0.01$) (Fig. 10).

The results show that the predominant receptor subtype in skin is the AT$_1$, confirming our previous study. Second, the stimulation of these AT$_1$ receptors leads to rapid metabolism of Ins(1,4,5)P$_3$, which is considered the critical second messenger molecule for mobilizing Ca^{2+} from intracellular stores.[23] Ins(1,4,5)P$_3$ is rapidly formed after ANG II stimulation. It is actively metabolized and increases at the same time as cytoplasmic Ca^{2+} concentrations increase in intact adrenal glomerulosa cells.[23] Our findings would suggest that the same is true in skin. The effect of PD 123319 at 10^{-3} M, potentiating ANG II action on PI hydrolysis, raises the question whether there are AT$_2$ receptors present in the tissue. Further, the result may be due to PD 123319 acting as an agonist or an evidence for an appropriate action between AT$_1$ and

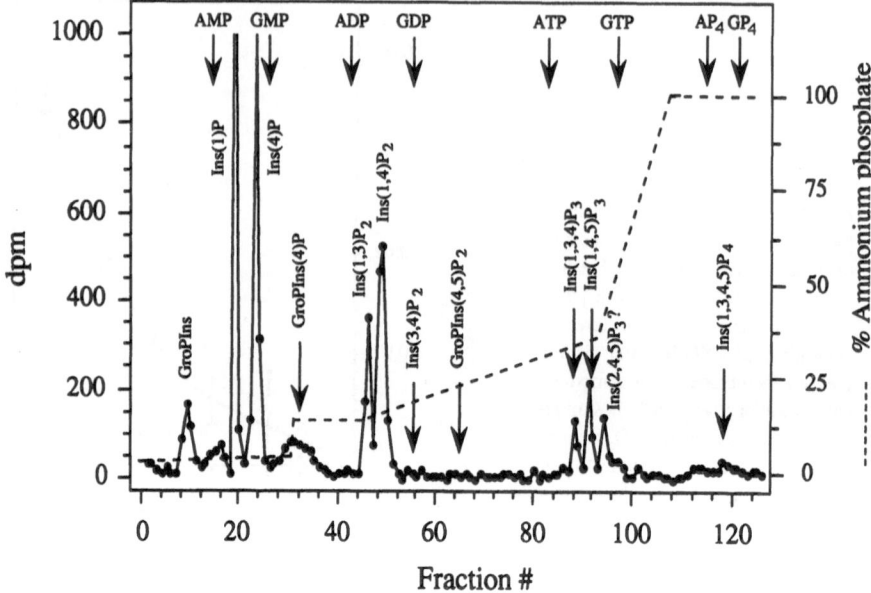

Figure 8. HPLC separation of [³H]inositol phosphates in rat skin slices. Myo-[³H]inositol and [³H]inositol phosphates were extracted and spiked with the indicated nucleotides and separated with SAX HPLC. Identification of samples were based on coelution with authentic ³H-labeled standards in separate runs (AP₄ and GP₄; adenosine and guanosine tetraphosphate, pH 3.7). (From Gyurko *et al.*[24] Reprinted with permission.)

AT_2 receptor functions. As noted above, we found very little evidence for AT_2 receptors in the adult skin. One could consider that the skin slices used in this experiment is a model for wounded skin rather than for intact skin because the experiments occur some hours after removal. Time is taken to cut the tissue, 1 hr is taken to wash the tissue, and 1 hr for incubation with myo-[³H]inositol. The incubations with angiotensin or angiotensin antagonists were 60 min so the total time was 3 hr after cutting. However, at 3 hr after cutting, AT_2 receptors were not evident.

6. ANG II INDUCED CALCIUM SIGNALING IN SKIN FIBROBLASTS

Since one of the effects of the second messenger $Ins(1,4,5)P_3$ is to mobilize calcium, it is necessary to show that Ca^{2+} can be stimulated by ANG II in skin tissue. Aguilera *et al.*[25] has shown changes in cytosolic calcium in fetal skin fibroblasts perfused with ANG II. Cells were preloaded with fluorescent indicator Fura-2 and nanomolar concentrations of ANG II caused a rapid increase in internal Ca^{2+}. The effect was specific because it could be abolished

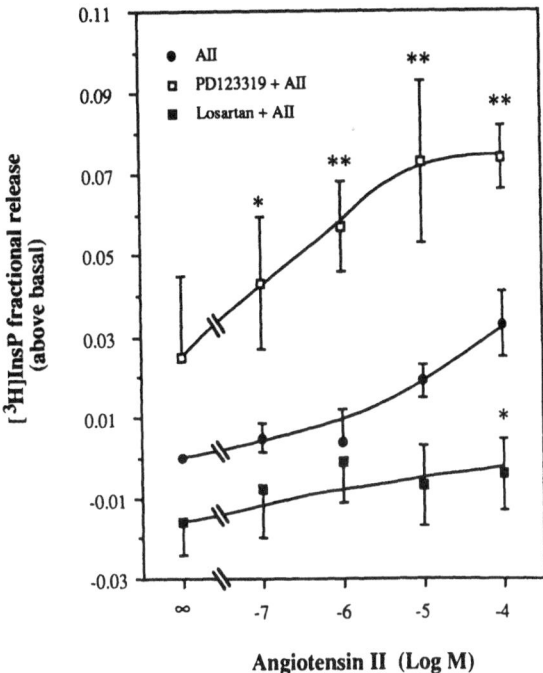

Figure 9. Inhibition of [³H]phosphate release to ANG II with 10⁻⁴M angiotensin type-1 receptor blocker losartan. PD123319, the ANG II type-2 receptor blocker at the dose of 10⁻⁴M, could not block the effect of ANG II on PI hydrolysis (*$p < 0.05$, **$p < 0.001$). (From Gyurko *et al.*[24] Reprinted with permission.)

by ANG II antagonist given 5 min prior to and during incubation with ANG II. However, the receptor subtype was unknown and since the study was carried out in fetal skin fibroblasts it is possible that the Ca^{2+} increase was mediated through AT_2 receptors. To test calcium responses to ANG II in adult skin fibroblasts we used an Attafluor (Zeiss) system and incubated dishes of fibroblasts with Fura-2.

Plates of cultured skin fibroblasts were loaded with Fura-2 at room temperature and 0.001% digitonin to release cytosolic Fura-2. The intracellular location of the intracellular Fura-2 was analyzed by digital imaging. Thirty minutes after calcium indicator loading, the plate was mounted on an inverted microscope and viewed with epifluorescence. An ultrasensitive camera was used to visualize the cells. Emissions of two wave lengths, one sensitive to calcium concentrations and the other insensitive to calcium concentrations, were monitored. A view of many of the cells in the dish was visualized and pixel squares placed over the cell images to identify which cells were responsive and which location of the cell was active. ANG II was delivered to the

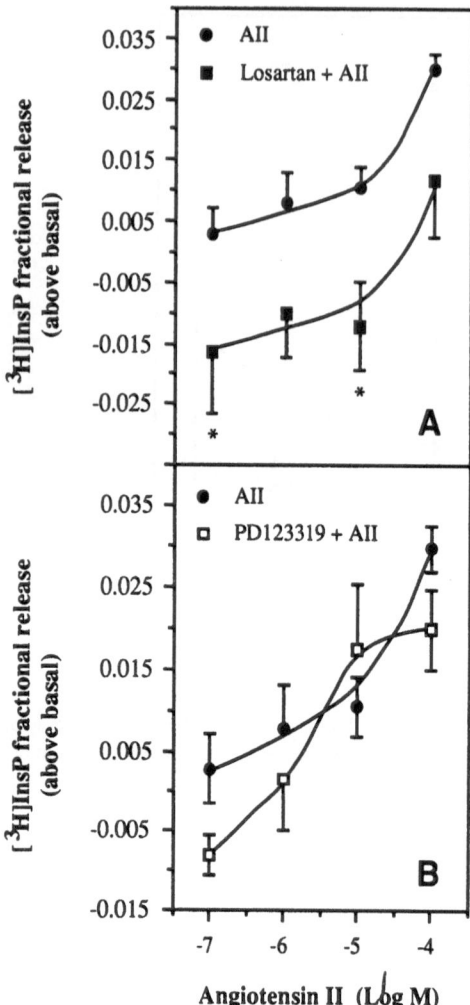

Figure 10. Effect of 10^{-3}M PD123319 and losartan on the dose-dependent curve of PI hydrolysis to increasing concentrations of ANG II. The enhancement produced by the type-2 receptor blockade suggests a blocking effect of type-2 receptors on PI hydrolysis. Alternatively, it can be explained as an agonistic effect of PD123319 on type-1 receptors. (From Gyurko *et al.*[24] Reprinted with permission.)

dish via a Hamilton syringe and the cytosolic calcium concentration was calculated at the dissociation constant for Fura-2 times the ratio of the calcium-sensitive wavelength to the calcium-insensitive wavelength ratio (340/380 nm) over time. Figure 11 shows a typical response to 1 nM ANG II. The response was biphasic with an initial decrease at about 50 sec after injection, followed by a pronounced increase in Ca_i2+ that repolarized after 30 sec and gradually came back toward baseline by 150 sec. It was noted that repeated stimulation was difficult to obtain. Although the study is incomplete, it shows that ANG II stimulates an increase in internal calcium of adult skin fibroblasts. Given the finding above of angiotensin stimulating AT_1 receptors, which then hydrolyze

Figure 11. Changes in cytosolic calcium in adult skin fibroblasts in response to ANG II added to the cells at zero time. The cultured adult skin fibroblasts were preloaded with fluorescent indicator Fura-2. Ten cells were recorded simultaneously and the ratio of 340/380 nM was calculated by an Attafluor Zeiss system. The illustration is the response in a single cell to 1 nM ANG II.

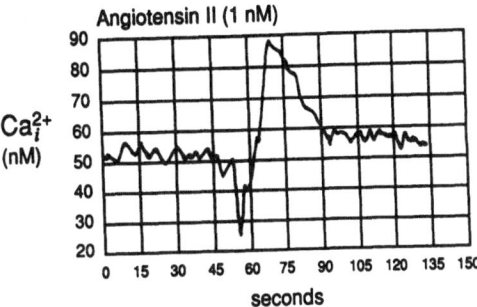

phosphatidylinositol, the results are supportive of the concept that Ins(1,4,5)P3 evokes rapid release of calcium from endoplasmic reticulum.[26] The Ca^{2+}-induced Ca^{2+} release by Ins(1,4,5)P3 would terminate the Ins(1,4,5)P3, contributing to the lack of responsiveness to a rapid restimulation of ANG II. The mobilization of intracellular Ca^{2+} by ANG II activates protein kinase C. This is apparently a critical step in the expression of *c-fos, c-jun,* and other early response genes.[23,27] Through this signal transduction mechanism, ANG II stimulates growth factors such as PDGF and TGF-β and vascular smooth muscle.[28] The next question is if this sequence of events is involved in TGF-β activity in adult skin fibroblasts.

7. ANG II STIMULATION OF TGF-β IN SKIN FIBROBLASTS

Skin fibroblasts were maintained in Dulbecco's modified essential medium with 10% fetal bovine serum (FBS) and 100 U/ml penicillin. The cells were plated in 100-mm plates and were used between passes 2–4. The confluent skin fibroblasts were treated with ANG II at 10^{-9}, 10^{-7}, and 10^{-5} M ANG II. Control plates were treated with buffer vehicle. After 16 hr incubation with ANG II, the cells were scraped, homogenized, and RNA extracted. Northern blotting was performed with a ^{32}P-labeled *in vitro* transcript of TGF-β cDNA clone. Figure 12 shows TGF-β mRNA in double lanes extracted from cells that had been stimulated by ANG II. The controls show that the quiescent levels of TGF-β mRNA expression is very low. It is induced by ANG II in a dose-related fashion. Having established that 10^{-7} M ANG II was a potent inducer of TGF-β mRNA expression in skin fibroblasts, we then tested the specificity of the response by the use of ANG II antagonists. Losartan (AT_1 antagonist) and PD 123319 (AT_2 antagonist) were both used at 10^{-5} M. They were combined with the 10^{-7} M ANG II and the experiment repeated. Figure 13 shows that losartan inhibited TGF-β mRNA expression, whereas PD 123319 did not inhibit expression but appeared to actually increase the mRNA abundance. The

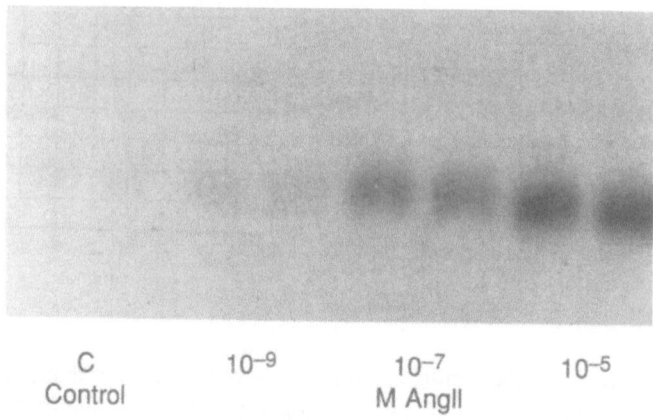

Figure 12. Northern blot analysis of RNA from fetal skin fibroblasts. Fibroblasts were prepared by enzymatic digestion from skin of fetal Sprague-Dawley rats at 18–20 days of gestation. Cells were grown to confluency in DMEM supplemented with glutamine (300 µg/ml), penicillin (400 µg/ml), streptomycin (50 µg/ml), and 10% FBS in a humidified atmosphere of 5% CO_2 and 95% O_2 at 37°C. On reaching confluence, the cells were replated at a dilution of 1 : 4 and used between passages 2 and 3. FBS was removed from the medium and the cells were treated with increasing concentrations of ANG II. After 24 hr incubation the cells were washed twice with ice cold phosphate-buffered saline and total RNA was extracted in a guanidium thiocyanate followed by phenol/chloroform extraction according to Chomczynski and Sacchi.[32] RNA was quantified by measuring the optical density at 260 nM. Samples containing 12 µg total RNA were separated on 1.2% agarose formaldehyde gel, transferred to a Genescreen (NEN) nylon membrane and hybridized with 10^6 cpm/ml ^{32}P-labeled *in vitro* transcripts of TGF-β1 cDNA clone.[33] The TGF-βa dCNA cloned in pBluescript II KS+ was a kind gift of Dr. Su Wen Qian. Membranes were hybridized overnight at 56°C, washed at a high stringency (0.1 × SSPE and 0.5% SDS at 70°C), and exposed to x-ray film for 3–7 days for autoradiography. The blot shown is representative of three similar experiments.

skin fibroblasts were tested for the presence of AT_1 and AT_2 receptors by competition curves and the presence of the AT_2 receptors in these fibroblasts was demonstrated by a 50% inhibition with PD 123319. Therefore the results show that while there are both AT_1 and AT_2 receptors in the adult skin fibroblasts, the AT_1 receptors stimulate the gene expression of TGF-β. AT_2 receptors, if they have a function in these cells, appear to act in the opposite direction and inhibit TGF-β. Blocking the AT_2 receptors with PD 123319 would allow only AT_1 receptors to be stimulated by the ANG II, leading to the observed increase response.

The β-type transforming growth factor is a member of a family of poly-peptides, which play a critical role in the regulation of cell proliferation during development.[29] TGF-β (5 isoforms have been identified) has numerous actions associated with mitogenesis, inhibition of endothelial cell proliferation, and as

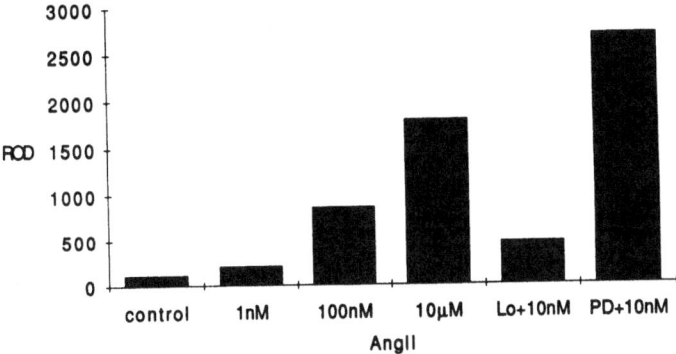

Figure 13. Densitometric analysis of Northern blots for TGF-β mRNA. Relative optical density (ROD) was determined with a video-based imaging system. 10^{-7} M ANG II elicited a severalfold increase in the expression of TGF-β gene.

a potent chemoattractant for macrophages, mononucleoleukocytes, and fibroblasts.[30,31] TGF-β is released from platelets, and since platelets aggregate at the site of wounds, TGF-β is released in wounds. These data support the hypothesis that ANG II has a paracrine action in skin to stimulate growth factors such as TGF-β in cell growth regulation and in repair.

8. CONCLUSIONS

Almost all physiological systems can be approached from several perspectives. A complex process, such as wound healing, becomes unbearably complicated to analyze by taking into account all of the multiple factors involved. What we have presented in this chapter is a systematic approach to understanding the mechanism by which one peptide, ANG II, might be involved in cutaneous wound healing. Our data suggest that there is a distinct tissue RAS in skin that is independent of the blood-borne RAS. The evidence for this is (1) the presence in skin of angiotensinogen mRNA, (2) high levels of authentic ANG II, and (3) increased levels of ANG II for several hours and days after wounding without change in the blood levels of angiotensin or in adjacent unwounded areas of skin. The rise in ANG II levels in skin is therefore independent of blood-borne ANG II. In addition, skin has angiotensin receptors and these receptors are predominately of the AT_1 type in the adult, in contrast to the predominance of AT_2 receptors in the fetus and very young rat. We suggest that the high levels of ANG II during wound healing are indicative of a mechanism by which skin tissue angiotensin is involved in the healing process. Specifically, we propose a model (Fig. 14) by which angiotensin is released close to the site of the wound. As the levels increase, the angiotensin

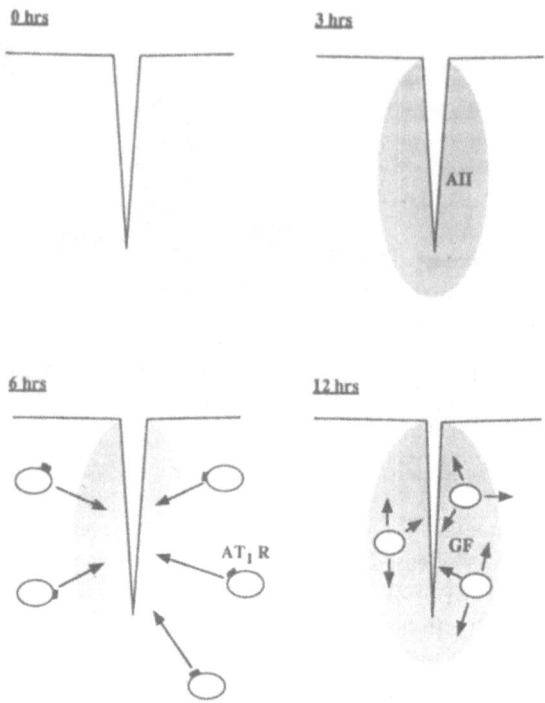

Figure 14. A model for the role of ANG II in wound healing. At 0 hr the wound is made in the skin. This stimulates the generation of ANG II in skin from as yet unidentified cells but is localized and not the result of blood-borne ANG II. From 3 hr on the concentration of local ANG II in wounded tissue but not in unwounded tissue begins to rise. It reaches a peak at 24 hr. The effect of the concentration of ANG II around the wound is to set up a gradient so that distant cells (neutrophils, platelets, fibroblasts) with AT_1 receptors are stimulated. The stimulated cells become motile and are drawn along the gradient to the greatest concentration which is at the site of wounding. The cells gathered at the site of the wound are stimulated through AT_1 receptors which mobilize internal calcium through a phosphatidylinositol pathway. The calcium presumably binds with calmodulin and acts on *c-fos* and *c-jun* to increase growth factor mRNA expression and the release of the growth factor into the site of wounding. In this way ANG II acts in a paracrine fashion and draws cells that secrete growth factors such as TGF-β to the injured site.

develops a diffusion gradient that stimulates multiple cell types, including platelets, fibroblasts, macrophages, and leukocytes at some distance from the wound. These motile cells are drawn along the angiotensin gradient toward the maximum concentration which is at the wound site. This migration is triggered by stimulation of their endogenous ANG II receptors which appear to be principally AT_1. The migration of platelets, macrophages, neutrophils, and other leukocytes to the high angiotensin in the wound results in the release of growth factors such as TGF-β through the intracellular signal-coupling mechanism of

IP hydrolysis and internal calcium mobilization, turning on early genes *c-fos* and *c-jun*. These proto-oncogenes interact at their leucine zipper regions to form heterodimers which then bind to a specific DNA sequence found in the 5′ regulatory regions of the early genes (the AP-1 site). This leads to the expression of growth factor mRNA and to the ultimate secretion into the tissue where the growth factor, such as TGF-β, stimulate the production of collagen and regulation of cellular proliferation and chemotaxis. Such a hypothesis leaves many questions unanswered but opens a new function of angiotensin to investigation, namely, the role of ANG II in skin and in the acceleration of wound healing.

ACKNOWLEDGMENT. We thank Greg Fossum, M.D., with whom we began studies in wound healing in the uterus that led to studies on skin. We acknowledge the support of NIH grant HL27334 and AHA Multidisciplinary Program Grant (Florida Affiliate). Robert Gyurko is a recipient of an American Heart Association Postdoctoral Fellowship Award.

REFERENCES

1. Naftilan RE, Dzau DVJ: Induction of platelet derived growth factor A-chain and C-myc gene expression by angiotensin in cultured rat vascular smooth muscle cells. *J Clin Invest* 83:419–1424, 1989.
2. Powell JS, Clozel JP, Muller RKM, et al: Inhibitors of angiotensin converting enzyme prevent myointimal proliferation after vascular injury. *Science* 245:186–188, 1989.
3. Osterreider W, Muller RKM, Powell JS, et al: Role of angiotensin II in injury induced neointima formation in rats. *Hypertension* 18(2):6–64, 1991.
4. Powell JS, Muller RKM, Baumgartner HR: Suppression of the vascular response to injury. The role of angiotensin converting enzyme inhibitors. *J Am Coll Cardiol* 17:137B–142B, 1991
5. Stouffer GA, Owens GK: Angiotensin II-induced mitogenesis of spontaneously hypertensive rat-derived cultured smooth muscle cells is dependent on autocrine production of transforming growth factor-β. *Circ Res* 70:820–828, 1992.
6. Majesky MW, Lidner V, Twardzik DR, et al: Production of transforming growth factor β-1 during repair of arterial injury. *J Clin Invest* 88:904–910, 1991.
7. Chiu AT, Herblin WF, McCall DE, et al: Identification of angiotensin II receptor subtypes. *Biochem Biophys Res Commun* 165:196–202, 1989.
8. Millan MA, Carvello P, Izumi SJ, et al: Novel sites of expression of functional angiotensin II receptors in the late gestation fetus. *Science* 244:1340–1342, 1989.
9. Tsutsumi K, Saavedra JM: Differential development of angiotensin II receptor subtypes in the rat brain. *Endocrinology* 128:630, 1992.
10. Johnson CM, Aguilera G: Angiotensin II receptor subtypes and coupling to signaling systems in cultured fetal fibroblasts. *Endocrinology* 129:1266–1274, 1991.
11. Phillips MI, Kimura B, Krim AJ, et al: Angiotensin II in skin may act as a growth factor and enhance wound healing. *FASEB J* 5:904, 1991.
12. Kimura B, Sumners C, Phillips MI: Changes in skin angiotensin II receptors in rats during wound healing. *Biochem Biophys Res Commun* 187:1083–1090, 1992.

13. Viswanathan M, Saavedra JM: Expression of angiotensin II AT_2 receptors in the rat skin during experimental wound healing. *Peptides* 13:783–786, 1992.

14. Dzau VJ, Burt DW, Pratt RE: Molecular biology of the renin–angiotensin system. *Am J Physiol* 255:F563, 1988.

15. Ekker M, Tronik D, Reugeon F: Extrarenal transcription of the renin genes in multiple tissues of mice and rats. *Proc Natl Acad Sci USA* 86:5155–5158, 1989.

16. Phillips MI, Speakman EA, Kimura B: Levels of angiotensin and molecular biology of the tissue renin-angiotensin systems. *Regul Pept* 43:1–20, 1993.

17. Phillips MI, Stenstrom B: Angiotensin II in rat brain comigrates with authentic angiotensin II in HPLC. *Circ Res* 56:212–219, 1985.

18. Phillips MI, Kimura B, Raizada MK: Measurements of brain peptides: Angiotensin and ANP in tissue and cell cultures. *Methods Neurosci* 6:177–206, 1991.

19. Hermann K, Ring J, Phillips MI: Presence of angiotensin peptides in human urine. *J Chrom Sci* 28:524–528, 1990.

20. Reuben E, Farber, JL: *Pathology.* Lippencott, Philadelphia, 1988.

21. Ignotz RA, Massague J: Transforming growth factor-β stimulates the expression of fibronectin and collagen and their incorporation into the extracellular matrix. *J Bio Chem* 261(9):4337–4345, 1986.

22. Mustoe TA, Pierce GF, Thomason A, et al: Accelerated healing of incisional wounds in rats induced by transforming growth factor-β. *Science* 237:1333–1336, 1987.

23. Catt KJ, Sandberg K, Balla T: Angiotensin II receptors and signal transduction mechanisms. In Raizada MK, Phillips MI, Sumners C (eds): *Cellular and Molecular Biology of the Renin–Angiotensin System.* CRC Press, Boca Raton, Florida, 1993, pp. 307–356.

24. Gyurko R, Kimura B, Kurian P, et al: Angiotensin II receptor subtypes play opposite roles in regulating phosphatidylinositol hydrolysis in rat skin slices. *Biochem Biophys Res Commun* 186:285–292, 1992.

25. Aguilera G, Johnson MC, Feuillan P, et al: Developmental expression of angiotensin receptors: Distribution, characterization and coupling mechanisms. In Raizada MK, Phillips MI, Sumners C (eds): *Cellular and Molecular Biology of Renin–Angiotensin System.* CRC Press, Boca Raton, Florida, 1993, pp. 413–431.

26. Wakui M, Osipchuk TV, Petersen OH: Receptor activated cytoplasmic Ca^{2+} induced Ca^2 spiking mediated by inositol triphosphate is due to Ca^{2+} induced Ca^2 release. *Cell* 63:1026, 1990.

27. Taubman MB, Burke BC, Izumo S, et al: Angiotensin II induces c-fos mRNA in aortic smooth muscle: Role of Ca^{2+} mobilization and protein kinase C activation. *J Biol Chem* 264:526, 1989.

28. Gibbons GH, Pratt RE, Dzau VJ: Vascular smooth muscle cell hypertrophy vs. hypoplasia. Autocrine transforming growth factor-β1 expression determines growth factor to angiotensin II. *J Clin Invest* 90:456–461, 1992.

29. Rizzino A: Transforming growth factor-β multiple effects on cell differentiation and extracellular metroses. *Dev Biol* 130:411–422, 1988.

30. Wahl SM, Hunt DA, Wakefield LM, et al: Transforming growth factor-β produces monocyte chemotaxis and growth factor production. *Proc Natl Acad Sci USA* 84:5788, 1987.

31. Postlethwaite AE, Keski-Oja J, Moses HL, et al: Stimulation of the chemotactic migration of human fibroblasts by transforming growth factor-β. *J Exp Med* 1675:251, 1987.

32. Chomczynski P, Sacchi M: Single step method of RNA isolation by acid guanidium thiocynate phenyl chloroform extraction. *Anal Biochem* 162:156, 1987.

33. Qian SW, Kondaiah P, Roberts AB, et al: cDNA cloning by PCR of rat transferring growth factor-β-1. *Nucleic Acids Res* 18(10):3059, 1990.

Index